Looking Ahead
Human Factors Challenges
in a Changing World

"Wise folk may or may not form expectations about what the future holds in store, but the foolish can be relied upon to predict with complete confidence that certain things will come about in the future or that others will not" (Peter Medawar). Alas.

Looking Ahead
Human Factors Challenges in a Changing World

by
Raymond S. Nickerson

LEA LAWRENCE ERLBAUM ASSOCIATES, PUBLISHERS

1992 Hillsdale, New Jersey   Hove and London

Lawrence Erlbaum Associates, Inc., Publishers
365 Broadway
Hillsdale, New Jersey 07642

ISBN 0-8058-1150-8 hardcover

ISBN 0-8058-1151-6 paperback

LC Card Number: 92-30953

Printed in the United States of America
10   9   8   7   6   5   4   3   2   1

To
> Amara
> Daniel
> Timothy
> Bryan
> Laura
> Rory
> Erin
> Danielle

and their generation

Contents

Preface   ix

Chapter 1
Looking Ahead   1

Chapter 2
Economics, Industry, and Productivity   9

Chapter 3
Energy   46

Chapter 4
Environmental Change   76

Chapter 5
Education and Training   138

Chapter 6
Transportation   160

Chapter 7
Space Exploration   184

Chapter 8
Biotechnology   195

Chapter 9
Information Technology   212

Chapter 10
Person–Computer Interaction   248

Chapter 11
Work   296

Chapter 12
Decision Making and Policy Setting   323

Chapter 13
Quality of Life   345

Chapter 14
Epilogue   370

References   375

Author Index   421

Subject Index   436

Preface

For the second time in my experience, a background paper prepared for an activity of the National Research Council's Committee on Human Factors has grown into a book. The immediate precursor to this book was a paper written for the committee's use in the planning of a report on Human Factors Research Needs, which was to be a sequel to a report on this topic issued by the committee in 1983. Involvement with the Committee on Human Factors has been immensely stimulating and rewarding, and I feel very fortunate to have been a member of it and to have participated in its activities.

As its title indicates, this is intended to be a forward-looking book. I am very interested—as I assume most of us are—in the question of what kind of a world we are moving toward. More particularly, as a parent and grandparent, I am concerned about some current trends and would like to understand better what can be done to increase the probability that the world we are bequeathing to future generations is one they will want to inhabit.

I believe that if one begins as a dyed-in-the-wool pessimist one will find plenty of support for that pessimism. On the other hand, I believe also that if one wants to be an optimist, one can find much support for that view as well. But we should not, I think, be easily persuaded to accept the inevitability of any particular scenario. The future is full of possibilities, some good, some bad. It will be what we make it.

The focus here is on the question of what major challenges and opportunities a world that is changing rapidly may have for the human factors profession. Or reversing the question, what does the human factors profession have to offer by way of solutions, partial solutions, or approaches to help solve problems that seem to be looming as we look ahead?

An obvious first step in beginning to think about this question is to try to understand what the problems are likely to be. To this end, I collected predictions, projections, forecasts, conjectures, and guesses about the near future, as well as data regarding recent developments and trends that struck me as having some relevance to the task of identifying challenges and opportunities that relate to human factors, broadly conceived.

I have not tried to define what I mean by human factors, but it will be clear that what I have in mind is considerably broader than the science and technology of designing equipment that is suitable for human use. If I offend some readers by stretching the bounds of human factors too far, I beg their indulgence and ask that they give more consideration to whether the problems being discussed are important and whether human factors, ergonomics, engineering psychology, or applied psychology more generally could have anything to contribute to solutions than to whether I have the turf and terminology just right..

A major risk in any book that focuses on the future is that, given the rapidity of change, much of its content is likely to be overtaken by events before the ink is dry. I am well aware that that is true of this book. I have had to make many changes in the evolving text because of unanticipated events that occurred during the time it was being written. Apologies in advance for projections that turn out to deal, accurately or inaccurately, with history by the time they are read.

My hope is that readers will find here a stimulus to think about certain problem areas in ways that they might not otherwise have thought and that some will see opportunities to address specific problems not yet addressed— or not often addressed—from a human factors perspective. If any of the specific suggestions pertaining to these problems in the book are pursued by researchers, that will be very gratifying of course; but even more rewarding would be the emergence and pursuit of better ideas that result from more thought and dialogue on these issues. What I have to say is directed primarily to people engaged in human factors research, or in the support of such research, and in the application of its results. But I have tried to write for a broad audience and will be pleased if others with an inclination to speculate about the future and to influence it for the better find it of some use.

I owe thanks to many people. Fellow members of the Committee on Human Factors and its technical staff over the period of time during which the book was written were Paul Attewell, Mohamed Ayoub, Jerome Elkind, John Gould, Miriam Graddick, Oscar Grusky, Douglas Harris, Robert Helmreich, Julian Hochberg, Beverly Huey, Roberta Klatzky, Thomas Landauer, Herschel Leibowitz, Neville Moray, William Rouse, Thomas Sheridan, Joyce Shields, Harold Van Cott, Christopher Wickens, Robert Williges, Frank Yates, and Lawrence Young.

I learned from all of these people and thank them for their inspiration and collegiality. In this context, I especially thank those who encouraged me to expand the background paper into this book. I have benefited greatly from discussions with committee members about problem areas addressed, but I do not attempt to represent the committee's views, or indeed anyone's but my own. The preparation of the committee's report has been proceeding on schedule, and the report should be issued in 1993.

Very helpful comments on parts of the manuscript were obtained from Erich Bender, Sanford Fidell, Tom Fortmann, Robert Glaser, Alex McKenzie, Robert Oliphant, and Sidney Smith. Carol Stillman read it from beginning to end and made numerous valued suggestions for improvement. My thanks to each of these people.

When I started working on the book, I was at Bolt Beranek and Newman Inc. I have retired from BBN after 25 years there but I want to express my gratitude to the company for its policy of encouraging members of its staff to participate in activities such as those sponsored by committees of the National Research Council and for making time available for them to do so.

I want to thank my friends and colleagues at BBN for an abundance of intellectual stimulation over the years. During the time this book was being written, I benefited especially from discussions with Sheldon Baron, Frank Heart, Richard Pew, and John Swets.

My thanks also to Lysa Stark, my secretary at Bolt Beranek and Newman, who patiently worked with more early drafts of the manuscript than either she or I care to remember, and to Marian Bremer, Maria Kneas, and Bridget Mooney, who helped me acquire numerous hard-to-obtain reports.

As always, my greatest debt of gratitude, by far, is to my wife, Doris, for her constant love and support, and to our family for being what it is.

—*Raymond S. Nickerson*
Bedford, Massachusetts

1

Looking Ahead

No one knows what the world will be like 50 or 100 years from now. Trying to project even 10, 20, or 30 years ahead is a fool's game, because there are so few ways to be right and so many ways to be wrong. The only truly safe prediction that one can make is that the future will be different from what we expect it to be, that it will be as full of surprises as it has been in the past. So, why should we play this game at all? I believe there are good reasons for doing so, the most compelling of which follows from the assumption that there are many possible futures, some much more desirable than others, and that the kind of future we will get depends to no small degree on the kind we work to achieve. The only alternatives to this view that I see are an insipid fatalism, which is singularly uninspiring, or a live-for-today let-the-future-take-care-of-itself form of egoism, which seems objectionable to me on moral grounds.

Most of us, I suspect, are futurists in the sense that we are interested in the kind of world that our children, their children, and their children's children will inhabit, and would like to play some part in making it better than it might otherwise be. To do so intelligently we need to have some idea of what the possibilities are so we can work toward those we prefer and against those we do not. As I think about what the world of tomorrow might be like, the thing that impresses me most is the enormous range of possibilities there appears to be, some very attractive and others quite frightening. I see justification for neither unbridled optimism nor petulant doomsaying, but many goals worth working toward as well as potential disasters to be avoided, and reasons to believe that thoughtful effort can make a difference.

My purpose in this book is to review some of the recent developments and trends that seem to me to be especially relevant to any attempt to understand,

1

even fuzzily, near-term future possibilities, to consider what a variety of knowledgeable people are saying about changes and developments that could occur, and to relate the possibilities to needs and opportunities for human factors research. Human factors, for this purpose, is taken to include not only the implications of human capabilities and limitations for the design of equipment and machines that are intended for human use, but applied psychology much more generally. In particular, it is taken to involve social systems as well as physical ones, the interaction of people with the environment as well as with machines, the facilitation of communication between people as well as between people and computers, and the design of policies and procedures as well as of equipment.

Although a consideration of recent developments and current trends is an obvious starting point for any attempt to understand what the possibilities for the future are, one must bear in mind that many of the twists and turns that the future will take will not be revealed by any analysis of the past. As Kay (1984) has pointed out, in warning against attempts to predict the future by looking at trends, "There is no trend that led from the railroad to the airplane. There is no trend that led from the horse and buggy to the car; no trend that led from the desk calculator to the pocket calculator; no trend that led from the ditto machine to the Xerox machine; no trend that led from the mainframe computer to the personal computer" (p. 111).

On the other hand, the automobile has evolved over this entire century; and there is every reason to believe that it will be a preferred means of transportation for the foreseeable future and that it will continue to evolve for some time to come. Electric power is generated in a variety of ways; each of these ways, as a technology in its own right, is undergoing development. How the various possibilities will compare in cost effectiveness in years ahead remains to be seen, but it seems reasonable to expect all of the ongoing efforts at improvement to continue in the future. Trends are apparent in the automation of industrial processes, the study of which should help improve our understanding of future labor needs. Long-term environmental changes are among the more worrisome trends that are attracting attention today, and we can ill afford to ignore them. I believe that Kay was right in suggesting that the study of trends will not reveal coming technological breakthroughs and innovations that can change qualitatively the way things are done, but it does not follow, nor did he suggest, that the study of trends is of no help in attempts to understand what some of the possibilities of the future are.

One reason why simple extrapolation of trends is risky is the fact that the growth history of many processes—biological, social, or technological—can be described by an s-shaped or logistic function. If one is observing such a process at a relatively early stage of its maturity, one can easily be misled into believing that the growth process is an exponential one. In the early part of the 20th century, the demand for wood for making railroad cross ties was growing sufficiently rapidly to cause concern among some leaders, including Theodore Roosevelt, that the forests could not continue indefinitely to meet the need and that a timber famine would result. In this case, the worst fears were not realized, and for several unforeseen reasons: Creosote and other wood-preserving techniques were used to extend the life of ties (apparently by a factor of about three), concrete was increasingly used for tie materials in place of wood, especially in Europe, and the rate at which new track was laid was eventually slowed (Ausubel, 1989).

There are these and many other reasons to be wary of projections, forecasts, and predictions, whether based on the extrapolation of trends or anything else. Perhaps one should be especially wary of one's own projections, given our penchant for convincing ourselves of the worthiness of our own ideas. None of these reasons, however, is a legitimate excuse, in my view, for not trying to understand, as best one can, what the possibilities for the future appear to be and how one might help shape things for the better.

My intention is to focus on anticipated problems (read *problems* here to include opportunities as well as troublesome difficulties) and then to ask how human factors research might contribute to their solutions, rather than to use the human factors research that is being done as the point of departure and then to look for future problems to which it might be applied. One risk of this approach is that of identifying problems for which such research has little if anything to offer. In fact, I do not limit attention only to problems for which experience indicates clearly that human factors research can be relevant. I assume, rather, that there are ways in which such research could be useful in addressing societal problems that the profession has not yet realized, and that they are more likely to be realized in the future if the community is actively seeking to identify them than if it is not.

Inasmuch as the impetus for this exercise came from an activity of the U.S. National Research Council, the focus is more on the United States than on any other country, although attention is also given to international and global issues. I have noted some research studies from the human factors and applied psychology literature that relate to various aspects of the problem areas discussed, but I have made no attempt at a comprehensive review of the relevant work. I do believe, however, that the studies cited are broadly representative of the relevant research that has been undertaken in the past few years and/or that is currently under way.

Projections, forecasts, and predictions were gathered from a variety of sources, which I identify when possible. In addition, I have not hesitated to include my own quite unauthoritative guesses. Whatever their origins, they are, for the most part, fairly conservative; that is, they are quite plausible, at least in my view. In their entirety, they are offered more as a stimulus to thought than as a serious attempt at forecasting. The time scale is the next few decades, 30 years at the outside in the vast majority of cases. The emphasis is on possibilities; the intent is to identify what could happen rather than to say what will happen. For convenience I have organized these thoughts under a few specific headings. There is a certain arbitrariness about these groupings: Many observations could as easily been put under some other heading.

When trying to anticipate what the future could hold, it is probably useful to remind ourselves how often people who have attempted to look ahead have been quite wrong. Each of us, undoubtedly, has a set of favorite prognostications that looked a little silly after the fact. The following are some of mine:

• In 1876, William Orton, president of the Western Union Telegraph Company, decided not to buy the rights to the telephone patent, with the comment: "What use could this company make of an electronic toy?" (S. Aronson, 1977).

• About 10 years later, in 1885, Arnold Morley, then Postmaster-General of Great Britain, told Parliament "the telephone could not, and never would be an advantage which could be enjoyed by the large mass of the people" (quoted by Marvin, 1988, p. 101).

• Late in the 19th century, Sir John Erichsen, a British surgeon, described the abdomen as "forever shut from the intrusions of the wise and humane surgeon." In 1930, George Moynihan, another British surgeon, judged it to be impossible to perfect surgery beyond its then-current state (Medawar, 1984).

• The 19th century Scottish mathematician and physicist, William Thompson (Lord) Kelvin ruled out the possibility of flight in heavier-than-air craft (Hardin, 1985). Early in the 20th century, the American astronomer Simon Newcomb did so, too: "The demonstration that no possible combination of known substances, known forms of machinery and known forms of force, can be united in a practical machine by which man shall fly long distances through the air, seems to the writer as complete as it is possible for the demonstration of any physical fact to be" (quoted in Clarke, 1962).

• In 1926, American radio pioneer Lee DeForest made the following assessment of the prospects for the commercial exploitation of television: "While theoretically and technically television may be feasible, commercially and financially I consider it an impossibility, a development of which we need waste little time dreaming." This pronouncement (p. 207) is one of many compiled by Cerf and Navasky (1984) in a compendium that makes for delightful, but somewhat sobering, browsing.

• As late as 1936, Ernest Rutherford ruled out the possibility of the use of nuclear energy, at least in this century (Eiseley, 1970).

Inventors of technologies that have changed the world typically have not foreseen the eventual uses of their inventions. Marconi, for example, saw radio not as a means of broadcasting speech and music to a wide audience, but as a wireless analogue to the telegraph and, in particular, as a means of providing point-to-point communication between one ship and another and between ship and shore. The American corporations that initially invested in radio also failed to see its potential as a qualitatively new form of communication and, treating it like a wireless telegraph, nearly went out of business as a consequence. The idea of using radio for broadcasting, in the modern sense, grew out of the activities of amateur ham operators whose numbers increased rapidly during the first decade or so of the 20th century. The amateurs used the radio for one-to-one communication, and many of the turn-of-the-century predictions about the future of radio assumed this to be the natural and continuing mode of operation. By the second decade of the century, however, amateur stations had begun to broadcast music and speech, and some of these stations became commercial in the 1920s (Douglas, 1986). As late as 1922, however, Thomas Edison predicted that the radio craze would die out in time (Cerf & Navasky, 1984).

Similarly, the pioneers of computer technology saw their creation, for the most part, as a device for mechanizing the arithmetic operations that were then being performed by cadres of human "computers." The then-current ideas regarding what these machines could do came primarily "from looking at the new invention strictly in the context of what it was replacing: calculating machines and their human operators" (Ceruzzi, 1986, p. 194).

These examples of views of the future were all too conservative in some way. One can easily find examples of views that erred on the side of extravagance as well. Often in the past, inventions and other scientific developments have prompted visions of idealistic and utopian change. The July 1899 issue of *Scientific American* carried the following prediction of how motor cars would affect urban life once mass production had brought their price low enough for people to acquire them: "The improvement in city conditions by the general adoption of the motor car can hardly be overestimated. Streets clean, dustless and odorless, with light rubber-tired vehicles moving swiftly and noiselessly over their smooth expanse, would eliminate a greater part of the nervousness, distraction and strain of modern metropolitan life" (Dubos, 1970, p. 95).

Most of us today, being sensitive to the automobile's contribution to air and noise pollution and aware of the nervousness, distraction, and strain that can be caused by the all-too-common urban traffic snarl, are likely to see only irony in this turn-of-the-century vision, but we should give the motor car its due. Citing some calculations made by Montroll and Badger (1974), and some plausible assumptions about a horse's per diem travel range and its solid and liquid waste generation, Ausubel (1989) has pointed out that a horse emits roughly 940 g of waste per mile, to the ground, whereas an automobile puts out only about 5 g (composed of .25 g of hydrocarbons, 4.7 of carbon monoxide and .4 of nitrous oxides) to the air.

There were many utopian visions, stimulated by the discovery in the middle of 20th century of the possibility of harnessing nuclear energy, of "power too cheap to meter," and of atomically powered airplanes and personal vehicles (Del Sesto, 1986). Work on the development of a nuclear-powered aircraft was carried on seriously for some years between 1946 and 1961, and was terminated by President Kennedy only after expenditures of more than $1 billion. Nuclear enthusiasts saw the potential benefits of nuclear energy but failed to foresee the problems—such as safety hazards and the disposal of radioactive waste—that would be encountered in attempting to realize that potential (Corn, 1986).

More generally, there was a good deal of utopian visionary writing late in the 19th century and early in the 20th based on what appears to have been an implacable faith in science and technology, coupled with a model of human nature not yet tarnished by two world wars and the major economic depression in between. Segal (1986) pointed out, for example, that "between 1883 and 1933, twenty-five individuals published works envisioning the United States as a technological utopia" (p. 119). Although most of the writers of these utopian visions were obscure, their values were representative of mainstream American thinking and in particular of the "belief in the inevitability of progress and the belief that progress was precisely technological progress" (p. 121).

Attempting to forecast the country's energy needs, or the way the needs would be met, has proved so far to be a particularly risky enterprise. According to a prediction by the Atomic Energy Commission in a 1962 report to the President on civilian nuclear power, by the year 2000, 30% of the demand for energy in the United States would be supplied by fission reactors. A report from the Interdepartmental Energy Study Group, issued in 1964, asserted "no ground for serious concern that the nation is using up any of its stock of fossil fuel too rapidly; rather there is the suspicion that we are using them up too slowly (A. M. Weinberg, 1988–1989). We are concerned for the day when the value of untapped fossil fuel resources might have tumbled and the nation will regret that it did not make greater use of these stocks when they were still precious" (quoted by A. M. Weinberg, 1988–1989, p. 81).

Thus one easily finds examples of past views of the future that were much too conservative in specific respects and examples of others that were, in retrospect, too radical. My sense however, is that, especially during recent years, we have been more often surprised by the ways in which technological advances have out-distanced our imaginations than by the ways they have lagged behind them. In any case, if the past gives us any clues at all as to what the future holds, the one thing that we can be quite sure of is that there will be surprises. No matter how carefully we try to anticipate all the possibilities, some of the most significant developments are very likely to be those that no one foresees. As C. Evans (1979) has pointed out, Alvin Toffler in his book *Future Shock,* which was published in 1970, painted a sufficiently radical picture of the near-term future that the book was criticized for being sensationalist, but he missed completely what has turned out to be among the most sensational of all developments—the microprocessor—and it came along within a very few years of the publication of the book.

In short, projections are just that; they can, and often do, prove to be wrong. We should, in keeping with Kay's admonition, be especially wary of projections that are based on simple extrapolations of past trends. None of the numbers in this book that refer to the future should be taken as more than somebody's opinion as to what is likely to happen. Before using any of them as the basis for decision making, one should have thought deeply enough about them to have an opinion of one's own as to their plausibility. Even numbers that are intended to represent historical data should not be accepted completely uncritically. The techniques by which such numbers are developed or derived are not error-proof; they often involve assumptions that are not made explicit, and different sources sometimes report different numbers for the same variables.

If the point needs illustration with a concrete example, a recent study by the National Research Council's Committee on National Statistics revealed that the number of scientists and engineers in the United States in 1984 was 1.9 million, 3.7 million, or 6.1 million, depending on whether one used the figures of the Bureau of Labor Statistics, the National Science Foundation, or the Bureau of the Census (Citro & Kalton, 1989). The discrepancies stemmed from various differences in counting policies—whether to count people with degrees in science or engineering who were working as managers or people with degrees from foreign institutions—and in sampling and measurement techniques. The Committee's study provides an unusual opportunity to compare estimates from three sources and it points up the importance of looking critically at the counting or estimating processes that produce any number on whose accuracy one wishes to depend.

Sometimes what appear to be discrepancies may be the result of sampling over nonidentical time periods or from different populations. What population is being described is not always clear in the way numbers are reported; counts that underlie unemployment statistics, for example, may or may not include unemployed people who, for one reason or another, are not actively looking for jobs. Despite these caveats, my general sense is that, although there are some discrepancies in the historical data in what follows, they do not, for the most part, negate the major points that are made.

There are, of course, many possibilities for the future that could have truly profound effects, should they occur: a world-wide economic depression, the discovery of intelligent extraterrestrial life, a natural disaster of global magnitude (such as collision with a sizable meteor), or an AIDS-like world-wide epidemic caused by a new virus transmittable by casual contact. Although such possibilities cannot be ruled out, no attention is paid to them here. Discussion of every imaginable event that could affect the future in a major way is clearly impossible and would not be very useful in any case; attention is limited here to what seem to me to be both reasonably likely possibilities and possibilities about which something can be done. My selection of topics on which to focus undoubtedly reflects my personal interests and biases, to a certain degree; I believe the topics included are especially significant for the future, but I do not wish to suggest that they are the only important ones. I have made no attempt to deal with ideological variables—philosophical, religious, or political—despite their obvious importance as causal factors in shaping local, national, and international events. Such variables lie farther outside the domain of human factors, as traditionally delimited, than any of those I have included, and some readers may consider a few of the latter already over the edge.

I had difficulty deciding how to organize this book. The problem is the interrelatedness of the topics discussed. Issues relating to energy and those relating to the environment are inextricably interwoven, because energy production and use are major causes of environmental problems and problems of the environment have profound implications for the cost—both present and deferred—of energy production and use; one cannot say much about economics without mentioning energy; further, any extended discussion of economics, energy, or the environment must, sooner or later, bring in transportation; one can hardly get started on a discussion of work, especially office work, without saying something about computer and communication technology. Growth in productivity is dependent in part on the availability of capital for investment in plant and equipment resources, the availability of capital is dependent in part on budget deficits which are affected by international trade balances, and so on. The organization that was finally adopted has a considerable degree of arbitrariness in it. It proved to be impossible to discuss any one of the selected topics without involving several others. Thus, some material in a given chapter could just as well have appeared in one or more of the other chapters, but I know of no way to avoid that, given the degree to which the topics overlap.

2

Economics, Industry, and Productivity

The U.S. economy has been the strongest and stablest economy in the world for many decades. Its continuing strength, stability and long-term growth are now threatened by a large federal deficit, an adverse balance of trade, and the country's recently acquired status as the world's leading debtor nation. The large deficit, coupled with a relatively low rate of private savings, means the country is dependent on foreign capital for investment in the new plants and equipment needed for economic growth. A continuing deficit probably means a weakening of the U.S. economy in a variety of ways. As D. Lewis, Hara, and Revis (1988) put it: "If continued over the long run, large budget deficits will either reduce domestic investment or be financed by increasingly uncertain, and potentially reversible, capital inflow from abroad. In either case, living standards of U.S. citizens would fall: a reduced level of investment would retard the growth of the economy, and continued heavy foreign investment would send a larger share of U.S. output abroad in payment of debt service or other returns to foreign investors" (p. 46).

The United States' position in the world economy is determined, to a large degree, by its ability to produce goods and services that can compete effectively in world markets, and this ability has been challenged severely in recent years. Our debt and trade balance depend on how what we consume relates to what we produce, and at the present time we appear to be better at consumption than at production. A simple extrapolation of current trends gives a picture of the future that is not very comforting. Changing the current course is certainly possible; whether we will do so is anything but certain.

9

WORRISOME ECONOMIC INDICATORS

As of 1986, the total public and private debt in the United States was over 200% of the gross domestic product (Landau, 1988). Due in large measure to the explosion of corporate debt—symptomized by junk bonds and leveraged buyouts—in the 1980s, the value of all debt issued by nonfinancial U.S. corporations now exceeds $2 trillion. This compares with a total market value for all U.S. corporations of $2.4 trillion (Dentzer, 1989). The total governmental debt (federal, state, and local) as of 1987 was $3.1 trillion, or about $12,600 per person; the federal debt accounted for about 77% of the total. The situation has worsened by more than a factor of two since 1980, when the total governmental debt was $1.25 trillion, or about $5,500 per person (Bureau of the Census, 1990). The federal debt tripled and the interest on it quadrupled during the decade of the 1980s (Nathan, 1991).

The U.S. balance of trade dropped steadily through most of the 1980s. The drop was precipitous for manufactured goods other than high-technology products; for high-technology products, the balance went negative in 1986, for the first time ever. (Although the total trade balance was still negative at the end of the decade—by about $100 billion—the balance at that time was again positive for high-technology products.) The interpretation of trade balance figures is complicated by the fact that they are influenced by changes in currency exchange rates, but a persisting trade deficit means that the United States is consuming more than it is producing (i.e., spending more than it is earning), a trend that cannot continue indefinitely. Landau (1988) summed it up this way: "The U.S. has been consuming too much, producing too little, and borrowing from abroad to maintain its standard of living" (p. 52). Hatsopoulos, Krugman, and Summers (1988) argue that the answer to the trade deficit must include some slowdown in the growth of national consumption, and that to ensure the future strength of the economy the United States must begin to produce more than it consumes.

Since 1960, the average annual rate of growth of the U.S. economy (3.1%) has been less than the average annual rate of growth of the world economy (3.9%; W. B. Johnston & Packer, 1987). Landau (1988) estimated 2.2% as the annual growth rate of the United States' real gross domestic product (GDP) since 1979, as compared with Japan's annual GDP growth rate of 3.8% during the same time period. Thurow (1987) gave 3.8% per year as the rate of growth of the gross national product of the United States in the late 1960s and 2% per year in the early 1980s, noting that it decreased by nearly 50% over a decade and a half. (*Gross national product* [GNP] is the total national output of goods and services valued at market prices; *gross domestic product* [GDP] is the output of all labor and property located within a country [Bureau of the Census, 1990]. For present purposes, the distinction is not important.)

D. Lewis et al., (1988) project a growth rate of about 2.6% per year, on average, over the quarter century from 1988 to 2013, which they consider insufficient to sustain the current U.S. standard of living. P. O. Roberts and Fauth (1988) see the GNP growing at about 2.3% per year until the turn of the century, and trending down to 2.0% approaching 2020. In the aggregate, these figures support the conclusion that the rate of growth of the American economy is currently lagging somewhat behind that of the world economy and the expectation that it will continue to do so for the near-term future at least. This means that our share of the world economy is on the decline and is likely to continue to be so.

Among the economic statistics that I find the most disturbing are those that appear to indicate a growing concentration of wealth in fewer and fewer hands. The proportion of total income in the United States that is going to the least affluent segment of the population (the bottom 60%) has been decreasing over the past couple of decades, while the proportion going to the most affluent segment (the top 20%) has been increasing. Excepting homes and real estate, 54% of all net financial assets in the nation are owned by the top 2% of all families; 86% of these assets are owned by the top 10% of all families; the bottom 55% of families have zero or negative financial assets (Thurow, 1987).

An increasingly visible and troublesome indication of the plight of people at the bottom of the distribution is the homeless population. Homelessness is a problem of growing proportions in the United States, although estimates of the number of people who are homeless vary over a very large range (from a quarter of a million to 3 million; Rossi, 1989). Whatever the correct number is, it is unacceptably large. The problem, undoubtedly, is more complex than is generally believed. Researchers do not entirely agree on what would constitute effective approaches toward solving it, but until more effective approaches are developed, the problem will remain a shame to the nation.

The robustness of an economy is a complex function of many variables. Especially important among those variables, however, is the efficiency with which goods and services are produced. When it costs more to produce goods and services domestically than to import them, the balance of trade suffers and the growth of the local economy is slowed. Human factors research aimed at increasing productivity is obviously relevant to this problem. Moreover, productivity can influence standard of living in two ways: When the productivity of one country increases relative to that of others, that country's products become more competitive on world markets; and increases in worldwide productivity mean more goods and services available generally for the same cost. We shall return to the topic of productivity shortly, because of its importance to economic well-being and its appropriateness as a target of human factors research.

Perhaps the other single most important determinant of the economic well-being of a country, according to many economists, is consumption. Consistent consumption by a population of more than it produces requires an outflow of cash, which cannot be sustained indefinitely by the selling of assets or by borrowing. The modification of consumer behavior, at either the individual or national level, has not been viewed traditionally as within the human factors domain. Perhaps the field has nothing to offer with respect to this aspect of the problem; on the other hand, the problem is related to other issues (e.g., manufacturing for durability and maintenance versus planned obsolescence) that are of interest to the field.

COMPETITIVENESS

Seventy percent of the goods produced in the United States compete directly with goods produced abroad (President's Commission on Industrial Competitiveness, 1985). Competitiveness in world markets is determined by the relative prices of the goods to be sold. Because labor represents a large fraction of the costs that must be recovered, an economy with relatively high wages will be competitive only if the productivity of the wage earners is commensurately high. Competitiveness is clearly related to balance of trade, but it should not be equated with it, inasmuch as it is possible to have a trade surplus and still be in poor shape financially, especially if the surplus is needed to service debt. The proper test of competitiveness is a country's ability to balance its trade while achieving an acceptable rate of improvement in its standard of living. It is possible to have either a positive balance of trade at the expense of a decreasing standard of living or an increasing standard of living at the expense of a negative trade balance, for a short time (Hatsopoulos et al., 1988).

The President's Commission identified four causes for the recent decline in U.S. competitiveness: failure to develop human resources as well as other nations, inadequate incentives for saving and investment, trade policies not adequate to today's international commerce, and shortcomings in commercialization of new technology. Other factors that can contribute to competitiveness are restraint in consumption (a willingness to live somewhat more frugally) and a strong work ethic (L. R. Klein, 1988).

The failure of the United States to develop its human resources as well as other countries is a particularly bothersome conclusion from the Commission's study. At the present time, the number of engineering doctorates awarded in the United States to non-U.S. citizens exceeds the number awarded to U.S. citizens, a fact that prompted W. E. Massey (1989) to observe that the country's "trade deficit" extends not only to goods and services, but to human capital as well. Although the problem of human resource development does not seem to fit comfortably within the domain of human factors, as usually conceived, the question again arises as to whether the human factors community can address it. I return to this question in discussing education and training.

The problem of inadequate saving and investment in the United States is closely related to that of overconsumption. Currently, the United States saves about 2% of its national income, compared with an average of 11% among other industrialized countries. The overall net savings rate in the United States has decreased from 7.9% in 1970–1979 to 2.1% in 1985–1987; the personal savings rate was 3.2% in 1987, compared to 8.0% in the 1970s (Dentzer, 1989). Hatsopoulos et al. (1988) argue that there is no reason to believe the United States can remain a first-rate economy unless it has an average savings rate close to that of its competitor nations. The situation is compounded by the fact that a significant percentage of savings are used, via the purchase of U.S. Treasury securities, to service the national debt and are therefore not available for investment in new business development. Investment in U.S. industry, therefore, is being financed to an increasing extent by foreign capital.

Complicating the problem of the low rate of personal saving in the United States is the scandalous mismanagement of a large segment of the banking industry. In 1987, the General Accounting Office's annual audit of the Federal Savings and Loan Insurance Corporation, which insures thrift institution deposits up to $100,000, found it to be insolvent. The federal financial exposure from the thrift industry was estimated to be $30 billion, a figure which now appears to have been low by an order of magnitude or more. According to the General Accounting Office (1987a), the problems with the nation's lending institutions developed over time as a consequence of institutions being allowed to hide, or delay recognition of, certain losses and to continue to operate with insufficient capital. The long-range implications of these problems for the economy, and of deeper troubles they may hint at, are not yet clear.

That the problems that have surfaced in the finance industry may be symptomatic of deeper troubles in the U.S. economy is a serious worry. Indicative of this possibility is what some observers see as the squandering of an unquantified, but probably large, percentage of the country's intellectual energy and other resources on a variety of financial games—what R. B. Reich (1983) called "paper entrepreneurism"—in which one wins by increasing one's personal wealth or improving a corporation's near-term earnings statements, but contributes nothing to the nation's productivity or the general standard of living.

One of the most striking aspects of the troubles that surfaced in the banking industry during is the extent to which they caught the American public, as well as many business people and financial professionals, by surprise. Was this because of a general lack of understanding of how the industry works? Was it because of the unavailability or inaccessibility of data that would have pointed to the fact that a serious problem was building?

What could be developed by way of information systems that would provide the average person with the kind of information needed to make enlightened decisions about savings and investments? One can imagine systems that would provide information, at various levels of detail, regarding investment options available to individuals with specific interests and assets. They would indicate not only what was available, but what the risks and contingencies would be in each case. Design, implementation, and evaluation of such systems would involve a variety of human factors issues relating not only to information representation and presentation but to such methodological issues as how to determine whether the information made available is understood well enough to be the basis for decision making.

MANUFACTURING

Manufacturing in the United States is also in trouble (Office of Technology Assessment [OTA], 1990a, 1990b). U.S. manufacturers have been losing market share to foreign competitors in many areas: steel, automobiles, textiles, appliances. Even in the case of semiconductor production equipment, the market share held by the United States dropped from over 75% in the early 1980s to less than 50% ten years later. "The history of consumer electronics is a history of successive retreats by American firms, with the result that foreign manufacturers have won an entire market without ever having to fight a pitched battle" (Dertouzos, Lester, Solow, et al., 1989, p. 12). Global economic pressures are likely to force significant changes in the long-standing U.S. approach to industrial production:

> The industries in which the United States can retain a competitive edge will be based not on huge volume and standardization, but on producing relatively smaller batches of more specialized, higher valued products—goods that are precision engineered, that are custom tailored to serve individual markets, or that embody rapidly evolving technologies. Such products will be found in high-value segments of more traditional industries (specialty steel and chemicals, computer-controlled machine tools, advanced automobile components) as well as in new high-technology industries (semiconductors, fiber optics, lasers, oil technology, and robotics). (Reich, 1983, p. 13)

The U.S. steel industry is one of the most obvious cases of a highly capitalized mass-production operation that is struggling, having decreased its employment roster by 60% (from 500,000 to 200,000) between 1975 and 1987 and having lost $6 billion from 1982 to 1987 (Szekely, 1987). Szekely suggested that hope for the long-term viability of this industry lies in its switching to the manufacture of novel high-value-added products that requires implementing innovative steel-making technologies for producing customized steels based on less energy- and capital-intensive processes. The products from such operations would acquire their value from the unusual processing involved.

R. B. Reich has warned that the organization of high-volume production is so different from that of flexible-system production that changing from the former to the latter will be difficult, but necessary. Others also have stressed the need for flexibility to enable a quick response to changing consumer tastes and facilitate the rapid application of new industrial technologies if this industry is to maintain competitiveness in the international arena (Cyert & Mowery, 1989; Landau, 1988; Lawrence & Dyer, 1983; OTA, 1988c). The importance of flexibility becomes apparent when one realizes that probably over half of the industrial production in the United States comes from production lots of 50 units or less (N. H. Cook, 1975). Carried to an extreme, flexible-system production can become customized manufacture where one-of-a-kind items can be produced for individual consumers. This is part of the vision of some futurists (Toffler, 1980).

An often-cited study by Dertouzos et al. (1989) took a "bottom-up" look at eight sectors of American industry: automobiles; chemicals; commercial aircraft; computers, semiconductors, and copiers; consumer electronics; machine tools; steel; and textiles. The study focused explicitly on production systems and productivity in manufacturing. No attention was paid to services, agriculture, mining, or construction. The primary conclusion drawn by these investigators was that American industry shows "worrisome signs of weakness". Although they explicitly noted that there is no cause for despair, inasmuch as many American firms are doing very well indeed, they saw in their results, in the aggregate, symptoms of systematic and pervasive ills. In their view, problems with American industry cannot be remedied simply by putting more energy into past practices. "The international business environment has changed irrevocably, and the United States must adapt its practices to this new world" (p. 8).

The need for such adaptation appears to be gaining increasing recognition. A recent report from the OTA (1990a) characterizes how organizational patterns in American industry are changing in terms of a set of contrasts relating to production, personnel practices, job ladders, training, and overall corporate strategies; these are shown in Table 2.1. The two central themes of these changing patterns, according to the OTA report, are "(1) reorganizing production so that lot sizes can be smaller and production runs shorter with little sacrifice in efficiency, and (2) transferring decisionmaking authority downward and outward to semiautonomous divisions and/or the shopfloor" (p. 6).

TABLE 2.1 Changing Organizational Patterns in U.S. Industry

Old model	*New model*
Mass production, *1950s and 1960s*	*Flexible decentralization,* *1980s and beyond*

Overall strategy

• Low cost through vertical integration, mass production, scale economies, long production runs. • Centralized corporate planning; rigid managerial hierarchies.	• Low cost with no sacrifice of quality, coupled with substantial flexibility, through partial vertical disintegration, greater reliance on purchased components and services. • Decentralization of decisionmaking; flatter hierarchies.

Production

• Fixed or hard automation. • Cost control focuses on direct labor. • Outside purchases based on arm's-length, price-based competition; many suppliers. • Off-line or end-of-line quality control. • Fragmentation of individual tasks, each specified in detail; many job classifications. • Shopfloor authority vested in first-line supervisors; sharp separation between labor and management.	• Flexible automation. • With direct costs low, reductions of indirect cost become critical. • Outside purchasing based on price, quality, delivery, technology; fewer suppliers. • Real-time, on-line quality control. • Selective use of work groups; multi-skilling, job rotation; few job classifications. • Delegation, within limits, of shopfloor responsibility and authority to individuals and groups; blurring of boundaries between labor and management encouraged.

Hiring and human relations practices

• Workforce mostly full-time, semi-skilled. • Minimal qualifications acceptable. • Layoffs and turnover a primary source of flexibility; workers, in the extreme, viewed as a variable cost.	• Smaller core of full-time employees, supplemented with contingent (part-time, temporary, and contract) workers, who can be easily brought in or let go, as a major source of flexibility. • Careful screening of prospective employees for basic and social skills, and trainability. • Core workforce viewed as an investment; management attention to quality-of-working life as a means of reducing turnover.

Job ladders

• Internal labor market; advancement through the ranks via seniority and informal on-the-job training.	• Limited internal labor market; entry or advancement may depend on credentials earned outside the workplace.

Training

• Minimal for production workers except for informal on-the-job training. • Specialized training (including apprenticeships) for grey-collar craft and technical workers.	• Short training sessions as needed for core workforce, sometimes motivational, sometimes intended to improve quality control practices or smooth the way for new technology. • Broader skills sought for both blue-and grey-collar workers.

From Office of Technology Assessment, 1990a.

16

What role automation will play in manufacturing in the future is a question that evokes much speculation. One thing that seems clear is that the nature of industrial automation is likely to change qualitatively. Robots with very limited versatility are used widely in high-volume operations (e.g., Unimate machines in auto assembly); it is expected that over the next few decades "smart" robots will become sufficiently versatile and inexpensive to be used in low-volume operations, as well. The goal of much current research on robotics and automation is to increase the flexibility of robotic devices so as to enhance the ability of manufacturers to respond rapidly to the changing demands of their markets (Wilfong, 1989). Abilities that robots are expected to have within the next few decades include: vision (sufficient to discriminate among various objects on an assembly-line conveyer belt or in a parts bin), mobility (permitting relatively free movement from place to place), much greater dexterity (grasping extremities modeled closely, in some cases, after the human hand with fingers, wrists, and many degrees of freedom of movement), and hand-off capability (selection of parts from a parts bin by a transporter robot, and delivery and hand-off to an assembler robot; Kinnucan, 1981).

In recent years, some attention from researchers has been given to human factors issues relating to manufacturing. Much of this work has focused on the control and operation of flexible and computer-aided manufacturing systems (Ammons, Govindaraj, & C. M. Mitchell, 1988; Hwang & Salvendy, 1984, 1988; G. I. Johnson & Wilson, 1988; Sanderson, 1989; Sharit, 1985; Sharit, Chang, & Salvendy, 1987; Wall, Clegg, & Kemp, 1987). Some of it has also focused on scheduling, which is especially important to manufacturing operations because inefficient use of either labor or equipment, and maintenance of unnecessarily large inventories of either parts or finished products, make for higher-than-necessary costs. According to Malone and Rockart (1991), the textile-apparel industry spends about $25 billion in inventory costs annually, about half of which might be saved through better coordination of the various activities involved in bringing clothing to the point of sale.

Sanderson (1989) has reviewed the literature on human scheduling and related it to the needs of modern manufacturing systems that use various degrees of automation for planning and scheduling purposes. Although attempts have been made to automate scheduling completely in some instances, and have human beings do it all in others, the evidence seems to suggest that most cases involving any significant degree of complexity are best handled by some combination of human and computer capabilities applied through an interactive system. Sanderson cautioned that generalizations based on the limitations of current scheduling algorithms could be invalidated as more adequate algorithms are developed, but she also noted that investigators tend to assume that interactive systems are unlikely to be replaced soon by fully automated ones and may be around indefinitely.

Manufacturing is, in my view, an aspect of U.S. industry on which human factors research could have a substantial impact, especially in view of the need for new and more flexible approaches to production and the increasing involvement of information technology in all aspects of production processes. Total automation is unlikely to be realized in many industries in the foreseeable future. What is likely is the increasing use of semiautomated and interactive processes in which people interact with semi-intelligent machines and systems in new and ever-changing ways. The types of problems that will have to be resolved to make flexible manufacturing systems truly cost-effective in today's world economy are precisely the kinds of problems on which human factors researchers have traditionally focused, although some of them are likely to appear in new forms.

MATERIALS

An area of very active research that could have far reaching implications for manufacturing in particular, and for the economy more generally, is that of materials science. A great deal of effort is going into the development of new materials and of methods for processing materials (OTA, 1988c), and this effort has resulted in considerable progress in understanding materials and their composition. The ability to control more accurately the chemical composition of materials, coupled with recently developed techniques for rapid solidification (bringing a material from a liquid to a solid state by lowering its temperature at rates of up to a million or even a billion degrees per second) provides an enormous range of possibilities for the development of materials with made-to-order characteristics (Chou, McCullough, & Pipes, 1986; Liedl, 1986; Steinberg, 1986). By controlling a material's microstructure, one can design that material to match the distribution of stresses that it is expected to endure. An increasing variety of new materials (ceramics, polymers, fiber composites, metal matrix composites) with made-to-order characteristics (high strength-to-density ratios, wear resistance, high or low electrical or thermal conductivity, flexibility) will become part of the array of "raw" materials for production and manufacturing.

The image of polymers (plastics), at one time associated with cheapness and goods of low quality, has changed for the better as these materials, sometimes reinforced with carbon fibers, are finding more and more uses in the production of ultradurable goods with many desirable properties. In 1976, plastics replaced steel as the nation's most widely used material. Now we use more plastics—over 10,000 varieties of them—than steel, aluminum, and copper combined (Childes, 1985). The production of plastics requires less energy than does the production of some of the metals they can replace;

unfortunately, plastics can cause serious waste management problems, about which more further on.

Interest in materials and how to make them more useful for specific purposes undoubtedly predates recorded history, but until recently, much of what was known about materials was based on empirical investigations, not on well-developed theories of atomic forces and energies (Slichter, 1988). The availability of supercomputers has helped the advance of materials science by making possible the kinds of calculations necessary to represent the atomic forces and energies within molecules and their implications for molecular structure and activity, and to represent the dynamics of collections of atoms or molecules. Computer simulation is proving to be an indispensable tool for synthetic chemists and materials scientists studying the interactions of materials at the atomic or molecular level. Such simulations allow researchers to observe on a computer display the motion of individual atoms as they interact in the growth of a crystal or the docking of a drug molecule at the receptor site of an enzyme. They permit chemists and materials scientists to see, literally, how the interactions of molecules depend on the details of their shapes. They also permit researchers to explore the consequences of adding, subtracting, shifting, or substituting atoms in synthetic materials without having to deal with the materials themselves (Goddard, 1988).

Composites currently being investigated include matrices of organic resins or metals with embedded high-strength fibers such as graphite, glass, or silicon carbide. (The use of carbon-epoxy composites to build Voyager, the first plane ever to fly around the world without refueling, gave it the ability to carry five times its weight in fuel and cargo [Childes, 1985]). Fibers of graphite, tungsten, or niobium may be embedded in copper to improve heat conduction; silicon carbide, silicon nitride, and aluminum fibers are being used to add strength to ceramics (Corcoran & Beardsley, 1990). In addition to providing high strength-to-weight and stiffness-to-weight ratios, such composites are typically resistant to fatigue and corrosion, can have a near-zero coefficient of expansion, and are relatively easy to mold into complex shapes (Economy, 1988). The integration of ceramic thin films with semiconductor technology is an active area of research that holds the potential for a new class of electronic devices with numerous applications in communication technology (Sayer & Sreenivas, 1990). The use of ceramics for a variety of purposes in industry has the attractive aspect that the elements from which ceramics are made are among the most abundant on earth.

Some scientists see the spectacular advances that are currently being made in materials science, especially with ceramics, composites, and semiconductors, as having profound implications for productivity and competitiveness in the world marketplace. One gets some feel for the potential of this impact from the fact that "on the average, every person in the U.S. requires the securing and processing of some 20,000 pounds of nonrenewable, non-fuel mineral resources each year" (J. P. Clark & Flemings, 1986, p. 51). The OTA (1989a) has estimated that the value of components produced in the United States from advanced materials will grow from about $2 billion in 1988 to nearly $20 billion by 2000. It cautions, however, that whether the United States will lead the world in commercializing these materials, as it has in developing them, is uncertain.

The U.S. government spends about $170 million per year for research and development on advanced materials, more than is spent by any other nation. In the past, the United States has spent more on advanced materials research and development than Japan, West Germany, France, and the United Kingdom combined. The magnitude of its lead, however, has diminished. Between 1965 and 1986, the United States' share of the total research and development of the five countries just mentioned dropped from 69% to 55%. As of 1987, Sweden, Japan, and West Germany spent a greater proportion of their GNP on research and development than did the United States. About 70% of the U.S. expenditures (and 50% and 34% of the expenditures of the United Kingdom and France, respectively) went to defense-related projects; in Japan and West Germany, almost all of it (95% and 87%, respectively) went to nondefense objectives (National Science Foundation, 1988). In addition, the United States has not been as aggressive as some other countries in initiating programs to commercialize the products resulting from this research (OTA, 1989a).

DEVELOPMENT VERSUS COMMERCIAL EXPLOITATION

The United States has led the world in developing new technologies, but it has been more effective at generating new knowledge than in applying that knowledge to the production of goods and services (Cyert & Mowery, 1989). It is in a very strong position in the science of high-temperature superconductivity, for example, but is not pursuing commercial applications very aggressively (Institute of Electrical and Electronics Engineers, 1988; OTA, 1989a). There is a great deal of concern, both in the government and in industry, that the United States will lose the race in the development and commercialization of high-definition television (HDTV), primarily because it is not running very hard (C. Norman, 1989). A variety of efforts to promote industry–government collaboration on the development of HDTV in this country have not met with much success so far (Corcoran, 1989).

The OTA (1989a) noted that similar observations can be made about many American industries. According to a recent report from the Technology Administration of the U.S. Department of Commerce (1990), by 2000 the United States will be behind Japan in most emerging technologies and behind Europe in several of them. Emerging technologies cited in the report include advanced materials, semiconductor devices, artificial intelligence, biotechnology, digital imaging technology, flexible computer integrated manufacturing, high-density data storage, high-performance computing, medical devices and diagnostics, optoelectronics, sensor technology, and superconductors.

Starting new companies around new technologies is risky business in the United States. If one needs evidence of how difficult it is to predict how new technologies will fare in the marketplace, the carcasses of countless technology-based start-ups that did not make it to the second round of capital funding should provide it (although it is quite amazing how rapidly the visible remains of defunct companies can disappear from view). Established corporations that have decided to put large amounts of money into new technology-based ventures have also guessed wrong on occasion. Wright (1990b) points out, for example, that "Knight-Ridder sank $50 million into its Viewtron project before folding it in 1986, while the IBM-Sears joint venture, Prodigy, has gathered only 200,000 subscribers. Scholastic, Inc., lost millions in its first bout with educational software in the early 1980s. No one has cashed in on electronic banking either; years after its introduction, fewer than 100,000 people nationwide bank by computer" (p. 85). What the government's role should be in nurturing the development of new high-technology industries has been, and continues to be, a matter of considerable debate (Corcoran, 1990b).

Some economists have been pointing to the importance of chance and of positive feedback processes in the economy as important factors in determining the success or failure of new ventures. Such processes can magnify small effects, and help to lock in and amplify an initial advantage that some product or nation has gained over competitors, perhaps strictly by chance. According to this view, two competing products that are equally good may start life with small and equal claims on the market. Eventually one product will begin to get ahead of the other, possibly as a consequence of some chance event. Once it is in the lead, the lead itself becomes an advantage that tends to increase the distance between that product and its nearest competitor. Arthur (1990) argues that knowledge-based parts of the economy are more likely than resource-based parts (agriculture, goods production, mining) to be subject to such positive feedbacks. "Standards that are established early (such as the 1950's vintage computer language Fortran) can be hard for later ones to dislodge no matter how superior would-be successors may be" (p. 99).

Such effects undoubtedly occur and complicate greatly the task of predicting how specific companies or countries will do in the marketplace with respect to specific products or technologies. One expects these effects to distribute themselves more or less randomly, however, and they do not account for consistent and systematic trends such as the regular loss of market share by the United States to foreign competitors in a large number of emerging technology areas.

Human factors is closely identified with the problem of designing machines and equipment in such a way that they are safe to use and that the requirements of their use match well the capabilities of their intended users. Referring to a piece of equipment as "well human-engineered" usually means that the device is well-designed from the user's point of view. This is appropriate, inasmuch as a primary objective of human factors work has been to help ensure the safeness and usability of devices intended for human use. Less attention, however, has been given to the problem of designing production procedures and processes that will yield usable products and that will do so in ways that are both cost-effective and desirable from the worker's point of view.

Of the various ways in which human factors as a discipline could have a direct and significant impact on the U.S. economy, easing the way from technological innovation to market-worthy products would appear to be a high-leverage possibility. If the United States cannot produce products that compete effectively in world markets, the U.S. human factors community must ask itself to what extent it shares the responsibility for that failure. More importantly, it should see the current situation as an opportunity for impact.

MICROFABRICATION

The ability to make components ever smaller has been, perhaps, the most obvious trend in computer technology since its inception. A natural extension of the continuing progress in miniaturization is interest in the development of ultra-small machines: for example, machines that are sufficiently tiny that they can function inside human organs, blood vessels, or, conceivably, even single cells. According to a recent news article in *Science*, "in the 1990s one of the major themes of physics, chemistry, and materials science is likely to be the study of how matter behaves at a scale of nanometers—billionths of a meter. The ability to design and manufacture devices that are only tens or hundreds of atoms across promises rich rewards in electronics, catalysis, and materials" (Pool, 1990a, p. 26). Goddard (1988) notes that about one sixth of the U.S. GNP requires catalytic manufacturing processes and that, in the search for new catalysts for processing petroleum, which he refers to as the hottest field of catalyst research, molecular engineering has practically replaced traditional research methods.

With the development of the scanning tunneling microscope, it is now possible to see the individual atoms that comprise a material's surface. In April, 1990, two IBM scientists reported in *Nature* their success in positioning individual xenon atoms on a crystal of super-cooled nickel so as to spell out the IBM logo with a total of 35 atoms. Each letter was 50 angstroms in height and the distance between adjacent atoms in the display was about

12 angstroms (Eigler & Schweizer, 1990). A team of scientists from Belcore and AT & T succeeded recently in etching almost 2 million lasers in an area less than one square centimeter; others have succeeded in building devices that can transmit electrons one at a time (Corcoran, 1990d).

This domain of interest is sometimes referred to as *nanotechnology*. If Drexler (1986), who has written extensively on this subject, is right, the technology of ultra-small machines will be a serious enterprise, and people who are less than about 65 in 2020 will have a good chance, thanks to this technology and "cell repair machines" that can slow the aging process, of living in good health for a very long time. As it happens, silicon at the scale of microns is stronger than steel and consequently is a preferred material for the building of ultra-small machines. Inasmuch as silicon has been so widely used in the construction of integrated computer circuits, much has been learned about how to shape it into structures with micron and even submicron feature sizes. With certain extensions and modifications, the technology that has been used in the fabrication of integrated circuits is beginning to be applied to the production of microminiature sensors, motors, pumps, valves, resonators, and various other mechanical devices. At MIT, for example, a motor with a 100-micrometer-diameter rotor has been built using such techniques (Howe, Muller, Gabriel, & Trimmer, 1990; Stewart, 1990).

Materials research and microfabrication make extensive use of computer-based tools for simulation and visualization. The design of such tools involves all the usual human factors interface issues. It has the added dimension that many of the structures that have to be represented visually are too small to be seen "in the flesh." Some of them are too small to be resolved by visible light, so the question of what they would "really look like" if we could only see them does not have a straightforward answer. How to represent such structures and the processes involving them is often a matter of intuition. There is room here for some research.

For the most part, when human factors researchers have attempted to design interfaces that are well-matched to the capabilities and limitations of their users, these machines have been approximately—say, within an order of magnitude—the same size as their users. When this has not be true, they have tended to be much larger than their users: jet aircraft, ocean-going vessels, power generation plants. In these cases the interaction occurs through display and control panels that are close to their users in size. This fact has obvious implications for the design of displays through which information is provided to the users and controls through which users affect the machines. The emergence of nanotechnology and the appearance of ultra-small machines adds a new dimension to the problem of machine design from a human factors point of view. What issues of usability, safety, and comfort are likely to arise vis-à-vis machines that are too small to see? How will one be able to tell whether a machine is functioning as it should? How will such devices be handled? Will there be problems of monitoring, repair, or replacement?

Perhaps the most significant effects the appearance of micromachines will have will be through their use as components in other machines and systems. Their proliferation will undoubtedly cause changes in the way familiar devices are constructed and operated. Just as microcomputers are beginning to appear everywhere—in automobiles, in cameras, in household appliances, in children's toys—so will micromachines, in time, surprise us with their many uses. What this will mean from a human factors point of view remains to be seen; what seems clear is that the development of this technology will broaden considerably the connotation of person–machine interaction and present some human factors problems that we can, at this point, only vaguely anticipate.

SERVICES

Future growth in the U.S. economy is expected to come more from the services sector than from manufacturing. Aggregate consumer expenditures on services have been increasing gradually but steadily since the mid-20th century. Some of the basic services that were traditionally provided at home by family members (e.g., care of children and the elderly) are now being purchased (OTA, 1989a). Service industries now account for over 70% of the GNP in the United States and about 75% of all jobs. From 1976 to 1986, services accounted for 85% of new private-sector jobs (Quinn, Baruch, & Paquette, 1987). Botkin, Dimancescu, and Stata (1984) characterized the economic changes that are currently taking place as a move from a capital-intensive, physical-resource-based economy to a knowledge-intensive, human-resource-based economy. The general shift from goods-producing to service-producing activities is a worldwide phenomenon (Klein, L. R. 1988).

Quinn et al. (1987) described the services sector as including "all economic activities whose output is not a physical product or construction, is generally consumed at the time it is produced, and provides added value in forms (such as convenience, amusement, timeliness, comfort or health) that are essentially intangible concerns of its first purchaser" (p. 50). They pointed out that it can sometimes be difficult to draw the line between products and services, because in many cases they are interchangeable (e.g., a home washing machine—a product, and use of a laundromat—a service), and the value of products often stems from the services they provide (e.g., transportation from an automobile, entertainment from a television set).

Manufacturing and product development are perhaps what come to mind first when one thinks of how human factors relates to the economy, because they suggest machines and devices; and the interaction of people with machines and devices has been a central concern of human factors research. In fact, a great deal of the work done by human factors researchers in the past has had at least as much relevance to service industries as to manufacturing and production. Problems pertaining to the layout and lighting of workspaces, the design of visual display terminals, the readability and intelligibility of instructions and printed forms, the effects of stress of various types on human performance, the development of decision-making aids, the study of human error in supervisory control situations, and numerous other subjects are encountered in the service sector as well as elsewhere, and research addressed to these problems has many applications in service industries.

The growing importance of services, as distinct from products, to the U.S. economy, and indeed to the world economy, should motivate even more thought on the question of how human factors know-how can be used to help improve the performance of the service sector. There must be many human factors issues pertaining to services that deserve attention and that represent opportunities for improvement. The challenge is to identify them and to develop the approaches necessary to work on them effectively. As the fraction of the population that is employed by service industries continues to grow, issues of job satisfaction in these industries will become increasingly important as well.

PRODUCTIVITY

A general answer to the question of what the human factors community might do to help improve the U.S. economy—or some sector thereof—is anything that would increase productivity. There appears to be a broad consensus (a) that the United States must improve its productivity if it is to stay competitive in world markets and (b) that only if productivity is increased globally will there be any hope of improving the standard of living everywhere.

Productivity is one of those words that most of us probably believe we understand until we have occasion to try to say precisely what it means. As applied to the production of material goods, productivity is usually expressed as a ratio, for which the numerator is some measure of output (what is produced or the value of same), and the denominator is some measure of input (what is used up in the production process or the cost of same). In theory, the same ratio may pertain when what is produced is a service or other intangible good, but in practice, quantification can be very difficult because of uncertainty about what exactly gets consumed in the production process or about what the "product" is really worth.

Closely related concepts are those of *effectiveness* and *efficiency*. A productive individual, system, industry, or nation is one that gets the intended job done with a minimum of wasted resources. If whatever one produces has value, increasing the efficiency of one's production is tantamount to increasing productivity. Whenever it is possible to accomplish the same task with fewer resources or to get more done for a given cost, we would probably say that productivity has increased. It suffices for our purposes to settle for this connotation of *productivity*, vague though it may be. In what follows, the word appears in a variety of contexts and undoubtedly does not always have precisely the same connotation. I pass it on as I found it, and assume that the intended meaning is usually adequately clear.

Input variables relevant to estimates of industrial productivity include labor, capital, energy, and materials. When focusing on competitiveness, profit margins constitute another factor in the equation, because they contribute to product prices. One definition of *labor productivity* is "the total market value of all goods and services produced, divided by the number of labor hours that went into the production process" (Baumol, 1989, p. 611), or dollars of output (adjusted for inflation) per hour worked. *Multi-factor productivity,* which takes into account inputs other than labor, is then defined as "the market value of total output divided by the market value of all pertinent inputs" (p. 611).

There seems to be broad agreement that a nation's productivity is a primary determinant of the general standard of living of its populace (Baumol, 1989; Hatsopoulos et al., 1988; L. R. Klein, 1988; Young, 1988). Some economists argue that the only way for the United States to deal effectively with the budget and trade deficits—which requires increasing savings and investment while reducing spending on consumption—without suffering a decrease in standard of living is to realize a greater rate of growth in productivity than it has in the recent past (Cyert & Mowery, 1989). Productivity per worker in the United States is expected by some observers to grow by about 1.4% per year, on the average, over the next quarter century, with a higher rate (3.4%) in manufacturing and a much lower one in the service industries (D. Lewis et al., 1988). D. Lewis et al. (1988) cite lower worker productivity as the greatest threat to the United States' long-term economic competitiveness and standard of living. (*Productivity* here is defined as GNP divided by the total number of people in the workforce.) A reversal of the recent trend and an increase in productivity in the service sector has been seen by some as the key to domestic economic growth in the future (Johnston & Packer, 1987).

SLOWING PRODUCTIVITY GAINS

The United States leads the world in overall economic output, as measured by GDP per employed person or by output per worker hour; however, several countries have been increasing productivity faster than the United States has since the 1950s, and, as of 1987, some of them had closed the gap considerably (Bureau of Labor Statistics, 1988a, 1988b). It has been claimed, for example, that between 1950 and 1983 the output per hour of U.S. workers increased by 129%, whereas that of Japanese workers increased by 1,624%; the increase in the productivity of Canadian, French, and West German workers fell between these extremes (Johnston & Packer, 1987). According to Thurow (1987), the rate of increase in productivity declined from 2.7% per year for the period from 1960 to 1970, to 0.9% per year from 1979 to 1985. Hatsopoulos et al. (1988) give 2.2% as the annual rate of growth in productivity between 1973 and 1985, compared with 3.8% for the period 1960 to 1973. They cite the annual rate of increase in output per worker as 0.3% and 1.9% for the same respective periods. Dertouzos et al. (1989) claim that increases in U.S. labor productivity averaged slightly under 3% from 1948 to 1973, a little over 0.5% from 1973 to 1979, and about 1.5% from 1979 to 1986. Small differences, they point out, can have very large cumulative effects: Had labor productivity continued to increase at 3% per year during the 1970s and 1980s the United States economy would now be almost 50% more productive than it is. The recent history of multifactor productivity, according to their estimates, is similar to that of labor productivity, averaging about 2% from 1948 to 1973, and going almost to 0% between 1973 and 1979. From 1979 to 1986 it bounced back up to roughly 0.5%.

Although most of the writing about recent trends in productivity in U.S. industry has been gloomy and much of it startlingly pessimistic, a more optimistic picture has been painted by Baumol (1989), who argues that despite legitimate grounds for concern, the situation is not ominous. Slowdown in productivity growth, particularly after 1965, has been universal in the industrial world and "probably attributable in good part to the exhaustion of the opportunities for spectacular productivity gains that had accumulated during the Great Depression and World War II" (p. 612). It is true that the productivity of most major industrial countries has been growing faster than that of the United States in recent years, Baumol argues, but this is mostly a matter of their playing catch-up. Given the ease with which technology can now be transferred between countries, it is to be expected that the productivity of leading industrial economies will converge, so that those that started behind will increase at a greater rate until they catch up. No other country has yet surpassed the United States in overall productivity level, he contends, and none is likely to do so in the near future.

As welcome and refreshing as this relatively optimistic view is, it is not a great source of comfort, because it fails to reflect what many economists who are writing about productivity are saying, and because among the factors that Baumol considers the determinants of productivity are some with which the

United States appears to be having considerable difficulty. One in particular, which is discussed further in chapters 5 and 11, is the skill and training of the country's labor force. Dale (1991), who, like Baumol, is generally optimistic about the U.S. economy and its prospects, also sees the problem of public education as a "potentially menacing" threat if not dealt with adequately, because of its implications for the quality of the future workforce.

In comparing productivity estimates from various sources, one is struck by the considerable differences among them, or at least by the appearance of such differences. Such a comparison is complicated both by the use of different time periods by different writers and by the uncertainty, in some cases, as to what type of productivity (e.g., labor, multifactor) is involved. What is significant is the general agreement among the estimates as to the slowing of the rate of productivity increase in the United States over the past few decades, especially relative to what has been happening on the world scene. Dertouzos et al., (1989) point out that a decline in productivity was experienced by every advanced industrial country beginning around 1970. What is most disquieting to U.S. economists is the fact that the United States has been trailing all other industrial countries in productivity growth since about that time. There are also indications that the country is not doing as well as it should with respect to the quality of its products and services or the speed with which its products are developed.

There is considerable variation in productivity growth from industry to industry and in many cases this growth has been negative in recent years. Manufacturing productivity growth increased sharply over the very recent past going from about 1.5% and 0.5% for labor productivity and multifactor productivity, respectively, during the 1973 to 1979 period to about 3.5% and 2.5%, respectively, during the 1979 to 1986 period. This is an encouraging development, but as Dertouzos et al. (1989) point out, the improvement was obtained, in part, by shutting down inefficient plants and laying off workers. By their calculations, the loss of jobs accounted for about 36% of the improvement in labor productivity. Also, because the improvement followed a recession, which is the normal course of events as factory output is increased to take up slack in the economy, they question whether the improving trend can be sustained. Mishel (1988) has argued that the appearance of a substantial increase in the output of U.S. manufacturing since 1973 is largely an artifact of an erroneous downward adjustment made to government statistical data on output and productivity in 1973.

Although many economists seem to take as a given the low productivity of the service sector relative to that of manufacturing, Quinn et al. (1987) concluded from a 3-year study that they should not do so. They noted that many service industries are as large-scale, as capital-intensive, and as grounded in technology as is manufacturing, and argued that technology, properly applied, can enhance the productivity of the service sector as it has that of manufacturing. They also pointed out the need, however, to be cautious in interpreting aggregate productivity data about services.

The more productivity figures one sees, the more likely one is to despair of getting a clear picture of productivity trends in which to put great confidence. In the aggregate, however, such figures provide a fairly strong signal that all is not well in this area. Increasing productivity in all economic sectors is among the major challenges facing the U.S. economy, and, indeed, the world economy, in the foreseeable future. An attractive aspect of the goal of improving productivity is the fact that success is not necessarily realizable only at someone else's expense. To be sure, more efficient production can result in greater competitiveness in specific marketplaces because it permits the selling of goods and services at lower prices, but in a larger frame of reference, it makes goods and services more widely available to consumers at relatively lower prices. Increased productivity, in the sense of more efficient, less wasteful, use of energy and other resources, is, in the long run, good for everybody; it is the best hope of being the rising tide that can lift all ships.

Finding ways to increase the productivity of individuals, systems, corporations, or other entities is an appropriate concern for human factors researchers and an objective that the human factors community is in a position to help attain. Germane to this issue are research studies on human capabilities and limitations; the cognitive, perceptual, and motor demands of tasks that people perform; motivation; physiological state; training; the way functions are allocated to people and machines; the design of machines (and especially person-machine interfaces); organizational and scheduling factors; and social and interpersonal variables that help define the character of the workplace.

Closely related to the issue of productivity is that of the quality of jobs. It is possible to increase the productivity of an industrial system or of an industry at the expense of degrading the quality of the individual worker's situation. From the human factors point of view, this is a poor trade. The objective should be to increase productivity and simultaneously enhance job situations, or at least not downgrade them.

DETERMINANTS OF PRODUCTIVITY

Determinants of productivity, according to Baumol (1989), include the "country's flow of inventions and innovations, the rate at which it learns to benefit from the flow of technology contributed by other economies, the rapidity with which it increases the capital stock per worker (including the plant and equipment at the worker's disposal), the skill and training of the country's labor force, and the incentives provided for the productive activities of entrepreneurs" (p. 614). Among the generally recognized causes of low productivity are failure of the government and industry to invest adequately in research and development, with a consequent shortage of innovations in the production process; failure to provide workers with adequate tools; inadequate training or motivation of the workforce; and inadequate quality control (resulting in wasted labor). Total private- and public-capital investment in plants and equipment and in infrastructure (highways, bridges, airports) in the United States in the late 1980s, came to about 14% of the GNP; the comparable figures for West Germany and Japan were 18% and 25%, respectively (Dentzer, 1989). Dentzer argued that this relatively low rate of capital investment, coupled with the fact that the U.S. labor force is expected to grow more slowly in the 1990s than at any time since the 1930s, bodes ill for American productivity in the future.

In its study of productivity in the United States, M.I.T.'s Commission on Industrial Productivity expanded the conventional notion of productivity to what it called *productive performance,* which encompassed "efficiency, product quality, innovativeness and adaptability, as well as the speed with which they [industries] put new products on the market" (Berger, Dertouzos, Lester, Solow, & Thurow, 1989, p. 40). This commission identified six "systematic weaknesses" that it considered to be hampering the ability of many U.S. firms to adapt to the changing international business environment:

- The use of outdated strategies: overemphasis on mass production of standard commodity goods and a parochialism that fails to recognize the worth of innovations originating elsewhere.
- Neglect of human resources: economically feasible in a system based on the mass production of standard goods where cost matters more than quality, but a serious problem when production requires flexibility in the workplace.
- Failures of cooperation, (within and among U.S. companies).
- Technological weaknesses in development and production: good at basic research, but poor at turning the results of that research into high-quality products that can compete in the market place.
- Government and industry working at cross-purposes, more so than in the countries that are our major competition.
- A short time horizon: preoccupation with short-term profits and steady growth in annual, or possibly even quarterly, earnings.

Technological innovation is seen by some economists as a primary vehicle for increasing productivity (Cyert & Mowery, 1989; L. R. Klein, 1988; Young, 1988). In the past, the application of technology in the workplace has typically had the effect of increasing the productivity of the individual worker. Perhaps the most striking illustration of this is seen in agriculture: In 1850, the average farm worker produced enough food for 4 people; now, with of the application of technology, one farmer can provide enough for 78 people; the number of person hours needed to produce 100 pounds of cotton in the United States went from about 42 in 1945 to about .07 in 1975 (W. D. Rasmussen, 1982). The application of technology includes mechanization and the use of fertilizers and pesticides, as well as application of scientifically gained knowledge about crop rotation, irrigation, breeding, and hybridization. (Unfortunately, some farming innovations have had costs, in terms of damage to the environment, that were not anticipated; see chapter 4.) A similar effect has been obtained in mining where, as a consequence of mechanization, less than 1% of the U.S. labor force now suffices to produce more than 80% of the country's mineral needs (Marovelli & Karhnak, 1982).

Although mechanization has had unquestioned positive effects on productivity, it has not always been the case that the greater the degree of mechanization the better. In coal mining, for example, it has been found that mines mechanized to an intermediate degree operate more efficiently than either the least or the most mechanized ones (Marovelli & Karhnak, 1982). Determining the appropriate degree of mechanization or automation for any given process is likely to be a continuing challenge to industry. This is one form of the problem of function allocation, and there appears to be no general solution to it. This is a problem, however, to which human factors researchers have paid a great deal of attention in various specific contexts, and it is one that will become increasingly important as the possibility of automation, to varying degrees, becomes a reality for more and more activities, including many in predominantly white-collar businesses.

The United States economy has been making a very large investment in information technology, in the belief that it will increase productivity. Information technology should, for example, contribute to increased productivity through information systems that facilitate the timely manufacture and movement of supplies and products, thus mitigating the need to maintain large, costly inventories. Inventories, including material and supplies, work-in-progress, and finished products, represent an enormous cost to manufacturing operations. Any steps that reduce the need for inventories without introducing delays into the manufacturing process or resulting in the unavailability of goods for a ready market should have an immediate beneficial impact on productivity by lowering capital costs. Electronic communication systems are beginning to be used in some industries to shorten the feedback loop between retailers and manufacturers so as to minimize the production of goods that do not sell and otherwise tune the production processes more finely to the desires and preferences of consumers (Corcoran, 1990d).

Gunn (1982) has argued that the major opportunity for improved productivity in manufacturing lies not in the direct work of making or assembling a product, which accounts for a relatively small fraction of the total cost of manufacturing, but rather in "organization, scheduling and managing the total manufacturing enterprise, from product design to fabrication, distribution and field service" (p. 115). The most important possible contribution to productivity of the factory from new information technology, he suggests, is the feasibility it represents for linking design, management, and manufacturing within a network of commonly-available information. "Information and the ability to transmit it quickly will come to be recognized as a resource as valuable as money in the bank or parts on the shelf" (p. 130).

Malone and Rockart (1991) made a similar point in observing that, unlike the industrial revolution, which was driven by changes in production and transportation, the revolution under way today will be driven by changes in coordination; furthermore, computers and computer networks, which make the coordination possible, "may well be remembered not as technology used primarily to compute but as coordination technology" (p. 128). In the office the use of information technology should increase productivity by reducing information "float," by making information more readily accessible to users when and where they need it (Giuliano, 1982).

In spite of these expectations, which seem very reasonable, numerous attempts to quantify the productivity gains that have been realized as a consequence of investment in information technology have yielded generally disappointing results. Although in some cases gains have been realized, in most they have not; in some instances productivity seems actually to have decreased. Attewell (1990), who reviewed the efforts to determine the effects of information technology on productivity, was careful to note that if it turns out that information has had no effect on productivity, in the sense of increasing the efficiency with which firms produce goods and services, it does not follow that it has left no mark on the world or on firms. Because it unquestionably has helped create new products and services that are bought in open and competitive markets, by definition, it has helped create value. There remains the interesting question, however, of whether the enormous investment that individual companies have made in information technology is paying off for those companies through resulting increases in either their productivity or their profitability. (Attewell pointed out that it is possible to increase profitability without increasing productivity; an increase in sales volume at a fixed level of productivity can generate an increase in profits. Similarly, an individual company can increase its profitability at the expense of competitors by getting a larger portion of market share, also without necessarily increasing the productivity of the industry or the nation as a whole. In this regard, Attewell warned that "if IT [information technology] investment is focused on the strategic goal of market share, and is shunted away from productivity-enhancing areas, the productivity stagnation currently experienced by American industry will continue" [p. 18].)

Attewell identified several mechanisms that could, in his view, negate or attenuate productivity gains that might be realized from the use of information technology. These include substituting slower channels of communication for faster ones (e.g., e-mail, which requires typing, for speech), the formalization of communication, the amount of time people must devote to learning to use new systems, increases in personal output that do not translate into greater cost-effectiveness for the firm (extra paper work or administration that does not contribute to improved corporate efficiency), and the expansive appetite of some managers for quantitative data that may or may not make for better managerial decision-making.

Clearly how the uses of information technology are affecting, or could affect, productivity requires more study. Among the questions that deserve attention are many of a human factors nature: What are the advantages and disadvantages of various methods of intracompany communication, and how does the answer to this question depend on the specifics of the communication situation? How should businesses go about determining what information their managers and other employees really need? (The assumption that managers need all the information they can get leads to very expensive management information systems that are wasteful of staff and other resources.) How can one distinguish between administrative procedures that contribute significantly to valid corporate goals and those that are simply consuming employees' time, or that are being done because of the illusion that technology makes them possible at no incremental cost?

White-collar productivity is of special interest and especially problematic. What is it? How should it be measured? How can it be improved? All of these questions lack definitive answers, but there is no lack of discussion of the subject. There seems to be general agreement among economists that, however defined and measured, white-collar productivity is not what it should be and the rate at which it has been increasing in recent years, in spite of office automation and its attendant wonders, is vanishingly small.

An inescapable conclusion of much of the discussion of white-collar productivity is that the ways in which specific business establishments are organized, the policies that guide their operations, the specific tasks that members of the organizations perform, and the ways in which they do them often lack any very clear rationale. Certain things are done, and they are done in certain ways, because of tradition, convention, or habit. Organizations, jobs, and tasks become self-perpetuating and the reasons for them, assuming there once were some, become lost. Too infrequently is it asked about any specific task, "What is its purpose? What important corporate objective does it serve? Who would suffer if it were not done?" Being busy is not necessarily the same as being productive, at least in any economically meaningful sense. Doing more efficiently something that should not be done at all is not necessarily progress, especially if all the greater efficiency accomplishes is to permit one to do more of the same.

Underlying many of the questions that can be raised about productivity, within white-collar industries and elsewhere, is the need to be explicit about goals and about the extent to which specific policies, practices, and procedures do or do not serve them. Systems analysis techniques that human factors practitioners have applied in many contexts should, perhaps with some adaptation, be useful in making clear how information technology serves, or fails to serve, corporate goals in specific companies and how it contributes to or impedes their productivity.

ORGANIZATIONAL AND CULTURAL EFFECTS ON PRODUCTIVITY

Although organizational and cultural determinants of productivity are not well understood, there is little doubt that there are some. It is known, for example, that many organizations become more productive with experience, and that different organizations "learn" at different rates. Studies have focused on such variables as organizational forgetting (e.g., following a strike or other disruption), turnover, and the degree to which knowledge transfers across products or across organizations. So far, a completely clear picture of the relationship between such variables and productivity has not emerged (Argote & Epple, 1990). Whatever the organizational effects on productivity have been in the past, they may well change in the future if, as a consequence of greater ease of communication, predominantly hierarchical organizations evolve into structures that are more in the nature of networks.

Culture, as the term is applied to organizations, has been defined by Schein (1990) as "(a) a pattern of basic assumptions, (b) invented, discovered, or developed by a given group, (c) as it learns to cope with its problems of external adaptation and internal integration, (d) that has worked well enough to be considered valid and, therefore (e) is to be taught to new members as the (f) correct way to perceive, think, and feel in relation to those problems" (p. 111). Bair (1987) defined *culture* simply as "the common awareness among people that forms a collective understanding for perception of events" (p. 178). Cultures vary in strength and degree of internal consistency with such factors as the stability of the group, the length of time it has existed, and the strength and clarity of the assumptions held by the group's founders and leaders. Schein argued that culture will become an increasingly important concept for organizational psychology because without it, it is not possible to understand change or resistance to change.

Competitive pressures are forcing organizational and cultural change in many companies. Such change is intended to increase productivity by increasing the efficiency of operations, but significant modifications of organizational structures and cultural patterns can be difficult to achieve (Beer & Walton, 1990). Among the major obstacles to organizational and cultural change in industry are the vested interests that some employees have in maintaining the status quo. Changes that are designed to eliminate layers of management, for example, are likely to encounter resistance among managers who perceive themselves as potential casualties of those changes (L. A. Schlesinger & Oshry, 1984). Pressures to make such changes, however, are likely to be strong. The M.I.T. Commission on Industrial Productivity found a trend for greater functional integration and less organizational stratification in essentially all of the successful firms among those it studied. Both of these characteristics, in the commission's view, facilitate quicker product development and greater responsiveness to changing markets (Berger et al., 1989).

Changes of high-level management and of corporate cultures as a consequence of mergers, buyouts, and takeovers have been a prominent feature of American industry in the recent past and the effects on productivity are not clearly known. During the 1980s, 28% of the 500 largest industrial corporations in the United States were acquired by new owners. During that decade, $1.3 trillion worth of assets changed hands. Given that many of the firms taken over were conglomerates that were then sold in pieces to buyers in the lines of business represented by those pieces, some economists have viewed the takeover wave of the 1980s as a return to specialized, focused companies, following years of diversification, a process they refer to as the "deconglomeration" of American business (Shleifer & Vishny, 1990). Although they note that the jury is still out on the question of the ultimate consequences of the takeover wave, Shleifer and Vishny argue that, in view of the disappointing experience with conglomerates, this move toward greater specialization could increase the efficiency of the resulting corporations.

Individual businesses increase their competitiveness by enhancing the quality of their products and by decreasing the costs of producing them. There are many ways to further these goals: enhancement of employee productivity through training and motivation, limitation of inventories through careful production scheduling, and control of quality to minimize waste are a few of them. Unfortunately, recognizing the importance of such objectives is not the same as realizing them. Krafcik (1989) made the thought-provoking observation that since 1981, General Motors spent more than $40 billion—enough to buy the Toyota Motor Company—on the modernization of its production technology, only to see its share of the U.S. car market drop from 45% to 37% and its profitability decline as well. Krafcik also noted, by way of contrast, that in 1984, General Motors invested $200 million (a small number in this context) in New United Motor Manufacturing, Inc (NUMMI), a "medium-tech" joint venture with Toyota. The

Fremont, California plant resulting from this venture "has achieved productivity levels 40% better than typical GM assembly plants, as well as the highest quality levels GM has ever known" (p. 29).

The production management style of the NUMMI plant is characterized as a "lean" approach because it features small parts inventories, small repair areas, multi-skilled workers, and work teams. This is to be distinguished from a "buffered" approach, which uses large inventories and repair areas, and narrowly specialized workers. Krafcik concluded from a study of 60 assembly plants in 15 countries that lean operations are consistently more productive and flexible, and more apt to produce high-quality products, than buffered operations. In fact, lean operations must be more efficient and have better quality control if they are to function at all, because, by design, they leave—in their small parts inventories and repair areas, for example—less margin for imprecise scheduling and error. Krafcik argued that the high expectations inherent in this approach tend to infect suppliers so that they gradually improve their own quality, delivery, and cost levels as well.

Clearly, there is a need to understand much better the effects of organizational and cultural variables on both individual and organizational productivity. That there are such effects is hardly in doubt. We know very little, however, regarding exactly what they are. Similarly, it is important to understand more completely how productivity at any given level of organization relates to productivity at a different level. How, for example, does the productivity of individuals in a particular corporate situation relate to the productivity of the corporation as a whole? How does corporate productivity relate to regional or economic-sector productivity, and how do these relate to the productivity of a nation or the world? It is not enough to assume that the productivity of a composite entity is determined in a linear fashion by the productivity of it parts, because different metrics are used at different levels of analysis and it is possible for behavior that appears to be productive at one level to work at cross-purposes with the goals at a higher level.

I have a friend who tells the story of motivating his children to dig dandelions from his lawn by paying them 10¢ a pound for what they dug. As the story goes, the incentive worked very well and the children dug lots of dandelions. In fact, over time they increased the amount of dandelions they dug very considerably, because what they did was buy dandelion seeds with the proceeds from their initial efforts. The story is apocryphal, I assume, but it makes a good point. It is important that the goals that individuals are encouraged to work toward within an organization be goals whose attainment will further the purposes of the organization as a whole. I suspect that situations in the business world analogous to my friend's dandelion-digging incentive program are not difficult to find.

PARTICIPATION, RESPONSIBILITY, AND QUALITY

An aspect of the lean approach that has significant implications for the individual in manufacturing operations is its use of teams of workers with multiple skills, able, within limits, to do each other's jobs. This is a definite step away from the convention of having individuals on an assembly line, each of whom does only one narrowly defined task over and over ad nauseum. The new approach (a new form, perhaps, of the pre-industrial idea of craftsmanship) is intended to give each worker more scope, a broader view of the production process, a stronger sense of identity with the product, and greater control over its quality. The worker is seen also as a valued source of ideas regarding how to improve, continuously, the production process and its output. Greater flexibility and broader participation by workers in defining and improving their operations assume a better trained workforce, so it is not surprising to discover that the lean approach puts considerable emphasis on worker training on a continuing basis.

The lean approach is in keeping with, and perhaps a version of, what is becoming known in this country as total quality management, or TQM. TQM probably began when modern quality-control techniques were introduced to Japanese industry in the early 1950s by American engineers, in particular W. E. Deming and J. M. Juran. The ideas and methods were adopted with considerable enthusiasm by Japanese engineers and businessmen and became institutionalized in a variety of forms, including quality-control circles and the Deming prize. Quality-control circles (QCs), small groups of workers engaging in quality-control activities on a continuing basis, have been introduced in about one quarter of the companies in Japan that employ 30 or more people. (Although participation in these circles is said to be voluntary, apparently *voluntarily* does not have quite the same meaning in Japan as it does in the West; as Shiba [1989f, p. 33] put it, "The Japanese believe that it will be helpful to their fellow workers' personal growth if they persuade them to join QCs activities." This helps explain why in Japan, so it is claimed, all workers join their section QC circles.)

The total quality control (TQC) movement, as it is referred to in Japan, is seen as a mass movement. A basic tenet of the philosophy is that for an implementation to be successful within a company, it must involve all members of the company, from the topmost executives to the most junior workers. The total commitment of the chief executive officer is seen as a must. Implementation is usually described as a distinctively top–down process; "from top to bottom" in this context connotes not only all-inclusiveness but also the direction of the enabling energy flow. In Japan, the top-to-bottom approach has been evident, in another sense, in the fact that the program has been promoted extensively over several decades at the national level through such organizations as the Japanese Union of Scientists and Engineers and the Japanese Standards Association (Shiba, 1989a, 1989b, 1989c, 1989d, 1989e).

Considerable importance is attached to adherence to certain specific "steps for improvement": identification of the problem, collection and analysis of facts, identification of main causes, planning and implementation of improvements, confirmation of effects, standardization of the process, review of the activities and planning for future work. These steps are sometimes grouped to comprise a "Plan, Do, Check, and Act" cycle. The idea is that the steps of this cycle can be applied iteratively to essentially any process that one wishes to improve, and that it should be applied routinely in one's everyday work. Various quality-control tools have been identified to help accomplish these steps. These include check sheets, Pareto diagrams, and cause–effect diagrams. All members of an organization committed to TQM or TQC are expected to be familiar with and able to apply the tools in the execution of the steps. Not surprisingly, employee training is a major aspect of the implementation of a total quality program in a corporation. Additional tools, referred to as *management tools for quality control,* are intended to help management evaluate quality in non-quantitative ways. These include affinity diagrams, process decision program charts, and so on (Shiba, 1989c).

The approach has been embraced by a number of American companies and by the U.S. Department of Defense (DoD). The DoD recently issued a directive (DoD Directive 5000.51) on total quality management and published a two-volume guide on its implementation (U.S. Department of Defense, 1990). The guide begins with an acknowledgment that the United States, even while continuing the longest peacetime expansion in its history, is "beseiged by serious concerns that threaten our very industrial existence." (p. xxx). What is needed, if we are to cope with such problems as the trade and budget deficits and the rapid loss of economic and technological standing, the guide goes on to say, is a change in our quality culture. The intent of the implementation of a total quality management program is to effect such a change within DoD organizations.

How effective is the total quality approach? There are many claims of increases in productivity (especially as a consequence of improvements in processes leading to larger yields and smaller rejection rates) by companies that have adopted it. I am not aware of any very careful scientifically valid studies on the question. Proponents credit it with having propelled some Japanese companies to leadership positions in their industries, and believe that it can work effectively in the United States as well. Whether it will be easy to implement in U.S. plants, and whether it will necessarily always accomplish what it is intended to accomplish when it is implemented remain to be seen. Implementation in any particular instance may require some customizing to the specific situation if it is to work well from both management's and the workers' points of view (L. Turner, 1989).

I find much of the total quality philosophy and approach very attractive: attaching great importance to listening carefully to one's customers and trying to satisfy them, setting improvement as a constant goal, recognizing that all members of an organization are important to its success, an emphasis on non-confrontational consensus building, the perception of tasks performed in the workplace as means to ends and not ends in themselves, and a stress on putting improvement in the hands of workers. Some of the methods and tools undoubtedly can be useful for specific purposes, but I also have reservations about some aspects of the TQM movement, at least as I have encountered it. These include a considerable rigidity in some aspects of the approach, a degree of precision implied by some of the language that does not seem to be justified, conceptual distinctions that are somewhat arbitrary, insistence that adherents accept the approach uncritically, and a devaluing of creativity and individual genius.

A serious risk is that of believing that one has accomplished more than one really has as a consequence of applying a TQM tool in a specific instance, of convincing oneself that one understands a problem—that one has gained deep insights regarding cause–effect relationships—when one has not. It is the risk of oversimplifying inherently complex problems—and their solutions—that can follow from unquestioning confidence in the power of the tools in one's kit.

There is a need for solid objective data regarding the effectiveness of the TQM approach, especially in view of the fact that businesses are committing nontrivial resources to its implementation. No one, of course, will question the desirability of improving the quality of goods and services or of improving the processes by which they are produced or delivered. The question is whether the TQM approach accomplishes these goals. I suspect that the answer will be that some aspects of the approach work better than others and that how well they work depends on how and where they are applied. Only well-conceived and -executed studies can give us the details that will provide an understanding of the merits of the approach and the ways to tailor it effectively to specific situations.

PRODUCTIVITY MEASUREMENT

Worker efficiency has been a focus of industrial and engineering psychology for decades (F. W. Taylor, 1913; C. B. Thompson, 1914). For better or worse, time and motion analysis is among the first things that come to mind when one thinks of the earliest applications of psychology to industrial settings. In recent years, psychologists have sought to influence productivity less by the study of specific repetitive motions required in assembly-line jobs than by improving the designs of tools and other machines that workers use and with which they interact. Increasing attention has also been given to the use of techniques to promote cooperation among workers, to build team spirit, and to encourage workers and groups of workers to take greater responsibility for the quality of their output (Cole, 1982; R. B. Reich, 1987). Teams are often responsible for products or specific services and are given considerable autonomy and self-management authority (E. Sundstrom, DeMeuse, & Futrell, 1990). Studies of efficiency should aim not only for the identification of ways to improve the design of machines and tools from a human factors point of view, but also for a better understanding of the determinants of team effectiveness and of how to facilitate team performance.

An especially difficult challenge for the future is to find ways to improve productivity in the performance of service-sector jobs. In some ways, productivity is a strange concept to apply to the service sector. The very word incorporates the term *product*, which usually connotes a material good that results from a production process. Productivity in this context can be quantified in terms of the cost of production and the monetary worth of the goods produced. Services can be thought of as products, in the abstract, but the variables involved tend to be less tangible and the identification of appropriate costs and values is less straightforward. Attempts to apply to the service sector productivity measurement techniques that were developed for application to manufacturing have yielded somewhat ambiguous results.

A nontrivial aspect of the challenge to improve productivity of the service sector is the need for better ways to measure productivity of workers in non-manufacturing jobs. Wallich (1990a) captures the problem of measuring the productivity of information workers nicely:

> Even if one focuses on the information-age equivalents of assembly-line workers, useful numbers are hard to come by. The number of checks processed per hour in a bank may be a useful indication of productivity, but other measures, apparently equally objective, have proved disastrous in practice. Counting the key strokes per hour of word processors does not guarantee that those are the right key strokes. Interviews with some telephone operators have revealed that they occasionally resort to hanging up on rafts of callers at a time to meet their hourly quotas. . . . One telephone company found that the increased stress from computer monitoring drove absenteeism to a level that negated any gains from increased productivity when workers were on the job. (p. 94)

In the absence of a better measure of productivity for information workers, statisticians have used the ratio of money paid for wages (adjusted for inflation) to the number of hours worked, which means that information workers' productivity is likely to remain constant almost by definition. Wallich (1990a) put the importance of more meaningful measures of productivity for information workers this way: "Until the NAS [National Academy of Sciences] or someone else develops accurate measures of productivity for the information age, initiatives to increase the productivity number could hurt a nation's economy rather than help it" (p. 95).

Clearly, there is a need for new ways to think about productivity, for new concepts in terms of which to define it, and for new approaches to its assessment. Measurement techniques must take account of quality of work as well as quantity of output, and ideally they should have some sensitivity to the long-term implications of an individual's work. How do we measure the productivity of a Thomas Edison, a Henry Ford, a Wilbur or Orville Wright, a John von Neumann? The long-term economic consequences of what such people produced are incalculable. Their work spawned industries and changed the course of the future in profound ways. How do we quantify the productivity of an Isaac Newton, a Gottfried Liebniz, a Karl Friedrich Gauss, a Niels Bohr, and countless others who made basic scientific discoveries that were critical to the development of modern technology and our present way of life? How do we assess, in terms of productivity, the work of the composers, poets, painters, and other artists who have made immeasurable contributions to the quality of our lives?

As already noted, Attewell (1990) and others have raised serious questions as to how the widespread introduction of information technology in the workplace has increased the productivity of the companies that have acquired and used it. Suppose, for the sake of argument, that the net effect of information technology on productivity has been zero or negative. (I do not mean to imply that I believe this actually to be the case, but it is useful to consider the possibility.) The production and operation of the computers and other equipment that comprise information technology are factors in the national productivity equation and, given the rapid growth of the computer and related industries, presumably have a strong positive impact, but this seems strange. Why should we take comfort or satisfaction in the rapid growth of one industry if it is proving to be more of a hindrance than a help to the other industries it is intended to serve?

In the past, ideas about productivity and its measurement have come primarily from economists. Although economists are likely to continue to take great interest in this concept and to worry about how to apply it more effectively outside the context of manufacturing, there is room for some thinking by human factors researchers and others who are interested in how people relate to their jobs. There is a need for some good new ideas; where they come from does not matter. There is a need for some fresh thinking, in general, about economic indicators and the various metrics we use to tell how well we are doing as a nation, or as a species. Waste disposal is a valuable service, whose performance, appropriately, adds to the GNP. In the interest of improving the quality of life, however, attempting to minimize waste and thereby reduce the need for waste disposal services are also of value. It is odd, however, to think that the elimination of waste should show up in economic statistics as a decrease in GNP.

The same point may be made with respect to equipment maintenance and repair. It is certainly in the interest of efficient use of materials and labor to build equipment that is maintenance-free. Maintenance and repair services, however, are also part of the GNP. To make the point with an extreme example, if I smash up my car in a highway accident, I generate work for the auto repairman, for the manufacturers of automotive replacement parts, perhaps for medical service providers, and possibly even for the local undertaker. Surely it would have been better from some point of view—at least from my own—had I not had that accident. It, thus, seems odd that its occurrence should have a positive effect on a statistic that is used as a primary indicator of the country's economic health. We perhaps are too willing to accept economic indicators uncritically as reliable reflections of our national well-being.

I do not wish to suggest that such indicators are useless, because I do not believe that to be the case, but I do think they are often given more status than they deserve and are interpreted in overly simplistic ways. If our productivity, or GNP, is growing faster than that of other countries, things are fine; if it is not, we are in trouble. The bottom line, in my view, is the individual's quality of life, and that is not something that is a simple function of growth in productivity or GNP. Productivity and GNP statistics do not tell us whether people are enjoying life more or less than they used to, whether they are more or less satisfied with their situations, whether they find their work more or less interesting and fulfilling. With respect to material possessions, one would expect the degree of contentment to vary inversely with the difference between what one has and what one wants. Because economic indicators give us only one of the terms of this equation, they reveal little about the state of people's minds.

The desire to acquire material possessions is a primary motivating force in our culture. Critics are quick to decry this as evidence of a crassly materialistic value system, but one suspects that few of the critics are prepared to carry their objections to the point of foregoing all the conveniences of modern living themselves. The fruits of technology—refrigerators, telephones, thermostats, washing machines, microwave ovens, and the countless other devices that support our lifestyles—are so much a part of our daily existence that it is difficult for most of us who live in developed countries to imagine what life would be like without them. There is, however, a legitimate question of perspective. One must wonder whether we are failing to distinguish means from ends. It is not clear that the abundance of material goods that technology has made available to us has made us more content as individuals or as a society. What is initially perceived as a luxury quickly becomes a convenience, and then a necessity. If our wants grow faster than our ability to satisfy them, as they appear to, then increasing productivity or GNP, or both, by whatever amounts, will not guarantee a happier populace. Is there any possibility that we could learn to get more pleasure out of what we have and to spend less energy worrying about how to get more? Aristotle was convinced that human avarice is insatiable. One must hope that he was wrong and that we have a greater capacity to learn in this regard than he gave us credit for.

* * *

Human factors relates to economics in a variety of more and less direct ways. The most direct link is via the objective of developing and applying knowledge that will increase the usability of products and enhance the safety, comfort and satisfaction of their users. Realization of this objective should be reflected, to some degree, in increased product values and marketability. Conversely, if a product that has benefitted from the application of human factors know-how is no more marketable than one serving the same purpose that has not, we must conclude that the application of the human factors know-how has not added any value to the product, at least in the consumer's view. Of course the consumer may be unable to distinguish a better from a poorer design and be unduly influenced by cosmetic issues, clever advertising, and so on, but however one accounts for consumer behavior, if human factors research and engineering are to have a positive impact on the economy, they must add value to products, and the added value must translate, one way or another, into greater acceptance in the marketplace.

Usability and safety have been key targets of human factors design activities in the past and will undoubtedly remain so in the future. However, as a consequence of increasing concerns about energy and resource conservation and about the environmental problem of waste disposal (see chapters 3 and 4) it seems probable that society will demand that industry put more emphasis on such issues as durability, maintainability, disposability, and recyclability in the future and will attempt to make the costs associated with a product's total existence span (and disposal) more visible so that they can be reduced and at least partially recovered from product sales. Human factors designers must take such issues into account in order to work out acceptable compromises when the need to design for disposability, say, has implications for usability or user safety. Further, designing for such criteria as maintainability and recyclability holds challenges for human factors research inasmuch as main-tenance, recycling, and related activities are likely themselves to involve the interaction of human beings with the products in order to effect the maintenance, recycling, or whatever is to be accomplished.

Perhaps the most obvious possibility for human factors to have a significant positive impact on the economy is in the area of productivity. Any increase in the efficiency of a production process means an increase in productivity. Success in the improvement of person–machine interfaces and operating procedures should translate into such economically significant consequences as increased output for given input, less waste and duplication of effort, and higher quality products. In the view of many economists, the need to increase productivity is among the most serious economic challenges facing the country in the near-term future. Increasing productivity in a way that is sensitive to the importance of job quality and worker satisfaction will be an especially difficult objective, given the need to produce goods that can compete with those produced in parts of the world where the cost of labor is extremely low.

There is a need also for better ways to quantify the effects on productivity of the many variables that are assumed to have such effects, including equipment design, operating procedures, and organizational factors. Measuring white-collar productivity remains an unsolved problem, and one that will increase in importance as the percentage of information-handling and other white-collar jobs continue to grow. There is, in particular, a need for a better understanding of how to increase productivity through the effective application of information technology in the workplace and for measuring productivity changes that occur. Increasing the throughput or output of a worker is no contribution to productivity, in an economically meaningful sense, if what is put through or out does not serve, however indirectly, some useful purpose.

There are numerous other challenges relating to the economy that represent opportunities for human factors research in the future. These include the discovery of ways to shorten the time required to turn new technology into marketable products; the development of better approaches to scheduling manufacturing operations and, more generally, the improvement of flexible-system manufacturing techniques; the design of interfaces for ultra-small machines; the establishment of a body of knowledge that can serve as guidelines for the design and operation of systems that involve the interaction of people with increasingly versatile robots and semiautomated devices; and the development of consumer information systems that can help consumers and small investors make better informed decisions about personal finances.

Economics is a very broad topic. It is very difficult to understand the present situation, let alone trying to get a reasonably clear picture of what lies ahead. Economists are not even agreed as to whether the United States is currently on the decline as an economic power (Dale, 1991; Nathan, 1991).

There can be no question, however, that many economists and other social scientists are very concerned as they discuss what they see as recent and current trends. My sense-although I have made no attempt to quantify it—is that the majority who have written on this topic believe that unless we manage to deviate significantly from the course we are on, we are in for a rough ride. My intent in this chapter was to note some of the reasons for concern as we try to look ahead, and to make the point that human factors, as a discipline, has something to contribute to improving the economic outlook. Most of the subsequent chapters relate to economics in one way or another. To the extent that human factors research can contribute to the solution of problems of energy use, environmental change, transportation, and others considered in those chapters, it will serve economic goals as well.

3

Energy

Strictly speaking, energy is neither produced nor consumed; it is converted from one form to another. The chemical energy in coal or oil is converted to the kinetic energy of a rotating turbine, which is converted to the electrical energy that services our homes. The electrical energy is again converted to kinetic energy in appliance motors, to radiant energy in heating and lighting systems, and so on. According to the second law of thermodynamics, in any closed system, such transformations, in the aggregate, always mean going from more highly structured to less highly structured forms and that, in turn, means a decrease in the amount of work that the existing energy can be made to do.

In other words, any conversion of energy from one form to another involves the loss of some of that energy for further use. The lost energy is not destroyed, but "escapes," as it were, and is not contained in the new form. The goal, in converting energy from one form to another, is to minimize the losses and retain as much as possible of the energy contained in the original form in order to apply it to the doing of useful work.

In what follows I speak of energy production and consumption, because that is the language that is almost always used, except perhaps in the more theoretical physics journals. It should be understood, however, that production, in this context, really means transforming energy into a more useful form and consumption means applying energy that has been put into a useful form to the accomplishment of some work and transforming it into a less useful form in the process.

WORLD-WIDE DEMAND AND DISTRIBUTION

The total amount of energy delivered to end users from all sources, excepting firewood, increased almost 60-fold from 1860 to 1985. Given that there were significant improvements in the efficiency of end use during that time, energy services increased at an even greater rate than did the amount of energy delivered. Global energy consumption has increased by a factor of about 15 since the beginning of the 20th century, and most of that energy comes from the burning of coal, oil, or natural gas (Gibbons, Blair, & Gwin, 1989); during the same period the world's population increased by a factor of about 3.5.

The energy demands of the industrialized countries far outweigh those of the underdeveloped countries: As of 1979 the average rate of energy consumption worldwide was about 2 kilowatt years per person per year; the average rate of energy consumption in the United States was about 11 kilowatts per person, as compared to about 1 kilowatt per person in the Third World (Sassin, 1980). (In the United States about one fourth of the electric energy that is produced is used for lighting purposes. [Fickett, Gellings, & Lovins, 1990]. Other major users of residential energy are furnaces, air conditioners, water heaters, and refrigerators [C. Seligman, Becker, & Darley, 1981].) At the end of the 1980s, 70% of the world's commercial energy went to one fifth of its population (Gibbons et al., 1989), but the demands of the underdeveloped countries can be expected to grow much more rapidly than those of the developed countries in the future.

Energy growth has tended to increase in the United States at about the same rate as has disposable income, the amount of money spent on consumables of all kinds. This relationship between energy consumption and disposable income suggests that energy demand is likely to grow enormously as poorer countries become more affluent (Hill, 1977).

TRADITIONAL ENERGY SOURCES

Fossil fuels have provided the energy to bring much of the world through the industrial revolution and into the current era. Until fairly recently there have been no serious alternatives. Such fuels have been relatively easy to obtain, process, store, transport, and convert from one form to another. Moreover, until relatively recent times, the prevailing assumption appears to have been that the supply was essentially inexhaustible relative to the demand. The demand has increased at a phenomenal rate during this century, however, and the world has been awakening to the fact that the global supply is limited.

Today, renewable resources provide about 18% of the world's total energy needs, nuclear power provides 4%, and the remaining 78% is provided by fossil fuels (G. R. Davis, 1990). There seems to be general agreement that the amount of time the world can depend on unrenewable energy sources is limited (although there is disagreement as to how much time there is), and that the challenge is to transition smoothly to sustainable energy sources (fusion power, direct solar power) in sufficient time to avoid crises arising from supplies being unable to meet growing demands. At the current rate of use, fossil fuel is being depleted about 100,000 times as fast as it is being formed, which means that, for practical purposes, what there is now is all there is going to be.

Independently of the question of supply, increasing evidence of the deleterious environmental effects of the burning of fossil fuels has made clear the inadvisability of continuing their use indefinitely. (This problem is discussed in chapter 4.) Among the fossil fuels, natural gas has at least two advantages over petroleum: It burns more cleanly, and the estimated recoverable reserves are greater. Until recently, gas has been less attractive than petroleum because of its relatively high cost of production, but as a consequence of research and technology these costs have been coming down. One technique that has increased the efficiency of drilling is the use of *real-time seismic profiling*, or *bore-hole tomography*, which involves three-dimensional modeling of a drilling area based on seismic data transmitted from the drill bit to signal-processing centers at the surface (Burnett & Ban, 1989).

The claim that natural gas is a cleaner-burning fossil fuel than petroleum is based on the fact that, as the least carbon-intensive of the fossil fuels, natural gas emits the least amount of carbon dioxide into the atmosphere per unit of energy generated. (In terms of carbon dioxide emissions, oil is about 40% dirtier than natural gas, and coal about 20% dirtier than oil [Gibbons et al., 1989].) Natural gas does emit methane, however, which is one of the greenhouse gases. Presumably, the methane emitted is not enough to offset the reduction in carbon dioxide emissions; however, decreases in carbon dioxide emissions resulting from the burning of natural gas in place of coal or oil could be offset by increased emissions of methane due to spillage during recovery and transportation. Calculations by Rodhe (1990) suggest that if natural gas spillage can be held to no more than 3% to 6%, there would be a net gain in switching from oil to natural gas, but if the spillage is greater than this there would be a net loss in terms of greenhouse gas emissions. To date, most of the attention has been focused on carbon dioxide as the primary greenhouse culprit, but, in fact, the relative importance of the various trace gases to the greenhouse effect is not well understood, and further research on this issue is needed.

G. R. Davis (1990) distinguishes between the consensus view and the sustainable-world view of how future energy needs will be met. According to the *consensus view,* patterns of energy use will continue more or less as they are at present, and consumption will increase by 50% or 60% in the next couple of decades, as will the amount of carbon dioxide emitted into the atmosphere. According to the sustainable-world view, environmental concerns will motivate both a slower rate of growth in energy demands and certain changes in the pattern of use (e.g., much smaller increases in the use of coal and oil) accompanied by greater increases in natural gas and renewable energy sources. Which view will turn out to be correct remains to be seen.

THE UNITED STATES SITUATION

Until the last couple of decades of the 19th century, the primary energy fuel in the United States was wood. From the last part of the 19th century through the first half of the 20th, it was coal. Oil has been number one since the 1950s, but the use of natural gas is increasing rapidly and is expected by some observers to become more important than oil by the end of the 20th century. The movement from wood to coal to oil to natural gas as primary energy sources represents a trend in the direction of increasingly hydrogen-rich fuels. The ratio of hydrogen to carbon is about 0.1 for wood, 1.0 for coal, 2.0 for oil, and 4.0 for natural gas (Lee, 1989). A continuation, or acceleration, of this trend would be a good thing with respect to the problem of carbon dioxide emissions into the atmosphere.

At the present time, the United States gets about 90% of its energy from oil, coal and natural gas, and almost half of that from oil (see Table 3.1). The country is strongly dependent for much of its current energy needs on foreign oil, and the fact that it has generally been cheaper to import oil and petroleum products than to produce oil or other liquid fuels domestically has provided a disincentive to make the investment necessary to decrease this dependence. Consequently, production of, and exploration for, oil in the United States have proceeded at relatively low levels. The United States' strategic petroleum

TABLE 3.1 Energy Consumption in the United States in 1987, by Percent of total Consumption

Source	Percent
Petroleum	42.9
Coal	23.7
Natural gas	22.6
Nuclear electric	6.5
Hydroelectric	4.0
Geothermal	0.3

Note: From Hoffman (1989).

reserve consists of about 600 billion barrels of oil, which is the amount of oil imported in about 3½ months.

Production costs in the United States are relatively high, even though wells are already in operation, because most U.S. wells are "strippers," some of which yield as little as 2 or 3 barrels a day; the 640,000 wells that were active in the United States in 1984 were producing an average output of 14 barrels a day. Mexico's wells, in contrast, averaged 801 barrels a day, Norway's 4,100, the United Kingdom's 5341, and Saudi Arabia's 12,011 (Abelson, 1987). As Abelson put it, "the relative abundance and inexpensiveness of petroleum have led to increasing reliance on this single source and left us vulnerable when supplies dwindle and prices rise" (p. 584). At least one projection has the oil import bill of the United States reaching about $200 billion per year by the year 2,000, which could be more than the country will be able to afford. Abelson warns that it would take a price-escalating crisis to motivate the action necessary to lessen the United States' dependence on foreign oil. The 1990–1991 crisis in the Middle East again illustrated the importance that much of the industrialized world attaches to accessibility to oil from this region. Whether it and subsequent events will provide the motivation for the United States to increase its efforts to develop alternative energy sources remains to be seen.

Estimates of how much oil and gas remain undiscovered in American reserves have varied considerably over the years and must still be viewed as highly uncertain. In 1972, U.S. Geological Survey estimates put the amounts of undiscovered oil and gas in the United States at 450 billion barrels and 2100 trillion cubic feet, respectively. In 1981, these estimates were revised downward to 83 billion barrels and 594 trillion cubic feet. In 1989 they were reduced still further to 35 billion barrels and 263 trillion cubic feet. These latest estimates, when added to what is believed to remain in known fields, imply a 16-year supply of oil and a 35-year supply of gas at current rates of use (Kerr, 1989c).

H. Kahn, Brown, and Martel (1976), on the other hand, list a sequence of estimates of oil reserves in the United States and predictions regarding the future ability to meet demands dating back to 1866, all of which, in retrospect, proved to be much too conservative. The estimated size of fossil fuel reserves is itself a function of technology; estimates have changed over time both with the discovery of new deposits and with advances in technology that have implications for what is and is not considered recoverable (*recoverable* typically means recoverable by means of current technology; T. H. Lee, 1989; Schurr, 1963). By way of illustrating the point, Lee cites the Kern River story, as described by Adelman (1987). After 43 years of producing oil, the Kern River in California was estimated to hold remaining reserves of 54 million barrels; in the following 43 years it produced 730 million barrels and was estimated at the end of that time—1986—to have remaining reserves of about 900 million barrels.

Reserve estimates are also affected by economic considerations. When the U.S. Geological Survey revised its estimates in 1988, the world price of oil was $18 per barrel; much of the oil in U.S. deposits is in forms (heavy oil, oil shale, tar sands) that make it unrecoverable at that price, given current recovery technology. A higher price of oil on world markets could change the equation. According to Abelson (1987), known U.S. reserves of oil include several hundred billion barrels that, because of location or viscosity, are not recoverable by conventional methods but require the use of heat, carbon dioxide, polymers, or other facilitative agents. A recent report from the Committee on Production Technologies for Liquid Transportation Fuels (1990) called for the support of research and development on new recovery methods and on techniques for making liquid fuel from coal and shale.

Coal, in abundant reserves in the United States and several other countries, is being seen as a major source of energy for the future. There are believed to be enough coal reserves to last perhaps 1500 years at the current world-wide rate of consumption. Techniques for burning it cleanly—in particular, with drastically reduced emissions of sulfur dioxide and oxides of nitrogen—are being developed, as are methods of converting it to gas and various liquid fuels. Scrubber technology is now claimed to be capable of removing as much as 97% of the sulfur dioxide from fuel emissions (Fulkerson, Judkins, & Sanghvi, 1990), although this technology is not being sufficiently widely applied yet to make such reductions imminent. The addition of magnetohydrodynamic (MHO) topping cycles to coal-fired power generation plants could increase significantly the efficiency of coal-to-electricity energy conversion.

The availability of electric power depends, of course, not only on the availability of fuel to run the generating plants, but on the existence of those plants themselves. Electric power generation is a capital-intensive business because plants are expensive to build and operate. In order to recover the capital costs of construction, a plant must be viable for a fairly long time. Uncertainties about how the technology of electric power production will change in the future, about the continuing availability of various types of fuel, and about the ability to handle environmental problems associated with power production can dampen investors' enthusiasm for putting money into new plants and equipment. Some economists worry about the possibility that the supply of electric power could be inadequate to meet demand within the next few decades.

It has been predicted, for example, that if the demand for power generating capacity compounds even at the modest rate of 1% per year, the demand could exceed generating capacity in the United States by about 2005, and serious problems are likely to be encountered during periods of peak use long before that time. Over half the plant capacity for burning fossil fuels will be more than 30 years old by 2000 and the operating licenses of about one third of U.S. nuclear capacity will have expired by 2010 (Balzhiser, 1989). The monitoring of plants and equipment involved in the generation of energy becomes increasingly important as those facilities age. Improvement in the techniques for detecting problems or, better yet, anticipating incipient problems is an important objective of all energy production programs.

The energy strategy that seems to be unfolding in the United States has been described by A. M. Weinberg (1988–1989) as *incrementalism*. The strategy involves what the word suggests: continuing to depend on known technologies and existing plants, adding capacity in small increments to meet needs as they develop. Weinberg argued that incrementalism has some benefits and is probably inevitable anyway, but noted also that it has some dangers in the long run and called for government supported efforts to find further ways to reduce demand and to develop cost effective means of supplying energy so as to end dependence on fossil fuels altogether.

It is clear that continuing to satisfy the growing energy demands of the country and the world will be a major problem, and a problem of increasing complexity, in the future. How can the human factors community participate in the development of solutions to this problem? In the past, perhaps the most obvious connection human factors researchers have had with this area has been in the design of power plant operating systems and procedures and the study of human error in the context of the operation of power generating facilities. Human factors issues in nuclear power plant operation have received considerable attention, especially since the Three Mile Island and Chernobyl incidents provided graphic examples of the importance of the human element in power plant operations (Moray & Huey, 1988). This will continue to be an area deserving of attention from the human factors research community, and, unfortunately, one fraught with liability issues.

The development and exploitation of alternative energy sources—wind, solar, geothermal, tidal—will require the design of new power generation operating facilities that will include various types of displays and operator consoles. The usual types of interface problems will have to be addressed, some of which will have novel aspects because of the newness of the operations that must be monitored and controlled.

Another challenge to human factors researchers relating to energy generation is the need for better techniques to assess risks, and costs and benefits more generally, both short and long term, associated with the various approaches that can be taken to energy production. At least as important as the problem of assessing risk is that of helping both policy makers and the general public to understand those risks and the various tradeoffs that are involved in the options that are available. This is an especially difficult challenge, because the tradeoffs are complex and not fully understood even by the experts, because not everyone approaches the problem with the same set of values, and because the issues evoke strong emotional reactions from many people, which inhibit rational decision making.

There is a need also for a better understanding of how policy decisions get made, of the variables that policy makers take into account, and of the ways in which long-term costs and benefits get traded off against short-term effects. When it is apparent that long-term effects are not being considered in policy making, how do we determine whether they are being overlooked or have been considered but discounted? To the extent that long-term effects are being overlooked, the challenge is to develop the kinds of decision aids that would bring such effects to the attention of policy makers and the general public. To the extent that the problem is one of inappropriately discounting long-term effects, what are needed are decision aids that will help make clear the relationship between such effects and present-day actions, and the kind of public education efforts that will make decisions that make a long-range view politically feasible.

ENERGY CONSERVATION

The problem of limited fossil fuel energy reserves can be addressed primarily in two ways: by increasing the efficiency of energy use and by developing alternative energy sources. Increasing the efficiency of use can be accomplished by reducing waste and by decreasing unreasonably high demands. Alternative energy sources include sources as yet untried or even unknown, but probably more important for the foreseeable future is development of the technology and know-how that will make economically feasible the widespread commercial use of alternative sources that are already being employed experimentally or on a small scale.

The efficiency with which energy is used differs greatly among countries, among industries, and undoubtedly among individuals. Energy inefficiency is particularly severe in the (former) Soviet Union, Eastern Europe, and China. Whereas these countries contain about one third of the world's population and consume about one third of the world's energy, they produce only one fifth of the world's economic output (Chandler, Makarov, & Dadi, 1990).

Energy conservation is a goal that seems to be desirable from just about any point of view: It reduces production costs, conserves raw materials and lessens the unfavorable impact of production on the environment. Emission of particulates into the atmosphere as a consequence of fuel burning can be greatly reduced, for example, by increasing the efficiency of the burning process (Shaw, 1987). Moreover, improvement in the efficiency of energy use is possible when motivation is high. This is seen in the aftermath of OPEC oil price increases in the early 1970s: Since then, the United States' GNP has increased about 35% without any increase in energy consumption (D. Hamilton, 1990; Rosenfeld & Hafemeister, 1988).

Another way to view this effect is to note that over this period energy intensity, defined as the amount of energy required to produce a dollar's worth of U.S. GNP, has decreased by 28% (Fickett et al., 1990). On the whole, American industry reduced the energy intensity of industrial processes by between 1.5% and 2.0% a year over that period; since about 1960, the amount of fuel consumed by U.S. industry per unit of output decreased by over 50% (M. H. Ross & Steinmeyer, 1990). In the United States, the energy efficiency of automobiles, as indicated by miles per gallon, has approximately doubled over the same period and tailpipe emissions have decreased by 60% (Wright, 1990a). More generally, between 1973 and 1985 the amount of energy used to produce a fixed unit of product decreased by about one fifth throughout the developed world (Gibbons et al., 1989; Greene, Sperling & McNutt, 1988).

Although, as already noted, the worldwide demand for energy has grown much faster than the population, so has the gross world product. For several decades, in particular from 1945 to 1975, the ratio of energy use to GNP in the developed countries was relatively constant, but from 1920 to 1940 and again from the mid-1970s until the late 1980s, the ratio decreased fairly regularly (A. M. Weinberg, 1988–1989). Explanations of why economic growth has been able, during some periods, to outdistance increases in energy use point to lower demand resulting from increased energy costs, the shift from manufacturing and mining to service industries, the effectiveness of conservation practices, and the conversion of a large fraction of primary energy to electricity.

Impressive as the recent gains in the efficiency of energy use are, some observers believe that much greater improvements are possible (U.S. Department of Energy, 1988b; M. H. Ross & Steinmeyer, 1990; M. H. Ross & Williams, 1981). Ross and his colleagues have pointed out that the energy efficiency increases that were realized in manufacturing were great, but did not come close to exhausting the potential for improvement; even in the most efficient plants, basic operations expend between 4 and 6 times as much energy as the minimum possible, as constrained by the laws of thermodynamics. Much greater savings could be realized in the future by the application of technology (e.g., greater use of energy-efficient lighting and of control systems to reduce or eliminate energy demands during periods of low production or nonproduction, the use of variable-speed motors, and the use of heat exchangers in place of the combination of heaters and coolers). The potential for energy savings is especially great in existing commercial buildings. Such buildings tend to be highly inefficient in their use of energy, and the cost of the energy they use can be as much as 30% of their total operating costs (Bevington & Rosenfeld, 1990). It has been estimated that monitoring and control systems that can adjust indoor heating, lighting, and air conditioning, depending on outdoor temperatures, direction of sunlight, and location of people can reduce energy use by as much as 10% to 20% (Gibbons et al., 1989).

A general strategy that has great potential for saving energy in the future is that of finding ways to substitute the transmission of information—electronically or photonically—for the movement of people and material, and this is a strategy to which the human factors community should be able to contribute. The technology already exists to permit much more to be done in this regard than is currently being done. Electronic mail and teleconferencing capabilities are perhaps the most obvious examples of existing technologies that have much greater potential to reduce the need for transporting material and people than has yet been realized. How to increase the effectiveness of these technologies is, in part, a human factors problem. More will be said about these technologies in chapters 9 and 10.

Related topics that are relevant to the problem of making more efficient use of energy and that encompass challenges and opportunities for human factors work include electronic publishing and the electronic distribution of "publications," the use of "virtual offices" connected by computer networks to decrease the need for travel to centralized work locations, and the exploitation of "artificial realities" that might be acceptable substitutes, in some instances, for trips to museums, galleries, and other sites of interest. Whether the fuller exploitation of such possibilities would decrease the amount of travel and transportation remains to be seen. Conceivably it might increase it, but it should at least mean more services delivered per unit of energy expended. The televising of major sporting events seems not to have decreased the numbers of people who attend them live, but it has made it possible for much greater numbers to be spectators from a distance. The use of artificial-reality technology to bring the Louvre, the Egyptian pyramids, or the Carlsbad Caverns to our living rooms whenever we wish to visit them might not decrease the number of people who would like to visit these places, but it would greatly enrich the lives of the many of us who will never have the opportunity to do so. Each of these topics is mentioned elsewhere in this book; all of them represent partially developed technologies that have human factors issues needing attention.

Efficiency of Electric Power Use

There appears to be general agreement within the electric power industry that with the appropriate application of technology, considerable improvements in the efficiency with which electricity is used could be realized. This is very important, given that the consumption of electrical energy has been increasing faster worldwide than total energy consumption (Hill, 1977). Estimates of what is possible range from 25% to 75% savings (Fickett et al., 1990). Fickett et al. argue that significant savings could be realized, given the intelligent application of technology, without sacrificing the quality of services the electricity provides, because many of the new energy-efficient devices function better and provide better services than the old ones they can replace; estimates of how much of the electric energy that is used for lighting could be saved by conversion to the most efficient lighting systems available today go as high as 90%.

Some analysts believe that the energy efficiency of buildings could be doubled over the next two decades, and that this improvement would represent a saving of $100 billion a year and a reduction of carbon emissions by half (Bevington & Rosenfeld, 1990). The simple expedient of using very low-emissivity windows and thick insulation on new homes can cut heating costs to a fraction of what they typically have been. Better load management, improved coal and gas combustion techniques, fuel cells, and photovoltaic technology all offer the potential for more efficient production of electricity in the future, as do greater use of solar, wind, and geothermal power (OTA, 1985d).

Automotive Efficiency

Automotive efficiency could also be improved through the use of lighter construction materials and more aerodynamically efficient designs. Composites are already being used fairly extensively along with light-weight metals such as titanium and aluminum in various automotive parts. Of course, one does not want to sacrifice crashworthiness for the sake of fuel efficiencies that might be realized from the construction of cars from lighter materials, but this tradeoff is probably not necessary. Research is producing new materials that are not only lighter but, in some cases, stronger than those currently in use. If all vehicles were lighter, this would be less of an issue, because how a vehicle fares in a collision with another one depends to a large degree on the relative weights of the vehicles involved. (Eighty percent of the people killed in accidents involving tractor trailers are occupants of the other vehicle, which is usually a car.) Moreover, the issue is not simply whether a vehicle can withstand a crash with the minimum of damage; the issue is occupant safety, and occupant safety is a function of many variables, including the effectiveness of seatbelts, airbags, and other features designed as protection in the event of a crash, features intended to prevent accidents, and safe driving practices.

Automotive safety has been a focus of human factors research for a long time and will undoubtedly continue to be for the foreseeable future. This topic is touched upon again in chapter 6. A special concern over the next few decades will be to determine when the goals of increased energy efficiency and increased highway safety are in concert and when they are in conflict, to understand the tradeoffs involved, and to find ways to make progress on both agendas simultaneously.

Quality Control and Recycling as Energy Conservation

Another major opportunity for energy conservation is in improved quality control techniques. The production of defective products requires as much energy as the production of quality products, and the energy used to produce defective products is wasted energy. A major tenet of the TQM philosophy that is attracting interest is that practices that do not put a premium on quality are more costly in the long run than those that do. Doing it right the first time is in the interest of energy conservation as well as that of saving unnecessary costs. Quality control is a matter, in part, of industrial procedures and, in part, of worker attitudes. How to improve this seems an appropriate focus of human factors research.

Recycling of used materials is also beneficial from the point of view of energy conservation as well as that of waste management, because it permits the bypassing of the most energy-intensive step of manufacturing, namely the conversion of ores and feed stocks into basic materials. Although it takes only half as much energy to make recycled materials as it does to make new ones, only about 20% of the paper, plastic, glass, and metal goods in the United States are made from recycled materials, whereas roughly 50% could be (M. H. Ross & Steinmeyer 1990). This topic is revisited in chapter 4.

* * *

Despite the fact that, according to numerous polls, about half of the U.S. population believes (or believed, shortly after the 1970s oil crises) that there is a real and serious problem of energy supply (M. E. Olsen, 1981), the importance of energy conservation is proving to be very difficult for the national psyche to acknowledge. Perhaps this should not be surprising. The United States is a very large country relative to the number of people who have inhabited it, even up to the present time. It has always been perceived, rightly, as a land with an abundance of natural resources. The tapping of those resources made possible an unprecedented rate of growth and industrial development. The abundance and the obvious advantages that have accrued from exploiting it, however, have also blinded us as a nation to the fact that it is not infinite. We have become so accustomed to plenty that warnings of problems that could arise from continued lack of restraint in consumption are often heard as obstructionist if not un-American.

Forty years ago, N. Wiener (1950) characterized the situation, somewhat harshly perhaps, this way. "So long as anything remained of the rich endowment of nature with which we started, our national hero has been the exploiter who has done the most to turn this endowment into ready cash. In our theories of free enterprise, we have exalted him as if he had been the creator of the riches which he has stolen and squandered. We have lived for the day of our prosperity, and we have hoped that some benevolent heaven would forgive our excesses and make life possible for our impoverished grandchildren" (p. 37). The fact is that many of the natural resources of this country, and indeed of the world, that were once abundant, relative to the present and probable future demands on them, are no longer so. Conservation is not an option; it is a necessity. The question we, as a society, have to answer is whether we will attend to this matter while there is still time, perhaps, to apply effective nondisruptive approaches to the problem or will ignore it until nature forces us to function in a crisis mode. The choice, for the moment, is ours.

There is little evidence to date that the need to conserve energy in order to avoid the depletion of energy sources is a very great concern to the American public (Milstein, 1976, 1977). Within government and industry, interest in energy efficiency has waned during the last few years because of the abundance of inexpensive oil. There has been relatively little investment in plant and equipment designed to minimize energy waste over the long term. When the capital costs of an industry far outweigh the costs of energy for operations, there is little incentive to spend more capital to be more energy efficient. As the energy costs increase relative to the capital costs, however, the economics becomes more favorable to the investment in energy conservation. U.S. companies are not noted for taking a long-range view; typically, they only make investments that will pay for themselves within two to four years (M. H. Ross & Steinmeyer, 1990). Unfortunately, much of the cost of current energy production policies will be borne by future generations, who will have to cope with both the problem of a smaller supply for an increased demand and the environmental problems that excessive energy consumption creates.

Reddy and Goldemberg (1990) make the interesting observation that, because developing countries do not yet have many of their vital industries, buildings, roads, and transportation systems in place, they have an opportunity to engage in "technological leapfrogging." Industrialized countries, which have already invested large amounts of capital in their infrastructures, however, have a disincentive to invest more money in upgrading to more energy-efficient systems. Developing countries, without such a disincentive may, therefore, be in a position to adopt new energy-efficient technologies before industrialized countries do.

The development of energy-efficient products and production techniques will be motivated, worldwide, by the costs of energy and by concern for the detrimental environmental effects of conventional methods of generating energy, which are likely to be expressed with increasing frequency in governmental regulations and strictures against environmental degradation. Therefore, another compelling reason for businesses to be interested in developing energy-efficient technology is the possibility of making money at it. The market for energy-efficient devices and processes is a global one and is likely to grow very rapidly as the disadvantages associated with operating inefficiently—in the factory, in the office, in the home—increase. Energy-inefficient products will not be able to compete with energy-efficient products that are equally good in other respects; companies that bring to market quality products that deliver the same services as their competitors with significantly less energy consumption could do very well indeed.

Several approaches have been tried to encourage individuals and households to be conservative in their use of energy. These include the use of persuasive communication, the offering of various types of inducements and incentives, the provision of models of conservation behavior, and the provision of feedback regarding the effectiveness of conservation efforts (S. W. Cook & Berrenberg, 1981). In addition, considerable research has been aimed at getting a better understanding of what determines people's attitudes and behavior regarding energy use and conservation. The results of this research must be interpreted cautiously, however, because almost all of it has been done with volunteer subjects, and people who volunteer for such studies may differ in a variety of ways from people who do not (Nitzel & Winett, 1977). If, for example, people who volunteer to participate in energy conservation studies have more positive or proactive attitudes toward energy conservation than people who do not volunteer for such studies, then the findings have limited generality.

Much of the experimentation on behavior that is germane to energy conservation and related issues, such as waste reduction and material recycling, has used college students as subjects. Experimenters have pointed out the need for more research directed at noncollege populations and done in noncollege environments, because college populations may differ from noncollege populations in a variety of ways that could affect the interpretation of results (Geller, 1981; Reichel & Geller, 1981). Also, just as strong opinions indicated by polls do not necessarily translate into political action (R. E. Dunlap, 1987), it is not safe to assume that what people say as respondents on surveys dealing with socially sensitive issues reflects accurately what they do in their private lives (Milstein, 1977); it is much easier to express socially responsible attitudes in response to a questionnaire than to behave consistently in socially responsible ways.

A related problem has to do with the reliability and accuracy of self-reports pertaining to energy consumption. This has been a matter of concern among researchers studying the effects of various variables on energy use patterns. Warriner, McDougall, and Claxton (1984) have argued that usage estimates tend to be about equally likely to err on either the high or low side and not to be related systematically to independent variables of interest and that, therefore, they should be viewed as useful data. Still, the development of methodologies that would produce more reliable and accurate self-reports would be a welcome advance.

Yet another challenge for research is the development of much more effective ways of sensitizing consumers to the implications of their moment-to-moment behavior with respect to energy use. We all know that we are using energy when we drive an automobile, watch television, use an air conditioner, or turn on an electric light. Most of us have little sense, however, of how much energy is required to power the various devices that we use or how that amount may be affected by the ways in which we use them. Studies that have provided feedback to people regarding their energy use have shown that such feedback can be effective in moderating use. The feedback methods used in these studies have been fairly crude, however, and have not provided the kind of detailed instantaneous information that would permit one to develop a precise understanding of the effects on energy use of small modifications in behavior.

As computing power becomes ever more inexpensive and widely available, it should be possible to equip appliances, automobiles, and residences with monitoring systems that can provide continuous instantaneous feedback to consumers regarding the amount of energy they are using at any given time. One can imagine such systems having the capability of determining when devices are being used in a sub-optimal way and informing users of how the efficiency of operation could be improved. In using such systems, people would develop more accurate models of the relationship between their behavior and energy consumption. They would learn what types of activities and device usage consumed the largest amounts of energy and how their energy consumption could be modulated by the specific ways in which they use those devices. They would discover what types of behavior change could have the greatest impact on consumption and would develop a "feel" for the energy-use implications of their moment-to-moment behavior.

SUPERCONDUCTIVITY

Among the most exciting hopes for very large improvements in efficiency of energy use is the phenomenon of superconductivity, which has been getting a great deal of attention recently. The phenomenon has been known since 1911, when it was discovered that mercury superconducts at $4.1°K$ (Kelvin). Between then and 1975 a few other materials were found to superconduct at temperatures up to about 23°K. Experimentation with oxide superconductors started around 1964, and beginning around 1986 a series of spectacular discoveries of oxides that superconduct at much higher temperatures were made (Sleight, 1988). To date, compounds have been found that will superconduct reliably at 120 to 125°K, and several researchers have reported at least temporary losses of resistance in some materials at temperatures as high as 250°K (Cava, 1990a, 1990b; Pool, 1989c). A material that superconducts at 125°K is a compound of thallium, barium, calcium, copper, and oxygen ($Tl_2 Ba_2 Ca_2 Cu_3 O_{10}$). Unfortunately, the fact that this thallium oxide is toxic makes it less than ideal for commercial applications. There are a few other known compounds that superconduct at temperatures above 77°K, the boiling point of liquid nitrogen, and about a dozen that do so above 40°K.

There is as yet no well-developed theory that will permit the accurate prediction of materials that will superconduct at higher temperatures, but enough has been learned that the search for new materials is not completely random; the expectation is that materials will be discovered that superconduct at still higher temperatures than those demonstrated to date. The one known feature that all high-temperature superconductors have in common is the sandwiching of planes of copper and oxygen atoms between layers of the other elements that make up the compound. The copper-oxygen planes, which are the electronically active regions of the crystal structure, move negative or positive charges they obtain from other layers of the structure (Cava, 1990a, 1990b). A better understanding of superconductivity could drastically alter the economics of power production and use.

Among the anticipated eventual uses of superconducting materials are "long distance electrical power transmission, levitated trains that ride on a magnetic field, powerful magnets in nuclear fusion reactors or medical imaging machines, and energy storage" (Wrighton, 1988, p. 113). The development of materials that will superconduct at room temperature is viewed by many researchers as a possibility, and the race is on to accomplish this. Some observers expect that superconductivity will be a $20 billion industry worldwide by 2000 (U.S. Industrial Outlook, 1989). There is also the possibility, however, that the time that will pass before superconductivity has a significant impact on energy use will be much greater than anticipated, so the prospect of the development of this technology is not a justification for putting less effort into other approaches to the problems of energy production and use.

ALTERNATIVE ENERGY SOURCES

Energy conservation and improvements in the efficiency with which energy is used must be among the highest priority goals for the United States and the world over the next few decades, but conservation and improvements in efficiency of use are not likely, by themselves, to constitute the solution to long-term global energy needs. As Hill (1977) pointed out, conservation measures that are effective in reducing demand in the short term only delay the problem unless they change the rate of demand growth for the long term. If, for example, the present energy use were reduced by 50%, but the demand continued to grow at the rate of 5% per year, it would take only 14 years to make up for the 50% reduction and bring us back to where we were before the reduction occurred. It will be necessary to develop new energy sources or to find ways to make already-known alternatives to fossil fuel cost effective and environmentally benign.

Public interest in finding alternatives to fossil fuel is currently very high; one recent poll found that 79% of Americans believe that renewable energy should be a top priority for federal research funding. In fact, however, relatively little is being done at present to push the country's renewable energy capabilities. Renewable energy now represents only about 10% of the nation's total energy production and about 14% of its electrical energy; hydroelectric power accounts for most of the latter (S. Williams, Fenn, & Clausen, 1990). Grubb (1990) noted that there have been two main attitudes toward non-hydro renewable energy: "One, widely expressed throughout the environmental community is that in the long run renewable energy will save us all from the unsustainable consequences of relying upon fossil fuels and nuclear power. . . . The other common attitude is that in a short to intermediate time horizon relevant to the real world of industrial and political policy formation and investment, non-hydro renewable sources are essentially irrelevant: that for the foreseeable future their contribution will remain marginal" (p. 525). Grubb took issue with the latter attitude, arguing that renewable sources could make sizable contributions to the world's needs even in the very near-term future. C. J. Weinberg and Williams (1990) have predicted that the use of wind, solar, thermal, and biomass technologies to produce electricity is likely to become cost-competitive during the 1990s.

Wind

The wind has been used as a source of power, especially for moving boats, throughout recorded history. Although it has never been a serious competitor with fossil fuels as a source of the power needs of an industrial society, interest in windmills as a means of generating electricity increased considerably as a consequence of the energy crises of the early 1970s, and significant improvements in this technology were made in the decade following (Moretti & Divone, 1986). There now are over 15,000 wind turbines operating in California, producing, in the aggregate, over 1,500 megawatts of electric power (Moore, 1990). The cost of generating electricity by wind energy decreased by almost a factor of 10 during the 1980s, and is now less than 150% the cost of generating electricity with coal-burning plants; it has been estimated that the cost of producing electricity by wind, at least in some locations, could decrease by another 50% between the 1990s and to 2010s (C. J. Weinberg & Williams, 1990).

Almost every country in Western Europe is experimenting with the use of wind for electricity generation. Wind turbine makers in several countries are developing new turbine designs. Possibilities for increased efficiency lie in the use of longer blades with variable-pitch capability and constructed from fiber-composite materials; the development of more effective variable-speed turbines also represents an opportunity for improvement (Moore, 1990).

A problem with wind as an energy source is its considerable variability, but this is primarily an argument against dependence on wind as the *sole* source and not against its use as an input into a power distribution facility that draws from other sources as well. As a supplement to a fossil-fuel burning plant, it could help to decrease the amount of fuel that would have to be burned, on the average, to deliver any given amount of power over a period of time.

Wind power obviously has the great advantage of not producing the atmospheric pollutants produced by many other primary energy sources. It is not entirely free of environmental problems, however: One concern that has been raised is the possibly unsightly effect of large windmill farms on the rural landscape. Another is the considerable noise that wind turbines can generate.

Water

Water has also been used as a source of power for centuries, especially through the medium of waterwheels of various types. Hydropower has long been used to generate electricity and is often promoted as an attractive alternative to fossil fuels because it does not involve the emission of gases into the atmosphere. The use of hydroelectric power has been opposed in many places, however, because it often requires the flooding of large areas of land, thus contributing to deforestation and perhaps some shifting of local ecological balances. The possibility of dam failures has also been a source of some concern. (The failure of the Lake Conemaugh Dam, above Johnstown, Pennsylvania, in 1889, and the consequential flooding of the Conemaugh Valley, resulted in nearly three times as many deaths—2,209—as did the great San Francisco earthquake in 1906 [Jackson, 1989].) Energy storage on a large scale is inherently dangerous, no matter what the storage form.

The Sun

The sun can be used to generate energy in a variety of ways. The use of solar panels and heat-storage systems to heat buildings is a relatively well-established technology. A variety of techniques are being investigated for using the sun to produce electricity. Among them are photovoltaic technology, in which photons create an electric current when absorbed in a semiconductor, and solar thermal electric technology, in which reflected solar radiation is used to heat a fluid to the point at which it can produce steam that can be used to drive a turbine generator.

Production of photovoltaic power has decreased in cost by about a factor of 5 to 10 since the 1980s. Further cost reductions, possibly by another order of magnitude, are expected in the 1990s. Some experts believe that megawatt power plants based on solar cells could be in use, and competitive with other means of energy production, by the turn of the century (Grubb, 1990; Hamakawa, 1987; Hubbard, 1989). Improvements in the technology are being made along several lines, including the development of low-cost photovoltaic materials, more effective device designs, automated manufacturing processes, and increased reliability and durability of devices and systems.

Most of the United States' satellites currently in space are powered by photovoltaic cells. Solar energy is also being used to power such devices as calculators and watches. Solar-powered plants can now produce electricity for between two and three times the cost of producing it with fossil-fuel plants. We should see much greater interest in solar technology as it becomes more cost effective, and especially if the price of oil and natural gas were to increase significantly. Sandia National Laboratories recently developed a photovoltaic cell that uses gallium arsenide and silicon and converts 31% of incident sunlight into electricity. This is seen as a milestone achievement that exceeded the expectations of many experts who doubted that photovoltaic cells could reach efficiencies comparable to those of more conventional energy sources (the average efficiency of coal- and oil-fired electric plants is 34%; Wollard & Zorpette, 1989).

Hubbard (1989) estimates that photovoltaic systems should be able to meet most of the electrical power requirements of the United States by the mid-2030s. He also argues that photovoltaic power has the potential to become the primary source of electricity worldwide by the end of the 21st century. Among the issues, in addition to cost, that are likely to affect the rate at which photovoltaic technology is adopted are the large land areas required for solar arrays, the intermittency of sunlight, and the resulting variability in the amount of power generated.

Inasmuch as vehicle emissions are among the worst sources of atmospheric pollution, the possibility of practical sun-powered vehicles is especially attractive from an environmental point of view. Prospects for the development of automobiles that can run primarily on solar power were brightened recently when a totally sun-powered vehicle won the 1,867-mile Pentax World Solar Challenge race by going from Darwin, Australia, to Adelaide in

44 hours and 54 minutes of running time at an average speed of 41.6 mph (H. J. Wilson, MacCready, & Kyle, 1989).

Geothermal

We know from the activity of volcanos, geysers, hot springs, and related phenomena that there is considerable heat stored not far below the earth's surface. If cost-effective ways could be developed to tap that heat, it could be a source of useful and relatively clean energy. One idea that has gotten some attention is using the "hot dry rock" (granite) beds that lie a mile or more below the surface in various parts of the country to heat water into steam that can then be used to generate electricity. A method for doing this involves the drilling of pairs of wells into a bed, then using one of the wells to carry cold water to the bed, where it can be heated by circulating through the fissures in fractured rock, and using the other to return the heated water or steam to the surface ("Hot rocks," 1990).

Biomass

Biomass has been estimated to account for about 14% of the world's primary energy supplies, most of this coming in the form of noncommercial fuels for open-hearth combustion, especially in developing countries (Grubb, 1990). Although biomass, in the forms of combustible waste, crops produced specifically for combustion, and gas produced from biomass by pyrolysis, can be used also for the generation of electric power, such use has not yet reached a very significant level.

Fusion

Nuclear fusion is an attractive potential source of energy because of its relative safety and environmental advantages over fission, and the greater supply of fuel. Fusion reactors would not emit carbon dioxide or pollutants to the atmosphere, nor would they produce high-level, long-lived radioactive waste. Fusion technology, in its current state of development, is not radioactively clean because neutrons escape during the process, but there is reason to hope that the technology can be improved to the point of solving this problem (Häfele, 1990). Unfortunately, the fusion process that is most likely to be practically feasible in the foreseeable future involves deuterium and tritium, the two- and three-neutron isotopes of hydrogen; further in the future is the hope of a practical deuterium–deuterium process. Deuterium–tritium reactors are not likely to be used widely because of the limited supply and the nasty nature of tritium. Because deuterium is abundant in the oceans, successful development of the deuterium–deuterium technology would essentially solve the problem of limited resources, but this technology is at a relatively primitive stage of development.

Although steady progress has been made in fusion research over the past few decades, it has been slow, and the OTA (1989a) has predicted that at least three more decades of research and development will be necessary before a prototype commercial fusion reactor could be operated and evaluated, and that fusion is unlikely to provide a significant fraction of electricity in the United States before the middle of the 21st century. On the other hand, recent developments in the use of short-wavelength lasers to heat heavy hydrogen have led some scientists to speculate that it might be possible to demonstrate the feasibility of harnessing fusion power by laser by the beginning of the century (Craxton, McCrory, & Soures, 1986).

The United States has been involved in a joint effort with Europe, Japan, and the (former) Soviet Union to produce a conceptual design of the International Thermonuclear Experimental Reactor (ITER), a fusion reactor with a thermal energy capacity of at least 1,000 megawatts. The conceptual design phase of this effort began under the auspices of the International Atomic Energy Agency of the United Nations in 1988. Some scientists have argued strongly that the international collaboration should be continued through subsequent phases of engineering design, construction, and opera-tion (Gilleland, Nevins, & Kaiper, 1990). The future of this project is uncertain, however, largely because of funding difficulties (Beardsley, 1990e). It appears that the European community may be stepping up its research on nuclear fusion, while the United States is retrenching (Cherfas, 1990).

The news in 1989 of observations of nuclear fusion reactions sustainable at room temperature raised hopes among some observers, at least for a short time, that the line of research that produced these observations could lead to a "virtually limitless source of clean, inexpensive power" (Pool, 1989a). The research quickly became highly controversial, however, and most experts believe that its implications for the near-term future are small or nonexistent. (For an account of this extremely interesting and somewhat bizarre chapter in the history of modern science, see Close, 1991.) The possibility of cold fusion catalyzed by negative muons has been the subject of some interest for nearly 50 years. Except for a short time following the first experimental observation of muon-catalyzed fusion by Luis Alvarez in the late 1950s, when investiga-tors thought the process might lead to inexpensive power, it has been generally believed to be too slow for practical use. Recently, however, as a consequence of both theoretical and experimental progress, the prospects for developing muon-catalyzed fusion into an economically viable energy source have been looking somewhat better (Rafelski & Jones, 1987).

Fission

The public debate about nuclear (fission) power has focused primarily on the issue of safety, although several other concerns have contributed to its intensity (Bronfman & Mattingly, 1976). Clearly, nuclear power is greatly feared by the American public (Fischhoff, Slovic, Lichtenstein, Read, & Combs, 1978). Within the industry itself, however, there is the belief that the technology exists to produce fission reactors that are more cost effective and safer than those currently in use (Agnew, 1981; P. E. Gray, 1989; J. J. Taylor, 1989). (A. M. Weinberg, 1989–1990, has estimated from probability–risk analysis that the mean core melt probability of light-water reactors currently in use is between 10^{-5} and 10^{-4} per reactor year. This means, that with about 400 reactors now in place, the probability of at least one core melt between now and the end of the century is between .04 and .4.)

Human error is among the most important vulnerabilities of present-day nuclear power plants (Golay & Todreas, 1990). As already noted, this is a topic that has received considerable attention from human factors researchers, and it will continue to be a very important focus for human factors research in the future. Some designers believe it is possible to design nuclear reactors in such a way that, in the event of a failure of the normal cooling process, the heat generated by fission products in the core can be dissipated passively and safely, thus eliminating the risk of a core meltdown and the need for external backup or cooling systems that have to be activated by humans (Lester, 1986), but fully automated plants appear unlikely to be built in the near future. Even if this were not the case, automated plants and their operating procedures are designed by human beings and the possibility of human error is no less real in this context than in that of hands-on plant operation.

As of 1989, about 16% of the world's electricity was produced by nuclear power plants (Häfele, 1990). France and Belgium produce nearly 80% of their electricity this way; in contrast, the United States produces only about 20% of its electricity from nuclear energy. Although nuclear power technology was pioneered in the United States, nuclear power is used to generate electricity far more in Europe and Japan than here: Of the approximately 560 nuclear power plants in the world, about one fifth are in the United States (Golay & Todreas, 1990). Lester (1986) argues that abandoning nuclear power in this country could have the effect of making the nation's electricity supply less efficient and less environmentally benign than it otherwise could be; a major advantage of fission power is the fact that its generation does not emit carbon dioxide and other greenhouse gases into the atmosphere.

Concern regarding the use of petroleum as an energy source has focused both on the possibility of exhausting the supply and on the detrimental effects on the environment of burning oil. Resistance to the use of nuclear power has come primarily from environmentalists and others concerned about the possibility of nuclear accidents and the problem of disposing of nuclear wastes. Very little attention has been given to the finiteness of the supply of uranium, the primary fuel for fission reactors, but uranium reserves *are* limited and could be exhausted within a century if the light-water reactors currently in use and those under construction were to produce electricity at the rate of 400 gigawatts as long as the supply lasts (Häfele, 1990). This is so because light-water reactors fission only about .5% of the available uranium atoms. The same limitation does not apply to fast-breeder reactors, which produce more fuel than they consume by converting non-fissile uranium into fissile plutonium.

Large-scale utilization of alternative means of generating electric power, if or when, it occurs, will of course have significant implications for existing industries. Over 80% of the coal that is mined in this country is now used for electric power generation. A sudden unplanned decrease in this demand for coal could have detrimental effects not only on the mining industry but on the industries involved in transporting, processing, and converting the coal to electric power. It is expected that this demand will not decrease precipitously soon, but that coal-fired plants could provide up to 70% of the country's electric power by the end of the century (Balzhiser & Yaeger, 1987). Considerable effort is currently being put into the development of new techniques for burning coal that are more cost effective and conducive to pollution control than are existing methods.

An aspect of the continuing space program that is expected to be controversial is the use of nuclear reactors as power generators both for vehicle propulsion systems and for energy-intensive activities such as mining and manufacturing on the moon or Mars (Kiernan, 1990). The search for more effective propulsion and power systems for use in aerospace operations is currently on the priority research agenda for the U.S. Air Force. Among the reasons it is there is the desire to develop an aerospace plane that will provide access to space from ordinary runways. (Compared to the publicity that has been given to the space station program, relatively little has been heard about plans for the national aerospace plane, but this aspect of the space program has its champions and it could, in time, move closer to center stage [Keyworth & Abell, 1989].) A nuclear propulsion system, called a *particle bed nuclear reactor,* that is under consideration for powering vehicles like orbital transfer vehicles would get its thrust from hydrogen that is heated by being passed over nuclear fuel encapsulated in small ceramic pellets. It is believed that such a system could be developed that would be safe and that an engine about the size of an oil barrel could produce a 50,000 pound thrust (U.S. Air Force, undated). Other propulsion possibilities include the use of electrostatic or magnetic fields to accelerate ionized gas; a more remote possibility would make use of antimatter (Corcoran & Beardsley, 1990).

ALTERNATIVE AUTOMOTIVE FUELS

In view of the fact that automobile transportation accounts for a very sizable fraction of the consumption of gasoline in this country, the search for alternative automotive fuels must be a high priority activity, for both conserving fossil fuels and decreasing deleterious atmospheric emissions. The most likely candidate substitutes for gasoline on a large scale are ethanol (grain alcohol) and methanol (wood alcohol). Some scientists have argued strongly in favor of making methanol the primary fuel for automobiles as quickly as possible, although the benefits of methanol over gasoline with respect to air pollution are still somewhat open to question (Seinfeld, 1989).

Among the reasons advanced for early conversion to methanol for automotive fuel as a means of achieving desired reduction in vehicle emissions are the following: Only modest changes would have to be made by the automotive and energy industries; methanol can be produced from a variety of abundant sources including natural gas, coal, wood, and organic garbage; methanol has a higher energy content per unit volume than other alternative fuels; it could be distributed by the existing distribution network for motor fuel with only minor changes in the system; and the use of methanol would dramatically reduce emissions of ozone and toxic substances into the atmosphere (C. L. Gray & Alson, 1989). The emission reduction would result from the fact that, with only modest engine changes, methanol-powered vehicles could be made to operate considerably more efficiently than gasoline-powered vehicles. Chrysler, Ford, and General Motors are all currently developing methanol-powered vehicles. At least one will probably bring such a vehicle to market, at least in a limited way, by the time this book goes to press.

One concern that has been expressed about methanol-powered vehicles is that they would emit twice as much formaldehyde as do vehicles powered by gasoline. Gray and Alson (1989) argue that only a third of the formaldehyde related to vehicle emissions comes directly from the tailpipe, the rest being formed by photochemical conversion of hydrocarbon emissions; because methanol-powered vehicles would generate only about one-tenth the amount of hydrocarbons generated by gasoline vehicles, conversion to methanol could actually mean a decrease in total formaldehyde levels in the atmosphere. They argue also, that the effect of formaldehyde would be dwarfed by the favorable impact the use of methanol would have on other carcinogenic air pollutants. (How much carbon dioxide methanol emits when burned depends on how the methanol was made: e.g., methanol derived from coal emits twice as much carbon dioxide as does gasoline [C. J. Weinberg & Williams, 1990].)

Another possible automotive fuel for the future is hydrogen. One of hydrogen's major advantages is its abundance. It is also relatively easy to transport; indeed, C. J. Weinberg and Williams (1990) claim that on a lifetime cost basis, transportation of hydrogen by pipeline costs less than transmission of electricity by wire. This is a great advantage because it means that hydrogen production facilities can be located wherever it is most convenient to locate them, and not necessarily near where the hydrogen is to be used. A major disadvantage of hydrogen at present is the cost of isolating it; another is the concern for safety in storing it. Weinberg and Williams estimated that it should be possible to produce electrolytic hydrogen with solar technologies by about 2010 at a cost about double what American consumers have been paying for gasoline on an energy-equivalent basis. James J. MacKenzie, of the World Resources Institute, has suggested that if the process of producing hydrogen by electrolyzing water were powered by solar energy, hydrogen could be cost competitive with methanol by the turn of the century (cited by Beardsley, 1989a). This seems like a low-probability possibility, however; few observers expect hydrogen-powered cars to be generally available until well into the next century, although some manufacturers are currently working on their development (Nadis, 1990).

Hybrid fuel systems can be effective both for the heating of buildings and the powering of vehicles. The use of solar energy as a primary source of heat in the home, backed up with a fossil fuel system or electric heating, is already a common combination. Many office buildings are being heated, insofar as possible, with waste heat produced by computers and other machinery, and draw on conventional sources only when these other sources prove inadequate to maintain a desired comfort level. It is possible, also, to build a power-generating facility that depends on sunlight as its primary energy source, but has a secondary fossil-fuel system for use when the sunlight is inadequate. Hybrid systems have not been used much in automobiles yet, but they are a possibility, and work is being done on their development. Volkswagen, for example, has designed a car that can run on either electricity or diesel fuel. Because of the limited distribution system for methanol, the first methanol-powered cars will probably be designed so they can run on different mixtures of methanol and gasoline (Moffat, 1991).

INDIVIDUAL BEHAVIOR AND ENERGY USE

When we think of increasing the efficiency of energy use, most of us probably tend to focus on the large energy consumers, such as industrial manufacturers. We should not, however, overlook the fact that any development that increases the efficiency with which individuals use fossil fuels and electric power in their day-to-day lives can have a significant impact because of the large numbers of people involved. Most of us probably make greater demands on energy resources than necessary. I do not wish to argue here for the enforcement, or even the urging, of ascetic lifestyles, but only for the promotion of energy-saving measures that are consistent with the high standard of living that middle-class Americans take for granted. My assumption is that there are ways that most of us could decrease our energy demands without decreasing either our standard of living or the quality of our lives.

It is claimed that people tend not to consider energy efficiency when selecting a home for purchase (Bevington & Rosenfeld, 1990). One suspects what is true of home buying in this regard is true also of the purchase of household appliances. Household appliances vary significantly with respect to the efficiency with which they use electricity. The variance is likely to increase in the future as manufacturers put more emphasis on the issue of efficiency. How can consumers be motivated to make efficiency of energy use a priority concern when purchasing such appliances? It appears that a very significant fraction of the variance in residential energy efficiency is attributable to the behavior of residents and appliance users (Woteki, 1977). What can be done to give consumers the necessary knowledge and to effect the attitude and behavior changes that would lead to reductions by individuals of their energy consumption? These are not usually thought of as human factors problems, but perhaps they should be.

The fuel efficiency of automobiles can be, and has been, improved very considerably through the use of lighter materials in construction and styling for aerodynamic efficiency. Further improvements are possible through technological innovation. The fuel efficiency of a particular vehicle also depends, however, on keeping the air pressure in its tires at an appropriate level, driving at moderate speeds, minimizing rapid acceleration and deceleration, keeping the vehicle properly maintained, and other variables under the driver's control. There seems to be very little evidence on the question of how important fuel efficiency is to the average driver. It might become more important if the cost of fuel were to plateau at a level substantially higher than the levels of the past few years, as many economists are saying it must. The challenge, then, would be to inform drivers as to what constitutes optimal behavior from a fuel-efficiency point of view. There is also the possibility of electronic monitoring systems that can provide drivers with feedback as to specific ways in which their driving behavior is contributing to inefficient use of fuel.

A considerable amount of psychological research has been done on beliefs, attitudes, and behavior relating to energy use and conservation (Baum & Singer, 1981). Approaches that have been taken to encourage energy conservation by individuals have been reviewed by McClelland and Cantor (1981). They include providing information regarding energy conserving behavior, attempting to persuade people that desired benefits can sometimes be achieved with less energy, providing information regarding energy costs, giving feedback regarding costs of energy use behavior, and explaining the long-term societal costs of energy use. Studies of the effect of having feedback regarding the amount or cost of energy used on further energy consumption have provided consumers with information that allows them to compare their current energy use with their past use, with expected use (given weather conditions, etc.), or with use by other consumers.

Such studies have looked at the effect of providing the feedback at much more frequent intervals than represented by the customary monthly or quarterly bill from the utility company. The results of numerous studies indicate that such feedback can be effective in reducing energy consumption in the home. Provision of continuous feedback on the use of electricity, for example, can lead to a reduction in consumption as compared with level of use by consumers who do not get such feedback (McClelland & Cook, 1979, 1980). For information to be effective in modifying behavior it must, of course, be believable. The credibility of commercial advertising in the United States is sufficiently low that promotions of energy-saving measures (e.g., installation of insulated windows) are likely to be received with a great deal of skepticism when they are made by companies with a vested interest in their adoption.

A number of investigators have explored the possibility of motivating energy conservation behavior through the provision of monetary incentives, such as cost rebates or cash prizes for consumers who do particularly well at it. These studies have shown that such measures can increase conservation efforts at least for short times (a few weeks); however, the effects have not been large and often the value of the energy saved has been less than the cost of the incentives provided (McClelland & Cantor, 1981). Not surprisingly, there is some evidence that people are somewhat more conservation minded when they pay their utility bills directly than when they occupy apartments in which the cost of utilities is included in the rent and does not vary with use (McClelland, 1980).

Seligman and colleagues have done a number of studies of the effects of feedback of various sorts on residential energy use (Becker & Seligman, 1978; Seligman et al., 1981; Seligman & Darley, 1977). They and several other investigators have shown that provision of explicit feedback at frequent intervals regarding energy use, especially for heating and air conditioning, can reduce consumption. The results of these studies were reviewed in Seligman et al. (1981).

Several experimental efforts have been made to modify driver behavior and, in particular, to reduce the number of vehicle miles driven per unit time, using modest cash prizes as incentives. As in the case of energy consumption in the home, these efforts have usually yielded reductions, but only of small amounts. It is difficult to know from the results of these studies, what to believe about the possibility of modifying long-term behavior in similar ways (Reichel & Geller, 1981).

State and municipal governments have instituted a variety of measures to provide incentives for carpooling or the use of public transportation, such as provision of special express lanes on crowded highways for cars with more than one occupant, provision of express bus service from city perimeter parking lots, and establishment of special toll rates for cars with more than one occupant. Corporations, often under government pressure, have instituted similar measures, providing preferential parking for carpoolers, free van transportation to places of business from relatively nearby public transportation terminals, facilities for bikers, company-sponsored services for matching up carpool participants, and recognition, bonuses, and financial rewards of various sorts. The evidence indicates that such measures can be effective, although many people find carpooling to be inconvenient or undesirable; it appears that relatively few carpools are formed without some well-organized incentive program to promote them. Factors affecting the acceptability of carpooling are as much psychological and social as economic and practical (Reichel & Geller, 1981).

Reichel and Geller (1981) drew a number of conclusions from a review of research on carpooling, among which are the following: Community interventions, such as the assignment of priority lanes for multi-occupant vehicles, can be effective, but enforcement of priority lane restrictions remains a problem. Efforts on the part of corporations and community agencies to facilitate carpooling by providing matching services have had mixed results. Reduced transportation costs are not sufficient incentives for initiating a carpool, but may serve as rewards for maintaining already established carpools. Social factors are important influences in the initiation and maintenance of carpooling. Although experimentation has shown that transportation behavior can be modified toward energy conservation by the use of incentives, the interventions are rarely very cost effective, and evidence is lacking that whatever effectiveness they do achieve would continue for very long periods of time. Reichel and Geller argue that creation of a truly energy-conserving community will require the evolution of social norms that value and reward energy-conserving behavior.

Some investigators have found that, with proper instruction, incentives, and feedback, drivers can increase significantly the miles-per-gallon they realize (Lauridsen, 1977; Runnion, Watson, & McWahorter, 1978). In one study (Lauridson, 1977), automobiles were equipped with meters that gave miles-per-gallon feedback on a continuous basis. Most drivers who were given this type of feedback were able to increase the number of miles-per-gallon they obtained by about 5% to 10%. This seems like a promising lead to follow. It should be possible, with modern technology and computer-based monitoring systems, to provide drivers with displays showing fuel efficiency where performance is averaged over very short intervals, so that at any instant the indicator represents the more-or-less instantaneous situation. Another possibility would be to display a continuous record of the accumulating cost of a motor trip. In this case, cost could reflect not only fuel consumed, but the assumed or measured wear on tires and other replaceable automotive components, and so on.

* * *

A fundamental fact about energy in the United States that is frequently overlooked is that energy here is, and for a long time has been, relatively inexpensive. One might argue that this has been a major deterrent to conservation at all levels. Although economic studies show consumption to be somewhat sensitive to price changes, the effects tend to be small, so that when the price of energy increases consumption falls off but not by much (Lehman & Warren, 1978; A. M.). It appears that energy costs are sufficiently low that significant price swings have relatively little effect on consumer behavior. The fact is, however, that the total costs of energy production and use are very difficult to quantify; they are not represented by what the consumer pays for gasoline, electricity, or home heating oil.

The direct costs of producing energy are relatively visible and easy to understand. These include the costs of drilling, mining, refining, transforming (e.g., chemical energy to electricity), and distribution. Not all of these costs are reflected in prices paid directly by consumers, however, because some are recovered indirectly through various types of subsidies and credits paid for with tax dollars. Much less obvious than the direct costs of energy production and distribution, but no less real, are the costs of dealing with the detrimental effects of various forms of energy use on the environment. These costs are less obvious because, in many cases, they are delayed: They are not borne by the primary energy suppliers or users, and they can be difficult to quantify, but any reckoning of energy costs that does not take them into account is incomplete.

We need a better understanding of the total costs of energy production and use, and better ways of informing the public of them and of the tradeoffs involved in the various approaches that might be taken to satisfying our energy requirements. Happily, there is some indication of a growing interest among policy makers in estimating the cost of the environmental effects of various methods of energy production, as well as the immediate production costs. Oregon, Illinois, Massachusetts, and Virginia, for example, now require their utilities to consider the environmental costs of energy production in their future resource planning (S. Williams et al., 1990).

There is a need also for some new perspectives and ways of viewing energy requirements. As Reddy and Goldemberg (1990) have argued, energy should not be viewed as a commodity, but as a means of providing services. If improved services are the objective, realizing it as a consequence of improved efficiency in the use of energy is as good as increasing the amount of energy available. The role of energy utilities, they suggest, should not be simply that of supplying energy, but that of providing heating, cooling, lighting, and other energy-based services. From this perspective, the appropriate indicator of a country's degree of development would be the level of services provided rather than the amount of energy used, or, better yet, degree of development might be judged on the basis of the efficiency with which services are delivered, which would mean thinking in terms of services delivered per unit of energy consumed.

Finding more efficient and effective ways to produce, distribute, and use energy is, and will continue to be, a major challenge facing the nation and the world. I believe there are opportunities for the human factors community to help meet this challenge in many ways. Some of the more obvious ways include the design of displays for power-plant operator consoles and control stations; the assessment of perceived risks associated with various approaches to energy generation and the communication of objective risks, to the extent they can be quantified; the study of decision making as it pertains to energy generation and use; finding ways to stretch resources—substituting low-energy limited-material goods and services for those that require large amounts of energy and material—without negatively impacting quality of life; increasing the usability and user acceptance of electronic means of information transmission; the finding of ways to have the electronic transmission of information substitute for the transportation of people and material; helping to bring the potential of electronic publishing, the virtual office, and artificial realities to practical applicability; improving the safety of lighter automobiles; helping to find ways to increase the quality of goods and services and to decrease the incidence of product rejection and duplication of effort; the study of individual and institutional participation and nonparticipation in energy conservation and recycling efforts and ways to increase participation; and development of better techniques for monitoring energy use and providing useful feedback to consumers. There are, undoubtedly, many other possibilities for applications of human factors research to the problem of meeting future energy demands that are less obvious than those listed here. The challenge, in this case, is to search them out.

4

Environmental Change

The earth is a dynamic, changing, ever-evolving system, but we have come to realize this fact only quite recently. Local changes resulting from earthquakes, volcanos, erosion, and similar natural causes have been recognized for a very long time, but, historically, the cumulative global effects of such forces have not received much attention. The discovery that the mean surface temperature of the globe has varied over a considerable range, producing a series of ice ages interspersed with periods of relatively warm climate, was made during this century. The theory of plate tectonics, according to which the earth's crust is composed of about a dozen continually shifting plates, was formulated by Alfred Wegener in 1912 but has become generally accepted only since the 1960s.

Only now are we becoming aware of how delicately the various components of the global ecosystem are balanced and how tenuous are the conditions that make the planet hospitable to human life. We are becoming aware also—although much too slowly, some would say—that human behavior has the potential to upset this balance in a variety of ways (Carson, 1962; W. C. Clark, 1989a, 1989b; Goudie, 1982). A major difference between the past and the present is the fact that humankind has become sufficiently numerous and powerful to have a significant effect on the global ecosystem. Not only have we amassed an unimaginable amount of power explicitly designed for destructive purposes, but we produce and use large enough amounts of energy to be in danger of affecting the ecosystem in ways that we do not understand. A recent report from the National Research Council (1989c) states the problem succinctly and well: "Modern technology, by making it possible for humans to alter natural processes at the level of the geosphere, has made possible disasters that could not even be fantasized a few generations ago" (p. 59).

A growing concern in relation to this is reflected in the considerable attention that environmental issues have been receiving of late both in the technical literature and in the press. It seems safe to assume that that concern will continue to increase as more is learned about the many ways in which the behavior of individuals, institutions, industries, and nations can have long-term detrimental, and possibly catastrophic, effects. Discussions need to focus on such topics as global warming, acid rain, urban smog, thinning of stratospheric ozone, water contamination, deforestation, desertification, decreasing biodiversity, waste disposal, contamination from industrial accidents, and natural disasters.

The topics are not mutually exclusive; they interrelate in a variety of ways. Moreover, most of the phenomena of interest can be viewed as both effects and causes. Global warming, for example, is usually discussed as an effect of the accumulation of greenhouse gases in the atmosphere; it can be seen also, however, as a cause, or potential cause, of desertification, a rise in ocean levels, and changes in the earth's biota. The accumulation of greenhouse gases in the atmosphere is probably thought of first as a cause of global warming, but it is also an effect of deforestation. Acid rain, which is an effect of the burning of high-sulfur fuels, is the cause of the death of fresh-water lakes and perhaps some forests. And so on. Nevertheless, these topics are convenient pegs on which to hang a discussion of environmental problems and are used here for that purpose. Many of the following comments are adapted from Nickerson (1991).

The order in which the topics are discussed is not intended to reflect my view of their relative importance. If asked to put these topics in such an order, I would find that a difficult task indeed. Each of these problems is potentially very serious, and, if not addressed effectively, could easily become catastrophic. Judging solely from the volume of printed material that is available on all of these topics, however, one must believe that global warming is the focus of greatest concern, so it is with that one that this discussion begins.

GLOBAL WARMING

The Greenhouse Effect

Most of the solar radiation received at the earth's surface is in the visible wavelength region, .4 to about 1 μ (micron). This solar energy fuels photosynthesis, and much of it is also reflected back into space. The earth itself emits radiation over a fairly broad range of wavelengths in the infrared region, with the intensity peaking at about 10 μ, falling off fairly sharply at lower wavelengths and more gradually at higher ones. Most of the infrared radiation immediately around 10 μ is emitted into outer space. At shorter and longer wavelengths, however, much of the earth's radiation is absorbed by water vapor (especially in the form of clouds) and certain trace chemicals, especially carbon dioxide, in the atmosphere.

The earth's average temperature is believed to have varied considerably over the ages (Budyko, Ronovn, & Yanshin, 1987; Burner & Lasaga, 1989), but consistently to have been subject to some warming from this trapping of infrared radiation in the atmosphere. This warming has become known as the *greenhouse effect*. A greenhouse effect, in itself, is not a bad thing; indeed, such an effect of the right magnitude is essential to a life-supporting environment. At present, the mean surface temperature of the earth is about 16°C (61°F), which is believed to be about 33°C (59°F) higher than it would be if there were no greenhouse effect (S. H. Schneider, 1989). Venus, which has an atmosphere that is about 96% carbon dioxide and about 90 times as massive as the earth's has a very strong greenhouse effect and a surface temperature of about 400°C; Mars, in contrast, has very little atmosphere and thus a weak greenhouse effect, and it has a surface temperature of − 50°C (Revelle, 1982).

The concern of scientists is not for the fact that a greenhouse effect exists— no one, to my knowledge, has argued that the world would be a more pleasant place if the effect disappeared entirely, and the mean temperature dropped by 33°C. The concern is for the possibility that, as a consequence of increasing concentrations of carbon dioxide and other greenhouse gases in the atmosphere, the magnitude of the effect will become inhospitably large. This possibility was pointed nearly a century ago by Svante Arrhenius (Asimov, 1972). In 1957, Revelle and Suess characterized the burning of fossil fuels and the consequential release of carbon dioxide into the atmosphere as a global experiment with the system that has been responsible for the relative stability of the earth's climate. Increased global warming, with attendant severe environmental and economic consequences, is now widely recognized as a possibility deserving serious attention (Bolin & Doos, 1986; Dickinson & Cicerone, 1986; Houghton & Woodwell, 1989; Kerr, 1990; National Academy of Sciences, 1983; Ramanathan, 1988; Rosenberg, Easterling, Crosson, & Darmstadter, 1989).

Although scientists are far from unanimity on the question, many believe there has been a rise in mean temperature over the past century of about 0.5°C. (The rate of increase has not been uniform over that period, and from about 1940 to about 1970 the average temperature actually fell almost 0.2°C before resuming an upward trend.) This would be consistent with an estimated increase of carbon dioxide in the atmosphere of about 43 parts per million over the same time (Revelle, 1982). Perhaps the most extensive effort to document temperature change was made by P. D. Jones, Wigley, and

Wright (1986; P. D. Jones & Wigley, 1990), who attempted to collect as many records as they could find of temperature measurements around the world since the beginning of such recording some 300 years ago and to account, as far as possible, for variations due to such factors as changes in thermometer design, the appearance in more recent times of urban heat islands, and changes in techniques for measuring the temperature of sea water. On the basis of their results, Jones and Wigley concluded that there has been a rise in global temperature over the past century on the order of 0.5°C. They did not attempt to pinpoint the cause of the rise, but did note that "the changes in the levels of greenhouse gases that have already occurred should eventually cause a global mean temperature increase between .8 and 2.6 degrees Celsius" (p. 91), the uncertainty in these figures being due to incomplete understanding of how the earth will respond to various types of external forces of climate change.

Not all scientists are in agreement with the prevailing assumption that greenhouse warming is a serious threat; a few have argued that the expressed concerns greatly exaggerate the problem or even that the atmosphere is capable of negating the effects of increased greenhouse gases (Kerr, 1989a, 1989d). To get an indication of the controversial nature of the question of global warming, one need only read the Marshall Institute report on the subject (Seitz, Jastrow, & Nierenberg, 1989), the *Science* article in which that report was discussed (L. Roberts, 1989c), and several letters to the editor in response to this article (Jastrow, 1990; Lindzen, 1990; Nierenberg, 1990). The whole dispute was recently reviewed by R. M. White (1990).

Understanding of the greenhouse effect is further complicated by the fact that the earth's average temperature may fluctuate, showing up-and-down trends over periods of decades as a consequence of other natural factors such as fluctuations in the amount of solar energy per unit area reaching the earth (possibly resulting from variations in solar activity such as those associated with changes in sunspot frequency). Long-term trends in global temperature can be obscured by significant short-term variation due to volcanic eruptions (the eruption of Krakatau in 1883 is believed to have lowered the temperature by 0.5°C, and the El Nino–La Nina cycle: The first part warms the atmosphere as a consequence of abnormally warm waters in the Equatorial Pacific, and the second part cools it because of the abnormally cold waters in that half of the cycle (Kerr, 1989b).

Also, although the global carbon cycle is understood at a general level (Baes, Goeller, Olson, & Rotty, 1977), recent research has shown the details to be more complex than was previously realized; efforts to balance all of the carbon fluxes over the last two centuries have not been successful and different mathematical models have produced inconsistent results (W. M. Post et al., 1990). In particular, how carbon dioxide is distributed among the atmosphere, the oceans, and land remains uncertain. The oceans are believed to hold many times as much carbon dioxide as the atmosphere (Broecker & Denton, 1990), but recent studies suggest that land is a greater repository of fossil fuel carbon dioxide than was previously thought, whereas the oceans are a lesser one (Tans, Fung, & Takahashi, 1990; Wallich, 1990).

The effect of land use by human beings on the global carbon cycle has proved to be especially difficult to estimate. Post et al. (1990) have pointed out that two of the more common approaches to making this estimate give diametrically opposite results, one showing a marked increase in the net flow of carbon from the biosphere due to land use and the other yielding a net decrease over the same period. Post et al. have noted further that although it is possible to document the growing human contribution of carbon dioxide to the atmosphere and the potential for additional increases, uncertainty about the effects of feedback processes—which could either moderate the increase in carbon dioxide or accelerate it—makes it impossible to predict with confidence the ultimate effects of further increases in atmospheric carbon dioxide on the global carbon cycle.

The Oceans

The importance of the oceans to an understanding of the carbon cycle is seen in the fact that, irrespective of whose estimates one uses, they are, by far, the greatest source of the carbon that participates in that cycle. According to Post et al., about 93% is stored in the oceans, about 5% in the terrestrial system (trees and soil), and only about 2% in the atmosphere. These numbers do not include the carbon stored in fossil fuel deposits—which is estimated to be about twice as much as is stored in trees and soil—because, except as a consequence of mining and burning, this carbon does not participate in the carbon cycle as does carbon in the other repositories. The processes involved in the mixing of carbon in the oceans include upwelling, downwelling, vertical diffusion, advection, gravitational drift, and biological pumping; the details of these processes and precisely how they interact are only poorly understood. The role of plankton as a major carbon dioxide consumer and counterbalance to its atmospheric accumulation also needs to be researched more thoroughly.

The accuracy of the predictions that climate models make of increases in temperature resulting from increases in the concentration of carbon dioxide is unknown. A major ingredient of this uncertainty is lack of knowledge of the effects of clouds and of the role of the oceans (La Brecque, 1989). Because they are heat sinks, the oceans can be expected to delay climate change somewhat, but given the current lack of understanding of the details of their interaction with the atmosphere, the delay could be anywhere from 10 to several hundred years. La Brecque has pointed out that if this delay is long, that could mean that only a small fraction of the heating due to the greenhouse gases already added to the atmosphere has yet been realized.

Carbon Dioxide and Other Greenhouse Gas Accumulation

Although the effects of further increases in greenhouse gases in the atmosphere have been the subject of much, and sometimes acrimonious, debate, there seems to be little question that the concentration of carbon dioxide and other greenhouse gases is, indeed, on the increase. Carbon dioxide is produced by the burning of organic compounds, whether rapidly, as when fossil fuels are burned to produce energy, or more slowly, when organisms, including humans, metabolize what they eat.

A major source of information regarding changes in CO_2 concentrations in the atmosphere over periods of time ranging from a few centuries to many millennia is air trapped as bubbles in the ice of glaciers (Barnola, Raynaud, Korotkevich, & Lorius, 1987; Friedli, Lütscher, Oeschger, Siegenthaler, & Stauffer, 1986). Attempts have been made to recover ancient air from other repositories—bubbles in amber, sealed chambers in the pyramids, airtight old artifacts such as compasses, hourglasses, telescopes, hollow buttons—but none of these alternatives has proved to be as reliable and useful as glacial ice (Weiner, 1989). Ice core data show the CO_2 concentration varying between about 200 and 300 parts per million over the last 160,000 years. For the first 850 of the last 1000 years it was quite constant at about 280 parts per million, but since the middle of the 19th century it has steadily increased (Post et al., 1990). Estimates of CO_2 concentration derived from ice core samples have forced a rethinking of assumptions underlying conventional ocean models, in some cases, because of incompatibilities in the results of the two methods (Enting & Mansbridg, 1987).

The longest continuous record of directly measured atmospheric carbon dioxide concentration was obtained at the Mauna Loa Observatory in Hawaii, and extends from 1958 to the late 1980s. It shows the concentration increasing from about 315 parts per million to about 350 parts per million over that period (Keeling, 1986). The current level of carbon dioxide in the atmosphere is estimated to be about 25% greater than it was in 1850, and the increase has been attributed primarily to the burning of fossil fuel and a decrease in the percentage of the earth's surface that is covered by forests (S. H. Schneider, 1989). (A ton of carbon in fossil fuel becomes 3.67 tons of carbon dioxide in the atmosphere when the fuel is burned [Ayres, 1989].) It has been predicted that the concentration of carbon dioxide in the atmosphere will double by the middle of the 21st century and probably quadruple before the use of fossil fuel peaks (Ingersoll, 1983).

Trace greenhouse gases other than carbon dioxide, including methane and the chlorofluorocarbons (CFCs), which exist in the atmosphere in even smaller concentrations, are believed to have increased by even larger percentages than has carbon dioxide and to be continuing to increase as well. Exactly what the effects of a doubling or quadrupling in carbon dioxide concentration, or of comparable increases in other greenhouse gases, would be on average global temperature is not known with certainty. Models of global climate consistently predict temperature increases, but by different amounts. Most predictions range from about 2° to about 6°C over the next century (J. F. B. Mitchell, 1989; S. H. Schneider, 1989).

Although most modeling to date suggests a doubling of carbon dioxide in the atmosphere before the end of the 21st century, there is no reason to assume increases will stop at this level; Houghton and Woodwell (1989) note that estimated reserves of recoverable fossil fuels alone are enough to increase the concentration by a factor of from 5 to 10. They contend that any resulting changes in climate are essentially irreversible over any time span of interest to us or our children. Possible preventive measures include lessening dependence on fossil fuels, halting deforestation, and promoting reforestation. An Environmental Protection Agency (1989b) report released in draft form to the U.S. Congress in March, 1989 (cited in Marshall, 1989a) concluded that the rate of increase of greenhouse gases in the atmosphere could be slowed down with appropriate action. Marshall described the action as including "price increases for coal and oil; lots of solar power devices; more use of nuclear and biomass energy; new forests all around the planet; sharp cutback in the manufacture of chlorofluorocarbons and related products; a lower per capita demand for cement; new ways of producing rice, meat, and milk; gas capture systems and landfills, including those in the developing world; and an unusual degree of cooperation among the developed and developing nations. If all these things begin happening in the early 1990s, the rate of gas buildup may level off in the 22nd Century" (p. 1544).

Water Vapor

Although most of the discussion of the greenhouse effect has focused on the role of carbon dioxide and other gases or particulates that are found in the atmosphere in small quantities, these components may be less important than water vapor, which often is not mentioned in discussions of the threat of increased greenhouse warming. How the concentration of water vapor might change in response to a significant rise in temperature resulting from an increase in greenhouse gas accumulation, and how the change would contribute to further modifications in climate, are not known. Neither an exacerbation of the problem and an acceleration of the warming trend nor some counteraction to warming through an increase in high-level cloud cover can be ruled out with certainty.

The role of clouds is among the least understood aspects of climate determination. Clouds help to cool the earth by shading it from the sun, but they also block the escape of radiation of infrared wavelength from the earth's surface. How these two effects trade off against each other is not known precisely, nor is it certain whether an increase in surface temperature would produce more clouds as a result of increased water evaporation, or fewer clouds as a result of the lower relative humidity of the warmer air.

Currently, about half the earth is covered by clouds at any given time; small changes in cloud cover could conceivably have a large influence on the greenhouse effect—one way or the other. Indeed, it has been estimated that the greenhouse effect of clouds could be larger than that resulting from a 100-fold increase in the carbon dioxide concentration of the atmosphere (Ramanathan et al., 1989). A major reason for large differences in the predictions of climatologists regarding changes in the earth's temperature over the next several decades is the result of different assumptions about these matters (H. S. Schneider, 1989; Seitz et al., 1989). The earth's radiation budget experiment (ERBE) is an attempt to collect observational data from three satellites launched into three different orbits between 1984 and 1986 on cloud cover and differences between incoming and reflected radiation, in an effort to quantify better the effects of clouds.

Ice Sheets

What role the ice sheets that cover much of Greenland and Antarctica could play in any future warming trend also is not well understood (Zwally, in press). Significant melting of these sheets could contribute to sea-level rise, adding to the effects of thermal expansion of the oceans (Meier, 1984); on the other hand, increased precipitation resulting from higher temperatures can accelerate the growth of glaciers, and this could counteract, to some degree, the increased rate of melting. Reduction in the extensiveness of sea ice in the polar regions as a consequence of atmospheric warming could also have the effect of increasing biological production in those areas and decreasing atmospheric carbon as a consequence. (Measurements of the ratios of ^{18}O to ^{16}O in ocean sediments has led to the conclusion that global ice volume has peaked about every 100,000 years over the last several hundred thousand [Broecker & Denton, 1990].)

Climate Models

MacDonald (1989) distinguishes three types of models that have been used to predict climate change. At one extreme of a complexity continuum are *energy-balance models,* which treat the entire earth's surface as having a single temperature. At the other extreme are *three-dimensional global circulation models* (GCMs), which partition the atmosphere into tens of thousands of regional cells and solve, for each cell, equations that represent how climate is affected by several dependent variables. Intermediate on the continuum are *radiative connective models* (RCMs), which recognize vertical structure (radiative effects) but do not deal with horizontal structure.

Several GCMs have been developed and are currently in use. One report from the National Research Council (1987) identified eight trends that are being projected by these models:

- Stratospheric cooling: lower temperatures in the upper stratosphere due to reduced ozone concentrations accompanied by increased concentrations of radioactively active trace gases.
- Global mean surface warming: generally higher surface temperature worldwide.
- Increase in global mean precipitation, following from increase in evaporation due to warming.
- Reduction in sea ice, as a direct consequence of warming.
- Disproportionate warming at the poles, due in part to reduction in sea ice.
- Increased midcontinent summer dryness: reduction in midcontinent soil moisture in summer due to warming.
- Increased precipitation at high latitudes, due to poleward movement of warm moist air.
- A rise in mean sea level, due to thermal expansion of water and to melting of glaciers.

A major complaint that climate modelers have is the lack of available time on supercomputers for running their simulations. Weather models are massively parallel and require enormous amounts of processing capacity. Typically, they divide the atmosphere into a three-dimensional grid that wraps around the globe. The conditions in each unit of the grid are updated at short intervals in accordance with a mathematical model that incorporates simple principles of atmospheric physics. How much computing is required depends on the sizes of the polyhedrons into which the atmosphere is divided, the number of variables in the model, and the frequency with which the variables have to be updated. Ingersoll (1983) described the situation this way:

> Keeping track of seven atmospheric variables (temperature, pressure, water vapor, cloud cover and the wind speed along three axes) in a grid consisting of cubes 200 kilometers square and ten layers deep actually means keeping track of

a million variables. Interactions with nearby variables largely determine how each variable changes with time. Some 500 arithmetic operations are required to compute the interactions to which one variable is subject, so that a total of about .5 billion operations must be done for each ten-minute time step of the model. A ten-fold increase in the spatial resolution of the three-dimensional grid would increase the number of variables by three factors of ten, and the number of time steps per hour would have to increase by a comparable factor. The number of arithmetic operations would therefore increase 10,000-fold, and yet such a model would still not resolve atmospheric phenomena smaller than 20 kilometers. (p. 168)

The climate models that have been developed to date represent only a subset of the variables that are known to be important, and their predictive accuracy is often unknown because the necessary validation studies have not been done. Such studies can be extremely difficult, or in some cases impossible, to do. The models are too complicated to permit their validation with controlled experiments, and inasmuch as they are used to make long-range predictions, determining the accuracy of those predictions by checking them against future climate data will take a long time. One technique that has been used for validation purposes is that of initializing the models' parameters to conditions of several years ago and checking their predictions against data that have been obtained from monitoring activities in the past.

The projections of climate models are consistent with data obtained from climate monitoring over the recent past in some respects, such as the fact of overall global surface warming and precipitation, but are problematic in other respects. With respect to the amount of warming that should occur, the models tend to disagree with each other and with actual trend data for the past 100 years by as much as a factor of 2 or 3 (Grotch, 1988; Schlesinger & Mitchell, 1987; Wigley & Schlesinger, 1985). Also, the match is less good with respect to temperature trends in specific zones; whereas the models predict more warming at higher latitudes, the data suggest less warming there or possibly even some cooling. The monitoring data also fail to show the trends predicted by the models with respect to sea surface temperature, sea ice, and snow cover. Such discrepancies illustrate the need for the maintenance of accurate monitoring databases and the frequent validation of predictive models against the data they hold. Numerous government reports have called for improvement in techniques and resources for monitoring global climate (National Research Council, 1983). Monitoring of global temperature by means of microwave radiometry from satellites promises to yield better information regarding atmospheric temperature than is obtained by monitoring from fixed sensors distributed over the earth's surface (R. W. Spencer & Christy, 1990).

The questionable validity of long-term climate models has been a major source of concern among scientists and policymakers alike, especially in view of the fact that some of their predictions are dire indeed, and have prompted the proposal of preventive or corrective action that would be extremely costly, at least in some parts of the world. The proposal to stabilize the emissions of carbon dioxide into the atmosphere could probably be realized, for example, only at the cost of slowing the development of the less developed countries. Clearly this is not a cost one would want to defend on the basis of predictions generated by models of doubtful validity, and there appears to be considerable doubt as to the validity of the models that currently exist (W. T. Brookes, 1989; Seitz et al., 1989). Manzer (1990) has argued that chlorofluorocarbons (CFCs) are sufficiently important to modern society that it would be irresponsible to cease their production immediately. The CFCs that are used to produce goods and services in the United States alone have been estimated to be worth about $28 billion a year and to involve 725,000 jobs (Hively, 1989). Production of substitutes, such as hydrochlorofluorocarbons (HCFCs) and hydrofluorocarbons (HFCs), is feasible, but the selection and manufacture of suitable alternatives require the solution of several technical problems.

The U.S. Department of Energy (1990) is planning two new research initiatives for the 1990s: the Atmospheric Radiation and Measurement program, which is aimed at improving understanding of cloud-radiation interactions, and the Computer Hardware, Advanced Mathematics and Model Physics initiative, which is intended to develop a climate model that increases the throughput capability (i.e., the amount of data able to be processed in a given period of time) over existing models by a factor of at least 10,000 over 10 years. The target model will improve on existing models in being able to represent slower changes in atmospheric composition, in representing the atmosphere with a finer grid (resolution of less than 100 km, latitudinally and longitudinally), in having a better representation of ocean and atmospheric chemistry, and in being able to simulate climatic events for longer periods of time.

Modeling the earth and its climate as a dynamic system is a very complicated undertaking. Considerable effort is being devoted to this objective, but the models that currently exist must be considered crude compared to what will be required to resolve many of the uncertainties about the determinants of climate change. The science is young, and the tools (e.g., computing resources) are only beginning to be close to adequate. Global modeling efforts must take many factors into account. These include emissions of gases and particulates into the atmosphere, changes in cloud densities and distributions, the roles of oceans as carbon and heat storage systems, circulation processes in the atmosphere and oceans, heat and chemical exchanges at the ocean–atmosphere boundary, changes in the density and distribution of forests, changes in the earth's albedo, the role of biota as consumers and emitters of substances important to earth system processes and balances, and positive and negative feedback effects from disruptions of steady-state dynamics.

Is There Cause to Worry?

The full implications of a sustained rise in global temperature are not completely clear, but predictions include a partial melting of the polar ice caps and thermal expansion of sea water, with a resulting rise in sea levels and inundation of low-lying land, an increase in the proportion of the earth in which the conditions are tropical, and probably major changes in the earth's biota, as well as in both regional and global climate and weather. (There is at least suggestive evidence that the global sea level has risen from 10 to 20 cm during the past century [Peltier & Tushingham, 1989]). A possible consequence of long-term climate change that has not received the attention it deserves is the potential for amplified effects of extreme conditions such as storms and floods. A 1 m global rise in sea level does not sound like a very significant change, but regional effects of such an increase could be severe under hurricane or flood conditions.

Given the high degree of uncertainty about the long-range effects of possible actions that might be taken in the interest of ensuring a hospitable future climate, and the costliness of some of these possibilities, it is not surprising that proposals often encounter considerable resistance. In one recent article in *The New York Times* that focused on the cost of controlling greenhouse emissions, the writer reported as "pessimistic, but not implausible" one estimate of the annual cost to the United States of holding the line on carbon-dioxide production as rivaling the current level of military spending (Passell, 1989). Greene et al. (1988) claimed that there is not a strong commitment to mitigate this effect at the present time either in the United States or elsewhere, because of uncertainty about the severity, location, and timing of its consequences.

On the other hand, in spite of the fact that climatologists are not totally agreed as to either the immediacy or the magnitude of the threat of global warming, there seems to be a fairly widely shared opinion that postponing action aimed at averting an increasing greenhouse effect until we can obtain incontrovertible evidence that the effect is already increasing—which could take decades—is not prudent (Kerr, 1988a, 1988b). As Holden (1990a) put it: "If we wait our knowledge will improve, but the effectiveness of our actions may shrink; damage may become irreversible, dangerous trends more entrenched, our technologies and institutions even harder to steer and reshape" (p. 162). A petition urging that the possibility of global warming be taken seriously was delivered to President Bush in February, 1990 with the signatures of about 700 scientists, including almost half the members of the National Academy of Sciences. Abelson (1990) cautioned against a "half-baked political response" to the threat of global warming, but argued that in spite of uncertainties about the magnitude of the threat, actions "based on well-thought-out, long-range goals" should be taken (p. 1529). As examples of such actions, he noted the promotion of conservation and more efficient use of energy through increased taxes on fuels—especially gasoline—and greatly expanded efforts to develop renewable energy resources, including biomass.

The "No-Regrets" Policy

As a strategy for addressing the possibility of a serious problem with global warming, while recognizing the uncertainty as to the reality or magnitude of the problem, some have advocated a *no-regrets policy,* a policy that includes actions that will be effective against a real threat, and that are desirable (or at least not detrimental) even if the threat proves not to be a serious one. Sometimes the no-regrets policy is described as actions that will have "tie-in" benefits even if the desired primary effect is not obtained (H. S. Schneider, 1989). Actions that would be included in such a policy are development of better monitoring and modeling capabilities so as to reduce the degree of uncertainty about the problem, steps aimed at conserving and making more efficient use of energy, phase-out of CFC production, development of alternatives to fossil fuel as energy sources, reforestation and forest preservation, and research on stress- and disease-resistant crops (R. M. White, 1990).

Making more efficient use of fossil fuel in the production of energy, for example, has a tie-in benefit because even if the accumulation of carbon dioxide in the atmosphere turns out to be a less serious problem than is currently believed, conserving fossil fuel is desirable for purely economic reasons. Increasing the tendency of people to rely on "manual" modes of transportation (bicycle riding, walking), when feasible, in preference to using motorized forms of transportation, has the triply beneficial effect of conserving energy, lessening pollution, and contributing to people's physical fitness. Another strategy that analysts sometimes advocate is the "insurance-policy" strategy. An *insurance-policy strategy* is one that seeks protection against catastrophic dangers for modest costs, recognizing that those costs might, in retrospect, prove to have been unnecessary, like having fire insurance on a house that never burns down.

White cautions, however, that desirable as such steps would be, the public should not be misled to believe that they will suffice to solve the climate warming problem, if the problem is as severe as many scientists think it is. In particular, hopes of reducing emissions of carbon dioxide into the atmosphere over the next few decades by a combination of improved efficiency of energy use and increased use of non-fossil energy sources are unrealistic to the degree that they fail to take the needs of the non-industrialized countries into account. Although about 80% of the greenhouse gases are now produced by the industrialized nations, which make up about one third of the world's population (Fulkerson et al., 1990), adequate control in industrialized countries will not suffice to solve the global problem, inasmuch as the CO_2 emissions from developing countries could exceed those from industrialized countries by around 2000 (Fulkerson et al., 1989). The development of contingency plans for accelerating the rate of reducing the emissions of greenhouse gases, should further research reveal the problem of global climate change to be more severe than is currently recognized, ought to be part of the general approach to the problem of climate change (Holdren, 1990).

The U.S. government has proposed to bring free-market pressures to bear on the problem of emissions control by the establishment of a system for allocating and trading emissions credits. Allocated credits would determine the amount of greenhouse gases that each participating nation could emit into the atmosphere over a specified period of time, but these credits could be traded among nations. Thus, a nation that could get its emissions below its allocated amount would have a commodity that it could sell; conversely, a nation that failed to limit its emissions to its allocated amount would be forced to incur the expense of buying further credits (Sun, 1989). This is the international version of a similar plan to help reduce acid rain in the United States. In this case, the U.S. government would set a national limit on sulfur-dioxide emissions and allocate to individual companies their fair shares of this limit. Companies could then buy and sell credits, the assumption being that this would provide a financial incentive for companies to emit less than their allocated share.

* * *

A specific difficulty associated with global warming (and with other atmospheric problems, such as acid rain and smog) stems from the fact that the problematic gases exist in extremely small concentrations but appear to be able to have very sizable effects when those concentrations are changed. Together, nitrogen (about 78%) and oxygen (about 21%) make up about 99% of the earth's atmosphere; most of the remainder is argon. The several gases that have been receiving so much attention in discussions of the threat of global warming and other environmental problems exist, by comparison, in minute quantities, accounting, in the aggregate, for much less than .01% of the atmosphere. Table 4.1 shows the concentrations of these trace gases in parts per billion by volume (ppbv) and estimates of their current rates of change.

TABLE 4.1 Concentrations of Greenhouse Gases in the Atmosphere, in Parts per Billion by Volume, and Estimated Rates of Increase in Them (after Rodhe, 1990)

Gas	Concentration (ppbv)	Rate of Increase (% per year)
Carbon dioxide	353,000	0.5
Methane	1,700	1.0
Nitrous oxide	310	0.2
Tropospheric ozone	10–50	0.5
CFC-11	0.28	4.0
CFC-12	0.48	4.0

The magnitude of the effect of increasing the atmospheric concentration of a greenhouse gas depends both on the gas and on its prevailing concentration. The addition of a specific amount of carbon dioxide, for example, has considerably less effect than the addition of the same amount of several other gases, in part, because the prevailing concentration of carbon dioxide is relatively high. At the present time an increase of a fixed amount of CFC-12 has about 6,000 times the effect of the same amount of increase (by weight) of CO_2 and about 15,000 times the effect of the same amount measured in moles. The effect of other greenhouse gases—methane, nitrous oxide, ozone—fall between these extremes. Although carbon dioxide has much less effect per unit weight, or per mole, than do most of the other greenhouse gases, it is responsible for about 60% of the effect because of its relatively high concentration. Methane is believed to be responsible for about 15% of the effect, tropospheric ozone for 8%, N_2O for about 5%, and CFC-11 and CFC-12, together, for about 12% (Rodhe, 1990).

Calculating the combined radiation-trapping effect of atmospheric constituents is complex, because the contributions of individual absorbers do not add linearly (MacDonald, 1989). Because the ranges of wavelengths that different constituents absorb overlap, a decrease in the concentration of one absorber could be offset to some degree by an increase in the amount absorbed by a remaining absorber, simply because there would be more radiation available for it to absorb. Also, increased concentration of CO_2 in the atmosphere could promote growth of certain biota, and the result could be a larger biological sink for CO_2.

It should be apparent that the subject of climate change generally, and of global warming in particular, is an exceedingly complex one. Considerable knowledge has been gained in recent years about the numerous variables involved, but as the knowledge has increased so has an appreciation of the complexity of the issues. Among the most obvious needs for the future is a better understanding of the long-range consequences of changes in variables that affect the climate and effect climate change. Especially important is a better understanding of the various ways in which human behavior plays a causal role in climate modification. It is likely to be a considerable time, however, before the picture becomes completely clear, if it ever does. In the aggregate, the evidence in hand strongly suggests that postponing action until the picture becomes clearer than it currently is would be running the risk of letting the problem grow so much that effective corrective action would be very difficult, if not impossible, to apply or even prescribe.

ACID RAIN

Acid deposition ("acid rain") consists primarily of sulfuric and nitric acids that are formed when sulfur dioxide, nitrogen oxides, and other materials emitted by industrial processes and road vehicles combine with water in the atmosphere. The main cause of these emissions is the burning of fossil fuels, especially coal. Acid rain contributes not only to the acidification of lakes and streams and the destruction of trees but also to the corrosion of materials, including irreplaceable historical and cultural treasures (Mohnen, 1988; S. E. Schwartz, 1989). Extreme acidification of lake water can be lethal, not only to fish, crustaceans, and marine invertebrates that live in the lakes, but also to birds such as loons that depend on them for food (Alvo, 1986). Human activity is believed to account for about 90% of all sulfur emissions into the atmosphere (Likens, Wright, Galloway, & Butler, 1979).

In the United States, acid rain is most problematic in the northeast, especially New England, New York, and Pennsylvania. It is also a severe problem in eastern Canada and in Europe. Comparison of the acidity of old ice with the acidity of snow or rain that falls today has led to the conclusion that, especially in the eastern United States and western Europe, the acidity of precipitation has increased from about neutral to mildly acidic over the past 200 years. Cases of extreme acidity have been reported, however, one of the most notable being a 1974 storm in Scotland in which the rain was the acidic equivalent of vinegar (pH of 2.4; Likens et al., 1979). In the greater Los Angeles area, the pH of fog gets as low as 2, which is about the acidity of lemon juice (Mohnen, 1988). Technological progress on the problem has been made: Sulfur dioxide emissions decreased by about one third from about 1975 to 1990, during a period when the use of coal increased about 50% (Fulkerson et al., 1990).

The goal of much research on acid deposition is to determine not only the amount deposited at specific sites, but the origins of the emissions that cause those depositions. Ultimately, researchers would like to be able to specify a set of source–receptor relations that specify, for any given region (receptor), the sources of the emissions that determine its depositions (Seinfeld, 1989). Because emissions travel considerable distances in the atmosphere, this is a very ambitious goal. Various simulation modeling techniques are being used on this problem, but the source–receptor relationships are still not understood.

The difficulty of empirically determining specific source–receptor connections has been increased by local responses to the acid rain problem; these have often increased the area over which emissions are dispersed so as to minimize the effects of the emissions on the air in the local areas in which they are produced. An example of such an approach is that of making emission stacks very tall. This, of course, mitigates the local effects of the emissions at the expense of exacerbating the effects over a larger region. By venting the emissions to higher levels of the atmosphere, one simply assures that the pollutants will be transported over larger distances, but it does not prevent their return to the earth's surface in the form of acid rain or their accumulation and contribution to other long-term effects.

AIR POLLUTION AND URBAN SMOG

Health-based air quality standards exist in the United States for six pollutants: lead, sulfur dioxide, ozone, carbon monoxide, nitrogen dioxide, and particulates. Improvements were made in the concentrations of all six from 1977 to 1986, but ozone and carbon monoxide remain serious problems: An estimated 100 million Americans live in areas that violate Clean Air Act standards for these two pollutants (Suhrbier & Deakin, 1988). From 1983 to 1985, the standard for ozone was exceeded in 76 urban areas in the United States (OTA, 1988d). In 1988, according to one estimate, one third of the U.S. population lived in areas that did not meet the federal safe standard for ozone (Marshall, 1988), which is proving to be the most difficult to control of the six officially recognized pollutants (Seinfeld, 1989). C. L. Gray and Alson (1989) estimated that 100 U.S. cities had ambient levels of carbon monoxide particulate matter and ozone exceeding the health levels established by the government.

The problem of poor air quality, like most other environmental problems, is international in scope. As of 1989, 88 Soviet cities, with a combined population of 42 million people, had toxic air pollution levels 10 times the "maximum permissible" (Chandler et al., 1990). The severity of the problems of both carbon monoxide and ozone was expected to increase in the subsequent decade or two. Carbon monoxide does not stay in the air very long, but it contributes to the greenhouse effect indirectly, because some of it is converted to carbon dioxide. Also, because carbon monoxide reacts readily with hydroxyl, which normally contributes to the breakdown of methane, it can contribute indirectly to methane accumulation.

Pollutant emission estimates are based in part on measurements, but in part also on models of how emission sources (e.g., traffic) are distributed in space and time. The estimates can only be as good as the models are accurate, and there is some evidence that hydrocarbon emissions in urban areas have been underestimated. If that is the case, ozone reductions may be less than originally believed (Seinfeld, 1989).

Before the availability of data collected on recent NASA space flights, carbon monoxide in the atmosphere was believed to come primarily from the burning of fossil fuels. The NASA data have revealed that equally as much atmospheric carbon monoxide comes from the burning of tropical rain forests and savannas (R. E. Newell, Reichle, & Seiler, 1989). This finding has also changed the prevailing assumptions about the distribution of carbon monoxide in the atmosphere, which previously located most of it in the northern hemisphere. It is now believed that the burning of fossil fuels is the major source of carbon monoxide in the northern hemisphere, whereas the burning of biomass is the major source of it in the southern hemisphere.

Although human behavior is both a cause of air pollution and is affected by it, much more appears to be known about the effects of pollution than about how behavior causes it and what might be done to change that behavior to lessen its detrimental consequences (G. W. Evans & Jacobs, 1981). The cost of air pollution is difficult to determine with any certainty, but undoubtedly it is very high. According to some estimates, the United States spends an estimated $16 billion annually on health care costs associated with air pollution, another $40 billion in decreased worker productivity, and about $7 billion as a consequence of damage to buildings, monuments, and other materials (T. M. Mitchell, 1987).

Most of the concern regarding possible health hazards from energy production has been focused on the use of fossil fuels and nuclear reactors. However, the traditional fuels widely used for cooking and water heating in the developing world can also be hazardous to their users' health: "Perhaps 80% of global exposure to particulate air pollution occurs indoors in developing countries, where the smoke from primitive stoves is heavily laden with carcinogenic benzopyrene and other dangerous hydrocarbons" (Holdren, 1990, p. 160).

STRATOSPHERIC OZONE THINNING

The "ozone problem" can be confusing, because sometimes it is described as one of excess accumulation and sometimes as one of depletion. In fact, there are two problems. One, mentioned earlier, is an excess accumulation of ozone in the lower troposphere (the air we breathe); the other is a decrease of ozone in the stratosphere. The latter is bad because stratospheric ozone absorbs much of the ultraviolet radiation from the sun, thus shielding the earth from this radiation, which can cause a number of undesirable effects.

Although the ozone layer in the stratosphere, centered between 25 and 35 km altitude, is about 20 km thick, the concentration of the ozone in this layer is extremely thin; if compressed to the atmospheric pressure found at the earth's surface, it would all fit within a layer about 3 mm thick. In spite of the minuteness of the quantity of stratospheric ozone, the protection that the ozone blanket provides is critical to the well being of living creatures on earth. Both ozone formation and ozone decomposition involve the absorption of ultraviolet radiation. The absorption process is so efficient that only one part in 10^{30} of the solar radiation at wavelengths near 250 nanometers gets through the ozone layer.

The most dramatic manifestation of the loss of ozone in the stratosphere is the ozone "hole" that has appeared over Antarctica every spring since about 1970 (Farman, Gardiner, & Shanklin, 1985; Hofmann, Harker, Rolf, & Rosen, 1987; Stolarski, 1988; R. T. Watson, Prather, & Kurylom, 1988), but there is evidence that less severe losses have occurred elsewhere around the globe, especially in the higher latitudes of the northern hemisphere (Proffitt, Fahey, Kelly, & Tuck, 1989). What effect the ozone hole is having on Antarctic life is not yet clear, and there are wide differences of opinion among scientists as to what the results of investigations will eventually reveal (L. Roberts, 1989c).

Chlorofluorocarbons (CFCs), which normally constitute only about one part per billion of the atmosphere, are believed to be the primary agents responsible for the thinning of the stratospheric ozone layer. (Somewhat paradoxically perhaps, the same technology that threatens the environment in so many ways has made it possible for us to detect some environmental changes with a precision undreamed of even a few decades ago; with modern gas chromatography, it is possible to measure trace elements in air that have concentrations of a few parts per trillion.) At high altitudes, CFCs are broken down by ultraviolet radiation, and the freed chlorine atoms catalyze the conversion of ozone to molecular oxygen. Small amounts of chlorine are able to destroy large amounts of ozone. Atmospheric chemistry is still not well understood and new reactions are being discovered all the time (Glas, 1989). Projections of changes in ozone concentration in the atmosphere as a consequence of effects of nitrogen oxides or CFCs have varied considerably since attention was first drawn to this problem in the early 1970s.

The concentration of CFCs in the stratosphere has increased rapidly since they began being used as refrigerants, aerosol propellants, and solvents several decades ago (Graedel & Crutzen, 1989). Concern about the possibility of stratospheric ozone depletion is an interesting case of a problem growing out of what was believed to be a solution to an earlier concern. When CFCs were invented, around 1930, they were seen as a most welcome alternative to toxic, flammable, and corrosive chemicals, such as ammonia and sulfur dioxide, that were then being used in home refrigerators, and they were quickly adopted for that purpose (Glas, 1989). The need to reduce CFC emissions is now widely recognized. Dupont, which manufactures one fourth of the world's CFCs, has committed itself to stopping production no later than the end of the century (Hively, 1989). The United States and several other countries banned the use of CFCs as aerosol propellants in 1978. So far, however, demand for CFCs for non-aerosol uses has continued to increase and is more than enough to offset the decrease as a consequence of the aerosol ban. Currently, in the United States alone, there is more than $135 billion of installed equipment dependent on CFC products (Glas, 1989).

WATER CONTAMINATION AND DEPLETION

The global water cycle involving evaporation, transpiration, vapor transport, precipitation, surface runoff, percolation, and groundwater flow makes an estimated 9,000 cubic kilometers of fresh water available at any given time for human use; this is believed to be enough, in principle, to sustain 20 billion people (la Riviere, 1989). It does not follow that all of the 5 billion people currently living on earth have an adequate water supply, however, because the world population is not distributed optimally with respect to where the available water is located. For millions of people today in certain regions of the world, insufficient water is a serious and growing problem, so prevention of contamination of available water supplies and cleanup of already contaminated supplies are significant worldwide challenges.

Water shortages due to drought have been common in central Africa in recent years and have been sufficiently grave to cause severe food shortages and major dislocations of people from their home territories. In some places where fresh water is plentiful, its availability for domestic use has been greatly diminished as a result of contamination, and even when it is available it is not readily so. According to 1980 estimates of the World Health Organization, 71% of the developing world's population does not have drinking water piped to its homes (Postel, 1985). Protection of groundwater from further contamination is a major national concern in the United States (OTA, 1984b).

The Safe Drinking Water Act of 1974 (PL93–523) requires the Environmental Protection Agency (EPA) to promulgate national standards for drinking water and regulations for enforcing them, and it directs the EPA administrator to arrange for the study of adverse effects on health attributable to contaminants in drinking water. Responses to this mandate include a series of reports from the National Research Council, the first of which (National Research Council, 1977) reviewed what is known about various types of contaminants of drinking water (microorganisms, particulate matter, inorganic and organic solutes, and radionuclides) and their effects on human health. Among the many possible contaminants mentioned in that report are those shown in Table 4.2. Some particulates that may be found in drinking water (e.g., asbestos fibers) can affect human health directly, as a consequence of being ingested; others are not harmful in themselves, but may facilitate the survival or multiplication of other harmful agents, such as microorganisms that cluster on the surfaces of such particles.

Although public drinking water is relatively safe in the United States and has been for a considerable time—the average citizen does not worry much about the possibility of contracting cholera or typhoid fever, two scourges of the past that were carried in drinking water—it is not totally devoid of disease-causing agents. In 1975, for example, over 10,000 cases of water-borne enteric disease were reported in the United States, and there are believed to have been many other cases that went unreported (National Research Council, 1977). Causal agents were identified in only about 10% of the reported cases. Further improvements in the safety of drinking water will require the development of even more effective systems for detecting and reporting water-borne diseases than now exist.

The unparalleled productivity of agriculture in the United States has been due in part to the mechanization of farm work and in part to the extensive use of fertilizers and pesticides to increase crop yields and limit losses from insects and other biological spoilers, but both fertilizers and pesticides have unfortunate side effects, and agriculture has become one of the major sources of surface- and ground-water contamination. Pesticides are also a health hazard because of the residues they leave in foods. Runoff of toxic chemicals from farming activities are stressing the Illinois River to a greater degree now than

TABLE 4.2 Some of the Drinking Water Contaminants Identified in the National Research Council's 1977 Report on Drinking Water and Health

Various genera of enteric-disease causing bacteria
Various enteric viruses
Pathogenic protozoa and helminths (intestinal worms)
Heavy metal cations (lead, chromium, copper, zinc, cobalt, manganese, nickel, mercury, and cadmium)
Anions (phosphates, arsenate, borate, and nitrates)
Many insoluble compounds (hydroxides and hydrous oxides of iron, manganese, and aluminum)
Particles formed by a variety of interactions of chemical and organic materials with clays
Asbestos in its many forms, and other fibrous minerals
Oil- and petroleum-derived hydrocarbons
Organic colloids
Numerous herbicides, fungicides, insecticides, and fumigants
Many compounds used as solvents, cleaning agents, refrigerants, and lubricants, and for other home or industrial purposes
Radioactive isotopes of potassium, tritium, carbon, rubidium, cesium, radium, iodine, strontium, and several other elements.

did the dumping of raw sewage from Chicago some years ago (Havera & Belorose, 1985). To date, most of the efforts to clean up the nation's waterways have been focused on controlling "point sources" of pollution. The effects of "non-point sources," such as urban and agricultural run-off, have been very difficult to determine, although they are assumed to be substantial (R. A. Smith, Alexander, & Wolman, 1987).

Monitoring of river water quality has increased greatly since the passage of the Clean Water Act in 1972. In the aggregate, the data suggest a considerable improvement attributable to point-source controls in fecal bacterial counts and somewhat less improvement in total phosphorus concentrations. Concentrations of nitrates, on the other hand, have often increased. Nitrate increases are believed to be due, in large part, to atmospheric deposition of nitrogen as well as to agricultural activity, especially the use of nitrogen-rich fertilizers. Salinity increases have also been observed in many cases and are attributed in part to the use of salt on the highways which, according to the Salt Institute (1980), increased by a factor of more than 12 between 1950 and 1980. In 1980, the amount of salt used on American roads for deicing came to almost 400 pounds per citizen (Hibbard, 1986).

How the concentrations of toxic substances have changed in recent years in not well known, although there is some indication of increasing amounts of dissolved arsenic and cadmium, which presumably come from fossil-fuel burning, certain manufacturing processes, and agricultural practices. Concentration of lead, on the other hand, is believed to have declined significantly in recent years, due largely to decreased use of leaded gasoline (R. A. Smith et al., 1987). A great deal more data from water-quality monitoring activities will be needed before the trends in surface-water quality and their determinants are thoroughly known.

Agriculture accounts for about 70% of the total use of water worldwide; industry accounts for roughly 25%, and residential and other municipal uses combined account for less than 10% (Postel, 1985). Data gathered by the World Bank indicate that between 50% and 60% of the increase in agricultural output of the developing countries from 1960 to 1980 was due to the increased use of irrigation (Crosson & Rosenberg, 1989). Postel (1985) noted that the amount of irrigated land worldwide tripled between 1950 and 1985. Nearly seven times as much water is used for agricultural irrigation in the United States as is used by all of the nation's city water systems combined (Udall, 1986). In California, about 85% of the water goes to agriculture, which accounts for about 2.5% of the state's economy; in New Mexico, agriculture uses about 92% of the water and represents about 18% of the economy (Reisner, 1988–1989).

As practiced in many countries, irrigation is highly inefficient (la Riviere, 1989), and unless properly controlled it can produce a build-up of salt in the soil and deplete the underlying aquifer. In some major agricultural regions in the United States, the annual withdrawal of ground water exceeds the natural recharge rate by large amounts, and this spells trouble for a number of states in the relatively near-term future. Texas, Udall (1986) suggested, could lose as much as half of its irrigated farm land by the year 2000.

Water management efforts are better focused on increasing efficiency of water use than on increasing the water supply, because increasing the supply is likely to be more expensive and only postpones crises (la Riviere, 1989). Moreover, sometimes increasing the supply may not be an option, but, given the inefficiency of traditional patterns of use, the potential for improving upon them is great. One approach to a dwindling water supply for farming— a technique that decreases dramatically the amount of water needed to produce a good crop—is *drip irrigation,* in which buried pipes deliver water directly to plant roots, thereby increasing efficiency of water use by minimizing runoffs and evaporation while reducing delivery to areas containing no plants. Unfortunately, the installation of drip irrigation systems is expensive.

The prevention of water pollution is likely to be far more economical and more effective than clean-up activities. Chemically polluted riverbeds can be extremely difficult to detoxify. Polluted sediments that have been covered with clean fill, either as a consequence of natural processes or by reclamation activities, can resurface as problems if the covering is subsequently disturbed.

The world's attention has been riveted on oil in recent years, and for good reason: It is a critical resource, and significant changes in its availability have far-reaching implications for the world's economy. As critical as oil is, however, its importance compared to that of water, to life and well-being, is minor. We are, by body weight, mostly water. As individual organisms, we can survive only a few days without this substance. There are substitutes for oil, however inconvenienced the world might be to have to rely on them too abruptly. There are no substitutes for water. No slightly-less-convenient, somewhat-more-expensive alternatives waiting to be developed. If our needs for water come to exceed the supply, or if we sufficiently foul the reservoirs we have, we will be in serious trouble, indeed.

Evidence that we are moving ever closer to that situation abounds. I have already mentioned the recurring problem of drought in various regions of the world. One need not look beyond one's daily experience, even in a highly industrialized society, however, to see unsettling signs of a water supply that is dwindling relative to the demands on it. My town buys water from a neighboring municipality because the wells from which it once got its supply have become contaminated. Municipal controls or bans on the use of water for nonessential purposes—watering lawns, washing cars—during summer months are common in many parts of the country. A couple of weeks before writing these words, I was reminded of California's problem by a card, posted in a conspicuous place in a San Francisco hotel room where I was staying, urging conservative use of water. Suggestions for how to conserve included using water in the shower only during the beginning of the shower (while soaping up) and at the end (to rinse off), using a similar approach for brushing teeth, making sure faucets were turned completely off, not flushing the toilet after every use, and going easy on the ice. The card referred optimistically— but not convincingly—to the situation that motivated this effort to decrease the hotel's demands on the water supply as a temporary one.

One of the problems associated with water conservation programs aimed at household consumers is that people appear not to be able to estimate very accurately how their behavior translates into usage figures. Reports of water saved by 471 participants in one water conservation campaign deviated considerably from the actual savings realized (D. C. Hamilton, 1985). The results led the investigator to suggest that, because knowledge about water use is generally poor, the provision of easily understood feedback about use might help increase conservation behavior.

Given that only about .01% of the earth's total water supply is contained in its fresh-water repositories, such as lakes and rivers (la Riviere, 1989), a long-range answer to fresh-water needs, at least for coastal areas, could be the development of effective and economical methods of desalinization. In spite of considerable efforts in this direction, however, only modest successes have so far been realized.

DEFORESTATION

Deforestation refers to the clearing of forest lands either to harvest timber or to make space for agriculture or industrial development. Harvesting timber need not mean a permanent reduction in forest land, because timber can be regrown; however, timbering operations have often been done in such a way as to make the land less conducive to forest growth or to delay reforestation by excessively long times. In many areas of the world, government policies, driven by economic pressures that afflict debt-ridden underdeveloped countries, contribute to deforestation by promoting commercial logging and conversion of forested areas to use for ranching and agriculture (Repetto, 1990).

If deforestation of Amazonia were to continue at its current rate, most of the tropical forests of the area would disappear within 50 to 100 years (Shukla, Nobre, & Sellers, 1990). On the basis of results obtained from an atmosphere–biosphere model used to investigate the consequences of deforestation in the Amazon region, Shukla et al. concluded that total destruction of the Amazon forests could be irreversible; the effect of the loss of vegetation on precipitation—because of the great reduction in evapotranspiration—could make it impossible for the destroyed forests to reestablish themselves.

Inasmuch as the wealth of many developing countries is in their natural resources, there is the growing risk that, as the world demand for these resources increases, this capital will be depleted further, in many cases without adequate provision for replenishment. Expansion of the populations of underdeveloped countries is expected to lead to further deforestation as the demand for building material, fuel, and farmland increases (Keyfitz, 1989).

The fact that economic pressures have been primary causes of forest depletion does not mean that countries that have harvested timber for this reason have done well economically; the capitalized value of the income derived from renewable nontimber forest products, such as animals, fruits, oils, nuts, and fibers, can, in some cases, greatly exceed the value of the timber that is harvested. Moreover, the income derived from the renewable products is the livelihood of local residents, whereas the profits from timber operations typically go elsewhere.

This suggests one way in which human factors expertise might be helpful. Innovations that would increase the cost-effectiveness of the harvesting of renewable forest products, relative to that of operations that deplete the forests, would increase the economic attractiveness of the first type of activity. This is not a problem on which human factors researchers have focused in the past, but the harvesting of renewable forest products involves people using tools and interfacing with machines. Presumably the issues that would arise, as efforts were made to increase the efficiency with which the necessary tasks could be performed in this context would resemble, in many ways, the issues that arise when the same objective is pursued in other work situations, and I suspect that approaches that have proven to be effective in the latter cases might be usefully applied to the former.

Repetto (1990) argues that deforestation is not inevitable but largely the consequence of inappropriate policies and inattention to certain social and economic problems from which these policies have derived. Even when there are official policies and regulations to protect forests, they are often unenforced because of powerful vested interests. He notes also, however, that there are some encouraging signs that awareness of the national and global significance of tropical forests is growing, and that countries are beginning to take steps to protect them or at least to use less wasteful means of exploiting them. For example, under the aegis of the Tropical Forestry Action Plans sponsored by several international organizations, over 50 countries are preparing plans to conserve and manage their forests.

Although there appears to be considerable agreement among scientists that air pollution can damage, and even kill, various forms of vegetation including trees—it is believed by some scientists, for example, that sulfate and nitrate depositions from combustion processes have contributed significantly to forest decline in central Europe in recent years (Schulze, 1989)—not a lot is known about the precise effects of specific pollutants. Moreover, whereas acid rain was once believed to be the primary cause of diseased forests, especially in Germany, more recent evidence suggests that ozone derived from nitrogen oxides from auto emissions is the more direct cause of this problem; this is still somewhat speculative, however, and the problem is being studied (Kiester, 1985). A workshop convened by the National Research Council noted the need to look for groups of "biologic markers" that, taken together, could shed some light on how forests respond to air pollutants. Among the approaches recommended were: measurement of nutrient use efficiency in leaves, correlation of tree ring growth with pollution records, and remote sensing of infrared radiation from distressed forests (Borchelt, 1989b).

Probably the aspect of deforestation that has attracted most attention is its effect on the amount of carbon dioxide in the atmosphere. Because trees and other vegetation remove carbon dioxide from the atmosphere and return oxygen, any significant reduction in the amount of the earth's surface that is covered with forests will contribute to a greater concentration of carbon dioxide in the air. Reddy and Goldemberg (1990) put the amount of global carbon dioxide emissions due to deforestation in developing countries at about 23%; trees use from 10 to 20 times as much carbon per unit area as is used by pastureland or land growing crops (Revelle, 1982); and trees planted in urban areas nearer the sources of increased CO_2 emissions are believed to be many times more effective in reducing the atmospheric accumulation of such emissions as trees in forests (Leahy, 1989). Clearly, reforestation programs could help limit further increases in atmospheric carbon dioxide; according to Merland (1989), however, reforestation by itself cannot be the total answer, because this would require doubling the net annual yield of the world's forests.

In addition to carbon dioxide, the burning of biomass in the tropics emits very significant amounts of other trace gases into the atmosphere, including carbon monoxide, several hydrocarbons (especially methane), and a variety of other compounds (Crutzen & Andreae, 1990). Deforestation also affects the water cycle, because water that normally would be returned to the atmosphere as water vapor from leaves becomes part of the water runoff. Because water vapor helps to moderate surface temperature, reduction in it could also contribute to general climate change.

DESERTIFICATION AND WETLAND LOSS

Desertification refers to the degradation of land, particularly once-arable land, as a consequence of soil erosion, salinization from irrigation, and other processes. It has been claimed that several ancient civilizations died because their lands became so salty as a consequence of irrigation that crops could no longer be grown on them (Pillsbury, 1981). The technology now exists to solve this problem and to permit continual irrigation of land without causing its salinization, but it is not being widely used.

Currently, about 12% of the earth's land surface is arid, and about 20% of it semi-arid. Semi-arid land can be degraded and permanently desertified by concentrated grazing or row-crop agriculture (W. H. Schlesinger et al., 1990). The United Nations Environment Program has compiled data showing that about 60% of the agricultural land outside humid regions of the world is experiencing desertification to some degree (Crosson & Rosenberg, 1989). According to the Worldwatch Institute, the world loses about 25 billion tons of topsoil per year, which is roughly the amount that covers Australia's wheatlands (MacNeill, 1989). A Panel on the Improvement of Tropical and Subtropical Rangelands (1990) of the National Research Council claimed that about 40% of the earth's land surface is rangeland and about 80% of that is degraded to some degree. Emmanuel, Shugart, and Stevenson (1985) have predicted that a doubling of atmospheric carbon dioxide would produce a 17% increase in the amount of the earth that is covered by desert. Crosson and Rosenberg (1989) caution, however, that global estimates of desertification can be of questionable validity for a variety of reasons, such as varying definitions of desertification and unavailability of accurate accounts of degradative processes in some areas.

Inasmuch as certain farming practices are major contributers to detrimental changes in soil quality, it is not surprising that these practices are now receiving considerable attention from environmentalists. *Sustainable agriculture* is distinguished from *conventional agriculture,* which makes heavy use of chemical fertilizers and pesticides with insufficient regard to long-term effects on soil quality and other environmental variables. Sustainable agriculture, as Reganold, Papendick, and Parr (1990) conceive it, "is not so much a specific farming strategy as it is a system-level approach to understanding the complex interactions within agricultural ecologies" (p. 112). It puts great emphasis on protecting and nurturing soil so as to ensure its long-term productivity, and is sensitive to other environmental problems, such as ground-water contamination, energy costs, and risks to health and wildlife habitats. Reganold, et al. also point out that sustainable agriculture does not necessarily mean a return to methods used before the industrial revolution. It makes use of many aspects of modern technology and of current scientific knowledge, but it emphasizes the diversification and rotation of crops, the build-up of soil, and natural methods of pest control. The use of "green manure crops," such as clover, to enrich the soil is encouraged. Also, experimentation with a variety of new crops is being done in order to find new sources of food and industrial materials that put less long-term stress on the soil (Hinman, 1986).

There is some evidence to indicate that it is possible for farms that grow a variety of crops with little use of chemicals to be as productive as farms using more conventional techniques. They might even be more profitable because of lower input costs, especially if the indirect costs associated with environmental degradation and health effects are taken into account (National Research Council, 1989b; Reganold et al., 1990). Unfortunately, monocropping—the practice of growing the same crop on a given field, with the help of fertilizers and pesticides—has been encouraged, in the United States, by the government's policy of subsidizing only a small number of specific crops, primarily wheat, corn, and other major grains. In addition to the problem of government subsidy policies that provide disincentives for farmers to diversify and rotate crops, an impediment to sustainable agriculture is the relative unavailability of information to farmers regarding practices that are conducive to it. It must be noted, too, that alternative agricultural techniques also have their critics (Beardsley, 1989b).

Until very recently, wetlands were widely considered to be wastelands, and the official government policy was to encourage reclamation; drainage of them was often subsidized. It is now recognized, however, that wetlands play a very important role in the total ecosystem and are among the most productive of all habitats of life in its countless forms. Nevertheless, it has been estimated that over half of the wetlands in the lower 48 states have been lost (Wallace, 1985). In spite of general agreement that remaining wetlands should be preserved, we continue to convert them to farms, shopping centers, highway corridors, airport runways, marinas, industrial parks, and housing tracts at a rate of 300,000 to 500,000 acres a year (Steinhart, 1990).

About 70% of America's wetlands are in Alaska, where they account for about 60% of the entire state. Here the tug-of-war between conservationists and people concerned that protecting the natural environment will stymie economic growth is intense (Lipske, 1990). President Bush has announced a "no-net-loss" policy regarding U.S. wetlands, whereby an acre of wetlands that is reclaimed for development must be offset by creation of an acre of new wetlands or restoration of an acre to wetlands status. Enforcement of the policy is meeting some stiff resistance.

DECREASING BIODIVERSITY

The total number of species of organisms on earth is not known; estimates vary from about 4 million to as many as 30 million, only about 1.4 million of which have been formally described. Biological diversity is believed to be, at present, at its lowest level since the end of the Mesozoic era, 65 million years ago, and to be decreasing (E. O. Wilson, 1989). The problem, believed by many biologists (among others) to be more serious than is generally realized (E. O. Wilson & Peter, 1988), is seen to be a direct consequence of the rapid increase in the human population. Fossil records and other evidence on which paleobiologists rely suggest that biological diversity has increased gradually over the millions of years since life first appeared on the earth, although on several occasions, mass extinctions have resulted in significant, but temporary, decreases.

E. O. Wilson (1989) notes that the diversity reached an all-time high just before the effects of human activity began to be noticeable—during the past 10,000 years or so. The loss due only to the clearing of large sections of the rain forests, the most richly and diversely populated regions of the world, he estimates to be between 4,000 and 6,000 species a year, which is about 10,000 times the natural rate of extinction that occurs in the absence of human interference. A task force of the National Science Board (1989) recently released a report warning that from one quarter to one half of the earth's species could become extinct within 30 years if current trends are not reversed. Preservation of biodiversity in the national forests is one objective of the National Forest Management Act of 1976.

The threat to certain species (e.g., whales and elephants) that results directly from their slaughter for financial motives has been well publicized. (Largely as a consequence of ivory-motivated poaching, the population of African elephants was reduced by more than half during the 1980s [Horgan, 1989b].) For the most part, however, the effects of human activity on biodiversity are much less direct and apparent, albeit not necessarily less significant.

The destruction of forests and wetlands can affect the fauna in areas far removed from where that destruction occurs. More than three quarters of North America's bird species are migratory. Many of them spend the winter in Central or South America. Some ornithologists believe that the migratory bird populations are shrinking, rapidly in some cases, as a consequence of the destruction of tropical forests (Conner, 1988). Similar observations have been made regarding various species of migratory water fowl with respect to the effects of loss of wetland habitat in Alaska (Lipske, 1990).

The OTA (1986a) issued a report describing the problem of decreasing biological diversity and identified a variety of technologies and strategies that could be useful in reversing that trend. One of the major needs appears to be for more complete biological databases and for more effective techniques for collecting, storing, and accessing biological data. Here, again, we see a need for attention to the human factors issues that relate to the problem of ensuring the usability of databases and information stores.

Although one can justify concern about decreasing biodiversity on ethical, aesthetic, or economic grounds, among the most alarming consequences of the extinction of populations or species is the loss of some of the free services ecosystems provide (Ehrlich, 1987). Ecosystems, of which all organisms are functional parts, are knitted together by flows of energy and cycles of materials, and can be thought of, in the aggregate, as the life-support apparatus of the planet. Such services include the production of oxygen for the atmosphere, maintenance of concentrations of gases that help control the climate, maintenance of a fresh water supply, and contributions both to soil fertility and to the pollination of crops.

WASTE

Waste comes in a variety of forms. Many of the gaseous and particulate emissions into the atmosphere from industrial, agricultural, and other types of processes can be considered waste from some point of view. This is especially true of emissions that are greater than necessary because of inefficiencies in the processes involved. In the following discussion, I focus, for the most part, on solid and liquid wastes, having addressed the problems of gaseous wastes in several of the foregoing sections.

Conventional Solid Waste

Solid waste disposal is becoming an increasingly serious concern, the magnitude of which has been dramatized, in the United States, by the observation that "every five years the average American discards, directly and indirectly, an amount of waste equal in weight to the Statue of Liberty" (O'Leary, Walsh, & Ham, 1988, p. 36). Rathje (1989) cited several other estimates of waste volume that make use of similarly graphic comparisons: "Catherine Kelly, in her book *Garbage* (1973), asserted that the amount of MSW [Municipal Solid Waste] produced in the United States annually would fill 5 million trucks; these, 'placed end to end, would stretch around the world twice.' In December of 1987 *Newsday* estimated that a year's worth of America's solid waste would fill the twin towers of 187 World Trade Center. In 1985, *The Baltimore Sun* claimed that Baltimore generates enough garbage every day to fill Memorial Stadium to a depth of 9 feet" (p. 100).

According to Ayres (1989), the American economy extracts more than 10 tons of mass (excluding atmospheric oxygen and fresh water) per person per year from the environment, roughly 75% of which is mineral and nonrenewable. Ayres claims further that only about 6% of this mass ends up in durable goods, the other 94% being converted into waste residuals almost immediately. Tonnages of waste residuals are greater than the tonnages of crops, timber, fuels, and minerals recorded by economic statistics. This is possible, in part, because certain non-priced and unmeasured ingredients, such as air and water, contribute both to industrial products and waste residuals. According to Tchobanoglous, Theisen, and Eliassen (1977), Americans were producing about 3,600 pounds of solid and liquid waste per capita per year, whereas the Japanese, who came closest to us, produced about 800 pounds, the Dutch 680, and the West Germans 500. Estimates from other sources do not show quite as large a disparity between production of waste material in the United States and elsewhere, but they indicate very sizable differences nonetheless (Herman, Ardekani, & Ausubel 1989).

Determining the volume or weight of waste for an entire country, or on a per capita basis, is difficult because there is no way to measure more than a fraction of all the waste that is produced (Rathje, 1989). Consequently, estimates vary over a considerable range: from 2.9 to 8.0 pounds per person per day, according to Rathje. (Estimates from the Bureau of the Census, 1990, are shown in Table 4.3.) On the basis of landfill excavation studies, Rathje sees the higher estimates as significant overstatements of the actual amount. More generally, he considers the problem of solid waste in the United States to be a serious one, but not beyond control, and one solvable with disposal methods that are safe and already available.

Approaches to the problem of waste vary in cost and impact on the environment, but none of them is ideal. One less-than-laudable strategy that developed countries have sometimes used to rid themselves of noxious waste is to pay less developed countries to take it off their hands. This approach has not always worked in the recent past, and there have been several well-publicized incidents of vessels laden with trash plying the seas looking in vain for a country that would accept their cargo.

Considerable emphasis is being put on reducing the creation of waste through resource conservation programs and the recycling and reprocessing of waste materials. Finding uses for waste products addresses both the problem of waste disposal and that of resource depletion. Effort is also going into the development of incinerator technologies that will reduce waste volume without aggravating the problem of air pollution. Waste incineration is being used as a means of energy generation in some instances. Because waste reduction and recycling are likely to be significant components of any long-term solution to this problem, a major task will be that of changing public attitudes toward consumption. It is difficult to overstate the importance of this task, especially in view of the possibility that consumer products

TABLE 4.3 Municipal Solid Waste Generation and Recovery in the United States (Adapted from Bureau of the Census, 1990, Table 355)

Year	Waste generated (in million of tons)	Waste generated (lbs/person/day)	Waste recovered (as % of generated)
1960	87.5	2.65	6.6
1965	102.3	2.88	6.1
1970	120.5	3.22	6.6
1975	125.3	3.18	7.3
1980	142.6	3.43	9.4
1985	152.5	3.49	10.0

may have more environmental consequences than does industrial activity (Friedlander, 1989).

Waste Recycling

The terms *consumption* and *consumer* convey an idea whose time may have passed. We do not really consume anything. We use things and transform them in the process. Many of the results of those transformations are useless, and they find their way to the trash heap, but this is not the end of them. Their continued existence in dumps or in atmospheric residues of their incineration stay around to cause problems of one type or another for a long time. We have to begin thinking more in terms of our role, not as consumers, but as users and processors of the goods we buy. This means, among other things, being more explicitly aware when designing, manufacturing, selling, or buying material goods of the fact that it will be necessary, at some time in the future, to dispose of them or, better, to recycle them.

To date, recycling efforts have been only marginally successful. Education and advertising campaigns have not been very effective in obtaining participation in paper recycling programs (Coach, Garber, & Karpus, 1979; Geller, Chaffee, & Ingram, 1975). Plastics, which represent about 7% by weight of municipal waste in the United States—roughly the same as glass and steel—are, in principle, recyclable essentially to their original form, but programs to recycle plastics have been neither extensive nor successful (Powell, 1990).

Geller (1981) distinguishes between high-technology and low-technology waste-recovery systems. A high technology system is one in which unseparated trash is collected from the consumer, and transported to a resource-recovery plant, where it is mechanically separated into paper, glass, aluminum, and other constituents for further processing. In the low-technology approach, the consumer separates trash into a variety of types, which are collected separately and transported to separate waste-processing facilities. The high-technology approach is most convenient for the consumer, but it tends to be more expensive. Geller lists several other disadvantages or problems with the high-technology approach, among which are the following: operation of resource-recovery plants can consume substantial energy and contribute to air pollution; refuse-derived fuel, when burned with coal and oil, as it often is for the production of electric energy, also contributes to air pollution; separation technology (especially for separating aluminum from other materials) is still not reliable; storage of shredded waste can be a fire hazard; reliable markets for products from resource-recovery plants may be difficult to find; and high-technology systems can be undercut by low-technology programs, which can divert valuable waste products from them.

Most of the sociobehavioral research on waste reduction and resource recovery has looked for correlations between attitudes and recycling-related behavior rather than studying community interventions for facilitating such behavior. Unfortunately, we know that the opinions people express regarding environmental issues are not good indications of their behavior relating to those issues (Bickman, 1972). There have been a few studies, mostly on university campuses, of efforts to stimulate participation in resource (predominantly paper) recovery programs. Most of these studies, which have been reviewed by Geller (1981), have shown that participation could be increased, at least for a period of weeks, by using some type of incentive program. Geller cautions, however, that none of them was sufficiently long-term or large-scale to provide the basis for specifying optimal conditions and contingencies for a community-wide resource-conservation program. Initial enthusiasm—as evidenced by a high degree of cooperation—for an environmental activity, such as sorting office waste paper for recycling, may wane over time, even among people who strongly favor environmental protection, if not bolstered in various ways (Humphrey, Bord, Hammond, & Mann, 1977).

Geller (1981) also points out that simple positive reinforcement procedures used to encourage specific types of behavior can sometimes evoke a variety of unanticipated and unwanted behaviors. He makes this point in the context of a discussion of experiments in which the investigators were looking for ways to increase participation in paper-recycling efforts. He recounts instances—faintly reminiscent of my friend's apocryphal dandelion-harvesting children—in which children who were rewarded for contributing bags of litter to a program sometimes filled their bags with litter that had already been collected.

An attempt to determine how people who voluntarily participate in community recycling programs differ demographically from those who do not was made by Vining and Ereo (1990). These investigators found, in a study of 197 Illinois households, that people who participated in recycling programs were more aware of publicity about recycling and more knowledgeable about recyclable material than people who did not, but they did not find clear demographic distinctions between recyclers and nonrecyclers.

An aspect of product design that deserves more attention than it gets is that of disposability or recyclability. All products have finite useful lives and they eventually wear out and become incapable of serving their intended purposes. How much of a problem they will create at that time should be considered an important component of the quality of their design. Other things being equal, a product that is easily recycled for other uses or, at least minimizes problems of waste disposal, is a better product than one that is not recyclable and is problematic as waste. We do not have much experience in evaluating products from this point of view, but as the problem of waste disposal grows it will become increasingly important to gain some.

Products that are not recyclable, in the usual sense of the word, should at least cause no harm as waste in landfills or when incinerated or disposed of in other ways. An issue that is often discussed in this regard is the question of material degradability. Although almost everything degrades in time, different materials take different amounts of time and decompose into end products that have different environmental effects. Decomposition is sometimes aided by biological or photochemical processes, and such materials are said to be *biodegradable* or *photodegradable*. Starch is now used as an ingredient in the production of some plastics in order to make them biodegradable; these materials are weakened and their fragmentation is facilitated when the starch particles are consumed by microorganisms in soil or water. Work is proceeding on the development of biodegradable plastics based on materials that would be more effective catalysts to decomposition than starch. Lactic acid is one possibility that is being explored. Related research is being directed at using plastics that are end-products of natural bacterial functions (biopolymers). Some of the possibilities offer the hope of low-cost production and very rapid in-soil decomposition (Studt, 1990).

Wastepaper Reduction

Obviously, to the extent that the creation of waste is reduced, the problem of disposal does not arise. Because paper accounts for about one third of all solid waste produced in the United States, the use of electronic or photonic (optical) information storage media instead of paper could help alleviate the solid waste problem considerably. As noted in the earlier discussion of energy, the widespread use of electronic mail, for example, could decrease the need to transport mail in hardcopy form. Electronic journals, magazines, newspapers, and books, should they be widely adopted, could also reduce the need for the use and transport of printed material. Replacement of "yellow-page" telephone books with electronic or photonic directories would, by itself, eliminate one quite significant source of demand for paper and for landfill space.

As of 1988, the circulation of daily newspapers in the United States was about 63 million (it has been roughly stable since 1970; Bureau of the Census, 1990). Given that every daily newspaper is obsolete the day following its date of issue, that most buyers of newspapers are interested in only a fraction of the information contained in them, and that the printing and delivery of newspapers are energy-intensive processes, this means of distributing information is inefficient in the extreme. There seems unlikely to be an effective alternative to the conventional newspaper as a method of news distribution in the near future, but electronic delivery should be feasible if suitable terminals become widely available in homes, perhaps as a consequence of the installation of television sets with built-in computers. Designing electronic newspapers that will be seen by the public as acceptable, if not preferred, alternatives to paper versions is a problem to which human factors researchers should have much to contribute.

The comment regarding the wastefulness of printing an individual newspaper for every reader applies equally to the printing of magazines and journals; in the case of journals, it seems reasonable to expect that some form of electronic publishing could be feasible on a significant scale fairly soon. One form the electronic publishing of scholarly material might take is that of providing to readers access to all that is published in the discipline, but delivering either electronically or in hard copy only what is requested by individual subscribers (Gardner, 1990). This would represent a very significant savings of paper and could have some other advantages as well.

Noting the nonexistence of electronic journals in psychology, despite years of discussion of the possibility, Gardner (1990) argues that most computer screens are not good for reading, electronically published material is not available to noncomputer-using members of a discipline, and personal computers are not as convenient as printed journals for storing and retrieving text. For electronic publishing of scholarly material to be successful, he suggests, it must solve all of these problems: "It must retain the highly evolved readability of a traditional printed journal. It must be accessible and attractive to all members of the discipline, whether they use computers or not. Most importantly, it must provide improved facilities for retrieving information, while continuing to serve as a permanent archive of the society" (p. 334). Realizing these desiderata will require addressing a variety of human factors issues having to do with matching information systems to the capabilities and preferences of their users. Assuming that these issues can be addressed effectively, increasing the use of electronic means of storing and distributing information has the potential to decrease the use of paper for these purposes.

A caveat is in order here. In spite of the vision of a "paperless society" that some futurists have promoted, I know of no evidence that the use of paper has decreased at all as a consequence of the increasing electronification of information technology. In fact, total paper consumption in the United States increased over 50% between 1970 and 1988, and consumption of paper for printing and writing more than doubled (Bureau of the Census, 1990). How this increase relates to information technology is not known; conceivably, greater use of paper has been stimulated by some of the word-processing and documentation tools, not to mention photocopying facilities, that are ubiquitous in the office workplace.

The management information system (MIS) of a high-technology company of which I am aware produces about 800,000 pages of printed material per month, or more than 3,000 pages per employee per year. These are pages produced only by the management information system; the count does not include memos from other internal sources, outgoing correspondence, technical reports and copies thereof, xeroxes of journal articles, hard-copy listings of electronic mail, and a variety of other uses of paper. I would be surprised to discover that this company is unusual among similar high-technology companies, either in its use of paper or with respect to the role that information technology has played in stimulating that use. One suspects there is a principle at work here whereby if a number can be produced by a management information system, someone will imagine that it is needed, whether it is or not.

Observing that consumption of writing paper in the United States tripled between 1959 and 1986, Herman et al. (1989) related the increase to information technology this way:

> Apparently nobody anticipated that the microchip would catalyze the burgeoning of paper to such an enormous extent. It would appear that the information age technicians did not understand that the amount of information was not fixed and that electronic information was not simply a substitute for paper. Computers are storing greater quantities of more kinds of information than ever before in extremely compact form, but people prefer reading from a printed page rather than the average computer screen, which in order to have excellent resolution must be improved by a factor of about 10. In addition, there is an increase in office workers compared to those in manufacturing jobs, and this shift leads to an increase in precisely the kind of people who generate paper. Note also that it is easy to produce photocopies compared to the old days when making carbon copies was indeed a great burden. (p. 62)

Whether the reliance on paper for information storage and distribution continues indefinitely, or will begin to diminish as people gain confidence in and become more accustomed to alternative media, only time will tell. With the proliferation of facsimile machines throughout the business community and elsewhere, it seems likely that, at least in the short term, the use of paper will get another incremental boost.

There can be no doubt that electronic word-processing and document-preparation tools, along with high-speed printers and copiers, make it extremely easy to use paper. Editing large documents is so easy with these facilities that they now may go through many more drafts than they did when editing was done with the help of scissors, tape, and other tools and techniques that were standardly used before computers entered the office. It does not follow from this fact alone that the preparation of documents should now require more paper than it did before; however, if paper copies of an evolving document are printed at frequent intervals, then the effect may indeed be a much greater use of paper. The ease with which multiple copies of a document can be made also seems likely to be a significant determinant of paper use.

We need a better understanding of why people use paper when the information is readily available electronically and is in a nonfinal or draft form. It would be useful to know, in particular, to what extent the use of paper is dictated by basic characteristics of human beings as information processors that are better matched to traditional paper media than to computer-driven video displays, by irrational preference for tradition or distrust of new media, by a preference for a tangible representation of one's efforts over an ephemeral—and easily erased—computer file, or something else. A specific question that deserves attention is whether the increased use of paper is a consequence, to some extent, of inappropriate or inefficient uses of computer technology. If the answer to that question is "yes," the obvious next question is how such inappropriate or inefficient practices might be changed.

Medical Waste

For some time there has been a trend in medicine to make more and more of the equipment and supplies used in patient care disposable. Bedpans, wash basins, and foam rubber bed pads go home with the patient following a hospital stay; syringes, diapers, and eating utensils are discarded after a single use. The preference for disposable materials is motivated, of course, by the intention to limit the spread of disease. Partly as a consequence of this practice, however, the problem of getting rid of medical waste is becoming very serious, as the unhappy experiences of bathers on several eastern U.S. beaches in the 1980s can testify, and the cost of disposal is becoming excessive.

There are, according to one estimate, about 6000 substandard medical-waste incinerators at hospitals throughout the United States (Hershkowitz, 1990): "Usually concentrated in populous areas, these facilities each year spew tons of toxic emissions—including dioxin, heavy metals, and acid gases—into the air at much higher rates than state-of-the-art incinerators in other countries" (p. 35). As of June, 1990, Hershkowitz noted, the United States had no federal regulations to control the high levels of heavy metals, acid gases, and toxic organic compounds that these incinerators emitted, nor did it have any statutes governing the transport of medical waste: "in this country refrigerated trucks legally can and do carry food after transporting medical wastes—without first being cleaned" (p. 39). He pointed to West Germany and Switzerland as countries that have particularly good medical-waste disposal facilities; in both countries air-quality regulations are sufficiently stringent to make it impractical for individual hospitals to incinerate their own waste, so they send it to regional facilities that have sufficient volume to make the use of advanced air-pollution control technologies cost effective.

Radioactive Waste

Disposal or storage of radioactive waste involves a special set of problems that have, at best, controversial solutions. The possibility of depositing such waste in steel drums in salt beds a half mile below the surface of the ground has been under consideration since the 1970s and is still being hotly debated (Borchelt, 1989a). Techniques for storing radioactive waste so that they are assuredly immune to catastrophic accidents for all future time simply do not exist (Krauskopf, 1988). Moreover, the psychological stress that comes from living near a hazardous-waste dump is as real, and conceivably as problematic, as the threat to physical health from contact with the waste materials or their products (Baum & Singer, 1981).

In the United States, high-level nuclear waste (which differs from low-level nuclear waste by its higher concentration of radioactive elements and the much greater duration of its threat to the environment) exists in the form of spent fuel rods and in liquid waste from the dissolving in acid of fuel rods for the production of plutonium. About one third of a nuclear reactor's fuel—uranium pellets encased in long thin metal rods—has to be disposed of every year. Although the fuel is spent, in the sense that it is no longer usable in the reactor, it is more radioactive than when it was initially inserted. Plans call for long-term storage of this material in a repository inside Yucca Mountain in Nevada, but for now, most of it is stored in cooling ponds near the reactors that produce it. The 22,500 tons of this waste is expected to double by 2000. The Yucca Mountain site that is being prepared as a repository for spent reactor fuel is also controversial, especially in view of its location between two prominent earthquake faults (S. Shulman, 1989).

Radioactive liquid waste is stored in steel tanks at the Hanford Military Reservation in Washington and at the Savannah River plant in South Carolina. The Hanford Reservation was established to produce weapons material for the Manhattan Project. It has had nine plutonium-producing reactors over the past 40 plus years, and is believed to have released millions of curies of radioactivity into the ground, water, and air. To put this in perspective, Shulman pointed out that the Three Mile Island accident released only 15 to 24 curies of radioactivity. A recent agreement between (the state of) Washington and the U.S. Department of Energy to undertake a 30-year cleanup of the area is expected to cost about $57 billion.

Finding a long-term solution to the problem of nuclear waste disposal that is acceptable to scientists, policy makers, and the general public has proved to be exceptionally difficult. The Nuclear Waste Policy Act of 1982 called for beginning the storage of nuclear waste in a mined repository in 1998, but the projected date has been pushed forward at least twice, and is now 2010. Although there seems to be general agreement that an underground repository is the only possibility, there is a great deal of skepticism as to the prospects of identifying a site and a containment technique that would pose no risk of eventual seepage of radioactive material into groundwater that could find its way to the surface. The skepticism is sufficiently strong and pervasive to ensure that every state in the union has taken a not-in-my-back-yard position with respect to where a nuclear waste storage facility, or facilities, should be located; this includes Nevada, where Yucca Mountain is located (Krauskopf, 1990). Assurances that the level of risk involved, given a carefully planned and constructed facility, is less than what we routinely accept from other sources have not affected public opinion on this issue very much. Disposal of high-level nuclear waste remains an unsolved problem for the nuclear power industry and the country as a whole.

In addition to the problem of disposing of the waste products of operating nuclear plants is the problem of disposing of the plants themselves when they are retired. The first commercial nuclear reactor to be decommissioned in the United States was the Shippingport reactor, located on the outskirts of Pittsburgh. This reactor, a relatively small one, with a capacity of 72 megawatts, was transported by barge via the Ohio and Mississippi Rivers, the Gulf of Mexico, the Panama Canal, and the Pacific Coast, to its burial site on the Hanford Reservation, at a cost of roughly $98 million. About 50 nuclear power plants in the Western world should be ready for retirement within a decade; about a dozen U.S. reactors are ready for decommissioning now (S. Shulman, 1989). The dismantling of decommissioned nuclear reactors and the cleanup of nuclear wastes are problems for which the use of robots and teleoperators would seem to make a great deal of sense, and in fact, some starts in this direction have been made. Human factors issues relating to the design and operation of teleoperators are receiving considerable attention from researchers.

The EPA is currently developing a strategic information system called the Resource Conservation and Recovery Act Information System (RCRIS) for intergovernmental administration of the hazardous-waste management program. The schedule, which has not been adhered to, called for completion of the system and phasing in of all states by the end of 1992. The design and development plans are controversial and have been criticized as representing an attempt to develop a technological solution to problems of trust, cooperation, and authority in intergovernmental administration, and skepticism has been expressed regarding the likelihood that it will be successful.

Radiological weapons, as well as chemical and biological ones, are not only a nightmarish threat to countries against which they could be used in conflicts, but pose a serious problem to the countries that have developed and stockpiled them as well, even if they are never used. The problem is that the stockpiles age and become increasingly hazardous; their mere existence becomes threatening to people who live anywhere near them, and disposing of them in environmentally benign ways is not easy to do.

More From Less

In 1946, Arnold Toynbee (1946) suggested that the history of technical development reveals a principle of *progressive etherialization* that seems to govern technical progress: That which is ponderous and bulky is replaced by that which is fast and light. Buckminster Fuller (1969) made much of a similar idea. His term was *comprehensive ephemeralization,* which he defined as "the doing of ever more with ever less, per given resource units of pounds, time, and energy" (p. 3). The idea is that of making more and more efficient use of nature's resources, and Fuller saw this as one of the main results of what he referred to as the *invention revolution,* the accelerated rate at which major scientific discoveries and technological advances have been made during this century. A similar idea, although in considerably more mystical form, appears in the writings of Teilhard de Chardin (1959). Here, the notion finds expression in the theory of the emergent spheres, each representing a more advanced stage of evolution and each characterized by more ethereal quality than the one from which it emerged.

Perhaps this principle is seen most clearly in the history of communication and information storage technology. Twenty-five years ago, James Miller (1965) pointed out how the matter-energy costs of storing the markers of information have decreased over the centuries: "Cuneiform tablets carried approximately of the order of 10^{-2} bits of information per gram; paper with typewritten messages carries approximately of the order of 10^3 bits of information per gram; and electronic magnetic tape storage carries approximately of the order of 10^6 bits of information per gram" (p. 195). Advances in information storage technology since 1965, especially in the ability to pack larger and larger amounts of information into smaller and smaller spaces, is nothing short of phenomenal. A CD ROM (read-only memory compact disk) packs information at the rate of more than 10^8 bits per gram, and that technology is already several years old. Recent successes in moving atoms one at a time have led to speculation that it may be possible in the foreseeable future to store a bit in a cluster of as few as 1,000 atoms; if this possibility comes to pass, it will mean that the entire contents of the Library of Congress will fit on a single 12-inch wide silicon disk (Yam, 1991).

Today, the term *dematerialization* is being used to characterize a decline over time in the weight of the materials used in industrial end-products or a decrease in the "energy embedded" in those products. Clearly, dematerialization, in the sense of getting more benefit from less material is a desirable trend from the point of view of nonwasteful uses of natural resources. Whether it is actually taking place on a broad scale depends on how it is defined (Herman et al., 1989). If reduction in the size and weight of products also means a decrease in their quality, then the ultimate result could be the production of a greater number of units (because of the greater need for replacements) and the consequent generation of unnecessary waste, which, from an environmental point of view is not dematerialization, at least in any positive sense. (From the point of view of waste control, an unfortunate side effect of improved manufacturing techniques and consequent decreases in product costs can be an increase in the probability that a product will be replaced, rather than being repaired, when it is no longer functional.) Conversely, dematerialization can occur as a consequence of improved quality and hence longer life of products, even though those products are not necessarily smaller or lighter than their alternatives. Herman et al. pointed to automobile tires to illustrate this point: Although total tire production has risen over time in the United States, the number of registered vehicles and the total miles traveled have increased faster, so the number of tires per million vehicle miles has declined.

Whether we refer to it as etherialization, ephemeralization, or demateri-alization, getting more mileage out of a fixed amount of material or energy must be seen as a good thing, both environmentally and economically. Finding ways to stretch resources, to increase the efficiency of processes, to substitute low-energy limited-material goods and services for those that require the expenditure of large amounts of energy or material must be high on the list of societal objectives as we look to the future. There are opportunities and challenges here for human factors research aimed at furthering these objectives in ways that are consistent with the goal of improving the quality of life.

CONTAMINATION FROM INDUSTRIAL ACCIDENTS

The possibility of environmental contamination resulting from industrial accidents is a continuing concern. Three Mile Island, Chernobyl, Bophal, and the *Exxon Valdez* are notable among the more recent reminders of the reality of this threat. The possibility of accidental release of radioactive material into the atmosphere as a result of mishaps at nuclear power generating facilities has been, and continues to be, a primary focus of attention, but other types of industrial accidents can also have long-term environmental effects in large regions of the world, if not globally.

Human error is a major factor in industrial accidents, and the errors that people make often appear to be related causally to the ways in which the systems and operating procedures they use are designed. The 1979 accident at the Three Mile Island nuclear power plant, for example, was judged to have been a consequence, at least in part, of errors made by the plant's control room operators, and these errors were seen to have had their origins in such factors as "inadequate training, a control room poorly designed for people, questionable emergency operating procedures, and inadequate provisions for the monitoring of the basic parameters of plant functioning" (Moray & Huey, 1988).

The cumulative effects of many accidents that get little attention, because individually they are not considered newsworthy, should be a matter of concern. The frequency of such accidents can be surprisingly high even in a relatively limited area. The number of significant accidental spills of toxic contaminants in the Great Lakes Basin provinces and states, for example, has been estimated to be about 3,000 per year (Great Lakes Science Advisory Board's Technological Committee, 1988).

The problem of assessing the risk to a specific community at any given time from *possible* spills or other accidents involving toxic chemicals used in industrial processes is a very difficult one because, at present, EPA regulations call only for the reporting of toxins that have actually been released to the environment. Various proposals have been made for the implementation of policies or practices that would provide communities with more complete information regarding types and amounts of toxic chemicals used by local industries (Committee to Evaluate Mass Balance Information for Facilities Handling Toxic Substances, 1990).

One approach to the problem of industrial accidents that is sometimes discussed under the rubric of *inherent safety* is to design plants and processes in such a way that accidents are highly unlikely to occur, or so that if they do occur the consequences cannot be disastrous (Weinberg, 1989–1990). Preventing accidents, of course, is greatly preferable to having to deal with their consequences, but the cost of reducing their probability to near zero is undoubtedly higher than society is willing to bear. So, developing ways to identify and respond rapidly to such accidents and to contain their effects will continue to be an important way of dealing with this problem.

A particularly challenging aspect of this task is that of anticipating people's reactions to various incidents and taking adequate account of them in contingency plans. At the time of the Three Mile Island incident in 1979, Richard Thornburgh, then governor of Pennsylvania, issued an advisory that pregnant women and preschool children who lived within 5 miles of the facility might want to evacuate and that everyone within 10 miles should consider staying indoors. The total number of pregnant women and preschoolers living within 5 miles of the plant—the number of people advised to consider temporary relocation—was estimated to have been about 3,500. In fact, about 200,000 people fled the area and went, on the average, 100 miles (Erikson, 1990).

NATURAL DISASTERS

"Natural disasters" such as earthquakes, floods, hurricanes, volcanoes, and collisions with meteors antedate humanity. Except for those very few, widely spaced catastrophes that have been hypothesized to have caused the extinction of species, the effects of natural disasters on the environment have tended to be transient and confined to relatively small regions of the globe. They take on increasing significance, however, as the world population grows; when such an event occurs in a densely populated area, its effects can be disastrous and can extend considerably beyond the point of occurrence in both space and time. Compelling reasons for taking an interest in natural disasters are the possibilities of predicting them, preparing for them, controlling their damage, and providing effective assistance to their victims when they occur.

The ability to predict earthquakes, at least probabilistically, has improved greatly over the past few years (Beardsley, 1990a). As seismic monitoring methods improve, it will become possible to make predictions about time of occurrence with much greater certainty. A major problem relating to earthquake prediction that has not received a great deal of attention, however, relates to the question of the use that should be made of predictive information. How effectively can individuals and communities respond to the knowledge that an earthquake in their locale is highly probable within some specified period of time, measured in months or years?

On a statistical basis it is now possible to identify areas around the world that are highly likely to experience specific types of natural disasters sometime—in many cases probably more than once—within the next few years. It requires no special prophetic powers to be aware of the high probability of continuing problems of flooding in Bangladesh and other low-lying regions. Earthquake-prone areas all over the world are well mapped, where hurricanes tend to originate and the paths they typically take are quite well known, and so on. The question is how, as individuals, communities and countries can we be reasonably ready for these events when they occur, without being neurotically apprehensive about them all the time? Numerous pieces of evidence, especially but not only in the Third World, of the ineffectiveness of organizations and governments to provide adequately for survivors of major natural disasters immediately, and in some cases for a considerable time, following their occurrence testify to the need to use more foresight in planning for specific contingencies and for research on crisis management, in general.

NEED FOR BETTER UNDERSTANDING OF PROBLEM

Much of the rhetoric about environmental pollution seems to rest on the assumption that the major problem is one of political will: We know how to clean up the environment, or at least how to stop polluting it further, but we lack the will to do it. Without denying that will power may be part of the problem, it is also the case that what *should* be done is often far from clear (Russell, 1988). The involved variables interact, frequently in unpredictable ways. A much better understanding of the causes and consequences of environmental change is needed to guide policy decisions and to ensure that measures taken in the interest of solving specific problems will not create worse ones in the process.

More than once in recent years, we have been reminded how very difficult it is to back out of a bad situation that should have been avoided in the first place, but anticipating bad situations is not always an easy task. Commoner (1966) has pointed out several instances in which science has failed to foresee the undesirable or even potentially disastrous consequences of technological advances. He noted, for example, that the massive atomic weapons testing program was launched before the biological risks of fallout were fully understood. The increase of carbon 14 in the troposphere, now a well-known result of the above-ground nuclear testing that was carried out in the mid 1950s, apparently was not anticipated at the time the testing was conducted, and the amount of only one of the fallout components—strontium 90— released by these tests before the enactment of the Test Ban Treaty of 1963 added radioactivity to the environment equivalent to about 1 billion grams of radium.

A less spectacular, but perhaps no less serious, failure to anticipate the deleterious effects on the environment of a product of technology (also pointed out by Commoner) is that of the effect of household detergents on water supplies. Detergents, which are synthesized from raw materials found in petroleum, are not broken down by bacteria in waste disposal systems. Impelled by the alarm regarding the rapidly increasing degree of contamination of water supplies in the early 1960s, attention was focused on the chemical composition of detergents and the reason for their resistance to natural degradation. Research led to the development of more readily degradable detergents by the mid-1960s. It is not clear how much damage was done to the environment before this problem was solved. Moreover, the "solution" is not without problems itself, and the long-range effects of biodegradable detergents are yet to be determined. Another example of a technological innovation that had an unfortunate consequence was the addition of lead to gasoline in the 1930s to decrease engine knock and increase fuel efficiency. I have already mentioned the switch, also in the 1930s, to CFCs—which were later found to be ozone-depleting compounds— as refrigerants.

Examples of unpleasant surprises involving the discovery of detrimental environmental effects of products or practices only after considerable damage has been done are easily multiplied. In the aggregate, they make a compelling case for the need for much more research on the great many ways in which the environment can be modified and for the development of more effective techniques for anticipating the future consequences (especially the undesirable ones) of specific policies and activities. A specific limitation for planning is incomplete knowledge of how much, and what kinds of, change the world can tolerate (Clark, 1989b). Much has been learned in the past few decades about changes that are now, and have been, taking place, and the monitoring of changes of various types is improving, but critical cause–effect relationships are still only partially understood, and predictive capability is very limited.

ENVIRONMENTAL DATABASES AND RESEARCH TOOLS

There already exist numerous databases that relate in one way or another to the problem of controlling environmental change. Several databases are maintained in the United States that record accidents involving chemical spills and other incidents involving hazardous materials. These include the Hazardous Materials Information System (U.S. Department of Transportation), the Pollution Incident Reporting System (U.S. Coast Guard), the Chemical Hazard Response Information System (U.S. Coast Guard), the National Transportation Safety Board File (National Transportation Safety Board), and the U.S. Department of Energy Database (Sandia National Laboratories). The National Center for Disease Control maintains a database of outbreaks of water-borne diseases reported to it by state health departments. The Carbon Dioxide Information Analysis center (CDIAC), which was established at the Oak Ridge National Laboratory in 1982, collects, stores, and distributes information regarding atmospheric carbon dioxide. It recently published a compendium of data on atmospheric carbon dioxide concentrations, carbon dioxide emissions, atmospheric methane concentrations, and recorded temperatures (Boden, Kanciruk, & Farrell, 1990). The compendium is beautifully organized with each data set—including tabular and graphical representations of the data, background information, a summary of noted trends, and references—occupying two facing pages of the document. It is a rich and highly accessible source of data for anyone working on any aspect of global climate change.

The computerized databases are of limited use to the random scientist working on environmental problems, however, because they are not interlinked and are not easily accessible from outside the government. What is needed is some "well-designed" front-end software and policies that will permit environmental scientists ready access to the diverse databases that hold the information they need in order to understand the problem with which they are involved (Bissett & Weaver, 1988). The enormity of the amount of data that will be gathered under the Global-Change Program represents a special challenge to the scientific community to figure out how to organize it, store it, make it available to individual researchers, and put it to good use. It is anticipated that the Earth Observing System satellite that the National Aeronautics and Space Administration (NASA) is scheduled to launch in the 1990s will transmit to earth every three weeks enough data to fill all the books of the Library of Congress (Kanciruk, 1990).

Several types of tools are needed: tools that will provide scientists with access to databases, tools to facilitate the processing of data, tools to help build and run predictive models, tools to help investigators share ideas and communicate results, and tools to help industries monitor their own performance vis-à-vis pollution regulations. Illustrative of the types of tools that will help scientists interact with environmental data in useful ways are the geographic information systems that are beginning to become available. Geographic information systems, which are designed specifically to represent spatial information, typically can provide users with cartographic representations of regions of interest (geographic and geopolitical maps) and non-cartographic information about those regions as well. Such systems are beginning to find uses by government agencies, public utilities, and private industries (Lang, 1988).

The National Center for Health Statistics and the U.S. Geological Survey are exploring the possibility of implementing an information system that will merge human health data with geophysical data. The goal is to facilitate the study of relationships between environmental conditions and human health (K. Taylor, 1990). The envisioned system would permit an investigator to examine, more or less simultaneously, data regarding the number of kidney cancer deaths and the location of polluted ground-water supplies in a given geographic region, to relate the number of melanoma deaths to the levels of ultraviolet radiation, and to explore similar relationships of interest.

Visual display design has long been a major focus of human factors research. There is much in the knowledge base that has been accrued as a result of this research that could benefit developers of geographical information systems. There undoubtedly remain many questions, however, about the optimal design of these systems, questions that could benefit from further research. The incorporation of cartographic-display generation capability, complex modeling tools, and powerful data analysis and representation procedures in the same software package in an integrated and readily accessible way will require continuing research and development efforts. Hopefully, whatever is learned from individual projects that has some generalizability can be codified in such a way as to provide useful guidance for future system development efforts.

PERCEIVED ENVIRONMENTAL RISKS

Until recently, the EPA had, in the words of a report from its Science Advisory Board, "made little effort to compare the relative seriousness of different problems" or "to anticipate environmental problems or to take preemptive actions that reduce the likelihood of an environmental problem occurring" (Loehr & Lash, 1990, p. 3). The agency has tended to react to environmental problems as identified in specific legislation, with each of its offices focusing on the particular problems it has a mandate to address. A consequence of this policy has been, again in the words of the Science Advisory Board report, "little correlation between the relative resources dedicated to different environmental problems and the relative risks posed by those problems" (p. 3).

In 1986, the Administrator of the EPA, Lee Thomas, asked a task force composed of members of EPA staff and of its Scientific Advisory Board "to examine relative risks to human health and the environment posed by various environmental problems." The results of the task force's work over a nine-month period were reported in an EPA (1987) document *Unfinished business: A comparative assessment of environmental problems,* in February, 1987. The task force rank-ordered problem areas with respect to each of four types of risk: cancer risks, noncancer health risks, ecological effects, and welfare effects (such as visibility impairment and damage to materials). Different groups ranked the problem areas with respect to the different types of risk, using somewhat different procedures to do so.

In reporting the results of its work, the task force stressed the incompleteness of the objective data at its disposal, the inexactness of the processes it used to produce its rankings, and the qualitative and judgmental nature of its findings. It also expressed general confidence, however, in the overall rankings that it produced. Severity of risk was judged in terms of risk to the entire U.S. population. The authors of the report pointed out that inasmuch as the rankings were intended to reflect severity of problems for the nation as a whole, they may not always reflect local situations or relative risks to individuals. None of the problems considered ranked relatively high or relatively low on all four types of risk. The highest ranking problems in each risk area are listed in Table 4.4.

The EPA task force attempted to compare its assessment of the relative importance of environmental problems with public opinion as reflected in polling data collected by the Roper organization over a two-year period. Because the questions asked in the Roper polls did not match exactly the problems identified in the EPA report, some judgment was required to relate the poll questions to the EPA problem areas. The 10 most important problems, as seen by the participants in the Roper polls, and recast somewhat to facilitate comparison with the EPA task force results, are shown in Table 4.5.

In spite of considerable attention to the problem, the assessment of perceived risk remains a delicate and difficult task. It is an important problem, however, because the willingness of policy makers and the general public to

TABLE 4.4 The Environmental Problems Classified as Relatively High in Importance in Each of the Four Risk Areas Considered by the EPA Task Force Working Groups, Listed in Rank Order of Importance Within Each Category

Cancer risks
    Worker exposure to chemicals ⎫
                                      ⎬ tied
    Indoor radon   ⎭
    Pesticide residues on food
    Indoor air pollutants other than radon ⎫
                                        ⎬ tied
    Consumer exposure to chemicals   ⎭
    Hazardous/toxic air pollutants
Non-cancer risks
    Criteria air pollutants
    Hazardous air pollutants
    Indoor air pollutants — not radon
    Drinking water
    Accidental releases — toxics
    Pesticide residues on food
    Application of pesticides
    Consumer product exposure
    Worker exposure to chemicals
Ecological problems
    Stratospheric ozone depletion
    Carbon dioxide and global warming
    Physical alteration of aquatic habitats
    Mining, gas, oil extraction, and processing of wastes
    Criteria air pollutants
    Point-source discharges
    Nonpoint-source discharges and in-place toxics in sediment
    Pesticides
Welfare effects
    Criteria air pollutants
    Nonpoint-source discharges to surface waters
    Indirect point-source discharges to surface waters
    To estuaries, coastal waters, and oceans from all sources
    Carbon dioxide and global warming
    Stratospheric ozone depletion
    Other air pollutants (odors and noise)
    Direct point-source discharges to surface waters

lead or participate in efforts to address environmental threats depends to a large degree on the perceived seriousness of those threats. Much more work is needed in this area. We return to this topic in chapter 12.

POLITICAL ACTION

The problem of environmental change is global in extent, and for that reason research and corrective action relating to it should stand a good chance of

TABLE 4.5 Ranking of Environmental
Problem Areas by Level of Public Concern,
as Reflected in EPA Task Force's Analysis of
Data Obtained in Roper Opinion Polls

Ranking	Problem Area
1	Chemical waste disposal
2	Water pollution
3	Chemical plant accidents
4	Air pollution
5	Oil tanker spillage
6	Exposure on the job
7	Eating pesticide-sprayed food
8	Pesticides in farming
9	Drinking water
10	Indoor air pollution

international collaboration. Already there are clear indications of growing worldwide concern, and there are even a few examples of international activity and information exchange. In 1983 the United Nations General Assembly called for the establishment of a special commission on the environment. What was established became known as the World Commission on Environment and Development. In 1987, this commission issued a report, *Our Common Future,* in which it proposed a variety of steps that should be taken in the interest of meeting the needs of today's world economy without jeopardizing the future. Another example of planned international collaboration on environmental problems is the International Geosphere–Biosphere Programme, a multi-decade project initiated in 1986 by the International Council of Scientific Unions. U.S. participation will involve such agencies as the National Science Foundation, NASA, the National Oceanic and Atmospheric Administration, the U.S. Department of Energy, the Environmental Protection Agency, the National Park Service, and the U.S. Navy (Science and the Citizen, 1987).

In 1987, the Montreal Protocol, which was signed by many nations, called for a 50% reduction in the production of CFCs by the end of the 20th century (United Nations Environment Program, 1987). In 1989, the European Economic Community announced its intention to ban *all* uses of CFCs by the end of the century (Dickson & Marshall, 1989). In May, 1989, a 10-day Forum on Global Change and Our Common Future was held in Washington, DC under the auspices of the National Academy of Sciences, the Smithsonian Institution, the American Association for the Advancement of Science, Sigma Xi, and the National Science Foundation. Speakers included scientists, business representatives, and government leaders from around the world. The conference had about 5,900 registrants from more than 80 countries, including the People's Republic of China, the (then)

Soviet Union, and all the (now former) Eastern Bloc countries (Abelson, 1989a).

The spotlight that has been turned on Eastern Europe because of interest in the amazing social and political events of the early 1990s has revealed some astonishingly serious environmental problems. It is conceivable that the spoiling of the environment could become the common enemy that would motivate truly collaborative efforts between East and West.

In the United States, a steady progression of environmental laws has been passed by the federal legislature since about 1965. These include the Water Resources Planning Act of 1965, the Clean Air Act of 1970, the Water Pollution Control Act of 1972, the Toxic Substances Control Act of 1976, the Clean Water Act of 1977, the Nuclear Waste Policy Act of 1982, the Superfund/Amendments and Reorganization Act of 1986, and numerous others. For a list of 33 such acts passed between 1965 and 1986 see Balzhiser (1989).

Shortly after President Bush took office in 1989, the National Academies of Sciences and Engineering and the National Institute of Medicine provided him with white papers containing policy recommendations in four areas in which critical decisions were deemed to be necessary within the succeeding four years. The areas addressed were: space (especially the Space Station), HIV infection and AIDS, environmental change, and technology and competitiveness. With respect to environmental change, recommendations included making global environmental change more prominent in the scientific, political, and foreign policy agendas of the United States; strengthening efforts to enhance energy efficiency and conservation, and developing alternatives to fossil-fuel energy sources; going beyond the provisions of the Montreal Protocol in reducing emissions of ozone-destroying chemicals; assessing the possible impacts of global warming (on sea level, agricultural systems, forestry, and water resources), and responses that could be made to them; implementation of efforts to mitigate the destruction of tropical forests; and a substantial investment in research and monitoring activities in the interest of improving the ability to predict the future of the global environment (National Research Council, 1989a).

The Committee on Earth Sciences was established by the Federal Coordinating Council for Science, Engineering, and Technology (FCCSET) in August, 1988. The committee's purpose is to increase the effectiveness of federally funded efforts to improve our understanding of the earth as a global system. Among other things, the committee identifies research and development needs and reviews federal programs in earth sciences. On the committee are representatives from the Departments of Agriculture, Commerce, Energy, Interior, State, and Transportation, the National Science Foundation, the EPA, NASA, the Office of Science and Technology Policy, the Office of Management and Budget, and the Council on Environmental Quality.

In a recent report to accompany the President's budget for the fiscal year 1990, the committee announced a U.S. Global Change Research Program, to establish the scientific basis for policymaking related to changes in the global earth system. The goal subsumes three parallel scientific objectives: monitoring, understanding, and predicting global change. The intent is to acquire information in such areas as: "ecosystem dynamics, the biological influence on the climate system, concentrations of significant atmospheric constituents, oceanic and atmospheric circulation, and regionally useful information such as predicted changes in growing seasons, precipitation, and soil moisture" (p. 11). The report identified seven interdisciplinary science elements on which the program is to focus: biogeochemical dynamics, ecological systems and dynamics, climate and hydrologic systems, human interactions, earth system history, solid earth processes, and solar influences. Federal funding for research and related activities in these areas will be channeled through several federal agencies.

One of the challenges identified by the Office of Science and Technology Policy of the Executive Office of the President for the recently announced Federal High Performance Computing Program is the prediction of weather, climate, and global change. Another is the development of a global ocean prediction model that will couple to atmosphere models for weather prediction purposes (Office of Science and Technology Policy, 1989).

Members of the 101st Congress introduced more than 700 pieces of environmental legislation (Foerstel, 1989). Among them was the World Environment Policy Act of 1989, introduced by Senator Albert Gore, Jr. (1989) as part of a proposed Strategic Environmental Initiative. The proposed act called for: a ban within five years on CFCs and other ozone-depleting chemicals, while promoting development of safer alternatives; massive reforestation programs; comprehensive recycling efforts; and radical reductions in carbon dioxide emissions through increased fuel efficiency. Senators John Chafee and Max Baucus introduced the Municipal Solid Waste Source Reduction and Recycling Act of 1989, which focused on reducing waste at its source and on recycling as the preferred approaches to solid-waste management. This bill called for measures that would provide consumers with information as to the recyclability of products and product containers, thus enabling them to buy selectively from producers that address the problem of waste management.

In testimony before the House of Representatives regarding the Global Warming Prevention Act, Representative Claudine Schneider, who introduced the bill, spoke of the "win–win strategy" of reducing global greenhouse gases and the costs of energy at the same time. The bill called for a 20% reduction of the carbon dioxide emission levels of 1988 by 2000, mandated the prioritization by the U.S. Department of Energy of its policies in accordance with least-cost energy options, authorized the establishment of 10 research centers to find ways to improve the efficiency of energy use, established national vehicle energy-efficiency performance standards, and established an Office of Recycling Research and Information within the Department of Commerce (C. Schneider, 1989). Unfortunately, government policies that give the traditional energy technologies a competitive advantage in the marketplace continue to be a significant deterrent to the development and use of alternative energy sources (S. Williams et al., 1990).

The military is also becoming more involved in environmental issues, both because it is a major contributor to environmental problems and must attempt to clean up its own facilities, and because it has technological and other resources that could be usefully applied to environmental problems more generally. These resources include detection and information gathering capabilities that can be used to monitor air, water, and soil contamination, as well as weather and meteorological phenomena. They also include research and engineering facilities that can be used to explore new toxic waste disposal and cleanup technologies on a relatively large scale. Both the Army and the Air Force are already involved in a variety of experimental efforts relating to decontamination of soil and water, processing of solvents for reuse, containment of paint fumes, large-scale toxic-waste cleanup activities, and development of alternative energy sources (Ackerman, 1990). An advantage of involvement of the military with environmental problems is the fact that in times of peace, the military represents a sizable workforce that probably does not need to spend all its time training or maintaining itself. Senator Sam Nunn, chairman of the Senate Armed Services Committee, and several of his colleagues have proposed that the concept of national security be expanded to recognize the threat to the country that a deteriorating environment represents, and that certain defense resources be redirected to help respond to that threat (Gilmartin, 1990).

Fred Krupp, writing as executive director of the Environmental Defense Fund for *The Wall Street Journal*, distinguished three stages of environmentalism in the United States: the conservation era, symbolized by the work of people like John Muir, Theodore Roosevelt, and Gifford Pinchot, the founding of the Sierra Club and the establishment of the U.S. Forest Service; the era of new laws and environmental-control regulations, a response, in part, to the publication in 1962 of Rachel Carson's *Silent Spring* and extending into the 1990s; and a third stage, not yet really entered, but, in Krupp's view, worth promoting. The third stage would see environmentalists recognizing the legitimacy of the societal needs out of which many, if not most, environmental problems grow, and working to find alternative and environmentally benign ways to meet those needs (K. Taylor, 1990). Conversely, with increasing recognition of the principle that neither individuals, industries, or nations have the right to disregard the implications of their behavior for the common good—a recognition that motivates movement toward regulation of activities that have significant impact on the environment—there is an increasing incentive to develop environmentally acceptable approaches to the pursuit of personal, corporate, and national goals. Such approaches will be economically attractive in the future by virtue of their insulation from regulatory problems.

RESPONSIBLE INDIVIDUAL BEHAVIOR

The problem of managing environmental change has as much to do with human perceptions, values, and activity as with atmospheric, earth, and ocean sciences. Ways to increase awareness, change attitudes, and motivate different behavior are among the most important of the issues that have to be considered. Planetary problems often appear remote to the individual. It is difficult for one to imagine that one's personal behavior can have any appreciable impact on the condition of the globe, but the behavior of cities, nations, or the world population as a whole is nothing more than the behavior of large numbers of individuals. If each person believes that his own behavior is insignificant in the grand scheme of things, the prospects of effecting voluntary change for the purpose of improving the environment are not good. The question is how to convince people that their own individual behavior does matter. How might we personalize the problems of the earth? How might attitudes that exacerbate the problem be changed?

The fact that individuals may see their own behavior as insignificant as it relates to global problems is compounded by short-term cost–benefit realities that are likely to motivate perpetuation of the wrong type of behavior. Hardin (1968) has referred to this problem as the *tragedy of the commons*. The metaphor is that of a herdsman who decides to add an animal to his herd. Because the herd is grazing on common land, he realizes a benefit of, say, one unit, while incurring a cost (in terms of the resulting decreased quality of the grazing land) of only a fraction of a unit. The fractional cost is also borne by all the other herdsman, but the single herdsman who has added the animal to his herd realizes a net gain. The problem, of course, is that when all the herdsman see the same opportunity to realize a net gain and decide to act in his own best short-term interest, they collectively ruin the land.

The tragedy of the commons is an example of what Platt (1973) has referred to as a *social trap:* a situation that, because of positive short-term benefits, entices individuals, organizations, or societies to move in a direction that will, sooner or later, lead to undesirable consequences that prove to be very difficult to avoid. Another characteristic of social traps is that the positive short-term consequences are typically enjoyed by individual members or subsets of some group, whereas the long-term negative consequences are shared by the group as a whole.

The willingness to obtain short-term gratification at the potential expense of great long-term costs seems to be a very strong human trait. Many people smoke, for example, in spite of incontrovertible evidence of the consequent increased likelihood of lung cancer and heart disease. Other illustrations of a willingness to make this tradeoff are easy to generate. In the case of spoiling the environment, the willingness is reinforced by the fact that the most serious of the long-term costs will be borne by generations other than the one that incurred them. People also may see the long-term cost as both inevitable and independent of their own individual behavior (Stern & Kirkpatrick, 1977).

Considerable experimentation has been done with laboratory situations that are intended to be analogous to social traps. Some of the results obtained provide at least suggestive evidence as to measures that could help people avoid such traps. The extent to which these results will transfer to real-world situations is not clear. Brechner and Linder (1981) have suggested that social-trap simulations might be useful for purposes of training aimed at making people better able to recognize situations where long-term consequences of their actions could be detrimental.

ENVIRONMENTAL PSYCHOLOGY

Environmental psychology has been recognized as a subfield within psychology at least since the late 1960s. There now are several journals that publish articles in this area, including *Environment and Behavior, The Journal of Environmental Psychology, The Journal of Environmental Education,* and *The Journal of Environmental Systems.* The *Handbook of Environmental Psychology* appeared in 1987, in two volumes of almost 1700 pages. The focus of the work has been somewhat different, for the most part, however, than that of this chapter. The predominant interest has been in the various ways in which one's environment can affect one's health, attitudes, and productivity (Stokols, 1990; Stokols & Altman, 1987). *Environment* in this context has usually referred to one's immediate workspace or living quarters: the "built environment." Considerable research has also been done on the effects of environmental stressors—heat, cold, altitude, noise—on human performance (Fidell, 1977). In all these cases, the emphasis has been on how the environment affects us; relatively little attention, in comparison, has been given to the question of how we affect it.

Environmental psychology has been concerned with the design of both living and working environments (Kasl, White, Will, & Marcuse, 1982), and especially the design of environments for people with special needs (Lawton, 1990). This work has undoubtedly been responsible for increasing sensitivity to the importance of taking into account the differing needs of individuals when designing living and working spaces; some researchers believe that environmental variables become more significant determinants of perceived quality of life when an individual's functional capacity is greatly diminished (Lawton & Nahemow, 1973; Rowles, 1978). The importance of providing special features in the interest of the safety, convenience, and comfort of people with functional limitations is now widely accepted. How to do this effectively, however, is not always apparent.

Lawton (1990) places the origin of environmental psychology in "classical theoretical psychology from which such constructs as *life space* (Lewin, 1935), *environmental press* (Murray, 1938), *competence* (R. W. White, 1959), and *adaptation level* (Helson, 1964) were drawn" (p. 638). Stokols (1990) suggests that psychologists became aware of the possible importance of the environment to human health and behavior as a consequence of the writing in the 1960s of such environmentalists as Carson (1962), Ehrlich (1968), and Hardin (1968). Before this time, Stokols claims, the prevailing assumption, which he refers to as the *minimalist view,* was that one's physical environment exerts little influence on one's behavior, health, or well-being. Although the tenability of this assumption is now being challenged, the results of research on the subject have not been as clear or consistent as one might wish.

Stokols (1990) contrasts what he refers to as the *instrumental* and *spiritual* views of people–environment relations. The instrumental view sees the environment as a tool—"a means for achieving behavioral and economic goals"; the spiritual view sees it as an end in itself—"a context in which human values can be cultivated." In the former case, the emphasis is on material features of the environment and environmental quality, seen primarily in such objective indicators as behavior, comfort, and health. In the latter case, the emphasis is on symbolic and affective features of the environment and environmental quality, seen not only in terms of comfort, health, and behavioral effects, but in the richness of its psychological and sociocultural meaning.

Much is known about the effects of the physical environment on human health and behavior (Stokols & Altman, 1987), but the relationship between environmental design and spiritual enrichment is less well understood. Moreover, the instrumental and spiritual views can be at odds with respect to their implications for environmental design, as, for example, when an emphasis on standardization and efficiency conflict with the desire for personalization and a concern for spiritual values. A key challenge, in Stokols's view, is to effect greater coordination between instrumental and spiritual views of environment and behavior. I suspect that few of us would doubt that different environments elicit different feelings and emotions. Some make us feel comfortable and at ease, and we find others depressing to varying degrees. It would be good to understand better the basis for these reactions.

On the basis of their review of the literature on the impact of the residential environment on mental health, Kasl et al. (1982) concluded that although there is a large set of more or less trustworthy findings, these findings do not permit any kind of closure. The major issues, such as the effects of crowding or density and those of work environments—which are the two most extensively researched classes of physical environment determinants of mental health—are still being debated. Studies of the effects of environmental variables on human attitudes and behavior are complicated by the fact that people in different environments (e.g., institutions, segregated facilities such as senior-citizen complexes, family homes, integrated apartments) may differ, on the average, in a variety of respects that could affect their attitudes and behavior independently of their residential situations.

In its May, 1990 issue, the *American Psychologist* devoted its "Psychology in the Public Forum" section to the role of psychological science in promoting environmental quality (Demick & Wapner, 1990). The articles in this issue focused on the question of how to measure the perceived quality of the natural environment, the problem of designing living environments for the elderly, contrasting philosophical views of how people relate to and are affected by their immediate environments, and roles psychologists have sometimes played in helping make or justify public policy with respect to environmental matters.

What is sometimes referred to as *environmental ergonomics* also focuses on how one's immediate environment—temperature, humidity, noise—affects one's bodily and cognitive functions and performance (Mekjavic, Banister, & Morrison, 1988). Of special interest are the affects of extreme conditions. Flach (1989) and Vicente (1990), for example, have described an approach to human factors based on an ecological perspective. The distinction is made between the *ecological approach,* which focuses on the interaction between individuals and their environments—which is to say, their immediate work situations and the constraints they represent for behavior—taking as its unit of analysis the human–machine system, and the *organismic approach,* which focuses on the human organism only. Proponents of the ecological approach argue that only by studying behavior as an interaction between an adaptive organism and a changing environment can one hope to produce results that are relatively invariant over tasks.

In short, the emphasis of environmental psychology has been on the implications of one's immediate environment for one's comfort, safety, productivity, feeling of well-being, and so forth. Comparatively little attention has been given to the effects of human behavior on the quality of the environment in a broad sense. I do not mean to discount the importance of studies of one's immediate physical environment and the effects it can have on one's performance and quality of life. This is an appropriate and very important focus for research. I mean simply to point out that most of the work that has been done by psychologists and human factors researchers to date under the rubric of environmental psychology or ecological human factors has not addressed itself to the question of how to make headway on the types of problems discussed in this chapter: climate change, air and water pollution, decertification, waste control, and so on.

One line of research that comes close to some of these problems has involved the application of classical psychophysical methods in an attempt to understand better the perception of environmental quality and, in particular, natural scenic beauty (Daniel, 1987, 1990; Daniel & Vining, 1983). Results from this work suggest a relatively high degree of agreement among observers even from different "interest groups" as to what constitutes scenic beauty, at least in forest contexts. Daniel and Boster (1976), for example, found a high degree of agreement among environmental activist groups, professional foresters, college students, and the general public in their assessments of scenic beauty. Similar results have been obtained by a number of other investigators. Studies in this domain have explored the effects on scenic beauty of numerous variables including forest harvest treatments, fire, recreation, road routes, insect damage, and air pollution (Daniel, 1990). Much of this work has focused on methodological issues and on efforts to develop and validate assessment procedures.

Efforts to find correlations between people's attitudes about environmental issues and sociodemographic variables have not found strong relationships; what seems to be a better predictor than such variables of attitudes on environmental issues, or at least on the desirability of environmental regulation, is political ideology (Samdahl & Robertson, 1989). Some attempts have also been made to determine linkages between people's behavior, insofar as it has implications for the environment, and their knowledge of environmental problems and their attitudes toward them; the results have not been very conclusive (Schahn & Holzer, 1990).

Although, as the foregoing examples show, there have been some attempts to bring psychological methods to the study of the relationship between human behavior and the environment, psychologists and human factors researchers have not yet seen the general problem of managing environmental change as one that represents a major challenge, opportunity, or responsibility for their fields. Noise pollution may represent an exception to this sweeping generalization, because this problem area has received a great deal of attention from psychologists and psychoacousticians, in particular. Excellent discussions of the problems and reviews of the work that has been done are readily available. (See, e.g., Broadbent, 1971; EPA, 1974; Fidell, 1978; Fidell & Schultz, 1980; D. M. Jones & Chapman, 1984; Kryter, 1985; Loeb, 1986.)

Paradoxically, noise is both the least threatening of the types of environmental problems considered here and the most thoroughly studied by psychologists. Noise, unquestionably, can be a form of environmental pollution. It can be highly annoying, distracting, and disruptive; in extreme circumstances, it can impair hearing. There is even some evidence that long-term exposure to high-intensity noise can have significant health effects in addition to hearing loss, such as increased risk of cardiovascular problems (S. Cohen & Weinstein, 1981). Unlike the other pollutants that are considered here, however, noise does not accumulate in the environment nor leave scars to offend future generations, nor does it have the same threat to the habitability of the planet as does, say, the unlimited build-up of carbon dioxide in the atmosphere.

ECOLOGICAL ECONOMICS

Computer models play a very important role in efforts to understand and predict environmental change. Most of these models represent the physics of environmental processes only. There is a need, also, for models that couple economics and environmental variables. Practices that affect the environment have economic implications, and, conversely, economic and fiscal policies can motivate behavior that further modifies the environment. That these cause–effect relationships exist is unquestionable, but precisely how they work is not well understood. It seems to be generally agreed that meeting the standards of the Clean Air Act is likely to be very expensive, for example, but how does one calculate the cost of not meeting them?

Attempts have been made to model the economic impact of a significant reduction in carbon emissions, and have resulted in predictions that the United States and several other countries will have to suffer a significant decline in GNP over the foreseeable future in order to accomplish it. Different models project different costs, however, and the only point on which there seems to be general agreement is that they involve assumptions about variables—like the cost of bringing alternative energy systems into use—that are highly uncertain (Corcoran & Wallich, 1990).

Ecological and economic management have generally been treated as separate problems and are the responsibility of different agencies at all levels of government. In reality, ecological and economic systems are interdependent. This is apparent, for example, in agriculture. Public policies intended to support agriculture, including subsidies amounting to over $300 billion per year in Western countries, can have the unintended effects of inducing farmers to clear forests for farmland and to use excessive amounts of pesticides and fertilizers (MacNeill, 1989). Similarly, nonemergency food aid to under-developed countries can, under some circumstances, inhibit efforts by those countries to produce their own food stocks, which would reduce the need to sell off natural resources for capital to buy imported food.

There are positive feedback loops in the processes that relate energy and environmental variables that exacerbate the environmental problems associated with energy use and that are not well understood. Fossil fuels, for example, are the main sources of energy for heavy industry, such as steel making and the manufacturing of heavy-duty machinery; and coal mining, oil drilling, and other industries involved in the recovery of fossil fuels are major consumers of steel and heavy machinery. So we have the somewhat paradoxical situation in which industries that are among the heaviest users of types of energy that are detrimental to the environment have among their primary customers the producers of that energy. Poland, for example, is the world's fourth largest producer of coal, which is used to fuel industry and to produce electricity, and in that country coal mining consumes 20% of the country's steel and nearly 10% of its electric power (Chandler et al., 1990).

To date, economists and environmentalists have not had the best of relationships, but have tended to be on opposite sides of arguments involving economics and environmental change. Indeed, economics has been viewed by some leaders of environmental groups and organizations as a tool of the opposition (K. Taylor, 1990). There is encouraging evidence, however, of a growing interest among some economists and ecologists in the possibility of a much better coupling of these disciplines. The International Society for Ecological Economics was formed in 1988 and now publishes its own journal, *Ecological Economics*. A meeting of the society in Washington, DC, in May, 1990 was attended by 372 participants instead of an expected 150. A major problem for this enterprise is to derive some consensus regarding how to determine the costs and values, in economic terms, of environmental variables, and, in particular, how to factor the long-term effects of today's activities and policies on future environmental capital into economic equations. Although efforts to make some progress on this problem are underway, they have not yet had much effect on the thinking of mainstream economists (Holden, 1990a).

An immediate challenge to ecological economists is the need for better methods of determining the costs and benefits to society of specific industrial and agricultural practices and of products and their means of production. As Frosch and Gallopoulos (1989) have pointed out, if a manufacturer produces nonrecyclable containers, taxpayers bear the landfill costs that this incurs. On the other side of the ledger, if a power plant reduces emissions that cause acid rain, the benefits are likely to be realized by communities elsewhere. Techniques are needed for fairly allocating to producers the real costs and benefits of their practices. There is a need, also, for much better approaches and techniques for evaluating the effectiveness of programs aimed at protecting or improving the quality of the environment. The evaluation information that has been available in the past has been described by the

General Accounting Office (1988) as "misleading, inadequate, or incomplete."

Judging the "greenness" of a product or, more appropriately, of the company that makes it, can be much more difficult than it may appear, because the problem is multifaceted and not just a matter of, say, whether the product is sent to market in a biodegradable package. Consumer guides and product or company certification programs that are beginning to appear must take the full gamut of considerations into account. The Green Cross Certification Company in Oakland, CA considers, when reviewing a company for environmental certification, the product, its packaging, the processes by which it was produced, the percentage of the company's resources that are sustainable, the percentage of its materials that are recycled, its adherence to governmental regulations regarding the environment, whether it has detectable residues of toxic effluents or emissions, its waste disposal and waste reduction plans, and whether it practices energy conservation (Welter, 1990).

*   *   *

One of the major challenges to psychology is helping us as individuals to understand what is at stake with respect to environmental change and how our behavior as individuals relates to it. As has been noted several times, many of the greatest threats to the environment are direct consequences of human behavior, and at least in some cases the behavior in question is that of individuals, as distinct from, say, corporations or governments. It seems safe to assume that often when people engage in such behavior, they are unaware of its environmentally detrimental effects. Although it would be unrealistic to believe that providing better information to the general public regarding the many ways personal behavior can have an environmental impact would cause everyone to behave only in environmentally protective ways, it seems likely that regular and continuing distribution of accurate and clearly presented information about how specific behaviors can damage or improve the environment would have some beneficial effect. Getting such information to the public and having it do some good requires, first, developing the information and, second, packaging it in attention-getting, understandable, and memorable forms.

As individual consumers, we have to learn to think more in terms of the environmental impact of the goods and services we purchase. We have to learn to distinguish between those that are environmentally benign and those that have a long-term cost to the environment that outweigh their short-term benefits to the purchaser. We must also develop the kinds of sensitivities and values that will dictate strong preferences for the former, but developing these sensitivities and values will require better information than we now have regarding the environmental implications of our purchases that are not reflected in what we pay at the counter. The human factors community can play some role in developing and communicating that information.

We need a much better understanding than we now have of attitudes and values that contribute to unnecessary waste: what they are, how they are formed, and why they are maintained. Presumably, few of us deliberately set out to be wasteful. On the other hand, vague or even graphic reminders, from time to time, of the fact that we are do not suffice to change behavior very dramatically. Wasteful behavior and the attitudinal variables that support or promote it are appropriate and practically important subjects for scientific study.

There are now sufficiently many laws and regulations relating to environmental protection to which industrial companies must adhere that ensuring compliance can be a significant problem. Computer-based systems that incorporate the laws and regulations that pertain to specific industries and that provide the guidance needed by individual businesses, not only to comply but to document properly their compliance, could be very useful to companies, especially those too small to maintain legal staffs that can be charged with the responsibility of ensuring compliance. The development of such systems requires attention not only to the problem of encoding the appropriate legal and regulatory knowledge bases, but also to the general problem of designing the interfaces so that the resulting systems will be easy for people other than their developers to use.

Much greater emphasis should be put on environmental issues in engineering education (Friedlander, 1989). There is a need also for greater sensitivity by the engineering profession for the environmental effects of its work:

> The engineer should bring to his or her work not only sound technical knowledge, disciplined technique, and a focused search for creative solutions to novel problems, but also a concern for the ecology of technology. Not all consequences of technological development can be anticipated; not all unfortunate extensions can be anticipated. Nevertheless, the imperative to understand the implications of a development in the broadest and most encompassing terms is a professional responsibility of the engineer, which must be incorporated into the task from the outset. (P. E. Gray, 1989, p. 202)

Clearly, this call is at least as appropriately addressed to human engineering, as a discipline, as to any other type of engineering.

The problem of environmental change has yet to be discovered as one to which human factors research has something important to contribute. This is unfortunate because there are many dimensions of this extremely complex problem that represent needs and opportunities for human factors involvement. Challenges include finding ways to make clear the long-term environmental implications of present behavior of individuals, organizations, and governments; development of approaches to assessing satisfaction with one's environment (a problem analogous to that of assessing job satisfaction); development of cost–benefit decision aids that take environmental impact into account; study of ways to improve the cost-effectiveness of methods to harvest renewable nontimber forest products; design of biological databases and development of tools that will facilitate the tracking of changes in species populations and habitats; development of more cost-effective methods of sorting trash for purposes of recycling; development of effective approaches to the evaluation of products in terms of disposability or recyclability; more generally, evaluation of products in terms of their total costs, taking factors such as the use of nonrenewable resources in their production and environmental implications of their residues into account; continued study of human error as a contributing factor in industrial accidents; development of more effective approaches for dealing with the aftermaths of natural disasters; design of interfaces for environmental databases, of front-end software that will facilitate access to different databases through a single front-end agent, and of tools that will facilitate useful interaction with the accessed data; design of compliance-monitoring tools and procedures for use by industries subject to environmental regulations; and development of environmental modeling tools and data displays. Many of the opportunities for helping to address the problem of energy generation and use, mentioned in chapter 3, relate also to the environment.

It would be easy to extend this list, but it is long enough to make the point. Moreover, if the problem of environmental change became a major focus of interest within the human factors community, if it became a topic of discussion at professional meetings, if human factors researchers began to look for ways to have an impact in this area, I believe that many possibilities would emerge that are difficult, or perhaps impossible, to identify now.

The problem of environmental change is real: Experts argue as to whether the magnitude of one or another aspect of it has been overstated, but there is a strong consensus that there is cause for concern and considered action. It is global in extent: Regional aspects of it can become global, and even while still regional can have indirect global effects. It is compounding: Ignoring it, or even postponing attending to it, is flirting with disaster. Much of the problem originates with human behavior, and progress toward a solution must involve modifying that behavior.

Among the long-standing goals of human factors research has been the design of working and living environments that are conducive to human safety, productivity, and comfort. A major effort to find ways to help ensure the continuing habitability and hospitability of the environment—writ large— would be in keeping with the best traditions of the discipline.

5

Education and Training

Education has long been viewed, in this country and elsewhere, as the doorway to opportunity for the individual, and an educated citizenry has been seen as essential to national prosperity and fundamental to a democratic way of life. Moreover, at least in this country, education is very big business; about 1 out of 4 people in the United States is a student or is employed by a school or college (Office of Educational Research and Improvement, 1990). According to estimates by the National Center for Education Statistics (1990b), expenditures in public elementary and secondary schools for the school year 1989–1990 were expected to reach $182 billion, or about $4,841 per pupil in average daily attendance. The same report projected a 10% increase in expenditures per pupil over the 5 years beginning with 1988–1989, which compares with a 20% increase during the preceding five years. Vaughan (1989) put the total annual U.S. expenditure—government and private—on education at all levels at about $600 billion.

Looking ahead, many observers have argued that the importance of education, both to the individual and to the country as a whole, will increase. Kasarda (1988) has put it this way: "With brain power replacing horsepower as the essential driving force of our emerging post-industrial economy, education is rapidly becoming the keystone on which future improvements in productivity and competitiveness will depend. At the same time, a good education will likely become a *sine qua non* for individual opportunity and social mobility in the 21st century" (p. 144). The OTA (1988a) also puts great emphasis on the importance of education as a key

138

determinant of the economic future of the United States; it stresses the need for research on learning and urges the development of new educational technologies. Many others have pointed to improvements in the educational system at various levels as fundamental to the United States' ability to be competitive in the world economy in the long term, and have suggested that the possible failure of the educational system to improve may be the country's greatest threat (Bartel, Lichtenberg, & Vaughan, 1989; Bloch, 1984; Botkin et al., 1984; National Commission on Excellence in Education, 1983; Young, 1984).

Technological change has increased the value of such basic skills as reading, writing, numerical reasoning, and problem solving, and some observers believe that the significance of such fundamental skills in the workplace will be even greater in the future than they are now. Cyert and Mowery (1989), for example, have noted the need for workers to have the ability to adapt when the demands of their jobs change, as they are expected to do with increasing frequency, and argued for the growing importance of a broad-based education that will prepare workers to adapt to rapidly changing employment situations. Mincer (1989) also has pointed out that the need for a well-educated workforce is greater during periods of rapid technological change than during periods of maintenance of the status quo, because tasks that depend on relatively mature technologies tend to become routinized and to require little adaptability. Another OTA report (1989a) described the challenge of preparing people for work opportunities this way: "The job market is too unpredictable to target specific personnel needs, so the goal of higher education, including that for science and engineering, should be to prepare students for an uncertain future by imparting a range of skills" (p. 32).

Education is also being recognized more and more as a life-long pursuit. Many professionals—physicians, scientists, engineers—find it essential to study and learn on a continuing basis in order to stay current in their specialties. Advancing technology, especially automation, frequently makes specific jobs obsolete and changes the skill requirements of others, thus necessitating the learning of new job skills by many workers. Increasing numbers of adults, including many older adults, have been enrolling in regular and special courses offered by colleges and community-sponsored adult-education programs either to acquire skills to serve specific practical objectives or to learn about some area of interest just for the intellectual satisfaction of doing so. Between 1969 and 1984, the number of people taking adult education courses increased at over twice the rate at which the population increased; during that time the percent of the population between the ages of 35 and 54 enrolled in such courses increased from 11% to 17% (Edmondson, 1988).

SIGNS OF TROUBLE

In spite of the acknowledged importance of education to the individual and to the country, and of the belief that its importance is likely to increase, there is substantial evidence that all is not well with the current state of education in the United States. To be sure, people get more schooling today than they did even in the relatively recent past—the median number of years of school completed by people in the 25- to 29-year-old age group was 12.9 in 1980 as compared with 10.3 in 1940, and the number of individuals with an undergraduate college degree or higher increased from 1 in 16 to almost 1 in 4 over the same time period (Ginzberg, 1982)—but such statistics provide limited comfort in the context of numerous indications of trouble:

• The problem of illiteracy is severe and appears to have been getting worse since the 1960s, at least in poor inner-city neighborhoods (K. S. Wilson, 1987). According to the widely cited Adult Performance Level Study conducted by the University of Texas at Austin in 1975, 23 million American adults at that time were sufficiently illiterate to have great difficulty functioning in our society, and an additional 40 million functioned, but not proficiently (Adult Performance Level Project, 1977). In 1981, a study by the National Assessment of Educational Progress led to the conclusion that only about half of the country's 17-year-olds could write a wholly satisfactory piece of explanatory prose, and only about 15% could defend a point of view effectively with a persuasive argument (National Assessment of Educational Progress, 1981). The Committee on Economic Development (1987) has estimated that less than 50% of high school seniors read at levels adequate to carrying out moderately complex tasks, and 80% have inadequate writing skills. Literacy means different things in different contexts, and how serious the problem of illiteracy in the United States is, relative to other problems the nation faces, is a matter of opinion (G. A. Miller, 1988); there seems little question, however, that it is a problem that deserves to be a focus of attention for research and merits continuing efforts to develop better approaches to the teaching of reading and writing to both children and adults.

• Education in science and mathematics has been the focus of much concern. A report prepared by the Educational Testing Service, based on 1986 data, indicated that only 7% of U.S. 17-year-olds are adequately prepared for college-level science courses, and that less than half of the country's 17-year-olds have enough scientific understanding to perform jobs that require technical skills or even to make informed decisions as citizens. The same report indicates that only one in thirteen 17-year-olds attains any degree of sophisticated understanding of science and warns that this proportion is substantially smaller than will be required by the work force of the future (Mullis & Jenkins, 1988). In an international comparison of math and science proficiency of 13-year-old students, Americans placed last in math and in the lowest of three tiers in science (Byrne, 1988, 1989). The teaching of science and math at secondary school level has been characterized as a national scandal (T. S. Perry, 1981).

• Former presidential science advisor Edward David, Jr. (1985) has pointed out that federal support of fundamental mathematics in the United States decreased by about 33% between 1968 and 1973, and remained at about the 1973 level for the next 10 years in constant dollars. The American Association for the Advancement of Science (AAAS, 1989), in its Project 2061 report, *Science for All Americans,* sounded the alarm about the current state of U.S. education in science, mathematics, and technology this way: "The fact is that general scientific literacy eludes us in the United States. A cascade of recent studies has made it abundantly clear that by both national standards and world norms, U.S. education is failing to educate too many students—and hence failing the nation. By all accounts, America has no more urgent priority than the reform of education in science, mathematics, and technology" (p. 3). As of 1991, only about one in three elementary science teachers had a college course in chemistry, and only about one in five had one in physics; almost one half of secondary school science teachers had not had college calculus.

• Both Walter Massey and Richard Atkinson, in their presidential addresses to the American Association for the Advancement of Science in 1989 and 1990, respectively, focused on the need to make improvements in science education a national priority. Each warned of the potential shortfall in technically trained people for the U.S. workforce in the near-term future and urged action by the government and the private sector to help address this problem. From 1966 to 1988, the percent of U.S. college freshmen planning to major in the sciences and mathematics fell from 11.5% to 5.8%, and the percent planning to major in mathematics went from 4.6% to 0.6%. Between 1972 and 1988, the percent planning to pursue careers in business went from 10.5% to 23.6% (Green, 1989). Between 1973 and 1987, the annual number of PhD's granted to U.S. citizens in mathematics fell from a little under 800 to about 350; during the same time the number granted to non-U.S. citizens increased from about 200 to about 400. In other words, in 1973 the ratio of U.S. to non-U.S. students getting PhDs in mathematics in the United States was about 4 to 1, and by 1987 the ratio had become roughly 1 to 1 (Atkinson, 1990). In 1980, Japan, with a population half the size of the United States, graduated about 87,000 engineers, whereas the United States graduated about 78,000. In Japan, the numbers of engineers, lawyers, and accountants for every 10,000 citizens were 35, 1, and 3, respectively; comparable numbers in the United States were 25, 20, and 40 (Botkin et al., 1984).

• Although what it means to be scientifically literate is not entirely clear, attempts to determine what the general public knows about general science have yielded some startlingly bothersome results: Over half of one sample of American interviewees did not know that the earth revolves around the sun once a year, and only 12% affirmed that astrology "is not at all scientific" (Culliton, 1989).

• According to a Bureau of the Census estimate (cited in Owens, 1988), an average of 682,000 students drop out of American schools each year: 3,789 per day. The Committee on Economic Development (1987) has estimated that one year's class of dropouts will cost the nation more than $240 billion in lost earnings and taxes, and this estimate does not include costs for crime control, welfare and health care, and other social services over the dropouts' lifetimes.

• The country's educational problems are disproportionately hard on children of low-income minorities. Such children frequently arrive at school inadequately prepared, either socially or intellectually, and often lack the home and social support to encourage and reinforce academic achievement; they can lag behind the national average by up to two years, and in large cities as many as 50% of them drop out of school (Comer, 1988; Tugend, 1984).

Unfortunately, it would be all to easy to make the list of worrisome indicators even longer, but the examples given here suffice to establish the existence of a serious problem. One assessment of the cost to the nation in lost output, due to the skills shortage indicated by the decline in test scores, approached $86 billion in 1987 and projected a doubling in constant dollars by 2000 (Dentzer, 1989). There are many specific suggestive evidences of the costliness of illiteracy: People who lack basic reading and writing skills account for the majority of the unemployed (Toch, 1984); 85% of the juveniles who appear in court are not literate (U.S. Department of Education, 1982); and an estimated 60% to 80% of prison inmates are functionally illiterate ("Ahead: A Nation", 1982; Boorstin, 1984). Protestations to the contrary notwithstanding, the country either is unconvinced of the importance of a first-rate educational system to its long-term well-being, or it is unwilling to make the investment necessary to ensure dividends that would be realized a generation hence. The latter possibility is consistent with much evidence that, as a country, we have an extremely short-range perspective and find it very difficult to make investments that have primarily long-term returns. Certainly we lack a vision of the exciting enterprise that education could be if there were a national commitment to develop the best learning environments that modern technology and current knowledge of cognition would permit.

This said, it is important to recognize that the problem is a multifaceted one and is not likely to have a simple and easily realized solution. In particular, there is more to education than whatever it takes to achieve high scores on standardized tests, and judging an educational system soley on the basis of such test scores is a questionable thing to do, at best. Conversely, one should not assume that a system that produces students who consistently get high scores on such tests is one that should be emulated in all respects. Apparently the Japanese education system, which does produce sudents that score well on standardized tests and is often held up as a model in the United States, has some warts of its own (Schoolland, 1990).

UNCERTAIN DIRECTION AND COMMITMENT

Recognizing that there is a problem and having the commitment and continuing resolve to address it effectively are quite different things. The AAAS (1989) warns that reform of U.S. education in science, mathematics, and technology, will take not only time, determination, collaboration, resources, and leadership, but also daring and experimentation. There is little evidence that the country is prepared to be very daring in this regard, and although there is general agreement that science education in the United States needs improving, there is no consensus on how best to attain it:

> How shall we change the curriculum? Concentrate on basic science literacy for all, or on training the best and brightest for tomorrow's research? Teach less? Teach more? Teach teachers to teach better? Make sure kids know readin' and 'ritin' before they tackle 'rithmetic? Forbid kids to use calculators? Force them to use computers? Reinstate drills in long division? Adopt the centralized Japanese system? Do research on how people learn? Get more science on TV? Set up mentoring and support systems for students who need help to stay in science? Provide more federal money? State money? Local money? (Powledge, 1989, p. 6)

There is a good reason why the answers to such questions are not known. The United States spends enormous sums of money on education, if one counts what comes from federal, state, and local taxes and from private sources, perhaps more than on any other single objective, including defense. Most of this money, however, is spent on educational facilities and the delivery of educational services; only a tiny fraction of it is used to underwrite educational research. Indeed, educational research has a shamefully low priority within the United States, at least as reflected in government funding patterns. A recent report in *Research and Development* (Cassidy, 1989), showed federal R & D funding for fiscal year (FY) 1989 (in millions of dollars) as follows: Defense—37,742; National Institutes of Health—7,152; Energy—6,179; NASA—4,192; National Science Foundation—1,885; Agriculture—1,064; Interior—396; EPA—335; Transportation—317; National Oceanic and Atmospheric Administration—176; and the National Institute of Standards and Technology—159. Research funded by the Department of Education, at less than $100 million, was apparently insufficiently significant to deserve mention.

In constant dollars, federal funding of educational research through the Department of Education has decreased considerably since the 1970s (General Accounting Office, 1987b). Also, in constant dollars, the total federal expenditures for education decreased by about 0.7% between FY 1980 and FY 1989 (National Center for Education Statistics, 1990a). In the early 1950s, the National Science Foundation spent almost nothing on education or educational research. From 1956 until 1965, its annual expenditures on education increased almost yearly until reaching nearly $500 million. It stayed at a little under $500 million per year until 1968, and fell off regularly thereafter, reaching a nadir of about $20 million in 1983. Happily, since 1983, its support of research on education in mathematics and science has been slowly climbing back up and was budgeted at about $200 million for 1990.

SOME NEW THRUSTS

In spite of the penurious funding of educational research, some research has been done, and it has yielded useful insights regarding how people learn and the kinds of environments that support learning. Emerging from this research is an appreciation of the importance of real student involvement—intellectual engagement—in the learning process. Researchers have argued, for example, that the poor performance of American students in science and their lack of interest in science subjects stem from an overreliance on textbooks and lectures as the main instruments of teaching, and have called for the adoption of teaching methods patterned after the methods of science itself. This means engaging students in the kinds of investigative activities that scientists engage in and doing so in the context of real-world problems of relevance to the students and their communities. Several major efforts to teach science or mathematics in more effective ways have been launched by organizations with a special concern for the future of science and mathematics as professions. The organizations involved include the AAAS (Project 2061), the National Science Teachers Association (Scope, Sequence and Coordinations; SSC), and the National Council of Teachers of Mathematics (Culotta, 1990). If Jenkins and MacDonald (1989) are right, however, we are unlikely to see much change in the way science is taught in schools until better techniques are developed and widely applied for assessing not only the factual information students have acquired about science, but their ability to think scientifically and to "do science" in a hands-on fashion.

Another noteworthy development in education in the recent past has been the rapidly growing interest in the teaching of thinking skills—reasoning, problem solving, decision making (J. Baron & Sternberg, 1986; Chipman, Segal, & Glaser, 1985; Nickerson, Perkins, & Smith, 1985). This interest is evident among both administrators and teachers and at all educational levels. State departments of education are beginning to mandate the teaching of thinking skills and to incorporate tests of thinking ability in their assessment programs. Conferences on the teaching of thinking, unheard of a few years ago, are becoming annual events in many states and regions. Numerous programs to teach thinking have been developed, and curriculum material for this purpose is becoming available in abundance.

In February, 1989 the American Association of Colleges for Teacher Education (1989) passed a resolution on critical thinking that says, in part: "Be it resolved that: AACTE encourage its membership, as a high priority, to implement within teacher preparation programs: (a) course work that requires those future teachers to enhance their own higher-order thinking skills, and (b) courses in pedagogy in which future teachers become proficient in applying strategies that will enable learners to acquire higher-order thinking skills of their own" The resolution calls also for the National Council for Accreditation of Teacher Education to incorporate the concept of higher order thinking in its assessment standards for teacher-education institutions. Clearly a greater emphasis in teacher-education programs on the development of thinking ability—of both teachers and students—is exactly right. The practical challenge is to figure out how best to accomplish this goal. This question has been getting some attention from educational researchers, but probably not nearly enough.

There is no more important objective of education, in my view, than the development of thinking ability, but in the quest to invent approaches toward attaining this objective, it is essential to recognize that thinking is complex, multifaceted, not yet very thoroughly understood. Thus, there are unlikely to be any simple short-cuts to learning to do it well. Unfortunately, the development of programs and materials, energized by the demand that is coming from prospective users of them, could be getting ahead of our knowledge of how thinking ability normally develops and how its development can be facilitated. The promotional character of the available documentation on many of the programs that teach thinking is a worry, as is the paucity of evaluative data. There is a need for much more research on thinking and especially on the question of how best to enhance it through instruction.

It is widely acknowledged that conventional multiple-choice tests of the kind that are used so extensively to measure academic achievement are much better at assessing what information students have assimilated than at determining their ability to think. Given that achievement tests are used to evaluate the effectiveness of educational systems, it is only reasonable to assume that as long as they measure primarily information assimilation, teachers will be motivated to use an information-dispensing approach to teaching. There is nothing very mysterious about this, and there has been much agitation, within the educational community and without, for the development and use of tests that will measure not only what information students have absorbed, but how deeply they understand that information and how effectively they can apply it in thoughtful ways. Everyone seems to know that this is what is needed, both to assess more meaningfully what students are learning and to drive the academic process toward a greater emphasis on teaching them—or helping them learn—how to think. The problem is that tests that do a good job of assessing thinking ability are neither easy to construct nor easy to administer.

Many investigators believe that some aspects of the process of learning can themselves be learned. According to this view, there are certain strategies and methods that are teachable and that, if learned, will facilitate further learning. Most of the strategies that have been studied to date relate to comprehension and retention of written material. There is a need for further research not only to identify and evaluate additional strategies, but to develop more effective methods for teaching them.

TECHNOLOGY AND EDUCATION

Although few people would want to promote technology as *the* answer to the complex multidimensional problem of education, some have seen in technology—especially computer technology—the potential for a significant beneficial impact (Kay, 1991; Kinnaman, 1990; Nickerson & Zodhiates, 1988). It must be said, however, that so far that impact has been vanishingly small, and experts are far from agreed as to how computer technology should be used in educational contexts (Holden, 1989a).

Although computers are appearing in more and more classrooms, so far the machines have been used primarily to teach about computers and have not been used extensively in traditional subject areas. In general, students who use computers in school do so for a little more than 1 hour per week, on the average (OTA, 1988a). It seems highly likely that the computer-to-student ratio will continue to increase and that, as we enter the 21st century, an unavailability of hardware will not be what limits effective utilization of computing resources in education. What is less certain is how long it will take to develop effective software and an understanding of how to integrate computing resources (hardware and software) with other educational tools.

Clearly, providing classrooms with an abundance of computer technology is not enough to ensure that learning will be enhanced, either quantitatively or qualitatively. There is still much to be discovered about how to use this technology effectively for educational purposes. It is not reasonable to expect individual teachers to make those discoveries on their own. Before this technology can be widely used to facilitate learning in the classroom, it will be necessary to provide teachers with much better preparation than they have had in the past or are now getting. In providing this preparation, we must resist the temptation, as Roszak (1986) put it, to lower the subject to the level of the computer when the computer cannot be brought up to the level of the subject.

That there is a real risk here can hardly be doubted. If one's ambition in life is to develop educational software, it is only natural to focus on aspects of any particular subject matter that lend themselves most readily to being taught by computer. This is all right provided one recognizes the limitations of what one is developing. Some aspects of a given subject are indeed more readily taught by computer than others, and perhaps there are aspects that one should not attempt to teach in this way even if it were possible to do so. The danger is that of equating what *can* be taught by computer with what *should* to be taught and with *all* that should be taught. Often people who attempt to assess the effectiveness of the use of technology, or any innovative approach, in the classroom focus on how well students have learned whatever the technology (or approach) has been used to teach; no such assessment is complete, however, unless it also considers any costs of lost opportunities that may have been incurred. If the new approach resulted in a much better understanding by the students of *x,* but left no time to cover equally important *y* and *z,* one may question whether a net gain has been achieved.

The OTA has estimated that, as of 1987b, there were more than 10,000 software products on the market intended for instructional use with stand-alone computers. Most of these products represented drill and practice approaches, and the quality was not judged to be uniformly high. It seems at least possible, however, that a lot of quite good educational software will become available eventually. There currently exist a few programs that are sufficiently powerful to provide glimpses of the potential that computer technology has for application in the educational arena. Although education is among the most conservative of institutions, there is some possibility that it will change dramatically over the next few decades; however, one cannot rule out the possibility that it will change very little (D. K. Cohen, 1988).

The development of intelligent computer-assisted instruction (ICAI) systems has been an objective of numerous researchers for several years. The systems that are targeted by this work would have capabilities similar in some respects to those of human tutors: an understanding of the domain on which the learning/teaching is centered, an ability to assess the student's knowledge of that domain and to tailor instruction so as to be appropriate to that knowledge, and an ability to ask and answer questions and to give explanations. They would also have some capabilities that human tutors do not have: for example, the ability to simulate a complex process and permit the student to do try-and-see experiments with it. The possibility of developing such systems was seen by some of the first investigators of the applicability of computer technology to education, and the idea of a tutorial system that could carry on a Socratic dialogue with a student emerged almost as soon as the earliest computer-assisted instruction programs (Feurzeig, 1964; Feurzeig, Munter, Swets, & Breen, 1964). Anything close to a full realization of the vision of a truly intelligent nonhuman tutorial system remains only a hope for the future, however, in spite of much effort and considerable progress by researchers in this field (Psotka, Massey, & Mutter, 1988; Resnick & Johnson, 1988; Sleeman & Brown, 1982).

There can be little doubt that computer video games of the Nintendo variety have a great deal of appeal and holding power with young people. Negroponte (1991), who points out that there is a Nintendo in more than 70% of all homes with a child between the ages of 8 and 12, calls it "America's largest domestic computer presence and potentially the country's major force for educational change" (p. 108). Some observers have been concerned that violence is a prominent aspect of many of the games that seem to have the greatest appeal, but there are exceptions, and among them are some that are believed to have the ability to teach useful skills, such as strategic planning and other forms of complex thinking (Bloom, 1990–1991).

Technology, I believe, is not now a major factor in limiting the effectiveness of education. The technology we currently have has the capacity to deliver far more effective learning tools and environments than have yet been realized. Our most serious lack is a national vision of what technology could mean for education if the potential that is already there were developed to even a modest degree. Visually rich, dynamic systems with which students can interactively explore and experiment with structures, relationships, principles, and processes could be very powerful educational instruments. How to ensure that this potential gets developed and applied to worthwhile educational goals is the hard question. The temptation to assume that this will happen in the natural course of events is easily resisted on a moment's reflection on the history of television, another technology with enormous— and still largely unrealized—educational potential. G. E. Brown (1982) has provided one vision of what information technology could mean to education in a broad sense:

> I see the possibility that, through information technology, the whole community will become a learning environment. Satellite transmission and cable television will bring a wide variety of cultural and educational programs into homes and schools. . . . Schools, museums, libraries, and governmental units

will be connected through computer and television networks and will have access to a wide variety of databases. Inexpensive microcomputers in homes will be able to access, through rapid, reliable networks, an almost unlimited range of learning resources. In the workplace, information technology will be used in a wide variety of activities, from routine tasks like electronic mail and electronic funds transfer to sophisticated applications of computer-aided design and robotics. Updating of skills and learning of new skills through satellite transmission and computer assisted instruction will be a standard feature of industrial and professional training. (p. 52)

Brown cautioned that, although it is easy to paint a rosy picture of what information technology could bring to our society, there is no guarantee that this technology will always be used in socially constructive and equitable ways. The vision of the entire community as a learning environment is an attractive one indeed and failure to realize such a vision will not be because the technology that would be needed to do so does not exist. The failure, if we do fail, will be one of intention and resolve.

The assumption that one learns best by doing, first made popular by John Dewey early in the 20th century, has considerable appeal among developers of software systems intended to facilitate the teaching and learning of science and mathematics. The explicit objective of many developers has been to build systems that will help students not to *learn* science or math, but to *do* science or math, in the spirit of the practitioners of these disciplines. The intent is not to teach students *about* science and mathematics, but to help them begin to think like scientists and mathematicians. Communication and collaboration have also been watchwords of much of the recent work on applications of technology to education. Cooperative learning, communities of inquiry, knowledge sharing, and collaborative writing are currently prominent themes.

Scardamalia and Bereiter (1990a, 1990b; Scardamalia, Bereiter, McLean, Swallow, & Woodruff, 1989) have promoted the development of "computer-supported intentional learning environments" in which "knowledge-building communities" in schools are supported in their learning and knowledge-generating activities by computer and communication technology. The technological supports are intended, in this case, to help students to "relate new knowledge to old, monitor their understanding, infer unstated information, and review, reorganize, and reconsider their knowledge" (1990a, p. 6), to do, in short, the various things research has shown that successful learners do. Central to this idea is the assumption that knowledge construction, and not just knowledge assimilation and retrieval, is the goal of education. In the computer-supported intentional learning classrooms that Scardamalia and Bereiter envisioned, students engage in a cooperative effort to build, extend, and refine a body of knowledge, represented in various forms—text, graphics, charts, simulations, and video 'clips.' Cooperation—

building on each other's knowledge—is a non-incidental aspect of this approach. So, also, is access to a wide variety of educational software and related learning tools.

A variety of software packages have been, or are being, developed to support inquiry and exploratory learning. These include the Geometric Supposer (in press), Inquire (Brunner, Hawkins, Mann, & Moeller, 1990), Function Probe (Confrey, 1990), Algebra Workbench (Feurzeig, 1987; N. Roberts, Carter, Davis & Feurzeig, 1989), and Physics Explorer (Richards, Barowy, & Levin, 1992), among others. All of these packages permit users to manipulate either mathematical constructions or simulated physical systems and see immediately the effects of those manipulations. Multimedia systems have been used for instructional purposes at Brown University for several years (Yankelovich, Haan, Meyrowitz, & Drucker, 1988). Pea, Boyle, and de Vogel (1990) described a software system under development that is intended to help students construct multimedia compositions involving text, photographs, graphics, video (tape and disk), sound, and animation.

The appearance of this kind of software and the fact that numerous efforts are ongoing to develop ever more versatile packages are exciting and encouraging. Much remains to be done, however, to ensure the pedagogical effectiveness of such facilities. Moreover, D. K. Cohen (1988) has warned that whereas technology enables educational change, it does not cause it. He and others (e.g., Cuban, 1986; Newman, 1990a, 1990b) have also noted that institutionalized education has been remarkably adept at assimilating technological innovations and using them in such a way as not to threaten the status quo. Ensuring that the potential that technology offers for the development and use of effective tools and environments for teaching and learning is realized to some significant degree will require that a variety of issues—social, psychological, organizational, political—be faced. Change has not come easily to the institution of education in the past—consider the degree to which technology has affected education during the 20th century, as compared to the degree to which it has affected transportation, communication, medicine, science, or almost any other major area of activity—and there is little reason to expect that it will come a lot easier in the future, but if change does not come, the cost will be dear indeed.

One idea of how to exploit information technology for educational purposes that is being promoted is that of a national education utility (Gooler, 1986; National Information and Education Utilities Corporation, 1990). The idea is to use computer networks to deliver to teachers and students, on demand, various educational tools and materials from a national resource center. One such utility, established by the National Information and Education Utilities Corporation, proposed to provide, under the control of the Taub Education Operating Environment, named for the corporation's conceiver and founder Jack Taub (who also was the founder of The Source, an information utility that offers a variety of information services to subscribers available via their home computers), courseware, databases, communication services (electronic mail and conferencing), and other resources from essentially all publishers, at relatively low cost, through interfaces that are easy to understand and use. The utility is expected to work with a variety of terminals, to provide access to the most current materials, and to protect against piracy of the materials it offers. According to one recent report, there are now over 10,000 packages, from over 700 publishers, available through this system (L. C. Oakley & R. C. Oakley, 1990). Usage is paid for on a metered basis and costs about 50 cents per access hour. Whether or not this particular venture proves to be a success, the idea of an education utility is a promising one. Availability is the key concept—ready availability to every teacher, to every student, to every individual who wishes to have it, of the best educational resources that exist. It is a vision well worth working toward.

Unfortunately, the new educational resources that are becoming, or are likely to become, available via national and regional computer networks will be of little use to many teachers and students in the classroom until more attention is paid to the development of communication infrastructures within schools (Newman, 1987, 1990a, 1990b). Although many schools have local area networks, which could be used to support cooperative learning environments that provide access to a rich assortment of educational resources, they typically are used to centralize management of individual computer-assisted instruction—which often means interaction with drill-and-practice software. Learning how to tap and use effectively the full range of educational resources that are available will be an increasingly demanding and important challenge to educators, as the set of resources that are available continues to expand.

As computer networks become more and more pervasive and provide ever greater access by people to information resources of an increasing variety of types, it seems inevitable that these developments will affect the ways in which people, in school and elsewhere, inform themselves and learn. If the potential that information technology seems to hold to increase manyfold the accessibility of information and to represent it in forms that are conducive to understanding are realized, even to a modest degree, the implications for education, in the broadest sense of the word, could be profound. One possible top-level effect could be that much of what people have traditionally learned as part of their formal schooling will, in the future, be learned outside of school. This is not to suggest that there will no longer be any need for schools—if it were now possible to acquire the equivalent of a high-school education by taking a pill, it would not follow that there would be no further need for schools—it is to suggest, however, that the ways that schools operate and the roles that they play in society, and especially in the education and socialization of the young, may have to be quite different in the future than they have been in the past.

INDUSTRIAL AND MILITARY NEEDS

There is some indication that industry will play a more important role in education and training in the future than it traditionally has played. The M.I.T. Commission on Industrial Productivity (Berger et al., 1989) stressed the importance of an adequately trained workforce for national productivity and noted the need for remedying the inadequacies of education at primary and secondary levels if real improvements in productivity are to be realized. It noted, too, the need for the United States to invest heavily in its future by investing not only in factories and machinery but in research and in human capital: "The most important investment in the long run is in the nation's schools. A better basic education will be crucial to the technological competence that will be required to raise the productivity of U.S. industry. Without major improvements in primary and secondary schooling, no amount of macroeconomic fine tuning or technological innovation will yield a rising standard of living" (p. 47).

The authors of a recent OTA (1990b) report cited worker training as a key to U.S. industrial productivity, competitiveness, and a continuing high standard of living. They worry that American workers are not well trained, as measured by international standards, and they anticipate an increasing need among both manufacturing and service firms for employees with good higher order skills, that is, reasoning and problem solving. According to the same report, nearly half of all investments for capital equipment now go for computers and related technologies. If these technologies are to be used effectively, the authors of the report argued, the workers who are to use them will have to learn some new and different skills. Workers will face new challenges also, they suggested, by virtue of new work organizations that will give more of the responsibility and authority that once belonged to managers to workers at lower echelons in the corporate hierarchy.

Exactly how much is spent on employee training by U.S. corporations is difficult to determine; one can find numbers spanning a considerable range. According to a survey conducted by *Training,* a magazine that focuses on industrial training needs, U.S. organizations with 100 or more employees were budgeted to spend $45.5 billion on formal training—not including on-the-job training—for their employees in 1990 (C. Lee, 1990). If this figure is accurate with respect to the domain covered, it must understate the total expenditures by a considerable amount, because not only does it exclude training costs incurred by organizations with fewer than 100 employees and the costs of on-the-job training, it includes only expenditures of training departments; training activities subsidized by other departments are not reflected. Vaughan and Berryman (1989), have estimated that investments in employer-sponsored training amounted to about $200 billion in the United States in 1985, or about half as much as investments in plant and equipment for that year. According to a recent article in *The Washington Post*, U.S. business is now spending about as much on training and education as is spent on public education at all levels in the United States (M. H. Hamilton, 1988).

Perhaps even the largest of these numbers should not be surprising in view of the fact that the half-life of a contemporary engineering education has been estimated to be about five years (Kaufman, 1978). According to the U.S. Department of Labor (1987), the increased and improved training that the U.S. labor force will need in the immediate future represents a challenge to all training institutions, but "most of the burden for upgrading the current labor force will fall on industry and its training capacity" (p. 45905). It is a burden, furthermore, that industry can ill afford not to carry, at least if one accepts Dentzer's (1989) assumption that because capital is so mobile in today's world, a country's future competitive advantage lies increasingly with a skilled and adaptive work force. The need for vocational training for workers whose jobs have become obsolete or who, for other reasons, lack the skills required to compete effectively in the job market seems likely to grow as the mix of job opportunities changes and the demands of many jobs increase.

Highlights from the *Training* magazine survey results (C. Lee, 1990) are shown in Table 5.1. The survey also listed the topics on which employee-sponsored training was focused. Not surprisingly, new employee orientation led the list. This was followed by performance appraisal, leadership, interpersonal skills, and new-equipment operation. At the very bottom was foreign language instruction, which, as J. Gordon (1990) pointed out, is somewhat surprising considering the emphasis that has been placed on the global

TABLE 5.1   Highlights from "*Training* Magazine's Industry Report, 1990"

Number of individuals to receive formal employer-sponsored training in 1990: 39.5 million.
Number of person-hours of training received: 1.3 billion.
Money budgeted for outside-company expenditures for training equipment and services: $9.2 billion. (Expenditures per employee do not necessarily translate into hours of training per employee, because an hour of training may cost more in some industries than in others.)
Types of organization with highest average training expenditure per employee: transportation, communications, utilities.
Types with lowest average: finance, insurance, banking.
Types of employees likely to get most formal training: salesperson (40.7 hours), professional, first-line supervisor and middle manager (35 hours). Middle managers are most likely of all to get *some* training, but more production workers than middle managers will receive training (because there are so many more production workers).
Percent of organizations using computer-based training: 41.
Percent using interactive video: 15.
Percent providing remedial education in the three Rs: 15.

Notes: Figures refer to U.S. organizations with 100 or more employees.
Adapted from C. Lee, 1990 and J. Gordon, 1990.

economy and the need for American industry to be able to compete effectively in world markets. To the survey question, "In your opinion, what general topic or trend will present the most critical challenge to your organization's training and development function over the next 2 to 5 years?," the most frequent response (16.2%) was "technological change." Second (14.9%) and third (14.4%) most frequent choices were "customer service" and "quality improvement."

Employer-sponsored training does not appear to have been solving the problems of people who enter the work force with an educational or other type of disadvantage, however; as a general rule, such training goes predominantly to employees who are among the better trained to begin with, so the effect is to widen the qualification gap between the better and more poorly trained workers. According to the OTA (1990b), only about $1 billion a year is spent on basic skills training by American industry. College graduates are more likely to receive employer-sponsored training than are high school graduates, who are more likely to receive it than high-school dropouts. Whites are more likely to receive it than non-Whites, and men are more likely than women (Vaughan & Berryman, 1989). Vaughan and Berryman described the positive feedback effect of education and employer-sponsored training the following way: "Well-educated people are the most likely to find employment and to receive training from their employers. Once trained, they earn more, switch jobs less frequently, and are rarely or briefly unemployed. If they change jobs, they find another one more easily and are likely to receive further training from their new employer. Those who enter the labor force lacking sound academic and problem-solving skills fall further and further behind" (p. 2).

The training problem of the armed forces also is severe. The military, in the aggregate, is the largest institution in the United States that recruits young people, and it employs them in a very large assortment of jobs (OTA, 1990b). The Army Technology Base Master Plan (U.S. Department of the Army, 1990) refers to training as the army's "primary peacetime mission" (p. V-4). What holds for the army in this regard is undoubtedly true of the other military services as well. One tends to associate military training primarily with training in the use of weapon systems and various support systems; however, there also is a need within the military for training and education in basic skills. The army established a Basic Skills Education Program in 1977. Ten years later, in 1987, 94,000 soldiers were enrolled in this program. Evaluations of the program have shown it to be valuable both to the individuals involved and to the army as a whole. Army plans call for continuing a skills-education program but with somewhat greater emphasis on job-related skills; with this new emphasis, the program is to be called the Jobs Skills Education Program. Some attention is also being given to the problem of identifying the academic skill requirements of Non-Commissioned

Officers (NCOs) and to help candidates for NCO training meet those needs (Simutis, Ward, Harman, Farr, & Kern, 1988). The military has also sponsored considerable research on the application of technology to education and training, and has been responsible for the development of several computer-based systems for the teaching of equipment maintenance, fault diagnosis, and repair (Psotka et al., 1988).

It seems unlikely that the need for either industrial or military training will decrease in the near future. Apparently, corporations are looking more and more to technology for the means of providing the training that is needed. According to the OTA (1990b), "large companies such as IBM, Ford Motor Co., and Motorola expect that by the late 1990s over half of their corporate training and education will be delivered outside the traditional classroom using some form of instructional technology" (p. 22). Interest in technology-based training is attributed to the flexibility and savings in time and money it is assumed to represent. To the extent that the purpose of some of the training that is needed is to prepare people to use computer-based systems, there is the possibility of providing those systems with the capability of training their users. The idea of embedded training has been around for some time now and has been of interest especially in the context of military systems. What has been accomplished so far in this regard is far from what proponents of embedded training believe to be feasible.

SIMULATION AND TRAINING

For a variety of reasons, not the least of which is the prohibitive cost of alternative approaches, training people to operate complex systems (aircraft, nuclear power plants, space vehicles) or to function in highly dangerous situations (war, outer space, natural disasters) depends increasingly on the use of simulated platforms and situations. A major concern relating to the use of simulation for training stems from the difficulty of determining, in many cases, the simulation's adequacy for training purposes. In what respects must a simulation be realistic, and how realistic must it be, in order to ensure effective training? The answers to these questions presumably differ from situation to situation, but the questions are appropriate in all instances, and in many cases the answers are unknown. Simulation has been used extensively for many years in flight training; but even in this area, where there is such a wealth of experience, there is a range of opinions regarding how realistic simulations must be in order to be effective for training purposes (Caro, 1988).

The veridicality of the simulation of some situations (the conditions of battle in a future war, the effects of a major natural disaster) is, in principle, impossible to establish, so the question of the effectiveness of the training that is done with simulations must remain in some doubt until those who have undergone the training have to perform in the actual situations. One hopes, of course, that performance in some of these actual situations is never necessary, but when that hope is realized, the question of the veridicality of a simulation may become increasingly troublesome over time. Imagine, for example, a period of, say, 50 years of peace, during which time the destructive power and technological sophistication of military weaponry increases manyfold, and the training of military personnel is done entirely with simulated systems and situations; how confident would we be that those personnel were adequately prepared for the situations they could face in an actual war?

P. K. Davis and Blumenthal (1990) have argued that although simulation, models, and war games are essential to the maintenance of an effective military, many of the models that are used extensively for training purposes are built on a "base of sand" and that their inadequacy is not generally recognized within the military or in the Department of Defense. If they are right, this is a cause for concern on two counts: first, because training done with flawed simulations may appear to be effective without, in fact, being so, and second, because the same simulations that are used for training are often used as decision aids in the context of planning and policy making; the riskiness of flawed models here is obvious.

*   *   *

Human factors researchers have paid a great deal of attention to the design of tools and workspaces, and rightly so. The efficiency, safety, and satisfaction of workers depend to no small degree on the usability of their tools and the organization of the space in which their work is done. Strangely, the design of educational tools and learning environments has not received much attention from the human factors community. The problems are analogous in many respects, however, and if human factors research has something to offer in the one domain, as it surely has, it probably has something to offer in the other. Educational tools, like work tools, may be more or less usable and more or less effective; and just as some spaces are more conducive to productivity than others, so some environments are more amenable than others to learning. The design of educational materials, procedures, and environments should be of much greater interest to human factors researchers than they appear to have been in the past.

It is a truism that, other things being equal, people who want to learn learn more effectively than people who do not, but what is it that makes people want to learn? Why will some children spend countless hours becoming experts at very demanding games while being unwilling to put much effort into mastering what they are expected to learn in school? How is it that we have managed to make academic learning such a chore to many youngsters instead of the exciting and enriching experience it could be? Or, to put the question more positively, what can be done to provide students with environments that not only facilitate learning by those who want to learn, but that are intrinsically motivating as well? Do the results of human factors research have anything to say on this question? Could future human factors research addressed to this issue be helpful?

Education involves, among other things, getting information from books—text, figures, diagrams, and pictures. Human factors researchers have given a great deal of attention to the problem of information representation and display. Considerable work has also been done on the interpretability of signs and messages and the intelligibility of instructions and expository writing more generally. Much of what has been learned as a consequence of this research clearly has applicability to the design of textbooks. As far as I know, however, human factors researchers have not focused explicitly on how information is represented in textbooks with a view to helping improve the educational process in this way, nor has there been a lot of communication between human factors researchers and textbook publishers. This is a missed opportunity. Significant improvements in the layout of information that students have to assimilate could be a substantial contribution to education.

We know relatively little about how people conceive of natural phenomena—what they believe is happening, for example, when water freezes, when wood burns, when a flashlight is turned on in the dark. Some recent research on this topic has revealed that people's understanding of such phenomena is often prescientific in fairly systematic ways. Research aimed at further elucidating conceptions that people acquire naturally should be useful in the design of educational programs that will foster more scientifically valid understanding of natural phenomena. This point pertains not only to natural phenomena, but to artifacts as well. In modern industrialized societies, people interact with many machines, appliances, and other devices, the operations of which they understand only superficially. People develop mental models, however, of how these devices operate. In many cases the models, although inaccurate, are functional, at least under normal conditions, but problems arise when the devices malfunction or when someone attempts to use one for a purpose for which it was not intended. In these cases, an inappropriate model can lead to behavior that is unproductive and perhaps unsafe. As machines become ever more sophisticated, and especially as some of them acquire cognitive capabilities of various sorts, it becomes increasingly important to understand the models that people spontaneously develop about their operation, to anticipate the conditions under which these models might lead to problems, and to find ways to help users of complex devices develop more accurate conceptualizations of their operations.

Another important educational concern for the human factors profession is the education of human factors specialists. Human factors is sometimes described as a bridging discipline that spans the chasm between the psychological research laboratory and engineering practice. To the extent that this is an accurate description, the human factors specialist, ideally, should have a fairly broad educational background, with training both in research methods and engineering. One of the themes of this book is that being prepared adequately to address some of the major human factors challenges and opportunities of the future will require a broad perspective and an education that goes considerably beyond what is required to develop expertise in research and engineering technique. The question of what precisely should constitute the education of human factors specialists, if they are to be prepared adequately to respond to the challenges and opportunities of the coming decades, is worth much serious thought and discussion.

Throughout this chapter, I have stressed the practical significance of education and training—the need of the country for a well-educated and well-trained work force, the importance of education and training to individuals who must compete for jobs in tomorrow's workplace. Although I think this is an appropriate emphasis for a book of this type, I do not mean to leave the impression that I believe that the only purpose, or even the primary purpose, of education should be to provide an adequate workforce for the country or to give people marketable skills, important as those objectives clearly are. Education, in my view, has intrinsic worth quite apart from its obvious practical value, because the ability to learn, to reflect, to understand, to wonder has a great deal to do with what it means to be human. It is the business of education, as I see it, to provide people—at all ages, but especially the young—with the opportunity to develop their potential in this regard as fully as possible. As a society, we do not yet know how to do this very well.

It is rather humbling, when one stops to think of it, to realize how ignorant most of us are with respect to most of what is known about the world in which we live. What is truly amazing, however, is how complacent we are with our ignorance. I suspect that we simply assume, although not necessarily explicitly, that the paucity of our knowledge as individuals reflects, for the most part, the limited capacities of our individual minds. As far as I know, the evidence for such an assumption, if there is any at all, is not compelling. An equally plausible assumption is that the fact that we know so little, relative to what there is to know, is a consequence of the limitations of the ways that have been available for centuries for representing information and for informing ourselves. Computer-based display technology, including hypertext and multimedia technology, expands the possibilities for the representation of information—visually and in other modalities—enormously, and for presenting it dynamically and interactively to a learner. The potential of this technology in education has barely begun to be developed, and we are only beginning to see vaguely what it could mean to education and the intellectual life of the individual. It is at least conceivable that we may discover, as this potential begins to be realized more fully, that the learning capacity of the average individual is very much greater than is generally assumed.

6

Transportation

Few aspects of life in the developed world have changed more extensively during this century than transportation. At the beginning of the century, world travel was possible only for the truly venturesome or very wealthy and required large amounts of time. Horse-and-buggy was still the main mode of land travel in the United States, other than foot, and a trip of a few hundred miles was a major undertaking. Now it is possible to go to nearly any major city in the world in a few hours, and the combination of railway systems, automobiles, and highways makes land travel in industrialized countries more convenient than anyone at the turn of the century could have imagined. American consumers now spend about 20% of their income on transportation, which is more than they spend for either food or shelter (Bureau of Labor Statistics, 1987).

AIR TRANSPORTATION

Air travel is a vital part of the national and world economies, and has increased rapidly in importance over the last few decades. The number of passengers on U.S. airlines increased fivefold, for example, between 1960 and 1980. Current airline passenger levels are expected to double by early in the 21st century (Transportation Research Board, 1990a). Ernst (1982) attributed such increases, in part, to the effectiveness and efficiency of electronic reservation systems that were introduced in the early 1960s and soon became adopted universally throughout the industry.

The distribution of passenger service over the airports in the United States is highly skewed: A small number of large metropolitan airports provide almost all the service, and the vast majority of airports provide very little by comparison. There are about 6,000 public-use airports in the country, about 500 of which service scheduled passengers on a regular basis. About 280 of these are designated as primary airports by the Federal Aviation Administration (FAA). Of this number, the 100 busiest serve more than 95% of all airline passengers. The 25 busiest serve almost 67% and the top 10 about 40% (Airport Network Study Panel, 1988).

A study panel of the National Research Council's Transportation Research Board has warned that the nation faces a rising crisis in aviation and projected that the strain on the airport and airways system is likely to increase dramatically in the coming decades (Airport Network Study Panel, 1988). "The FAA has estimated that 11 major airports now experience severe chronic delay in operations as a result of traffic congestion and that the number could reach 29 by 1996 and perhaps 47 by early in the next century" (p. 1). It has been estimated that airline passengers experience 12 million hours of delay annually at Chicago's O'Hare Airport alone (L. R. Johnson, 1990). The study panel attempted to project increases in demand in terms of revenue passenger miles into the foreseeable future, making use of econometric forecasting models, survey data on air travel propensity by age group, and Census Bureau demographic projections. Uncertainties inherent in the projection technique yielded a wide range of possible demand growth patterns, but the study group concluded that the demand was likely to at least double by 2050 and could increase by as much as a factor of 6.

The panel recommended three measures, all of which should be taken to enlarge the capacity of the airport system: (a) optimization of the present system (addition of new runways, introduction of improved traffic handling procedures, improvement in terminal facilities); (b) addition of new airports and, in particular, remote transfer airports (the latter being located outside highly populated areas and serving exclusively as passenger transfer points); and (c) development and introduction of new vehicle technology (large capacity—700- to 1,000-seat—aircraft, reduced take-off and landing and vertical take-off and landing aircraft, advanced engine technology—with reduced noise generation, permitting less intrusive use over densely populated neighborhoods—and high-speed ground transportation systems to supplement or substitute for short-hop air travel).

Accessibility to air travel and freight facilities is expected by some observers to be a primary consideration in business relocations or expansions that could produce employment growth during the near-term future (Kasarda, 1988; Massey, 1988). Kasarda (1988) pointed to Hartsfield International Airport in Atlanta, GA as an example of what good international airline access can mean to the economics of a region: "The virtually certain increased importance of large airports to the future economy of the nation should provide a unique

comparative advantage to major metropolitan areas in capturing the employment growth in the years ahead" (p. 129). An obvious way in which
transportation has affected land use in the recent past is seen in the
proliferation of shopping centers, restaurants, hotels, and other service
industries in the vicinity of interstate highway-access ramps.

International travel for purposes of business and pleasure is a multibillion
dollar industry. Some observers believe that in the future tourism could
become one of the largest industries in the world (H. Kahn et al., 1976). This
is both good news and bad. The thought that many more people will have
the opportunity to travel extensively and to see parts of the world that most
of their predecessors could never have hoped to visit is a positive one. On the
other hand, a great increase in the ease of access to remote places brings with
it the risk that those places will become less interesting to visit. The
attractiveness of many places to the would-be traveler comes in large part
from their uniqueness and their "unspoiled" character, and such properties
can quickly change when tourism becomes the tail that wags the dog.
Advertisements that invite people to avail themselves of excursions to
out-of-the-way, secluded, paradises contain the seeds of their own obsolescence. No secluded paradise that becomes easily accessible to the random
traveler will remain secluded or a paradise for long. A hard question for the
future, and one whose answer is not at all obvious, is that of how to increase
significantly the opportunities for people to travel and at the same time
preserve the integrity and uniqueness of the places to which they would like
to go.

France and Great Britain plan to build a successor to the Concorde, which
has been in transatlantic service for 20 years, that will have a range of 6,500
nautical miles (the Concorde's range is 3,700) and will provide 6-hour service
from Europe to Japan. In the longer term future, they see a hypersonic plane
that will carry passengers from Europe to Japan at Mach 5 in 2 to 3 hours
(French Embassy, 1990). Although it could be possible within a few decades
to travel from any major airport to any other major airport on earth in less
than 2 hours, the more significant technological challenges relating to air
travel in the future have to do with increasing the safety of air travel in spite
of more congested airways and improving transportation to and from
airports.

When most of us think of aviation, we probably think primarily of the
commercial airlines and the fleets of passenger jets they operate. In fact,
general aviation represents a very significant fraction of the total aviation
system, but in the United States, there are more private pilots (300,000 as of
1988) than airline transport pilots (97,000) and other commercial pilots
(143,000) combined (Bureau of the Census, 1990). There are also about
137,000 student pilots. As of 1987, there were a total of 222,000 active civil

aircraft in the country. Only about 5,000 of these were commercial air carriers; the remaining 217,000 were classified as general aviation aircraft.

AVIATION SAFETY

Aviation safety will be a continuing and growing concern as increasing numbers of people travel by air, airways become more crowded, and airline fleets age (OTA, 1988b). The OTA (1989a) has identified human performance, weather, aircraft component failure, and the air traffic environment as the primary causal factors in commercial aviation accidents. Noting that major improvements to air safety can come through solutions to problems of human performance, it has called for greater efforts to understand human error and to enhance the performance of controllers, mechanics, and cockpit crews. Currently, the aviation industry faces a critical shortage of aircraft maintenance personnel, and this at a time when aircraft require significantly more inspection and maintenance operations, in part, because of the aging of the airline fleets (Foushee, 1990).

The fatality rate is many times higher for general aviation than for commercial air carriers providing scheduled services; as of 1988, it was 14.9 and .001 fatalities per 1 million aircraft hours flown, respectively. Fatality rates for commuter air carriers and air taxis fall between these extremes (Bureau of the Census, 1990). The accident rate in general aviation is especially high during instrument approach and landing (Parker, Duffy, & Christensen, 1981).

Operators of general aviation aircraft are a much more heterogeneous group than operators of commercial aircraft with respect to the amount of training they have received, the amount of time they spend flying, the types of aircraft they fly, their age, and in numerous other ways. Moreover, general aviation pilots typically do not fly on a regular basis, usually have total responsibility for control of their aircraft, often fly into completely unfamiliar airports, and must perform their tasks in the context of a system not designed for them (Ritchie, 1988). For these reasons the human factors issues relating to general aviation may be even more challenging than those relating to commercial aviation.

The problems of general aviation are likely to become more pronounced and more visible in the future. If the population continues to increase, as it shows every sign of doing, and the economy remains relatively robust, the number of people who can afford to fly for pleasure is likely to increase as well. The demands of general aviation will have to be met in the context of a fixed amount of air space and of a total system, the capacity of which is already being severely strained. Maintaining the safety of the airways in the face of these growing demands will be an increasingly difficult challenge. To date most of the human factors work relating to general aviation has focused on the design of displays and layout of instrument panels. Attention is also being given now to problems of navigation, communication and the ability of general aviation pilots to deal with emergencies during flight (Ritchie, 1988). Work on these problems can only increase in importance as the numbers of general aviation pilots and aircraft continue to grow.

The Aviation Safety Research Act (Public Law 100–591) enacted by Congress in 1988 calls for an increase by the FAA in its research in the human factors of aviation as well as for coordination of its efforts with those of other agencies involved with aviation, such as NASA and the Department of Defense. As an initial response to this legislation, the FAA (1990) issued a two-volume draft *Plan for Aviation Human Factors,* which set forth a proposed 10-year program intended to alleviate many of the human performance problems that have troubled the aviation system.

According to this report, "human error has been identified as a causal factor in 66% of air carrier accidents, 79% of commuter fatal accidents, and 88% of general aviation fatal accidents" (Vol. I, p. 6). Human error has come to account for an increasingly greater proportion of aviation accidents, as equipment has become ever more reliable (Nagel, 1988). Foushee and his colleagues have pointed out the critical importance of crew performance, as opposed to performance of individual crew members, in assuring aviation safety (Foushee & Helmreich, 1988; Foushee, Lauber, Baetge, & Acomb, 1986; Foushee & Manos, 1981).

Accident analyses have often revealed some failure of communication among crew members as a significant causal factor. R. L. Sears, in an unpublished report cited by Nagel (1988), analyzed 93 airline accidents occurring between 1959 and 1983, and classified their causes (see Table 6.1). The Aviation Safety Reporting System, which collects confidential reports from aviation personnel regarding performance errors, mishaps, and "close calls" in the aviation industry, has recorded 180,000 reports since its establishment in the mid-1970s; approximately 70% of these reports involve some sort of information transfer problem (Foushee, 1990). On the basis of an extensive literature review, Graeber (1988) summarized the effects of fatigue and sleep loss on the performance of aviation flight crews in terms of four major categories of effects: increased reaction times, reduced attention, diminution of memory, and withdrawn mood.

The safety of military aviation is a special issue. C. O. Miller (1988) alludes to a reported loss by the Department of Defense of about 210 aircraft per year from accidents. This rather amazing number raises a variety of questions. One must wonder, for example, how many of the aircraft that are lost during a war are lost by accident. If we manage to lose 210 per year accidently during peacetime activities, the thought of what that number might turn into under

TABLE 6.1 Significant Accident Causes and Percentage Occurrence in 93 Major Accidents (from Nagel, 1988)

Cause of accident	Occurrence (%)
Pilot deviated from basic operational procedures	33
Inadequate crosscheck by second crew member	26
Design faults	13
Maintenance and inspection deficiencies	12
Absence of approach guidance	10
Captain ignored crew inputs	10
Air traffic control failures or errors	9
Improper crew response during abnormal conditions	9
Insufficient or incorrect weather information	8
Runway hazards	7
Air traffic control/crew communication deficiencies	6
Improper decision to land	6

war conditions, in which the opportunities for accidents must be much greater, is very sobering. It makes one wonder, more generally, about what percentage of casualties suffered during war time are really the consequences of preventable accidents.

Military aircraft are likely to depend more on automatic control systems in the future than they have in the past, and they will be able to remain airborne for much longer periods of time. Crews will monitor performance of automated systems and be able to override them in some situations. Crew performance itself will involve interaction with various complex electronic subsystems. Demands on the crews may change from very low to very high over short periods of time; one specific problem of some concern is that of maintaining readiness for high-demand situations over long periods of tedium (Southwest Research Institute, 1982). This is a problem that surfaces in many contexts in which the workload imposed on system operators can shift abruptly from extremely light to very heavy with little warning.

Helicopters have their own set of human factors issues, some of which are the same as those of fixed wing aircraft, but many of which are different (Hart, 1988). The possibility of movement in any direction, the need to fly at very low altitudes for certain types of purposes, the ability to hover and make vertical movements, problems of excessive noise and vibration, as well as large temperature variations in the cockpit and the complex flight dynamics of the vehicle, all contribute to the difficulty of the helicopter pilot's task. The workload imposed on helicopter pilots, especially in the demanding situations often encountered in military operations, is extremely high, and much of the attention of human factors research is addressed to the question of how to keep it within bounds. The army's effort to develop the LHX (Light Helicopter Experimental), which can be operated under all conditions by a single pilot, has required an effort to develop a variety of electronic aids that can relieve the pilot of some of the flight control tasks.

The FAA's Plan for Aviation Human Factors partitions the total problem space of human factors in aviation into five major areas: aircraft flightdeck, air traffic control, aircraft maintenance, airway facilities maintenance, and flight-deck/air traffic control integration. In each area the plan describes several projects, each composed of a number of specified tasks. Issues identified by the plan as needing attention include the problem of drastic changes in workload (from very low under some flight conditions to very high under others), inadequate awareness of flight situations, hesitancy or inability of humans to take over from an automated system, job dissatisfaction from lack of challenge, boredom, skill maintenance, human–machine interface design, personnel selection and training, effects of fatigue and circadian dysrhythmia on flight performance, influence of organizational and management culture on performance, training and assessment of crews, air-to-ground and ground-to-air information transfer, maintenance documentation, and intelligent job-performance aids.

The use of "command displays" that tell the pilot how to control an aircraft, as distinct from "status displays" that provide information regarding the current state of the aircraft, are becoming increasingly important as aircraft become larger, more complex, and slower to respond to control inputs. A variety of innovative approaches for providing situation information to the pilot have been developed. These include head-up displays (the projection of information onto a cockpit's windshield), displays that show the position of the pilot's own aircraft relative to the positions of other aircraft in the vicinity (such as the cockpit display of traffic information, or CDTI), and peripheral-vision displays that are intended to reduce the amount of visual clutter that results from the large assortment of instruments to which a pilot has to attend. These and other innovative display techniques have important advantages, but raise some questions for research as well (Stokes & Wickens, 1988).

Air traffic control is an extremely complicated process involving the coordination of activities of air crews and controllers, communications in many directions (between air and ground, and between various components of the air traffic control system: ground control, approach control, en route control), and the management of multidimensional dynamic information that is changing rapidly in real time. Moreover, the total activity is done under conditions in which the tolerance for errors is very low and the consequences of errors can be catastrophic. The job of an air traffic controller is well known to be a highly stressful one (Crump, 1979; Finkelman & Kirschner, 1981). With increasing air traffic and the concomitant congestion around airports, it seems inevitable that the stressfulness of this work will increase, unless much better electronic aids for the controller are developed.

On the other hand, Hopkin (1988a, 1988b) has pointed out that, in at least some air traffic control facilities, the demands on the controllers fluctuate greatly over time and that the periods of high demand can be separated by periods of relative inactivity and boredom. This problem of workload variation has received attention in several contexts, but it has been somewhat neglected in the air traffic control situation, Hopkin argued, and consequently it is not clear how best to deal with it.

Much of the technology on which air traffic control operations rely is several decades old. Upgrading is essential if the system is to be able to accommodate the expected continuing increase in air traffic. Upgrading plans, as outlined in the FAA's National Airspace System Plan, call for greater reliance on computers and on automated flight monitoring and control techniques. Much of the ground-to-air communication, such as weather data and forecasts, en route advisories, and collision avoidance warnings, will be generated and transmitted by computer.

A major concern associated with the planned modernization of air traffic control operations relates to the question of the adequacy of fallback procedures that can be used in the case of system failures. Until now it has been possible to handle most air traffic situations manually when it was necessary to do so; however, as automated systems become more powerful, traffic patterns will naturally become more complex, simply because of the systems' ability to handle the increased complexity. Then, in the case of system failure, fallback to manual control procedures will become a less viable option.

The role of automation in aviation, in general, and in flight control, in particular, has been controversial. Although the need for some degree of automation is widely recognized, the question of how much is enough is a hotly debated one (Peterson, 1984; E. L. Wiener, 1983, 1985). Some observers see signs that automation is making many jobs in the aviation industry less challenging and therefore less attractive to talented personnel than they used to be (Foushee, 1990). On the other hand, a study of airline crews who were flying an advanced technology aircraft, the Boeing 757, found mixed, but generally positive, attitudes toward flight automation (E. L. Wiener, 1989). Reasons given for the more positive attitudes included "workload reduction potential, the reliability of the systems, the ease of navigation, the advanced displays, and the EICAS [Engine Indicating and Crew Alerting System]" (p. 158). More negative attitudes were attributed to "perceived increases in workload at critical times, the difficulty of programming the CDU [Control Display Unit], and two fears: making a gross error, and loss of manual flying skills" (p. 158).

E. L. Wiener and Currey (1980) have argued that pilots should play a more active role in flying than they do with many modern aircraft, and that the most appropriate function of electronic systems is to provide decision support of various types, especially warnings of trends that could lead to trouble. An example of a high-priority need is for an automated system to provide instantaneous warning to pilots of windsheer problems. A type of windsheer that is particularly troublesome for takeoffs and landings, called a microburst, is produced when some atmospheric event, such as a thunderstorm, creates a rapid downdraft that spreads out horizontally when it strikes the ground. Windsheer has been blamed for 32 plane crashes and near crashes since 1964 (J. W. Wilson, 1986).

The focus of human factors research on problems of aviation goes back essentially to the beginning of human factors as a discipline (E. Edwards, 1988). In commenting on the long association of human factors with aviation, E. L. Wiener (1990) pointed out that work has related to many aspects of aviation, including aircraft design, pilot selection and training, accident prevention and investigation, and air traffic control. A major cause of the acceptance of human factors inputs by the aviation industry, Wiener suggested, was the realization by the early 1970s that a very large percentage of airline accidents were results of human error on the part of airplane crews, air traffic controllers, maintenance staff, or others directly or indirectly responsible for some aspect of flight safety. Wiener predicted that the aviation industry's interest in human factors will be greater in the future than it has been in the past, and that the main impediment to a satisfactory response to this interest will be lack of human factors scientists qualified to do the work.

GROUND TRANSPORTATION

Ground transportation is also a vital component of the economies of the nation and the world. In the United States, the cost of moving goods and people on the ground accounts for more than one sixth of the nation's GNP (Compton & Gjostein, 1986). Trucks account for more than $3 of every $4 spent on freight movement in the country (OTA, 1989a). For every $1 worth of goods exported from the United States, 25¢ is spent on transporting raw materials, semi-finished goods, and components between domestic manufacturers (P. O. Roberts & Fauth, 1988). About half of the world's oil is used to fuel road vehicles (Bleviss & Walzer, 1990). (Surprisingly, according to Bleviss and Walzer, light trucks now account for about half of the fuel used for personal transportation.)

Private vehicles are by far the preferred mode of transportation in the United States. C. L. Gray and Alson (1989) argued that the private automobile has shaped U.S. society more than any other product of the industrial age: "By providing mobility and convenience particularly attuned to the American desire for personal freedom, the automobile has come to dominate not only the nation's transportation network, but also its very culture" (p. 108). As of 1987, there were over 3.8 million miles of roads and highways in the United States. Every city, town, village, and hamlet is accessible by automobile; practically every dwelling and place of business in the nation has immediate access to a motor-vehicle road. Traveling as the driver or passenger in a private vehicle accounts for more than 80% of personal trips; public transportation of all forms accounts for less than 3% (Klinger & Kuzmyak, 1986). The number of rides provided by transit systems in the United States decreased from 17.2 billion in 1950 to a low of about 7.3 billion in 1970; in 1987, the number was at about 8.3 billion (American Public Transit Association, 1987).

In 1980, private conveyance—automobile, truck, van, motorcycle, or bicycle—accounted for 84% of all work trips in urban areas; work trips accounted for somewhat less than one third of local travel mileage. In 1983, U.S. residents over 5 years of age took an average of 974 trips per person, or 2.7 trips per day, not counting long-distance vacation trips. Less than one quarter of these trips were work related. Work trips averaged 8.5 miles; social and recreational trips averaged about the same distance, whereas family and personal errands were somewhat shorter (Lowry, 1988). Adults in the United States with driver's licenses average about 30 miles of local personal travel a day; adults without drivers' licenses average about 10 miles.

These figures represent the highest level of personal mobility in history and the highest level in the world today (Klinger & Kuzmyak, 1986; Reno, 1988). Travel by personal vehicle is expected to grow between 1.3% and 1.7% per year between the 1990s and the 2020s, with a somewhat higher rate during the immediate future and a lower one later in the period (Transportation Research Board, 1988b). The annual growth rate is expected to be considerably higher for large truck travel than for other vehicle types (Transportation Research Board, 1990b). Total vehicle travel has been projected to increase by approximately 50% between 1987 and 2005 (U.S. Congress, 1989).

The popularity of personal vehicle travel is no mystery: It has been estimated that an automobile user can reach more than 30 times as many jobs, stores, or friends in an "acceptable" period of time as could people of earlier generations who had to walk or depend on animal transport (Sobey, 1988). Of course, general dependence of a society on private transportation puts people who do not have such transportation at a disadvantage. A growing problem for inner-city residents, especially inner-city poor, is the inaccessibility of jobs in the suburbs because of the lack of public transportation and the prohibitive cost of owning and operating an automobile in the inner city (Kasarda, 1988).

As of 1988, there were an estimated 140 million cars, or about 1 car for every 1.75 people, in use in the United States, and an estimated 164 million licensed drivers (Bureau of the Census, 1990); the comparable figures for

1970 were 80 million and 108 million. Approximately 85% of all adults have drivers' licenses and about 96% of those between the ages of 25 and 35 do (Kostyniuk & Kitamura, 1987). The percent of people over 65 who have a driver's license went from 43% in 1969 to 62% in 1983 (Transportation Research Board, 1988a). It has been predicted that by 2000 there will be a private vehicle for every licensed driver in the United States (Lowry, 1988). Ervine and Chen (1988–1989), estimated that by 2005 there will be 50% more cars on American roads than there were in 1988, and that this will cause about a fivefold increase in congestion, as measured in vehicle hours of delay.

Worldwide there are about 500 million automobiles, or about 1 for every 10 people, which is approximately the ratio that prevailed in the United States in 1920 (Compton & Gjostein, 1986). Differences among countries in their automobile-to-people ratios are very large. Chandler et al. (1990) made the contrast in particularly striking terms: The United States, they noted, had enough cars for every American to be in an automobile at the same time with no one sitting in the back seat, whereas the (then) Soviet Union had 1 car for every 20 persons, and China had 1 for every 1,000. The automobile population worldwide is currently increasing faster than the human population and if current rates of increase continue, the number of automobiles will quadruple by 2025 (Keyfitz, 1989).

Given these figures, it is not surprising that the problem of traffic congestion is expected to worsen over the next few decades. Currently, the number of cars and miles driven in the United States is increasing much more rapidly than the capacity of the nation's highways. In California, the number of vehicle miles driven on the highways increased by almost 140% over the two decades ending in 1986, during which time the highways themselves expanded by less than a third (Wolkomir, 1986). Anticipated continuing growth in the size of the driver population and in the annual miles per driver will mean that more miles of highways will operate at poor levels of service during peak periods and that higher proportions of urban travel will occur under congested conditions (Bay, 1988; Reno, 1988). One estimate from the Federal Highway Administration (1987) has traffic delays nearly quadrupling, from about 2 billion vehicle hours to almost 8 billion vehicle hours between 1985 and 2000.

Everyone seems to agree that something must be done about congestion, but there is no strong consensus as to what. One response to this problem, especially in major cities, has been the development of computer-based traffic control systems that use information gathered from sensors embedded in roadways to regulate the traffic flow. This can help in the short term, but only postpones the inevitable grand clog. The answer to the problem must be found, at least in part, in a much greater use of mass transit systems. How to make such systems acceptable to, if not preferred by, the American public is a major question for the future, and it is one that has a variety of human factors ramifications.

Whatever happens with mass transit systems, though, the private automobile is likely to remain a major means of transportation for the foreseeable future, and much effort is being devoted to finding ways to improve automotive design and performance. Many of the expected innovations involve the application of information technology to this domain. Many of the components and subsystems of motor vehicles already contain computers to help increase the efficiency and effectiveness of their operation, and this application of computer technology will surely increase in the future. Unlike these uses of this technology, which are relatively invisible to the driver, others will be very much in evidence.

Automobiles and other personal vehicles will be equipped, for example, with such driver aids as automatic navigation and vehicle location systems, as well as more general information systems (Koltnow, 1988). One can imagine being able to call up on a visual display a map of a particular area showing, on request, point-to-point routes, the locations of specific spots of interest, and information regarding available accommodations. An electronic map could also show the vehicle's present location, and it could give information about accidents or other causes of congestion on one's intended route and suggest alternative routes. Because map reading is not an easy task for many drivers (Streeter & Vitello, 1986), the effectiveness of such systems might be enhanced if the maps were supplemented with verbal instructions regarding how to get from one specified point to another. One indication of the potential usefulness of better driver information systems is seen in the finding that the first trip to a new destination is likely to be 40% longer than necessary (Sobey, 1988). As Sobey pointed out, however, introduction of some types of driver aids are likely to lag their technological feasibility, because of liability concerns of manufacturers.

An experiment, called Pathfinder, is currently being conducted on the Santa Monica Freeway in Los Angeles in which motorists with specially equipped cars receive information regarding traffic delays ahead on a dashboard-mounted display and by computer-synthesized voice. Experiments with sophisticated navigation systems are being conducted in other countries as well. Among innovations that are now technically feasible and could soon be affordable, in addition to route guidance systems, are vision-enhancement systems that can look through fog with infrared sensors and show what is ahead on a windshield display similar to those used in some cockpits, and adaptive cruise-control systems that automatically cut back on the gas when a car is closing too rapidly on a leading vehicle (Sheldrick, 1990).

An ambitious ground transportation development effort currently going on in Europe is referred to as PROMETHEUS (Program for European Traffic with Highest Efficiency and Unprecedented Safety). A six-country collaboration involving 20 automotive manufacturers and 70 research institutes, this project's goals include reducing collisions by 50% by the year 2000 and increasing the efficiency of highway transportation through enhanced communications and traffic control schemes. Japan has similar efforts underway to apply computer and communication technology to the development of advanced ground transportation systems (Ervine & Chen, 1988–1989).

The development of automatic highway systems and automatically controlled vehicles is a focus of research by transportation planners as another means of relieving congestion and improving highway safety. This kind of technology, sometimes discussed under the rubric intelligent vehicle/highway systems (IVHS), could also help to improve fuel efficiency and to reduce automotive emissions. The concept includes a variety of computer-based capabilities for increasing the efficiency and safety of highway use. In addition to providing a driver with information on traffic patterns, congested areas, accidents, and highway accommodations, an IVHS might also include automatic vehicle control capabilities for optimizing the flow of traffic through particularly congested areas. Several efforts to develop such systems are underway in Europe, and the feasibility of establishing IVHS research in the United States is being investigated by the General Accounting Office and the National Academy of Sciences (Wright, 1990a).

Fully automatic highways are not expected to appear until the second quarter of the 21st century, but some aspects of such systems are likely to be introduced gradually over the next few decades (Koltnow, 1988). An electronic license plate, for example, is a small innovation that is probably feasible in the near future. It could help reduce traffic delays in some areas by making it unnecessary to stop at toll booths; such plates could also be the basis for automatic charging of gasoline purchases and parking lot usage and could be effective in helping to locate stolen vehicles (Sobey, 1988). Eventually all major vehicle subsystems (engine, transmission, suspension, steering) could be under electronic control. An integration of stand-alone control systems into a single unified system is a goal toward which designers are working (Compton & Gjostein, 1986).

The dream of a car that can fly, or, more accurately, a vehicle that can operate effectively either in the air or on the ground, is an old one—a car with wings, the Curtiss Autoplane, was exhibited at the Pan American Aeronautical Exposition in 1917—but it has proved impractical to date, in spite of the building and operation of numerous prototype aerocars. The dream, however, persists to this day (Childes, 1989).

HIGHWAY SAFETY

The annual number of fatalities in the United States from automobile accidents has varied around 50,000, plus or minus 5,000, for several decades, but it has been at the lower end of that range since 1982. When number of fatalities is normalized to the number of registered vehicles, the number of licensed drivers or the number of vehicle miles, a very considerable decrease in the death rate (nearly 50%) is seen over the period from 1970 to 1985 (Bureau of the Census, 1987). Contributing factors include lower speed limits, increased use of seat belts, energy-absorbing steering columns, antilacerative laminated windshields, antilocking brakes, improved vehicle lighting systems, infant and child restraint systems, padded dashboards, nonprojecting interior door handles, head restraints, forgiving highway barrier systems, break-away sign posts, and reflective signs and pavement markings. The effects of these improvements, along with other safety measures—such as efforts to decrease the incidents of driving while intoxicated—have been a reduction by about 50% of the number of deaths per 100 million vehicle miles traveled since the mid-1960s (Transportation Research Board, 1990b). An analysis of traffic accident and traffic fatality data for California for the entire period 1939 to 1973 showed that the frequency of accidents involving injuries stayed relatively constant over that time, whereas the probability that an injury-producing accident would involve a fatality decreased by about 50% (Stanislaw, 1987).

Kahane (1989) has estimated that the center high-mounted rear brakelights mandated by the National Highway Traffic Safety Administration on all new cars as of September, 1985, cost consumers about $1 million a year. As a consequence of a 17% reduction in rear-end collisions, however, they are saving about $910 million annually in property damages. The mandating of this safety measure resulted from about 20 years of research on rear lighting systems that cost the National Highway Traffic Safety Administration less than $5 million.

Still, as Table 13.3 (p. 357) shows, highway mishaps are not only the most frequent cause of accidental death in the United States, they equal, more or less, the number of deaths from all other accidental causes combined. In spite of the decrease in the rate of fatalities, this class of accidents has remained at the head of the list with a prevalence more than four times that of falls, the next most frequent cause of accidental death. It is a sobering thought, indeed, to realize that more U.S. citizens have been killed on U.S. highways than in *all* the wars in which the country has been involved combined. Most communities also experience greater losses from motor vehicle accidents than from crime (Streff & Molnar, 1990).

The vast majority of fatal motor vehicle accidents involve road vehicles; trains and airplanes together probably account for no more than about 2% of motor vehicle deaths. About 15% of motor vehicle fatalities are pedestrians, and over two thirds of the pedestrian fatalities occur at night. About 8% of traffic fatalities are motorcycle riders (National Highway Traffic Safety Administration, 1989). After the first few months of life, accidents are the major cause of death of children in the United States, and about 15% of children who die from accidents do so as passengers in motor vehicles. Older people are more vulnerable when involved in an automobile accident than are younger people; in fatal crashes involving both an over-65-year-old driver and a driver 20-years-old or younger, the older driver is four times more likely than the younger one to be the fatality (Partyka, 1983).

In addition to lives lost, motor vehicle accidents cause nearly 4 million injuries per year. The total annual cost to the United States in lost productivity, medical expenses, and property damage resulting from such accidents has been estimated at $70 billion, which is comparable to the costs of losses from cancer and heart disease (Transportation Research Board, 1990b). Federal and state governments combined spend about $70 million annually, or about .1% of the total cost of motor vehicle accidents (a little less than 30 cents per person) on research on motor vehicle safety.

Human error has been identified as the single major causal factor in motor vehicle accidents. Over 60% of vehicle accidents are attributable to this cause (Perchonok, 1972). Driving at a speed too fast for conditions and inadequate driver training are two of the most common factors associated with heavy-vehicle accidents; a third factor is vehicle age. In 1986, 52% of all fatal automobile accidents involved a driver or a pedestrian who had been drinking; the same year, teenagers accounted for 7% of licensed drivers and were involved in 17% of fatal automobile accidents (Lescohier, Gallagher, & Guyer, 1990).

Although driver error is the major contributing factor to accidents in all age groups, the kinds of errors that are most commonly made seem to differ for different groups (Allgier, 1965). At least when a fatality is involved, drivers over 65 are much more likely than young drivers to have failed to yield the right of way, and they are much less likely to have been speeding or driving while drinking. For the same number of miles driven, drivers over the age of 75 are about twice as likely as middle-aged drivers to be involved in a crash. By 2020 there will be about 22 million people 75 years of age or older still eligible for a driver's license (Transportation Research Board, 1988a).

Night driving is a special problem. Many drivers drive faster at night than is appropriate for the visibility provided by their headlights, so that unexpected obstacles in the road do not become visible in time for them to take appropriate evasive action (Shinar, 1978). One of the most common impairments associated with aging is declining vision; decreased static and dynamic visual acuity (especially under low illumination), and contraction of the visual field are commonly reported. Perhaps because of the gradualness of the loss, many people may be unaware of the fact that it is occurring.

Leibowitz (1988; Leibowitz & Owens, 1986) has suggested that vision plays a dual role in driving: It permits one to guide one's vehicle and also to recognize objects in the road that need to be avoided. The latter function is progressively impaired as darkness increases, whereas the former is not.

Because drivers' ability to steer their vehicles is unaffected by darkness, they are unaware of the impairment in recognition vision and do not decrease their driving speed to compensate for it. Leibowitz refers to the unjustified confidence drivers have in their ability to drive as well at night as in the daytime as one of a class of "silent killers," because it poses a threat to safety of which people are unaware.

As part of the Surface Transportation Assistance Act of 1987 (Transportation Research Board, 1988c), the U.S. Congress asked for a study of "(1) problems which may inhibit the safety and mobility of older drivers using the nation's roads and (2) means of addressing these problems" (p. 2). In response to this request, the Transportation Research Board (1988a), issued a report on the mobility and safety needs of older drivers, passengers, and pedestrians. One of the conclusions of this report was that "the assumptions about human performance used in signing, roadway marking, traffic control, and highway design are becoming increasingly inappropriate for an aging population of drivers" (p. 4). Legibility standards, for example, which assume visual acuity of roughly 20/25 are not adequate for a large percentage of older drivers. The report points out that the Manual on Uniform Traffic Control Devices suggests that traffic signalling systems at crosswalks be designed to accommodate a walking speed of 4 feet per second and many older persons walk more slowly than that.

The committee made numerous recommendations regarding how to improve transportation for the elderly. The recommendations concerned the design of highway and street signs (including greater use of advance-notice signs), roadway delineation markings, traffic signals (in particular, the timing of them at pedestrian crosswalks), the design of roadways (especially at intersections), seatbelts, consumer information, illumination, crash worthiness, licensing, vision screening, and cognitive performance as it relates to driving. The report also notes the need to improve alternatives to the automobile for transportation for the elderly (Transportation Research Board, 1988a).

People engage in risky behavior on the highways: They drive too fast, follow leading vehicles too closely, and drive under the influence of alcohol or drugs. Such behavior often has dire consequences. (According to L. Evans, Frick, & Schwing, 1990, the chance of being killed in an interstate highway accident is 1000 times greater for a young unbelted, intoxicated driver in a small vehicle, than for a middle-aged belted, alcohol-free driver in a heavier car.) A question of some practical significance is whether people typically underestimate the riskiness of their behavior or overestimate their ability to remain in control of the situation, or whether they assess the risks accurately and simply choose to take them. It is at least a plausible hypothesis that people often underestimate the risks that they are taking. There is suggestive evidence that drivers *do* underestimate the risks represented by substandard performance of certain vehicle components (D. G. MacGregor & Slovic, 1989).

Modern vehicle suspension systems, intended to give automobile occupants a smooth ride, plus power steering, power brakes, cruise control, and other features that simplify driving and enhance riding comfort, may also tend to detach drivers from an immediate appreciation of the physics of a situation. They may feel as though they are not going as fast as they really are and that they have the ability to stop in less time than they actually can. To the extent that this hypothesis is valid, the challenge is to provide drivers with the kind of information and feedback that will give them a more accurate appreciation of the risks they are taking in specific situations. The importance of this challenge is likely to increase as vehicles become more sophisticated and the "feel of the road" that they provide becomes even less pronounced. A particularly interesting aspect of the challenge relates to the question of how to increase the chances that an intellectual appreciation of risk will translate into appropriate risk-reducing behavior; one may realize full well that one is safer as an airline passenger than as an automobile driver, but feel safer in the automobile all the same. It is now feasible to equip cars with the ability to provide drivers with specific information regarding the risks they are taking in certain situations. How this provision will affect driver behavior is an open question.

Many of the advanced technological devices and systems that are being developed for use in vehicles and transportation systems are intended to improve highway safety. Whether they will do so remains to be seen. The Transportation Research Board (1990b) has warned that "If their design and development proceed without adequate attention to driver capabilities and limitations, the full safety potential of these systems may not be realized; poorly designed systems could actually degrade safety" (p. 78). The board urged that a broad program of human factors research be initiated soon so that its findings can influence the design of these new technologies.

More generally, the Transportation Research Board clearly stated the need for an expanded program of human factors research addressing highway safety problems that are likely to be even more severe in the future than they are now. It identified six areas deserving of research in the interest of mitigating problems anticipated as a consequence of increased traffic on the highways (i.e., more older drivers and pedestrians, more large-truck traffic, a greater range of vehicle sizes and weights, and an aging highway infrastructure). These areas are crash avoidance, occupant protection, highway safety design and operation, post-crash acute care and rehabilitation, management of highway safety, and driver information and vehicle control technologies (Transportation Research Board, 1990b). Highway safety is not a new area of concern for human factors researchers; it is an area, however, in which the need and opportunities for human factors research are likely to be even greater in the future than they have been in the past.

RAIL TRANSPORTATION

The idea of levitating and propelling trains by magnetic force has been around for some time. Prototype high-speed magnetic-levitation *(maglev)* systems (able to travel at 200+ mph) have been developed in West Germany and Japan, and the building of a high-speed line between Los Angeles and Las Vegas is planned for the mid-1990s (Murphy, 1989). Both Germany and Japan plan to have revenue-generating maglev systems in operation before the end of the 20th century (L. R. Johnson, 1990). Johnson contends that the development of maglev technology would be good for the environment, because these systems would emit smaller quantities of air pollutants, such as hydrocarbons, carbon monoxide, nitrogen oxide, and particulates, per passenger mile than more conventional forms of transportation. Noting that half of all airline flights are less than 500 miles, Johnson argues also that maglev train technology should be developed so as to provide a cost-effective and convenient alternative to air travel for trips of a few hundred miles, and, in particular, to serve as feeders into existing hub-and-spoke airline networks.

Wheel–rail trains going 190 mph are already in daily service in France; 200 mph lines are being built to connect Brussels to London and Paris, and a 225 mph track is expected to run from southwest Germany to Paris. Successful testing has been done with wheel–rail systems at 320 miles per hour (Duchemin, 1990b). Duchemin claims that, at 190 mph, the most recent high-speed trains can achieve a per unit energy consumption that is equivalent to about 235 miles per U.S. gallon per seat available, which is about half that of a private car at much lower speeds and one third that of a modern aircraft, such as an Air Bus 310 or Boeing 767.

Duchemin (1990a) argues also that, although less glamorous than magnetic levitation transport technology, wheel–rail systems may represent a more realistic and practical alternative for the development of high-speed transportation, especially in the United States. Because air drag is the most significant force to be overcome by any system traveling at speeds above 150 mph, all high-speed ground transportation passenger vehicles must be similar in shape and will require identical amounts of energy for propulsion at high speeds. Also, the noise problem is similar for maglev and rolling stock technology, because most of the noise that is generated during high-speed movement comes from air turbulence.

In the United States, rail transport has been on the decline for some time. Whether measured in terms of the numbers of companies in operation, miles of track operated, locomotives or cars in service, or passenger miles traveled, the trend has been down for several decades. Perhaps the fact that the number of passenger miles traveled by rail is small relative to the number traveled by air or by car accounts for the little attention that rail transportation has received from human factors researchers relative to that received by air and motor vehicle transportation. (Non-rail ground transportation accounts for about 10 times as many passenger miles as does air and about 100 times as many as does rail.) Rail transportation could become an increasingly attractive alternative in the future, however, if the quality of travel time with other modes declines as a consequence of congestion and if rail service is improved. Of special interest is the possibility of high-speed rail transport between cities separated by not more than a few hundred miles, and greatly enhanced intracommunity subway facilities. Ensuring the safety, convenience, and utility of rail systems and services will require attention to numerous human factors issues.

Even today, railway safety deserves much more attention than it gets. Although, as already noted, U.S. railways deliver only about one tenth as many passenger miles as do U.S. commercial air carriers, the annual number of railway fatalities is roughly 10 times as great as the number of fatalities in commercial airline accidents. This is not a completely fair comparison, of course, because many of the railway fatalities have nothing to do with passenger services (e.g., collisions of freight trains and automotive vehicles at railroad crossings), but nevertheless, people die unnecessarily every year as a consequence of avoidable accidents involving railroad operations. Collisions between trains continue to occur, with very significant tolls in human life and injury. Investigations of railway accidents have typically identified human error as at least a contributing factor, and sometimes the results of such investigations have revealed situations that make one wonder why accidents are as infrequent as they are. An account of one noteworthy and thoroughly investigated incident will make the point.

On the 12th of December, 1988, a collision involving three trains near Clapham Junction Railway Station between Waterloo and Wimbledon, England, claimed the lives of 33 people and seriously injured 69 others. The immediate cause of the accident was the failure of a signal light to show red to indicate the presence of a train on a track a short distance ahead of an approaching train. The signal failed because alterations in the wiring of the signaling system shortly before the accident had been done improperly. In particular, a wire that was supposed to be disabled was disconnected at one terminus but not at the other, and the disconnected end was not shortened, insulated, or otherwise altered so as to make inadvertent reconnection at the disconnected terminus impossible.

A thorough investigation by the British Department of Transport of the accident and the circumstances leading to it resulted in the identification of some 16 errors that were considered to be implicated in some way. These included the following: "working practices were permitted to slip to unacceptable and dangerous standards, . . . the quality of supervision was permitted to slip to an equivalent degree, . . . the quality of testing did not meet standards, . . . there was no proper system of training of installation and testing staff, . . . [and] there was no effective system of communicating to the workforce the proper standard required of installation and testing work" (Secretary of State for Transport, 1989, p. 165). It was determined also that the technician's work was not inspected by his supervisor, as it was supposed to be, nor was a test conducted of the altered circuit.

I cite this case because the investigation of it was so thorough and because the results of that investigation provide some insight into how multidimensional a problem situation can be. The findings should disabuse us of our tendency to assume that accidents always have simple explanations and one-factor causes. It is at least a plausible hypothesis that often the immediate or most obvious cause of an incident (in this case, the failure of a signal light to show red when it should have) is itself the effect of a system that is functioning below par in a variety of ways.

TRANSPORTATION AND ENERGY

One fourth of all energy (and almost two thirds of all oil) consumed in the United States is used for transportation. In 1987, the United States used more petroleum for transportation alone than it produced (Transportation Research Board, 1988b). More oil is used for transportation in this country than is used by all developing countries together for all purposes (Greene et al., 1988). These are compelling reasons for trying to develop more energy-efficient transportation systems and to begin making the transition to alternative fuels.

As measured in miles per gallon, the energy efficiency of automobiles has about doubled in the United States since the first OPEC oil price increases in 1973. This has been achieved with such innovations as "electronic engine controls, widespread use of frontwheel drive, fuel injection, lock-up torque conversion on automatic transmissions, friction reduction, improved aerodynamic designs, and other technologies, in addition to reduced weight and engine size" (Greene et al., 1988, p. 219). Thanks, in part, to the use of advanced high-strength lower-alloy steels in place of carbon steels and cast iron, the weight of the average automobile decreased 15% between 1974 and 1986, and was expected to decrease another 27% by 1992 (J. P. Clark & Flemings, 1986). Innovative vehicles, including half-width automobiles, could also play a significant role in changing the characteristics of personal transportation in the future. A half-width automobile developed by General

Motors is believed to be capable of 120 miles per gallon at 70 mph (Sobey, 1988).

Greene et al. (1988) anticipate a full-scale transition to alternative fuels for transportation being in progress by 2020, with methanol and natural gas as the primary alternatives; they also expect electric vehicles to be common by then. The idea of electrically powered vehicles is not new. At the beginning of the 20th century, 40% of the automobiles in the United States were steam-powered, 38% were electric, and 22% used gasoline (Wright, 1990a), but the ratios shifted dramatically as improvements in the gasoline engine increased the convenience and range of gasoline-powered cars. Of course, electric batteries have to be charged periodically. If that is done by means of power from a fossil-fuel utility plant, it is not clear that much will have been gained either in terms of energy conservation or benefit to the environment.

Efficiency improvements are also being sought in air transportation. Assuming research can solve the problems of brittleness and high processing costs, the use of ceramics in jet engines could increase their maximum operating temperature nearly 50%, thus improving fuel efficiency and permitting the use of lower quality fuels (Clark & Flemings, 1986). The use of ceramics could also reduce engine weight.

As noted in chapter 3, there has been some speculation regarding the possibility that future transportation needs might be offset to some degree by the substitution of the transmission of information for the transportation of people and material. Inasmuch as the transmission of information consumes far less energy than the movement of people and material, whenever one can substitute the former acceptably for the latter, it is in the interest of energy conservation—as well as environmental protection—to do so. Many forecasts regarding transportation, including some of those here, do not seem to reflect an expectation of much relaxation on the demands of transportation systems occurring in this way. In some cases, this may be an oversight; in others it may be a considered omission. Not everyone who has thought about it expects a decreased demand for transportation to be a consequence of better means of transmitting information. Herman et al. (1989), for example, argue that "demands for communication and transportation appear to increase in tandem as complementary goods rather than as substitutes for one another" (p. 51).

Although technology clearly has the potential to decrease the need for personal travel in a variety of ways, the past makes it equally clear that it also has the capability of increasing the opportunity for travel and thus the demand. On the average, people travel much more frequently and much greater distances today than in the past, largely because technology, in the form of automobiles, airplanes, and other powered vehicles, has made it possible to do so. Moreover, as transportation systems have become more accessible, people have used them more. Whether this trend will continue if technology manages to offer truly attractive substitutes for travel remains to be seen. The technologies that could make such substitution effective on a large scale are still young and will need considerable further development, including much attention to the human factors aspects of their user interfaces before they can be reasonably expected to be used enough to have a significant impact on transportation needs.

* * *

Human factors research relates to transportation in numerous ways. Work on highway, aviation, and railway safety is perhaps the most obvious connection. Design and evaluation of safety restraint systems, individual displays and indicators, dashboard and control panel layouts, and braking systems relate to safety as well as to other aspects of driving or piloting performance. Transportation safety will become an even greater challenge as pressures mount to make vehicles lighter and smaller to accomplish greater fuel efficiency and material conservation.

A continuing and serious problem is that of evaluating the effects on safety and comfort of specific changes that are made in transportation systems. The Transportation Research Board (1990b) described the problem this way:

> Despite nearly half a century of building highways to modern standards, relatively little is known about the safety consequences of design improvements (i.e., to what extent these improvements reduce crashes) or of traffic engineering decisions, such as the installation of traffic control devices and the location of signs. . . . The information that is available in policy design guidelines and traffic codes rarely provides adequate documentation and quantification of the associated costs and benefits of specific improvements or packages of improvements so that highway designers and traffic engineers can readily analyze alternatives. (p. 60)

Perhaps the most frustrating aspect of the excessively high rate of injury and death from automotive accidents is the realization that a large percentage of such accidents are due to human carelessness and, in principle, are avoidable. The question of how to motivate and effect safer driving behavior deserves a great deal of attention from human factors researchers, as well as from anyone else who is repulsed by the idea that 40,000 to 50,000 deaths and many more serious injuries are part of the annual toll Americans are willing to pay for the convenience of our ground transportation system. (How even to get people to use seat belts voluntarily remains an unanswered research question.) Just to keep the toll from increasing significantly over the next few years represents a considerable challenge. If the annual growth rate of travel slows from the 3.5% it averaged in the 1970s and 1980s, to between

2% and 3% over the following decade and a half, as it has been projected to do, keeping the annual number of fatalities from vehicle accidents below 50,000 by the year 2005 will require a reduction in fatalities from the 1988 rate of 2.3 deaths per 100 million vehicle miles traveled to a 2005 rate of between 1.45 and 1.75 deaths per 100 million vehicle miles traveled (Transportation Research Board, 1990b).

A great deal of work has been done on the modeling of human performance in complex control situations, such as the driving of automobiles or the piloting of aircraft (Baron, 1988; Pew, Baron, Feehrer, & Miller, 1977; Rouse, 1980). A specific challenge for the future is to develop models that incorporate both control-theoretic representations of manual performance and cognitive-process models that represent operators' perception and understanding of the situations in which they function. The development of optimal control models, with their emphasis on human information-processing characteristics, perceptual variables, and display requirements, is a step in this direction (Baron & Levison, 1980; Kleinman, Baron, & Levison, 1971).

Operators of complex vehicles like jet aircraft, have, for some time, been finding themselves interacting with the machine more and more through computerized control subsystems. The human's role in the operation of these vehicles is sometimes described as one of "supervisory control" (Sheridan, 1987, 1988), because rather than controlling the machine directly, what one does is supervise the operation of one or more automated control devices. The building of models designed explicitly to accommodate the evolution of the pilot's and crew's task into one of supervisory control is another important trend in modeling in the context of human performance in aviation (Baron, Muralidharan, Lancraft, & Zacharias, 1980). Development of effective approaches for maintaining skill levels of people in supervisory control jobs and for coping with the abrupt and unpredictable changes in workload that are sometimes encountered in these jobs is a need of increasing importance.

Vehicle handling quality and rider comfort will be other foci of attention. Increasingly sophisticated driver information systems and displays will benefit from design inputs from human factors engineering. The question of how to present information regarding vehicle state, traffic conditions, route options, and other facts of interest, without distracting attention from the driving task will require research. What can be done to make public transportation an attractive alternative to the use of private vehicles, especially in urban areas, is a human factors question, at least in part. The problem of making public transportation facilities accessible to people with disabilities has not been completely solved. The design of signs and symbols for highways, air terminals, train and bus stations, and other travelers' way stations will provide continuing opportunities for human factors research, as will the design and layout of travelers' accomodations more generally. Considerable effort has already been made to develop a set of iconic signs whose interpretation would be intuitively clear and relatively culture free, so that the same set could serve around the world.

Other opportunities for human factors research include continuing work on the development of more effective procedures for safely controlling the increasing flow of air traffic, both commercial and general; improvement in the procedures used to detect incipient problems in aircraft; continued study of human error in the context of flying and other aviation-related jobs; improvement in air crew communication and message reception verification procedures; improvement in situation monitoring and representation techniques for the operators of aircraft and other high-performance vehicles; continuing studies of driver performance and especially of the variables that contribute to unsafe driving; work on the development of techniques for giving motorists an accurate sense of the risks associated with specific driving situations and patterns of driving behavior; improvement of rail transport services in terms of passenger safety, convenience, and comfort; and enhancement—from the users' perspective—of telecommunication systems that, in some instances, could reduce the need for the transportation of people and material goods. In thinking about future research needs and opportunities, we should not lose sight of the fact that much of the basic research that has been, and is being, done on perception, cognition, and motor skills is directly relevant to pilot and driver performance, and safety more generally (Leibowitz, 1988; Wickens & Flach, 1988). The challenge with respect to this research is to get the results of it appropriately applied.

Multifaceted transportation systems are essential to modern life; without the ability to move goods and people from place to place quickly and in a variety of ways, civilization as we know it could not exist. The transportation capabilities we enjoy, however, come at a considerable cost, some portion of which will be paid by future generations. The three most apparent components of this cost are the drain on fossil-fuel resources, the environmental effects of vehicle emissions, and the toll of death, injury, and property damage on the highways. The challenge, as we look ahead, is to enhance our ability to transport goods and people quickly, safely, and comfortably, and at the same time to reduce the cost of doing so, especially those components of the cost that mortgage the future. There are many opportunities for human factors research to work toward that objective.

7

Space Exploration

Space exploration does not address an immediate and pressing national or international need in the same sense in which programs to develop new energy sources, to reverse environmentally destructive trends, or to improve education would do so, but space exploration is motivated, at least in part, by deeply rooted human curiosity and the need to explore the unknown. There are, of course, many frontiers today, metaphorically speaking, but unlike most others, space permits continued realization of the age-old quest of visiting, studying, and perhaps settling, new territories in a literal sense. Although the popularity of a space program may wax and wane over time among policymakers for a variety of reasons, a continuing effort to learn more about space and to extend the domain to which humankind has physical access is, in my view, inevitable.

For human factors researchers, space exploration represents a frontier also in the sense of giving rise to questions about human performance and well-being in environments and situations unlike those encountered on earth. Some of these questions can be answered by appropriate research in ground-based laboratories; some will require observations and experimentation in space. Some of the questions that will turn out to be most significant will probably not even be asked until they are prompted by situations that arise in space missions. Space exploration will, I believe, provide many opportunities, some of them unique, for human factors research for a long time to come. This is not to suggest that this work will, or should, be done in isolation from work addressed to problems on earth, because the results of much of the work that is done in direct support of the space program has applications in other contexts, and conversely.

PLANS AND PROSPECTS

If the recommendations of the National Commission on Space (1986) are implemented on schedule, the space station will be in operation before the end of this century; by 2020, space ports will be orbiting the earth and the moon, manufacturing operations will be ongoing on the moon, and a human outpost will have been established on Mars. Still, there are skeptics regarding the country's commitment to space exploration and its ability to meet an aggressive time schedule (Beardsley, 1987a, 1990b), and Congress' funding of the Space Station Program has been hesitant so far (Marshall & Crawford, 1988).

The National Commission's report identified as the main thrusts of the space program: advancing understanding of the planet, solar system, and universe; exploring, prospecting, and settling the solar system; and stimulating space enterprises for direct benefit of people on earth. Recommendations for research in the report include a sustained program to understand the evolution of the universe; studies of the solar system, sun, and earth; a continuing search for extraterrestrial life; and new research on the effects of different gravity levels on humans and other biological systems. The report calls for the development of state-of-the-art facilities for laboratory experiments on earth and in space and recommends special emphasis on the development of intelligent autonomous systems. It recommends also demonstration projects in seven critical technologies: aerospace plane propulsion and aerodynamics, advanced rocket vehicles, aerobraking for orbital transfer, long-duration closed ecosystems, electric launch and propulsion systems, nuclear-electric space power, and space tethers and artificial gravity.

In addition to the space shuttle and space station, the commission's vision involves several vehicles playing a variety of roles in a continuing program. These include cargo and passenger transports to low earth orbit; an earth spaceport in permanent low earth orbit; transfer vehicles for destinations beyond low earth orbit; an orbiting variable-g research facility; lunar, Mars, and libration-point spaceports; and cycling spaceships.

In July, 1989, President Bush reaffirmed the United States' intentions of playing a major role in space exploration, both in the immediate future and over the long term, by announcing a commitment to establish Space Station Freedom before the end of this century and then, after returning to the moon early in the next century, to undertake a manned mission to Mars. The plan, as described in a special task-force report (NASA Task Force, 1989), is in keeping, for the most part, with the National Commission's recommendations. Of special interest, for present purposes, is the report's mention of "behavior, performance, and human factors in an extraterrestrial environment" (p. 1) as an area requiring investigation and "about which we have very little data" (p. 5).

The Bush administration also launched the Outreach Program to seek new ideas for the Space Exploration Initiative that includes the return to the moon and the subsequent mission to Mars. Ideas, which have been solicited broadly from universities, national laboratories, and contractors, are to be studied by a Synthesis Group chartered by the White House and NASA. According to a recent report in *Aviation Week & Space Technology,* the Synthesis Group is expected to produce at least two alternative moon/Mars mission scenarios, and maybe as many as six or seven (Covault, 1990).

MOON BASE

One reason for interest in the establishment of a base on the moon relatively early in the future space program is the possibility of obtaining materials there that could support subsequent phases of the program. The lunar surface is rich in oxygen and silicon and contains also calcium, aluminum, iron, titanium, magnesium, manganese, and chromium. Ceramics and composites are likely to play a larger role in lunar manufacturing than metals because they require less processing (Mackenzie & Claridge, 1979). A special interest is the possibility of mining oxygen and perhaps hydrogen that, in liquified form, are major propellants used in rocket fuels (Arnold, 1980; Burt, 1989). If rocket propellant could be produced on the moon, this, coupled with the moon's relatively low gravity, would make the moon very attractive as a point of departure for vehicles headed to Mars or farther reaches of the solar system. Some thought has also been given to the possibility of mining other extraterrestrial bodies and in particular the near-earth asteroids (J. S. Lewis & R. A. Lewis, 1987).

The moon's almost total lack of an atmosphere, its very low levels of seismic activity (about 100 million times less energy in a typical moonquake than in a typical earthquake), and the relative lack of interference from light and radio waves, make it an especially well-suited spot for astronomical observation. Observatories there should be able to exceed the resolving power of earth-based observatories by perhaps as much as five orders of magnitude, and to enable the study of radio waves at frequencies below those that are accessible from the earth's surface (about 30 megaHz). On the negative side are the exposure to radiation that results from the absence of a magnetic field, bombardment by small meteoroids because of the lack of an atmosphere, and drastic and rapid temperature changes (also due to the lack of an atmosphere), combined with the moon's slow rate of rotation; but these are considered to be solvable problems. In addition to observatories located on the surface of the moon, consideration is being given to the construction of a moon–earth interferometer that would provide a resolving power of about 10^{-5} arcsec at a frequency of 10 gigaHz (Burns, Duric, Taylor, & Johnson, 1990).

THE SPACE STATION

The space station, which was mandated by President Reagan in 1984, is to be a permanently manned facility in low-earth (approximately 300-mile) orbit. Original plans called for it to be in orbit and operational by 1992. That date has slipped more than once. Currently the hope is for a launch some time around 1996. The station is expected to evolve over time; design plans call for modularity and expandability. The intent was that the project be an international one involving at least Europe and Japan, in addition to the United States. Space exploration is a major arena for international competition, but it also represents unprecedented opportunities for international cooperation (OTA, 1985c).

The station is expected to serve both scientific and commercial functions. In addition to being a platform from which parts of the solar system can be explored, it is to be a laboratory in which experiments can be performed on a variety of problems of commercial interest. It is to provide a means, for example, of studying the growth of crystals under conditions of microgravity and the alloying of metals in the absence of convection currents. It is to serve, also, as a staging platform for other space missions, as a servicing station for satellites and other space vehicles, and as a construction base for large structures (NASA/Johnson Space Center, 1979). Although there is now some question as to the role the space station will play in the earlier phases of the Space Exploration Initiative (Covault, 1990), it seems clear that it, or something very much like it, will be required before the program can proceed far beyond a return to the moon.

Because the station is to be as self-contained as possible and relatively independent from ground control, much attention is being given to the recycling of supplies, such as water, and to on-orbit maintenance and repair. Safety and habitability are major concerns for obvious reasons. Design plans have been modified several times and, at this writing, are not yet final. Living modules were originally to be pressurized cylinders 35 to 40 feet in length and about 15 feet in diameter, each intended to accommodate 6 to 8 people. An individual's private quarters were to contain about 150 cubic feet of space. The living modules, along with others of similar size that will serve as laboratories, are to be interconnected, but isolatable in the case of an emergency.

Both the habitability of the space station and the productivity of its occupants are topics of interest to human factors researchers, and they will become increasingly important foci of research as the space program moves along (Clearwater, 1985; Gillan, Burns, Nicodemus & Smith, 1987; Nickerson, 1987; Wise, 1986). The more general question of how to maintain health and psychological well-being on extended space missions will be a continuing matter of concern, especially as technology makes longer range missions feasible. Settling on a specific design for the space station is proving to be a difficult process and one that has fueled debate among the station's promoters and detractors, including members of Congress who control the purse strings (Marshall, 1989b).

A. Newell (1987) listed 15 "hard constraints" that can be expected to apply to the space station and that have human factors implications:

- Long lifetime of the station (decades).
- Medium-term crew residence on board (months).
- Small group of residents aloft (less than 10, to begin with).
- Large group of operators (nonresidents) aground (i.e., hundreds).
- Very small amounts of resources available per resident.
- Very small amounts of space available per resident.
- Infrequent physical communication (measured in months).
- Continuous but limited-bandwidth communication.
- Time delay of station communication of .5 to 2 seconds.
- Modest time constants of action (minutes to hours).
- Weightlessness.
- Continuous high task load.
- Continuous high threat level of many potential errors.
- Continuous public exposure.
- Completely artificial environment. (p. 19).

This list makes clear that the space station poses a great challenge to both designers and occupants, but less by virtue of any particular constraint than by the combination of many. Many of the individual constraints are not unique to the space station but have characterized other environments as well. Because of this, useful ideas about potential problems and how to prepare for them can probably be obtained from studies of extended submarine patrols (Weybrew, 1961, 1963), wintering-over parties in the Arctic and Anarctic (Gunderson, 1963, 1974; Gunderson & Nelson, 1963), and the like. Of course, the results of such studies should be extrapolated to the space station only with great caution, but they could provide at least suggestive evidence of some of the problems that should be addressed and perhaps some clues as to how to address them.

Consider, for example, the problem of living for long periods with other people in a very restricted space. There are data from other environments that are similar to the space station in this respect, and the results of studies of those environments provide at least some hints as to the types of problems that may arise and what might be done about them. Studies of the effects of crowding on naval vessels, for example, have sometimes found a positive correlation between perceived crowding and illness rates (Dean, Pugh, & Gunderson, 1975, 1978). Studies of crowding in prisons have yielded similar correlations (Cox, Paulus, McCain, & Karlovac, 1982; McCain, Cox, &

Paulus, 1976). It must be noted that crowding and perceived crowding are not necessarily the same, and it is perceived crowding that appears to have the negative health effects. Apparently, the perception of crowding can be lessened somewhat as a consequence of the way a given space, especially a living space, is laid out and partitioned (Cox et al., 1982). It is probably important to understand this relationship much better, given the long durations for which people will have to live in crowded situations in extended space flights.

In addition to the hard constraints listed by A. Newell (1987), there are many other characteristics of the space station, as planned, that could have implications for how well the station and its crew function. These include: a high degree of interactivity, and especially cognitive coupling, between crew and equipment; the fact that most control actions and information displays will be mediated by computer; the critical nature of information systems; the need for aiding or augmenting human thinking for troubleshooting and decision making; the need for continual concern for safety; the need for an ability to deal with unanticipated events; the shared responsibility of flight-control decisions between ground and flight crews; the need, sometimes, for operating procedures and principles to be negotiated with sponsors while in orbit; the heterogeneity of space station residents (different professions, different cultures, different amounts of technical and flight experience); the importance of having satisfactory ways for occupants to spend free time; and stress (Nickerson, 1987). There are, within these items, numerous human factors issues and problems that need to be addressed.

Artificial intelligence (AI) and expert-system technology are expected to play a role in the space station in two ways. First, further development of AI and expert-system technology is to be an integral part of the space station program, on the assumption that capabilities of these types will become increasingly important as space exploration projects become more ambitious. Second, specific AI and expert-system capabilities are to be incorporated later into the space station's operational procedures as these capabilities become sufficiently reliable to justify their use (Buchanan, 1987; T. M. Mitchell, 1987).

Remotely operated devices are expected to be used for several purposes in space station missions. The ability to control external devices from inside the station will reduce the need for extravehicular activity by humans, which is an especially costly aspect of the space program; and of course, much greater force can be exerted by remotely controlled devices than human beings can exert directly.

STATUS AND FUTURE OF THE U.S. SPACE PROGRAM

Since shortly after the launching of Sputnik in 1957, the United States and the Soviet Union were engaged in what became widely known as the "space race." Both countries have accomplished feats that would have been considered impossible just for a few decades ago. In recent years, several other nations have also established space programs. Before its dissolution the U.S.S.R. appeared to have the lead in this race.

The Soviet Union had an impressive array of reliable launch rockets: In 1987, it accounted for 86% of all space launches; as of 1989, Soviet cosmonauts had gained 143,000 person hours of experience in space, compared to 43,000 person hours of experience by U.S. astronauts, and had maintained an almost continuous presence in space since 1977 (Stever & Bodde, 1989). The U.S.S.R.'s launching of the space shuttle, *Buran,* its continuing attempts to develop heavier lift launch vehicles, and its aggressive marketing of launch vehicles and launch services to other countries were seen, until very recently, as indicative of a continuing commitment to the space program (Banks & Ride, 1989). The recent unprecedented political upheaval in the Soviet Union and the uncertainty as to the structure that will emerge from the ongoing turmoil leaves the future of the Soviet program in some doubt.

At least a half dozen countries—the United States, the U.S.S.R., France, Great Britain, Japan, and China—have been competing to make money launching commercial payloads into space. In 1986, the Soviets were offering launching services to the West for $25 million a shot, which compared with $70 million to $100 million per launch by comparable Western rockets. Private companies in several countries are now offering launch services for small- or modest-sized payloads; one might expect that market competition will help to improve this technology and to reduce its cost (Morgan, 1989). Because of almost total reliance on the space shuttle from 1972 until 1986, very little was spent on the development of alternative advanced propulsion and launch systems during that time in the United States (Corcoran & Beardsley, 1990).

The future of the U.S. space program is clouded by a variety of uncertainties and tensions. Some scientists have been critical of the emphasis on manned vehicles, which they believe to be impeding progress that could be made with much less costly unmanned flights; total dependence on the shuttle as the primary launch vehicle has often been criticized (Morgan, 1989; Van Allen, 1986). Logsdon and Williamson (1989) predicted that by early in the 21st century Europe, Japan, and perhaps China would have the ability, along with the United States and the U.S.S.R. to fly human beings to and from space and that "only the U.S. will be using for that purpose a vehicle initially designed with the technology of the late 1960s in mind" (p. 40). On the other hand, the United States' space program has shown some signs of revitalization after several years of relative dormancy.

In 1989, NASA launched the Cosmic Background Explorer, which will record radiation over four orders of magnitude of wavelengths over the entire sky (Gulkis, Lubin, Meyer, & Silverberg, 1990). The *Magellan* and *Galileo* spacecrafts were also launched in 1989 to explore Venus and Jupiter, respectively, and the long-awaited launching of the Hubble space telescope took place in the spring of 1990, although its operation since the launching has been disappointing. At a cost of more than $1.5 billion, the Hubble Space Telescope has been called "the most complex and expensive scientific instrument ever built" (Beardsley, 1990d). When fully functional it is supposed to produce a continuous flow of data that will total over 30 terabits a year (Herskovitz, 1990). Some 36 other major missions are on NASA's schedule through 1995 (Bell, 1989a). Plans call for the launching of three major satellite observatories, in addition to the Hubble telescope, by the end of the century: the Gamma Ray Observatory in 1990, the Advanced X-ray Astrophysics Facility in 1996, and the Space Infrared Telescope Facility in 1998, all to function in low earth orbit (500 to 600 kilometers above the earth; Burns et al., 1990).

Already our knowledge of the outer planets, Jupiter, Saturn, Uranus, and Neptune, has been increased manyfold by the two *Voyager* probes. For the first time, we have detailed photographs of the surfaces of these planets and their moons. As Kinoshita (1989b) has put it, "in the 12 years since they were launched, the *Voyager* spacecraft have contributed more to the understanding of the planets than 3 millennia of earth-bound observations" (p. 91). The *Magellan* spacecraft is expected to produce a radar map of the surface of Venus with a 0.25 to 0.5 kilometer resolution (Herskovitz, 1990). *Pioneer 10*, which was launched in 1972, and escaped the solar system in 1983 after returning the first close-up pictures of Jupiter, is still sending useful information back to earth from a distance of over 4.6 billion miles, more than 50 times the distance of the earth from the sun.

The National Academies of Sciences and Engineering and the National Institute of Medicine recently urged the President to support a balanced space program with two structural components: "a base program to assure access to space by a variety of manned and automated vehicles funded at about $10 billion per year and selected special projects (such as the space station) funded at $3 to $4 billion in peak years for each such initiative" (National Research Council, 1989a). Keyworth and Abell (1989) have drawn an analogy between the development of the space program and the history of progress of computer technology. They see a third generation of space policy, not yet implemented, that will be to the space program what the introduction of personal computers was to the computer industry. The third generation of the space program, they suggest, will be characterized by many applications— commercial, military, and scientific—making use of relatively inexpensive satellites launched by relatively small rockets. Such satellites will be crammed with microelectronics and will be launchable "on short notice with little fanfare" (p. 16). Other scientists have argued strongly in favor of greater use of unmanned satellites in space exploration. Dyson's (1989) vision of what could be done with miniature space probes represents one end of the continuum of space exploration vehicles, going from the very large to the very small.

Stever and Bodde (1989) suggest three reasonable national goals for the U.S. space program: (a) pre-eminence in space science (which could be attained with primarily unmanned flights and no permanently manned space station), (b) mission to planet Earth (a wide-ranging program of earth observation for the purpose of studying earth processes and monitoring global environmental change), and (c) human exploration (manned exploration of the solar system). Any one of these goals would be expensive, so pursuing all three of them at once is not an economically feasible option. It will be essential to the success of any future space program, Stever and Bodde argue, to establish some clear priorities: "Ahead of us is a greater number of worthy objectives than this nation can afford to pursue simultaneously and alone. What could limit us is not the inadequacy of our technology, but the inadequacy of our policy" (p. 71).

*   *   *

The human element is a critical aspect of the Human Exploration Initiative, and protecting and sustaining the lives of space explorers is of the utmost importance. The success of this bold and exciting interprise means accomodating human needs in environments that are hostile and exceptionally different from Earth, at distances never achieved by any space program. Fundamental differences between space and Earth—the lack of gravity, inadequate atmosphere, deep cold, and radiation hazards—challenge technological ability to protect, nurture, and sustain the crew members who will be the pioneers of the solar system. (NASA Task Force, 1989, p. 6–1)

The human's role in spaceflight has changed and expanded over the life of the space program. In the earliest flights, humans were primarily passengers in automated or ground-controlled vehicles. As the spacecraft became more complex and the missions more ambitious, the crews took on more of the responsibility of piloting the spacecraft and of performing numerous other tasks during spaceflight. Crew tasks have included monitoring the various spacecraft subsystems (guidance and control, propulsion, environmental control, and life support), guidance and control during rendezvous and docking, landing and taking off of lunar module, maintenance and repair, monitoring of data quality, and housekeeping (Loftus, Bond, & Patton, 1982). As the space program has progressed, an increasing fraction of the tasks performed have related to the performance of scientific experiments and observations (Garriott, 1974; Garriott, Parker, Lichtenberg, & Merbold, 1984).

Many of the tasks that crews have performed, and will perform in the future, involve activities outside the spacecraft. Extra-vehicular activities pose special problems because of severe constraints on mobility and dexterity imposed by pressurized space suits, limited visibility, greatly reduced tactile feedback to the hands with pressurized gloves, free floating or tethered (and easily tangled) tools, limited voice communication with in-station crew, problems associated with personal hygiene and comfort, and problems of eating and drinking (Nickerson, 1987). Finding ways to increase the safety, productivity, and comfort of extra-vehicular activities will be a continuing research objective.

Most of the functions that will have to be performed by people in operating the space station will involve interaction with computer-based systems of one sort or another. The typical mode of interaction, at least for the immediate future, will be via keyboards and video displays. (Loftus et al., 1982, have estimated that about 10,000 key strokes are required to complete all the elements of a lunar landing mission.) As the technologies mature to the point of being sufficiently reliable and robust to be used in high-risk contexts, automatic speech recognition and production, and other more esoteric input–output capabilities (eye movements and fixations, head and hand movements in virtual-reality environments) will probably be added.

Because so many of the functions in the space station will be performed by people and machines in interaction, the design of the various workstations and person–machine interfaces will be of critical importance. Much of the research that has been focused on interface design issues is relevant to the problem of designing interfaces for the space station and other spacecraft, but the physical and operational constraints imposed by the spaceflight context will undoubtedly necessitate some fine tuning.

Some of the human factors research that must be done in anticipation of extended human excursions in space can be done in simulation facilities on the ground; however, much of it must also be done in space facilities, such as the space shuttle or the space station, because it is not possible to simulate all the important properties of a space environment on earth. Conducting this space-based research will involve special challenges because of the severe limitations on the amount of equipment and number of personnel that can be included in space missions for the purpose of experimental data collection.

Life-support systems will have to be designed that will provide all the necessities for survival in whatever environments are to be explored. For missions involving excursions over lunar or planetary surfaces, this means portable equipment for providing air, food, and water, for disposing of body wastes, and for protecting individuals against inhospitable environments. All of this must be packaged in ways that do not unduly restrict movement and prohibit productive work. Vehicles, habitats, and storm shelters will be necessary for protection against galactic cosmic rays and radiation from solar flares. Solar flare prediction equipment will have to be developed and crew members trained in its use. Long-term consequences of living in a zero-gravity situation will have to be induced from the limited relevant data that exist and from theoretical considerations, and countermeasures will have to be designed to minimize its deleterious effects. Contingency plans and provisions will have to be made for the various types of emergency situations that could arise during a long-duration mission (e.g., serious injury or illness). Thought must be given to how to resolve conflicts that may arise among crew members, to how to deal with stresses of various sorts, and to the general problem of maintaining crew motivation at a productively high level. The challenges are many, and any effort to identify them all is very likely to miss some that will prove to be very important.

8

Biotechnology

Biotechnology has been making spectacular strides in the last few decades and will undoubtedly continue to do so in the foreseeable future. Like space exploration, biotechnology is a highly visible area of cutting-edge scientific activity, and it has been the focus of considerable debate among policymakers in terms of what research should be supported by public funds. Unlike the case of space exploration, however, no one doubts that work in this area will have very significant implications for life on this planet, and probably in the near future as well as in the longer term. Biology is rapidly changing from a primarily observational science to an experimental and manipulative one. Human genes have been inserted into several animal species for a variety of purposes: to stimulate growth of cattle and pigs; to support the study of certain human gene defect disorders in mice; and to stimulate the production of pharmaceutical proteins and hormones in goats, rabbits, and mice (Erickson, 1990c). Recipients of such genes are referred to as transgenic organisms.

It is now possible to manipulate the genetic code and effect specific changes in organisms to "create versions of life" as R. A. Weinberg (1985) put it, "that were never anticipated by natural evolution" (p. 48). Chemists are able not only to synthesize natural proteins, but to redesign them to improve their stability or modify their activity, and even to design completely new proteins, including some that contain amino acids other than the 20 that are used by nature (Barton, 1988; Kaiser, 1988; Petsko, 1988). The knowledge that is being generated by research in biotechnology has the potential of being applied in a wide variety of ways in medicine, in industry, in agriculture, and in mining. The long-range consequences of many of the possible applications are very difficult to anticipate, but it is hard not to believe that they will be substantial.

Unfortunately, the breathtaking rate of advance in biotechnology has not guaranteed the adequacy of the delivery of medical services in the United States. The skyrocketing costs of medical care—the United States currently spends about 12% of its GNP on health care, and medical costs are increasing at about twice the rate of all other costs (Dentzer, 1989)—and the fact that many limited-income Americans do not receive the care they should are serious national problems. Some of the statistics on the effectiveness of the medical system as a whole—such as the fact that we currently rank 17th among industrialized nations in infant mortality (OTA, 1989a)—are, and should be, an embarrassment to a country with the finest medical technology in the world.

How human factors research relates to this area is less clear than how it relates to space exploration. Perhaps this is because the connection is intrinsically more tenuous. On the other hand, it may simply reflect the fact that human factors researchers have not been, for whatever reasons, much involved with biotechnology in the past. Inclusion of a chapter on this topic in this book is dictated both by the belief that the social significance of work in this area is enormous and by the surmise that there are opportunities and needs for human factors involvement waiting to be identified.

TECHNOLOGY AND MEDICINE

The applications and potential applications of biotechnology that have attracted the most attention are those that relate to medicine. Computational techniques combined with various medical imaging processes are making it possible to see specific organs and internal body parts volumetrically and in ever-increasing detail. Computer-based imaging techniques include computed axial tomography, nuclear magnetic resonance imaging, positron emission tomography, and digital subtraction angiography. Some techniques, by assigning different degrees of transparency to different layers of bone and tissue, let the observer not only see them but look through them and see what lies within or behind them (Fitzgerald, 1989).

The question of the diagnostic power and cost effectiveness of these imaging techniques, relative to more conventional and less expensive x-ray technology, has been the focus of several psychological studies (Swets, 1988; Swets & Pickett, 1982). Signal detection theory and analysis techniques have proved to be very useful in this context (Metz, 1986; Swets, 1979). As medical imaging technology continues to develop, the need for studies evaluating the effectiveness of specific instruments and procedures will increase. This is so especially in view of the considerable cost of the equipment involved and the fact that the diagnostic images produced will not always be interpreted by image-interpreting specialists but by physicians, who view certain types of images only occasionally. There is a need, also, for the development and evaluation of procedures (including computer-aided procedures) to help clinicians, and especially those with limited experience in image interpretation, to extract from medical images the diagnostically relevant information they contain (Getty, Pickett, D'Orsi, & Swets, 1988; Swets et al., 1991).

It is somewhat paradoxical that the very high cost of health care is due in part to such spectacular advances in medical technology as those represented by imaging technology. The new computer-based techniques represent enormous advances over conventional x-ray technology in providing noninvasive windows to internal body structures; however, this technology makes use of expensive equipment, the purchase, operation, and maintenance of which adds significantly to a hospital's cost of doing business. A laboratory in a well-equipped, modern hospital can perform countless sophisticated tests that were not possible and, in many cases, not even dreamed of a few decades ago, but the equipment to perform these tests is also expensive.

We would not be willing to give up any of these capabilities, but as a society we have not yet quite learned how to use them in a way that ensures we will be able to continue to afford them. The availability of so many diagnostic procedures to choose from in medicine complicates the problem of the individual physician's deciding what is appropriate for any given circumstance. Cost–benefit relationships are not well understood in many cases. Moreover, the understandable tendency that physicians may have to overprescribe tests, because of a strong desire not to overlook any useful diagnostic information that might be obtained, can have the long-term effect of driving up the costs of medical services and, possibly, decreasing their cost-effectiveness.

A variety of new materials are now being developed for use in medicine, some of which are capable of chemical bonding with body tissue or bone. These include polymers, ceramics, bioglasses, and various types of composites (Fuller & Rosen, 1986). A line of work in biotechnology that has especially far reaching implications for medicine involves the development of immunotoxins in which toxic protein molecules are attached to antibody molecules, which can then carry the toxin to desired targets (Petsko, 1988). Such immunotoxins could be the key to major advances in the treatment of many diseases, including cancer and AIDS.

A development from medical research that has given rise to ethical dilemmas is the finding that embryonic brain tissue grafted into adult mammalian brains may help reverse damage from brain disease or injury (Fine, 1986). Difficult issues have also arisen regarding the patentability of genetically engineered organisms (OTA, 1989a) and about ownership of certain biological materials, such as human tissue and cells, used in research in medicine and, at least potentially, in industry (OTA, 1987d). The first patent on a living organism was issued, amidst a great deal of controversy, to A. M. Chakrabarty in 1980 for a genetically altered oil-eating bacterium.

Today's medical practice offers a spectacular range of technologies, medicines, and surgical procedures with which to treat illness and disease. This impressive assortment of tools and capabilities makes the medicine of just a few decades ago pale by comparison, and few of us would be willing to turn the clock back to the earlier times. However, in the eyes of some observers, the greatly increased technical sophistication has been accompanied by the loss or diminution of the human touch in doctoring. The medical profession has been accused, in this era of high technology and narrow specialization, of losing the ability to deal with patients as persons rather than as hearts, livers, bones, immune systems, or whatever the physician's specialty happens to be.

H. Leventhal, Nerenz, and E. Leventhal (1982) point out that dehumanization can occur in environments that are intended to promote the individual's well-being, such as medical institutions. One of several subjective aspects of dehumanization they discuss is a sense of isolation. A factor contributing to this sense in medical-care situations, they suggest, is the tendency of physicians to see patients as a collection of physical and physiological processes. Another factor that can contribute to the sense of helplessness and loss of control by a patient in a medical situation is the incomprehensibility of the language often used by medical professionals (Ley, 1977). Many people have written of the feeling of personal diminution that can come with a loss of one's independence and the control of one's daily life (deciding when to eat, when to go to bed) resulting from being temporarily incapacitated and under medical care. The experience of constantly encountering aversive events over which one has no control can produce not only a sense of powerlessness, but a variety of cognitive and affective disturbances and, can result in what M. P. Seligman (1975) has referred to as *learned helplessness*.

Pennebaker and Brittingham (1982) suggest that people are continuously bombarded with information both from within their own bodies and from outside of them, and that the processing of one type of information can inhibit the processing of the other type. Given this assumption, one should not be surprised to find that people become more aware of within-body stimulation when they are in situations in which the external stimulation is reduced, for example, when they are cut off from normal social contact or isolated from the usual hustle and bustle of the world, which is likely to be the case when one is confined to a hospital bed.

There is a need for research on the effects of the design of hospital environments on human behavior and affect. There is a literature on this topic, but it is widely dispersed among books and journals representing a variety of fields, and it includes relatively few empirical studies (Jaco, 1972; Reizenstein, 1982). There seems to be little doubt among people who have written on the subject that the layout and appointment of patient-care facilities can have significant effects on the behavior and attitudes of both patients and medical care providers, and quite possibly on the quality of the care that is given and the effectiveness with which patients respond. Although there are guidelines and recommendations for the design of hospital facilities, they appear not to be based very solidly on the results of empirical research (D. Cantor & S. Cantor, 1979; Jaco, 1972). R. Olsen (1978), for example, argues that, depending on the way they are designed and appointed, hospital environments can transmit to patients the message that they are sick and dependent and should behave passively, or that they are independent, competent, and in control of their recovery process. Hospital patients, visitors, and care providers perceive the hospital environment from different perspectives. Properties of a room that may be inconsequential to a visitor or a staff person who is in it only for short periods of time and is free to come and go, can be quite significant to a patient who is confined to it 24 hours a day (E. L. Brown, 1961; Sommer & Dewar, 1963).

I believe that there are many opportunities within medicine, especially vis-à-vis medicine's use of technology, for human factors research to help improve the delivery of medical services by making them more effective, more cost-effective and more responsive to human—as distinct from strictly physiological—needs. The design of patient-care facilities, the study of medical procedures—in operating rooms, emergency treatment centers, intensive care units, chronic care facilities—for the purpose of identifying error-inducing situations, the design of medical tools and patient-monitoring equipment, and the analysis of methods used to communicate medical information to patients are but a few examples of areas where human factors work could have an impact.

A current need within medicine, and one that the human factors community could help meet, is for effective information systems tailored for use by doctors and other medical personnel. Such a system would contain information on specific diseases, injuries, drugs, patients, and medical resources, with reference materials in areas of medical specialties (such as immunology, hematology, and radiology), as well as in more general areas, such as physiology, biochemistry, and nutrition. A truly versatile medical information system would be able to provide information not only as text and tables but in the forms of pictures and process simulations as well. Whether such a system would be used by medical personnel would depend in part, of course, on the extent to which it gave users access to the kind of information that they would find really useful in their daily work, but also, in part, on the adequacy of the interface design.

It is important to note that the mere existence of information services, even those tailored for specific user groups, does not ensure their use. An example of an on-line information system that was designed to meet the needs of medical practitioners is the Physician's Data Query (PDQ) system. This system contains information about state-of-the-art cancer treatment, a file of active cancer research protocols, and a directory of physicians and organizations that provide cancer care. The information on cancer treatment is updated monthly by an editorial board. The system is accessible 24 hours a day through the National Library of Medicine and private vendors. Somewhat surprisingly, although, as of 1989, the system had been in operation for several years and its use had been increasing, it was still not used very extensively: less than 400 hours per month by, or for, a few hundred physicians, according to estimates by the National Cancer Institute. This suggests a very small percentage of even the 12,000 physicians in the PDQ directory consult the system during an average month. Why the system is not consulted more extensively is not known (General Accounting Office, 1989), but a study of this situation might produce some useful insights that could be applied to the design of future systems.

GENE THERAPY AND THE HUMAN GENOME PROJECT

There are few, if any, areas of research and development that have more far-ranging implications for the future than genetic engineering. The application of recombinant DNA techniques for the purpose of inserting, replacing, or modifying specific genes in human beings for medical purposes is a very active quest of research biologists (Stahl, 1987). Theoretically, gene therapy can take one of three forms: replacement of defective genes, correction of such genes by modification of the relevant portions of their DNA sequences, or augmentation of the genetic make-up of an individual through the introduction of foreign normal genetic sequences that will counteract the effects of defective genes (Friedmann, 1989). The third technique is the easiest of the three to apply and is technically feasible in a limited way now. The first experiment involving gene transfer in humans was approved by the National Institutes of Health in January, 1989, and the results were reported in *The New England Journal of Medicine* in August, 1990 (Cournoyer & Caskey, 1990; Rosenberg et al., 1990).

We know that several thousand diseases are caused by single-gene defects and that many others result from the combined activities of several genes. The identification of disease-causing genes is a major goal of the field, and progress toward it is being made daily. The gene that causes neurofibromatosis, a disfiguring neurological disease that effects about 1 in 4,000 newborns, was recently isolated by two independent research groups (L. Roberts, 1990b).

Two other research groups isolated the gene that encodes a clotting protein that, when defective, results in hemophilia (Lawn & Vehar, 1986), and succeeded in splicing it into the DNA of blood cells drawn from hemophiliacs in laboratory tests. The clotting-protein gene is 186,000 base pairs long—as of 1986, the largest gene ever to be cloned and expressed in foreign cells—although somewhat over 95% of these base pairs are in intron (noncoding) segments that are interspersed among 26 exons that comprise the coding parts of the gene. Diseases caused by two or more genes working in concert include cancer, diabetes, hypertension, heart disease, and schizophrenia (Pines, 1987). Isolating the genes involved and learning exactly how they have their effects are among the main objectives of teams of genetic researchers.

New abilities to detect or predict incipient disease, including techniques using recombinant DNA technology, will open up new possibilities for prevention or early treatment; they will also produce new dilemmas relating to issues of disclosure (to insurers, employers) and information privacy, and they will necessitate difficult decisions that have not arisen heretofore. Government approval for the insertion of a foreign gene into a human being for medical purposes brought some of the issues relating to the use of genetic technologies in medicine into focus and triggered heated debate (L. Roberts, 1989a). With the promise of continuing rapid advances in research on gene therapy comes a host of very difficult ethical questions: "For instance, should therapy be applied simply to improve one's offspring, not only to prevent an inherited disease? Who would be empowered to decide? Is society willing to risk introducing changes into the gene pool that may ultimately prove detrimental to the species? Do we have the right to tamper with human evolution?" (Verma, 1990, p. 68).

Probably the most visible major undertaking in biotechnology at the present time is what has become known as the Human Genome Project. This project, the objective of which is to sequence the entire human genome, obviously could have very significant consequences, including, presumably, some that we can now only dimly imagine. One knowledgeable scientist has predicted that the entire human genome will be available to biologists for about $10 on a CD ROM by about the end of the 20th century (Branscomb, 1986).

A DNA sequence is a sequence of nucleotides, each of which is composed of a phosphate group, a sugar, and a base. For our purposes, the important part of this composite is the base. The DNA molecule uses only four bases: adenine, guanine, cytosine and thymine. Abbreviated as A, G, C, and T, respectively, they form the four-letter alphabet which is used to encode the characteristics of every individual. DNA is a double-stranded molecule, the two strands of which form the famous double-helix structure. Each base in one strand is linked to a corresponding base in the other. Sequencing a segment of DNA means identifying the sequence of base pairs in that segment. Because adenine always links with thymine, and cytosine always links with guanine, the information in one strand is completely redundant with the information in the other. Because knowledge of one member of each pair determines the other, a sequence is usually expressed by showing only one member of each pair: for example AATCGT.

Sequencing the entire human genome is an enormously ambitious project. Realization of Branscomb's prediction will require the development of effective automated sequencing techniques, inasmuch as, with currently available methods, DNA sequencing is a very slow manual process; an experienced sequencer "working flat out" can sequence perhaps as many as 100,000 base pairs per year (L. Roberts, 1988). Although 100,000 may seem like a large number, it is very small compared to 3 billion, which is the estimated number of base pairs in the genetic complement of a human being. (I have seen estimates as high as 6 billion, but I believe 3 billion to be a much more common one.) It is also an extremely expensive undertaking, with sequencing currently costing between $3 and $5 per base pair (J. D. Watson, 1990). C. R. Cantor (1990) has also noted the need to find more cost-effective techniques for DNA sequencing if the genome project is to be completed within a reasonable time.

Advances in state-of-the-art of DNA sequencing could come either from software or from hardware designed especially for pattern recognition. Such hardware would probably involve many pattern recognizers operating simultaneously on the same data set, perhaps along the lines of the Pandemonium pattern-recognition model described by Selfridge and Neisser (1960). Whether Branscomb's prediction proves to be accurate with respect to time is far less important, of course, than the fact that the effort to sequence the genome is underway and is likely to pick up momentum as useful results are obtained. As of 1989, about 30 million nucleotide base pairs had been sequenced, and the number was then growing at the rate of about 10 million pairs per year (L. Roberts, 1989d).

One goal of the genome project is to locate and sequence individual genes, each of which is composed of a large number of base pairs. The total number of genes that a human being has is not known precisely, but is estimated to be between 100,000 and 300,000 (Dervan, 1988). For purposes of this discussion, I will use 100,000. (One cannot simply infer an average of 30,000 base pairs per gene from the estimates of 3 billion base pairs and 100,000 genes, because a sizeable fraction of the base pairs are located in segments of the DNA molecule that do not carry protein-sequence information. These segments are called *introns* and their function is not well understood.)

Individual genes can come in many varieties called *alleles* (e.g., there are at least as many alleles for the gene that determines eye color as there are eye colors). So many different sequences of base pairs can occur, even in the small region of a chromosome that constitutes a single gene. A complete understanding of the human genome would involve knowing the nucleotide sequences of all the alleles of all the genes that could occur in the genetic makeup of an individual, and the locations of the genes on the chromosomes. The number of possible nucleotide sequences that could conceivably occur is unimaginably large. It is believed, however, that the DNA molecules of unrelated individuals are identical with respect to 99.9% of their base pairs; that is, the fully sequenced DNA of two unrelated people would be expected to differ with respect to only about 1 base pair in 1,000.

This leaves plenty of opportunity for the coding of individual differences—about 3 million base pairs—but it makes the genome sequencing problem many orders of magnitude simpler than it would be if individual sequences differed with respect to a large percentage of their base pairs. It means that determining the base-pair sequence of the DNA of a single person is tantamount to determining all but 0.1% of it for everyone. (That differences in a small percentage of the base pairs of two organisms are enough to make large differences between the organisms is seen in the fact that the genetic difference between humans and their closest neighbors among the primates is about 1 base pair in 100. A conclusion one might draw from this observation is that we are more likely to notice the differences between individuals, or between ourselves and other species, than we are to notice the similarities, despite the fact that the similarities far outnumber the differences.)

To someone outside the field, like myself, the terminology associated with the genome project can be confusing. In particular, in addition to base-pair sequences, several types of "maps" have to be distinguished. Genetic mapping, as distinct from base-pair sequencing, usually refers to the process of locating "markers" on the chromosomes that can be used to specify the approximate location of genes. The situation is further complicated by the fact that there are several types of maps.

A *genetic linkage map* shows the approximate distances between genes and markers, as inferred from the frequency with which specific genetically determined traits co-occur. During meiosis, when reproductive cells are formed, individual chromosomes often divide into segments, and these segments recombine. The closer two genes are on the same chromosome, the less chance there is that the chromosome will divide between them; hence, the greater the likelihood that they will end up together in the same germ cell and therefore be co-inherited if that egg or sperm is combined with its counterpart and develops into a new organism. Two genes that are separated by about one million base pairs—one centimorgan—will be separated during meiosis about 1% of the time. (The term *centimorgan* is used in honor of Thomas Morgan who, in genetically mapping the housefly, used the idea that the probability of the joint translocation of two genes between chromosomes should be a linear function of the proximity of their locations. A centimorgan is the distance between two genes that would be co-inherited 99 times out of

100. One million base pairs per centimorgan comes from an estimated 100 million base pairs per chromosome and the assumption that the likelihood of a translocating break is the same at any point on the chromosome as at any other.)

A *physical map* shows the precise distances, in terms of base pairs, between specific genes or markers on the chromosome. At a minimum, physically mapping the human genome means determining which chromosomes specific genes are located on; ultimately the goal is to determine the precise locations of all the genes on all the chromosomes. Locating a gene on a chromosome means determining its proximity to genes whose locations are already known or to other markers. A marker can be a gene or simply a specific, identifiable sequence of bases whose location on the DNA molecule is known. Typically, when a specific gene is the object of a search, its location becomes known with greater and greater precision as the search increases. The cystic fibrosis gene, for example, which was precisely located in 1990, has been known since 1985 to be in a particular small region of chromosome 7.

Physical mapping involves the process of slicing chromosomes, with the help of restriction endonucleases, into short segments and then determining the order of those segments. Restriction endonucleases cut DNA at specific sites based on their ability to recognize and bind to sequences of base pairs from 4 to 8 base pairs long. One of the challenges on which synthetic chemists are currently working is that of developing restriction endonucleases that can recognize sequences as many as 12 to 15 base pairs long, because it is this kind of recognition capability that would be required to isolate individual genes (Derban, 1988).

Mapping proceeds by a process of successive approximations. Markers are located on the individual chromosomes to which the locations of specific genes can be referenced. The goal is to increase continually the number of markers and thereby decrease the average distance between them. At the present time, that average is about 10 centimorgans. One oft-cited 5-year goal of the Human Genome Project is to produce a map with markers located at 1- to 2-centimorgan intervals.

The gene mapping effort clearly must involve synthesis of the work of many investigators and laboratories. A major problem to date has been the lack of agreement among investigators as to what to use for mapping landmarks; another has been non-uniformity in the recording of mapping results. There appears to be some reason for optimism that agreement on these questions may be attainable, following a meeting among mappers in August, 1989 (L. Roberts, 1989e). A status report on the effort to map the human genome was recently given in *Science* by J. C. Stephens, Cavanaugh, Gradie, Mador, and Kidd (1990). They reported that about 1,900 genes, or about 1.9% of the estimated 100,000 genes, have been located with respect to which chromosome they are on, but only about 170 of them have been located with the precision that is desired.

The genome project is an example of big science. It is expensive and it requires the collaboration of numerous scientists in many laboratories around the world. It requires, in addition to sophisticated biological research techniques, the support of powerful data management and communication facilities. Scientists working in widely separated laboratories must be able to contribute to and draw from a common database. The complexity of the database that will be required to support this project is probably not generally appreciated. Maintaining that database, and assuring its currency, accuracy, and constant availability to researchers is a major challenge for the project. There are many human factors issues that need to be resolved relating to data organization and representation and to the user interface.

J. D. Watson (1990), has suggested the year 2005 as a target for sequencing the entire human genome, but he has also cautioned that if this goal is to be attained, it will be necessary to plan carefully, so as to realize economies of scale, and probably to develop far larger sequencing facilities than exist today. The United Kingdom, the U.S.S.R., Italy, and Japan have also announced human genome programs. Other countries are also expected to do so. Efforts are being made to ensure cooperation and information sharing rather than competition. Only time will tell whether they will succeed.

The genome project is not supported by all biologists. Many have argued that what will be required to underwrite it would be better spent supporting those investigators who are looking for specific genes, like the gene that causes cystic fibrosis, which was recently found (B. D. Davis & The Department of Microbiology, 1990; Pines, 1987). Some are sufficiently concerned about its costs and the possibility of its diverting money from other biological research projects to have mounted letter-writing campaigns in an effort to stop it (L. Roberts, 1990a). In a recent *Science* editorial, Koshland (1989) pointed out, however, that relative to the cost of a supercollider or of a space station, the price tag of the genome project (currently estimated at $3 billion over 15 years) is really not very large, and that, in view of the amounts of money that are currently spent investigating the genetic bases of specific illnesses and the potential benefit to society of a better understanding of multigenic illnesses (such as manic depression, Alzheimer's disease, schizophrenia, and heart disease), the project is worth its expected cost. Watson (1990), on the other hand, has expressed concern for the ethical and social implications of acquiring the kind of knowledge the genome project will produce, in particular, the knowledge of genetic diseases, which could have obvious beneficial applications, but also great potential for misuse, as well as for inadvertently causing grief for individuals carrying genes for diseases for which there is no known remedy.

What will the consequences of a complete mapping and sequencing of the entire human genome be? Surely, no one knows. The possibility of developing effective therapies or preventive measures against genetically caused diseases is an obvious and desirable possibility. A better understanding of growth and human development seems highly likely. Whether we will be able to handle in a rational and humane way the awesome capability that this kind of knowledge will represent is an open and troubling question.

Precisely how human factors researchers can best relate to the Human Genome Project is not clear. Of course, there are problems of a human factors nature pertaining to the design of the tools and databases that sequencers, mappers, and data custodians use, and the questions of how to facilitate communication among the thousands of people working on various aspects of this project around the globe and of how to integrate the results of their work so as to limit the amount of unnecessary or duplicated effort, have human factors dimensions. My main reason for devoting space to the topic in this book, however, is not because I see so many opportunities for direct human factors involvement in this effort, but because I believe the effort to be enormously significant and one that should be of great and continuing interest to members of a profession which has as a main focus the matching of artifacts and environments to their human users and occupiers.

In the past, human characteristics have been taken as given, modifiable only in modest ways through training, and the burden has been on the equipment designer to shape what was built to match those characteristics. In the future, it may be possible, within ever-broadening limits, to modify those characteristics by design. I do not mean to suggest that anyone is eagerly waiting for the opportunity to use genetic engineering to modify the characteristics of individual human beings so that they better match the artifacts they may have to use, but I find the very fact that such a possibility is becoming less and less remote to be disconcerting. A major challenge of the future to the human factors community, and to all of us, I believe, is to maintain a clear distinction between what can be done and what should be done. As science and technology continually extend the realm of the possible, human values and purpose must increasingly guide explicit choices.

INDUSTRIAL BIOTECHNOLOGY

Although perhaps less apparent than the medical applications of biotechnology, there are a number of industrial applications of very considerable importance. For example, microorganisms (yeasts, molds, single-cell bacteria, and actinomycetes) are being used more and more widely as the worker agents in the manufacture of a variety of commercial products: enzymes (used in brewing, baking, cheese-making, and the manufacture of textiles, detergents, and leather), antibiotics and other pharmaceuticals, proteins, and bacteria that can detoxify and degrade industrial waste or facilitate the extraction of metal from low-grade ores (Aharonowitz & Cohen, 1981; Demain & Solomon, 1981; Gaden, 1981; Greenwell, 1987; Phaff, 1981).

Potential applications of genetic engineering that could have far-reaching consequences are those aimed at the modification of the genetic makeup of certain bacteria with the intention of improving their ability to perform these functions (Landow, Panopoulos, & McFarland, 1989). Products currently being manufactured, or being considered for manufacture, with the help of genetic engineering include insulin, interferon, urokinase and plasminogen activator, rennin, tumor-necrosis factor, hepatitis-B vaccine, human growth hormone, certain enzymes, and viral peptide antigens (Simmons, 1988; R. A. Weinberg, 1985). The utility of microbial cells for industrial production stems, to a large degree, from their extremely small size and consequential high surface-to-volume ratio; this permits rapid transport of nutrients into the cell, which in turn supports a high rate of metabolism: some microorganisms reproduce in 15 minutes (Demain & Solomon, 1981). Other potential uses of microorganisms include growing them on a large scale as a direct source of food for both animals and humans (Rose, 1981).

The insertion of cloned genes into microorganisms in order to induce high-volume production of specific gene products is the foundation of a new and rapidly growing industry. It has been estimated that by the year 2000, the use of recombinant DNA techniques will have resulted in the creation of $15 to $20 billion worth of new products (Good, 1988). Much biotechnology research is currently being done under the sponsorship of industries that are in a position to exploit the results of such research commercially. Dupont, for example, spends a $1 billion annually on research, about one third of which supports biotechnology research. This, according to Simmons (1988), is indicative of the trend throughout the chemical industry.

BIOTECHNOLOGY IN AGRICULTURE AND MINING

Genetic engineering is already playing an important role in crop agriculture (Gasser & Fraley, 1989) and is beginning to be applied to the modification of livestock, although here much work is needed to determine the long-term effects, especially any detrimental ones, of genetic alterations designed to produce specific immediate improvements (Pursel et al., 1989). Applications of biotechnology in agriculture include the development of plants that produce their own insecticides or nitrogen, or those that produce fruit with improved nutritional value (Simmons, 1988). The potential importance of biotechnology to energy and the environment is seen in the fact that the development, through genetic engineering, of plants that can fix their own nitrogen from the air would eliminate or reduce the need for nitrogen fertilizer for certain crops. This would result in a significant energy savings, because the production of nitrogen fertilizer currently consumes 2% of all industrial energy, and it would also be beneficial to the environment in terms of soil preservation and reduction of emissions of nitrous oxide into the atmosphere (M. H. Ross & Steinmeyer, 1990).

A number of genetically engineered crops (tomatoes that ripen but will not rot, disease-resistant soybeans and cotton) have been studied experimentally and are being readied for market. Getting them to market is a long and slow process, however, in part because a number of federal agencies, including the U.S. Department of Agriculture, the Federal Drug Administration, and the EPA, are involved in determining the safeness of the products both to consumers and to the environment (Erickson, 1990b, 1990c). Such applications make some people nervous because of the possibly disastrous nature of a serious mistake. Uncertainties about long-term ecological effects of the release of genetically engineered organisms into the environment are a source of considerable concern among some scientists (Tiegdje et al., 1989).

Debus (1990) points out that the effectiveness of the approach to the mining of minerals that has been used for millennia—digging rocks from the ground and smelting them to obtain the desired metals—works well under three conditions, none of which may hold much longer: the existence of high-grade ore, the availability of inexpensive energy, and dormant concerns for environmental degradation. An alternative approach to mining, which is currently being explored in a variety of contexts, involves the use of microorganisms to help separate metals from the substrates in which they are embedded; when this approach is successful, the freed metals can then be leached or pumped to where they can be gathered with a minimum of defacement of the landscape.

This approach is particularly attractive when the mineral of interest exists only, or primarily, in low-grade deposits. Copper is a case in point. The grade of copper ore in the United States has decreased by an order of magnitude (from 6% to 0.6%) since 1880; today when copper is mined in the conventional way, the production of 1 pound of copper is attended by the production of 198 pounds of waste, much of which is problematic from an environmental point of view (Debus, 1990).

Biohydrometallurgy is the term used to connote the application of biology to the mining and processing of metals or to the removal of unwanted metals from other substances. Microorganisms are being used experimentally to remove sulphur from coal so that the burning of the coal will produce less sulphur dioxide. Bacteria, algae, and fungi are being used to remove metals from industrial wastewater. Copper production is already being greatly aided by the use of a biological catalyst. A disadvantage of the use of biological agents in mining is their slowness relative to more conventional techniques;

although the recovery of metals can be less expensive, per pound, through biological leaching than by digging and smelting, the time required may sometimes be unacceptably long. The search is on, however, for ways to speed up the process, and it appears that the role of biotechnology in mining is likely to increase considerably in the years ahead.

* * *

Biotechnology is moving extremely rapidly, in part because of the wealth of new knowledge about biological structures and processes—in particular, about how individual molecules interact and how this interaction depends on the three-dimensional structures of molecules of interest—that has been acquired in the recent past, and in part because of the power of some of the tools that have become available as a result of advances in other fields. The technology of computer graphics is being used more and more by molecular biologists, chemists, and pharmacologists to study the structure and behavior of biological molecules. Unlike the space-filling models that have been used for this purpose since the 1940s, computer graphics representations permit the scientist to "move about" inside a molecule, observing its structure at arbitrary levels of magnification and from any perspective desired (Monmaney, 1985). Drug researchers are now using computer-based modeling techniques to help them visualize the atomic structure of molecules, especially enzymes, so they can design drugs that will couple with them and block their action. This approach to drug development, which is referred to as *rational design,* yielded Captopril from the Squibb Laboratories in 1975, and is now being widely used by other drug companies. Systems that support this work are extremely complex, and there are many human factors issues that relate to their design, and especially to the design of the displays and controls that constitute their user interfaces.

One of the direct consequences of advances in biotechnology has been the appearance of countless pieces of high-technology equipment in hospitals and doctors' offices. Such equipment is an undoubted boon to the practice of medicine and to the beneficiaries of that practice. However, as the practice of medicine has become more complicated, in part as a consequence of the use of sophisticated devices, the opportunities for errors and problems have also increased. Errors made in the use of equipment in the operating room or intensive care unit can have serious consequences, indeed. Even "routine" nursing care today often involves the use of complicated devices and equipment that represent opportunities for errors and malfunctions, and patients frequently have to use equipment they do not fully understand in caring for themselves at home. In 1984, the reporting of serious accidents caused by the misuse or malfunction of medical equipment became mandatory in the United States (Food and Drug Administration, 1984; Sind,

1990). In the following 2 years an estimated 10,000 to 20,000 accidents were reported, the primary causes of which were equipment design flaws, maintenance or calibration problems, and user error (Bassen, 1986). Accident and product-problem data are maintained in several publicly available computer databases (Sind, 1990).

A primary focus of human factors research has been the interaction of people with machines. The term *person–machine interaction,* in this context, conjures up a variety of images: a driver behind the wheel of an automobile, a pilot in an airplane cockpit, an office worker at a computer terminal. Less commonly, I suspect, does it evoke a mind's eye picture of a surgeon, an anesthesiologist, or a nurse with the tools of their trades, but the person–machine, or person–device, interaction is no less real in these cases than in the others, and the same kinds of issues pertaining to usability and safety apply.

Another image that is rarely evoked by the idea of person–machine interaction is that of a person connected to a variety of monitoring and life-supporting devices while confined to a hospital bed, but this is a form of person–machine interaction in which the coupling is extreme, and it is one that has implications not only for the individuals' physical health, but for their psychological state and sense of personal identity. H. Leventhal et al. (1982) pointed out, also, that technology can contribute to dehumanizing experiences by interposing itself between the practitioner and the patient: "The size and strangeness of the machinery and the need to modify one's residence and pattern of eating and sleeping to be looked at by a machine, stimulate a sense of self as an object handled by other objects" (p. 104).

The practice of medicine has not been a major focus of attention of human factors researchers in the past, but as medicine makes ever increasing use of technology, the need for such attention grows. As the technological devices that are used in medical contexts become increasingly sophisticated and as they are used more frequently in contexts other than operating rooms and intensive care units that are staffed by highly trained professionals, the importance of effective designs of interfaces and operating procedures will also increase. The application of human factors know-how to the design of medical devices, and to the layout of operating rooms, emergency rooms, and other hospital service units should be helpful in reducing the number of accidents and mishaps that occur in the provision of medical care, and in making the experience of being a patient—which is seldom a totally positive one—less dehumanizing than it otherwise might be.

Other opportunities for human factors research relating to biotechnology in the medical and health care domain include: further study of human factors issues in the design and interpretation of medical imaging displays; application of ease-of-use and failsafe principles to the design of medical tools and equipment; studies of decision making in medical and other diagnostic and care-giving contexts; cost–benefit and tradeoff analyses with respect to medical testing procedures; investigation of human error in operating rooms and other medical contexts, and the development of error-prevention and error-recovery techniques; studies of communication between medical professionals and patients; and design of information resource systems for medical practitioners.

Biotechnology promises to be an exciting and rapidly moving field of research and development during the next few decades. The long-range consequences of developments in this area, undoubtedly, will be profound, not only as they relate to medicine and health care but as they affect other domains as well. With respect to consequences that relate to traditional human factors concerns, some of the anticipated developments are likely to change drastically certain jobs that people do. The use of microorganisms, and other applications of biotechnology–in industry, agriculture, and mining–will alter the demands for human involvement and will change the nature of the tasks that will have to be performed. Many of the changes that will be effected more generally will have implications for the ways in which people interact with tools, systems, and environments.

9

Information Technology

Information technology has to do with making marks on a medium and moving those marks from place to place. The marks are not arbitrary, of course, but are meant to signify meaning according to some agreed-upon conventions or code. The purpose of moving them from place to place is to convey specific messages from the people who originate them to their intended receivers. Over the millennia the media have included stone, clay, animal hides, papyrus, paper, and, very recently, magnetizable and optically modifiable surfaces. Until modern times the only method of conveyance over long distances was transportation, that is, physical conveyance of the medium on which the marks were borne. Only relatively recently have various forms of transmission that do not require transportation been possible.

The latest chapter in the history of information technology—the one that is still being written—is a rapidly moving one, indeed. The last few decades have yielded numerous developments that have greatly increased the efficiency with which information can be gathered, processed, transmitted, and stored. Notable among these developments are the vacuum tube, the transistor, integrated circuitry, laser and fiber optics, and the electronic digital computer. We may summarize the consequences of developments in information technology by saying they have made it possible to store very large amounts of information in very small spaces, to transform that information quickly in many useful ways, and to transmit large amounts of information from place to place extremely rapidly and at low cost.

In principle, computers do nothing that could not be done with other media and methods of conveyance. Their power comes from their speed, both the speed with which they lay information down (make marks), and that with which they transmit it from point to point. It is important that both of these operations be fast, because the slowest of them will limit what the computer can do in a fixed amount of time. In the early days of computers, the marking speed was determined by the time it took to open and close a mechanical relay or turn a vacuum tube on or off. These were very slow operations compared to the time it took to get an electronic signal from one point to another over a copper wire. As relays and vacuum tubes have been replaced, first with discrete transistors and then with integrated circuits, the marking speed has increased by several orders of magnitude, while transmission speed has increased hardly at all. Because electrical signals travel at an appreciable fraction of the speed of light already, one cannot expect huge increases in performance here. This problem has been addressed, therefore, not by increasing transmission speed, but by decreasing the distance that signals have to travel, at least inside any given computing device.

TRENDS IN COMPUTING DEVICES

The major trends in computer technology over the last few decades are well known: steady and spectacular increases in speed and reliability of components, accompanied by equally spectacular decreases in component size and power requirements and in unit costs. All of these trends are due largely to advances in the technology of miniaturization, and such advances are expected to continue more or less at their current rate at least through the beginning of the 21st century. Switching speed increased by approximately three orders of magnitude (from a microsecond to a nanosecond) between 1960 and 1980, and is expected to increase by another three orders of magnitude, reaching about a picosecond, by 2000 (OTA, 1985b).

Electronic gates change state (open or close) as a consequence of the movement of electrons, so the speed of their action can be increased either by increasing the speed at which the electrons move through a medium or by decreasing the distance they have to travel. The medium has been silicon for some time, and the speed increases have come about as a consequence of miniaturization. The number of active elements that can be placed on a single silicon chip has increased by roughly three orders of magnitude every 10 years since about 1960 (Bromley, 1986). Increases in packing density cannot, of course, continue indefinitely. Linear extrapolation of past trends would make the number of transistors on a chip in about 250 years equal to the estimated number of atoms in the universe. There is little evidence that the technology will run up against fundamental limits in the immediate future, however. For a description of several beam-controlled processing technologies (using photons, electrons, ions, and molecules) that are likely to be involved in future efforts to increase the packing density of components on semiconductor chips, and thereby decrease further the distances electrons must travel, see Kern, Kuech, Oprysko, Wagner, and Eastman (1988).

New Materials

Further gains in switching speed in the future may be realized from the use of different building materials (e.g., electrons travel more rapidly through gallium arsenide than through silicon) and by going to devices that operate by moving photons rather than electrons. The effective mass of an electron in gallium arsenide is about one third the effective mass of an electron in silicon; consequently, gallium arsenide transistors are believed to be able to attain switching speeds from two to five times as fast as silicon transistors (Chaudhari, 1986). Also, because electrons move through gallium arsenide more easily than through silicon, gallium arsenide circuits consume less power and produce less waste heat than do silicon circuits. Furthermore, gallium arsenide can function under a broad range of temperatures and is highly resistant to high energy radiation, which means it is suitable for incorporation in devices that must operate under adverse temperature and radiation conditions. It also has the added advantage of being able to radiate and to detect radiation in the near-infrared range, and therefore can be used as a photo emitter or detector within the context of high-speed electronic circuitry (Brodsky, 1990). Unfortunately, large defect-free crystals of gallium arsenide are much more difficult to grow than are those of silicon, and gallium arsenide components cannot be packed as densely as can silicon components. (A planned activity for the space station is the attempt to grow such crystals in an environment that is relatively free of gravity, impurities, and convection forces.) Diamond is also being investigated as a possible basis for electronic chips because it, too, has some especially attractive features (Whiteside, 1988). Photonic switching devices, which are at a more preliminary stage of development, promise even greater speeds eventually.

Computer circuits of the future may also be built of devices that exploit quantum mechanical effects. The use of such quantum semiconductor devices may make it possible to put an entire supercomputer on a single chip (Bate, 1988). Gallium arsenide may provide the basis for the development of quantum-effect transistors for which substrate layers may be only a single atom thick, and switching is effected by electron tunneling. Texas Instruments has already built a quantum-tunneling transistor that can switch on and off a trillion times a second. Quantum-effect transistors are still in a very preliminary development stage, but considerable energy is being devoted to solving the problems that stand in the way of their practicality (Port, 1989).

New Architectures

Further increases in computational speed are being realized through the development of machine architectures that exploit parallelism and multiprocessor coordination. Some observers believe that machines capable of executing a trillion operations per second will be feasible before the end of the 20th century (Beardsley, 1987b). In any case, computer systems with varying degrees of parallel-processing capability will become increasingly common. Interest in the building of multiprocessor systems dates back at least to the design and construction of the ILLIAC-IV, a 64-processor machine, during the latter half of the 1960s. A few other parallel-architecture machines were designed and built in small quantities during the 1970s. Several more—among the better known of which are the Cosmic Cube, developed at the California Institute of Technology, the Connection Machine, built by Thinking Machines Inc., and the Butterfly and TC 2000, built by Bolt Beranek and Newman Inc.—were introduced during the 1980s. IBM is developing a parallel-architecture machine that will have 32,768 processors, each capable of 50 megaflops (50 million floating-point operations per second), and it is claimed that Thinking Machines Corporation is thinking about building a connection machine with over a million processors (Denning & Tichy, 1990).

Unlike single-processor machines, all of which are fairly similar to each other in basic design, parallel-architecture machines differ greatly from each other in many respects, including the sizes of the processors and how they are linked, the way in which memory is partitioned and is accessed by the different processors, and the degree to which the processors share instructions and data. In some cases, the individual processors are tightly coupled and work in concert, all executing the same instruction sequence in synchrony; at the other extreme, are systems of loosely coupled processors, each of which works independently of the others, producing results that can then be aggregated by other processors that have been programmed to perform that task. It may turn out to be more efficient to use machines of different designs for different problem areas than to attempt to develop one general-purpose parallel architecture that is suitable for all problems.

One hears much today about supercomputers. The meaning of *supercomputer* changes continually as computer technology advances; as La Brecque (1990) pointed out, a general definition would be something like "the most advanced computing technology available at a given time" (p. 523). The machines just mentioned qualify as supercomputers, and most supercomputer systems today have a significant degree of parallelism. Supercomputer applications include simulations and analyses of aerodynamic flow patterns; modeling of molecular structures; flight simulation; and modeling of stress and earthquake resilience of bridges, nuclear reactor containment vessels, skyscrapers, and other large structures (Erisman & Neves, 1987). Learning to exploit parallelism fully will be a major challenge because it requires the development of qualitatively different approaches to problem solving. What parallelism means in the context of multilevel, nested processes of varying durations is a complex matter that is not well understood.

Looking ahead, Corcoran (1991) describes the "decathlon of supercomputer designers" as the three Ts: "a trillion operations a second, a trillion bytes of memory, and data communications rates of a trillion bytes per second" (p. 102). Corcoran points out that each of these goals is about three orders of magnitude greater than the capability of most existing supercomputers and that realization of them will require abandonment of traditional approaches, such as serial computing. There seems to be little doubt, however, that the goals will be realized, and in the not very distant future. Indeed, on October 29, 1991 Thinking Machines Inc. announced the CM-5, a massively parallel machine that it claimed has the capability of reaching one trillion operations per second; the machine is modular and can accommodate up to 16,000 processors. The rate of advance in computer technology continues to be nothing short of phenomenal. What the implications are of the unimaginably enormous and ever-increasing amounts of computing power that the technology is making available remain to be seen. The fundamental question, of course, is whether we can learn to use this power wisely and well.

Storage Technology

Advances in the miniaturization of logic circuitry have been matched by similar progress in information storage technology. CD ROMs with a capacity of 540 megabytes have been available for several years. As D. C. Miller (1986) pointed out, given that these ROMs measure 4.72 inches across and weigh 0.7 ounces, they represent a storage capacity of about 12 billion bytes per pound. One CD ROM disk costing $10 or less, can store as much material as microfiche costing $150 to make, or books costing $1,000 to print (Laub, 1986). As a rough rule of thumb, 1 megabyte is about one book of 400 pages of text, with 350 to 400 words per page. So, a 500-megabyte disk will hold the equivalent of about 500 typical books. To date only a few books (encyclopedias and reference works) have been put on disks, but it is easy to imagine this technology becoming widely used if it is coupled with well-designed aids for retrieving specific information from, or browsing in, large information repositories.

The most important type of memory for builders of computers and computer systems has always been the high-speed random-access memory that is an integral part of the machine. Secondary, or auxiliary, memory has been important, too, but not nearly as important as the random-access memory, which is usually referred to as *primary memory* or *machine memory*. For many years random-access memories were commonly made of individual, magnetizable doughnut-shaped cores, each of which was capable of storing one bit of information, so for some time *core memory* was more or less a synonym for primary or machine memory. Today random-access memories, usually abbreviated RAMs (or DRAMs, for dynamic random-access memories), are made as integrated circuits that can accommodate many orders of magnitude more information per unit area.

Chips that can store 16 megabits of information have been state-of-the-art for a while. In June, 1990, Hitachi announced a prototype 64-megabit DRAM chip. This chip, which has circuit lines measuring about .35 μ, is very close to the limit of what can be obtained with optical lithography techniques. In fact, it is better than what was thought possible until very recently. To obtain smaller feature sizes, say .20 to .25 μ will require the use of wavelengths shorter than those of visible light. Expectations are that with x-ray lithography it will soon be possible to mass produce DRAM chips with 256 megabits of capacity. Japan is said to be spending about twice what the United States is on the development of this type of capability, however, and is therefore expected to lead the world in electronics production and trading within a few years (Corcoran, 1990c).

Random-access storage devices will continue to decrease in size and cost per unit of capacity and to increase in storage capacity per unit area (or volume). In an experimental storage system, IBM has succeeded in achieving a packing density of 1.8 megabits per square millimeter, or one gigabit (1 billion bits) per square inch (Herskovitz, 1990). New storage technologies that are being investigated for possible future use include technologies that mix magnetic and optical techniques in an effort to get the best of both worlds; such a hybrid system might, for example, use a laser beam to read and write data on a magnetic medium (Kryder, 1987). Optical techniques also offer the possibility of much greater packing density for information storage; experimental work is being done on three-dimensional optical storage devices with a theoretical maximum capacity of 6.5×10^{12} bits per cubic centimeter (Parthenopoulos & Rentzepis, 1989).

Costs

The effect of steady progress in miniaturization of computing and storage components has been to decrease, steadily and dramatically, the cost of computing operations. Between 1950 and 1980, the cost of a fixed number of calculations dropped by roughly an order of magnitude every 10 years (OTA, 1981). Entry-level workstations decreased 60-fold in price just during the 1980s (U.S. Industrial Outlook, 1989). For some time now, the bulk of the expense has not been in system components but in the interconnections among them (Mayo, 1986). The cost of connections between a computer chip and external wiring, for example, is several orders of magnitude greater than the cost of connections within the chip. Extrapolation of current trends leads to the expectation that in the not-distant future, very large amounts of both computing power and storage capacity will be available to the average person at very modest cost. It has been argued that we should work on the assumption that there will be enough computing resources to satisfy everyone's needs at costs that are within everyone's means (Giuliano, 1982).

Personal Computers and Integrated Computing Devices

Recently, the fastest growing aspect of the computer industry has been the market for small machines. The number of personal computers in use worldwide went from about 200,000 in 1977 to over 50 million in 1987. Over the same period, the value of the world market for personal computers went from a few million dollars to $27 billion. According to several surveys in 1983 and 1984, between one half and three quarters of home computers in the United States were being used primarily for game playing. Other uses include word processing, learning, information retrieval from computerized databases, checkbook balancing, and household budgeting (Forester, 1987). The user population, or at least the potential-user population, for which interfaces must be designed now includes essentially everyone.

One vision of what will be available by the end of the 20th century includes a battery-operated 1000 × 800-pixel high-contrast display, a fraction of a centimeter thick and weighing about 100 grams, driven by a low-power-consumption microprocessor that can execute 1 billion operations per second and that contains 16 megabytes of primary memory, augmented by matchbook-sized auxiliary storage devices that hold 60 megabytes each, and larger disk memories with multiple-gigabyte or even terabyte (trillion byte) capacity (Weiser, 1991).

The number of computers in the world that *look* like computers is increasing rapidly and is likely to continue to do so for the foreseeable future, but so is the number of computing devices that are *not* visible because they are incorporated as parts in other things that do not resemble computers at all. Indeed, microprocessor-based computing power will soon be everywhere: in vehicles, household appliances, television sets, hand tools, games and toys, clothing, and devices implanted in bodies for medical purposes, among other places.

NETWORKS

Accompanying the spectacular advances in the design and manufacture of computing machines has been an equally spectacular development of the technology for connecting computers in communication networks. Tesler (1991) has characterized the history of computing technology in terms of four paradigm shifts that have occurred at roughly 10 year intervals, each of which has introduced a new mode of computer use. The first such shift—to batch computing—occurred in the early 1960s, when the computer began to be used on a large scale by corporations for data processing. The second and third shifts led to widespread use of time-sharing and desktop computing in the 1970s and 1980s, respectively. The fourth shift—to networked computing—is just now occurring; its harbingers, Tesler suggests, "are the increasingly networked laptop devices and electronic pocket calendars— mobile machines I call pericomputers" (p. 88). The growth of networked computing will be visible in the proliferation of wireless terminals that are sufficiently compact to be carried essentially anywhere.

A source of many of the ideas behind the development of computer networking was Paul Baran of the Rand Corporation, who expressed these ideas in a series of memos in the early 1960s and in at least one published report (Baran, 1964; see also Boehm & Mobley, 1969). Shortly after Baran's report, small experimental networks were designed at the Computer Corporation of America (Marrill & Roberts, 1966) and at the National Physical Laboratory in Middlesex, England (Davies & Barber, 1973). What was to become both the largest and the dominant network for many years was launched under the auspices of the Advanced Research Projects Agency of the U.S. Department of Defense in 1969 (Heart, 1975; Heart, McKenzie, McQuillan, & Walden, 1978). The ARPANET, which was built and operated by Bolt Beranek and Newman Inc., started out as a 4-node network. Its successor, the Internet, supports several million users operating on over 5000 networks in 26 countries (V. G. Cerf, 1991). Wright (1990a) refers to the Internet as "the intangible eighth wonder of the world" (p. 92). Crucial to the success of the ARPANET was the new technology of packet switching that made possible the efficient use of telephone trunk lines already in place.

A *packet-switching network* gets its efficiency, in part, from the fact that it does not tie up transmission lines during times when nothing is being transmitted on them. Consider a long-distance telephone conversation. When a connection is made between the caller and the receiver of the call in a conventional circuit-switching network, the route over which that connection is made is reserved for the parties to that conversation until one or the other of them hangs up. Thus, that circuit is not available for other uses even during pauses in the conversation when no information is being transmitted over it. In a packet switching network, the information that is to be transmitted is divided up into small "packets," each of which has the ability to find its way to the ultimate destination by the fastest available route. Different packets from the same message may take different routes and may even arrive in an order different from that in which they were sent. Header information carried by each packet allows the packets to be arranged in the proper order at the receiving end. This approach permits more efficient use of the total capacity of a network, because it ties up any given node-to-node link only when it is actually transmitting information. (For further details regarding packet switching, see R. E. Kahn, 1978.)

There are now numerous networks in the United States and throughout the world serving a variety of user communities. In the United States, one user community that has been the focus of considerable attention, especially in recent years, is the academic world and, in particular, academic researchers. Several commissioned reports have pointed out the need for better access by this community to advanced computing resources (Curtis & Bardon, 1983; Lax, 1982; Press, 1981). One aspect of this need is for network facilities that will give researchers access to the limited number of supercomputers in the country and to each other. The U.S. government and, in particular, the National Science Foundation, have responded to this need by funding five supercomputer centers and establishing the NSFnet, which links these centers and through which scientists at other universities have access to them (Jennings, Landweber, Fuchs, Farber, & Adrion, 1986). The five centers, which were funded for a 10-year period, from 1985 to 1995, are at Carnegie-Mellon University, Cornell University, Princeton University, the University of California at San Diego, and the University of Illinois.

The NSFnet came into existence in 1985, and has been evolving since its launching, growing in size and increasing in speed. Beginning as an access route to supercomputers, it is becoming a high-speed channel linking major research computing facilities in the United States and in other parts of the world. Because communication among users is so easy, it greatly facilitates collaboration among colleagues at different universities and research establishments. Electronic mail is its primary use, but it also provides for the sharing of large amounts of data and remote use of programs and computing facilities. The capacity of this network is expected to expand rapidly over the next few years (La Brecque, 1990).

As networks began to multiply, the problem inevitably arose of establishing certain standards of operation and communication that would permit them to be interlinked. The need for such standards is universally recognized; getting agreement on what the standards should be, however, has required a great deal of effort, cooperation, and compromise. Network protocol standards in the United States have for some time been the TCP/IP (Transmission Control Protocol/Internet Protocol) developed by the Defense Advanced Research Projects Agency. International protocol standards represented by Open Systems Interconnect (OSI) are scheduled to become the lingua franca of networking worldwide, sometime in the 1990s.

Some observers see in networks the prospects of work environments that will enhance the productivity and effectiveness of their users greatly, through the immediate access they provide to remotely located resources and colleagues. Jennings et al. (1986) describe, their expectations for a research network intended to serve all research scientists and engineers as follows:

> Our vision of this network is of a vast network of networks interconnecting the scientist's local advanced graphics workstation environment to other local and national resources. Scientists and engineers will be able to work at such workstations, using tools that are both comfortable and familiar and interacting with an environment that reflects a model of their scientific world. Through the network, a researcher will be able to build programs, execute and modify models, and collect and analyze data without concern for where the tools or programs come from or models reside. The scientists will be able to bring powerful computational resources to bear on problems, without explicit knowledge of the physical machines and communications involved. The procedures and formats for accessing these resources and other services will be as uniform as possible. . . . Our vision is of a network integrating the computer resources available and presenting these resources to the user as a single interactive system. (p. 231)

Local, National and Global Networks

Networks designed to serve a single office building differ from those intended to cover the world. The bandwidth of both local-area and wide-area networks has been increasing dramatically and promises to continue to do so for the foreseeable future. For some years, 56 kilobits per second was considered a high bandwidth link for a wide-area network. Today the more common component is a T1 link operating at 1.5 megabits per second. Ethernet provides a 10-megabit transmission rate for local-area networks. Fiber Distributed Data Interface products that are now available offer 100-megabit rates. Microwaves have been used as the basis for regional networks operating at 10 megabits, and 1-gigabit fiber links are available in some areas. Multiple gigabit and terabit (trillion bit per second) long-distance networks based on laser and fiber-optic technology are probably on the way (Partridge, 1990). R. E. Kahn (1987) has suggested that transmission rates of 10 gigabits per second might be attained by the end of this century; an operating network with a capacity of at least 1 gigabit seems highly likely.

The development of an integrated national computer network system that would permit communication between any two computers in the country has recently been called for by the National Research Council's Computer Science and Technology Board (1988). A. Gore (1991) has called for a national commitment to the building of high-speed data highways, the absence of which he has referred to as "the largest single barrier to realizing the potential of the information age" (p. 152). A worldwide integrated-services digital network (ISDN), capable of handling digitally encoded information of any type (voice, data, facsimile, motion pictures) and linking homes, offices, and schools to information resources of various types around the world (libraries, museums, national and international data banks), is seen by some as a possibility in some form by the turn of the century (Forester, 1987). There is some chance that Eastern European countries will be connected to the West via Bitnet in the near future (Pelca, 1990).

Denning (1989a) gives a picture of an emerging worldwide network of computers that he refers to as Worldnet: "By the year 2000," he suggests, "Worldnet will be ubiquitous and pervasive. It will be as important for conducting business, distributing information, and coordinating work as are the existing transportation and telecommunications networks. Few enterprises, commercial or scientific, will succeed without mastery of this technology" (p. 432). Among the implications of the existence of a worldwide network, Denning notes, is the possibility of achieving the coordination of worldwide efforts on such problems as averting famines, fighting AIDS, mapping the human genome, and modeling global climate change. He notes too, the implications of such a network for individuals: "Computers with cellular telephone and FAX connections will be common, enabling people to maintain a link to Worldnet no matter where they are—at work, at home, or travelling" (p. 432).

Fiber Networks

One of the technologies that is enabling the development of such facilities as Worldnet is that of fiber optics. Optical fiber has been displacing copper wire in telephone networks since about 1980; in the United States there are now over 1.8 million miles of fiber in the telephone system (Shumate, 1989). The first transatlantic fiber-optic cable, TAT-8, went into commercial operation between the United States and Great Britain in December, 1988. It can carry 40,000 conversations simultaneously, which gives it a greater transmission bandwidth than all other existing transatlantic cables and satellite links combined (Bell, 1989b). Television is already being transmitted to a few homes by fiber on an experimental basis; Shumate (1989) estimated that most U.S. residents should have access to the broadband world through fiber networks in less than 20 years, and many of them will in less than 10.

It seems inevitable that more and more of the purposes that have been served in the past by the movement of electrons will be served in the future by the movement of photons. Optical fiber will replace copper wire as the primary conduit for point-to-point communication. Photonic devices will also become increasingly common as logic elements and storage components in computer systems. The advantages of photonics over electronics are several—greater transmission speed, and thus bandwidth, lower power requirements, no problems of electromagnetic radiation—but the technology is not yet as well developed.

Satellite Networks

Another important enabling technology is satellite communication. The transmission of information via satellite has already made possible instantaneous broadband communication around the globe, so that people in all parts of the world can simultaneously witness the same event as it occurs. Satellite TV dish antennas, of which there now are some 2 million in American yards, have been a special boon to areas like the hills of West Virginia, where conventional broadcast reception is unusually poor (Stewart, 1989). The total bandwidth of the global system of satellite communication will continue to grow rapidly: The number of 40-megaHz-equivalent satellite transponders in operation in the world is expected to increase from about 1,400 in 1985 to nearly 9,900 by 2000 (Dede, Sullivan, & Scace, 1988).

Terminals

Developing better terminals through which people can access computer networks will be a major challenge for information technology in the future. The cellular telephone is a step in the direction of making practical wireless terminals a reality. Growth in the cellular telephone market is strong in the United States and outpacing expectations: The number of subscribers increased by 82%—from about 880,000 to about 1.6 million—in a recent 12-month period (U.S. Industrial Outlook, 1989). Prices still put this technology out of reach of the average consumer, but they are dropping, and in time radiotelephone communication could become very common. Cellular phone technology, or some comparable wireless means of transmitting telephone conversations, could eventually replace much of the transmission now accomplished by wire or fiber. (Whether the use of telephones by automobile drivers while driving constitutes a traffic safety hazard and, if so, of what magnitude are not yet known.)

High-resolution color monitors with three-dimensional graphics should also become widely available, perhaps before the end of the century (Dede et al., 1988). High-definition, flat-screen TVs permitting large wall-mounted displays are likely to be in place by 2000. Digital TV will make the television set a qualitatively different type of instrument with a whole range of new applications. Visual display technology is likely to make significant advances on several fronts, involving liquid crystal displays, electro-luminescent displays, three-dimensional (including holographic and volumetric) displays, and head-mounted displays. Reflection Technology Inc., of Waltham, MA, has developed a light-emitting-diode (LED) display, dubbed "Private Eye," that measures 1.2 by 1.3 by 3.5 inches, weighs 2.5 ounces, can be mounted on a head-set or clipped onto eyeglasses, draws less than 0.5 watts, and is claimed to produce a picture that has the appearance of what one would see on a 12-inch monitor (Curran, 1990).

The use of computers to help scientists and engineers visualize complex structures and processes—earthquake migration patterns, weather and climate systems, molecular structures, wave patterns and turbulence, biological cell growth and multiplication, population dynamics, black holes—is a subject of considerable interest. This interest is sparked both by the growing realization of the importance of visualization to the understanding of complex structures and processes and by the availability, especially at the supercomputer centers, of enough computing power to produce high quality visual images and to support their rapid modification and manipulation by viewers in an interactive fashion (Cassidy, 1990).

Promises and Problems

The existence of such resources is likely to result in the spontaneous emergence of new forms of inter-person communication. Electronic mail is almost certain to be used by a growing percentage of the population both for business and personal correspondence. As of 1988, there were an estimated 1.5 million electronic mailboxes in the United States (U.S. Industrial Outlook, 1989). Voice mail is now a major market for digital speech technology (Hogan, 1987). The electronic bulletin board is already an established institution. There are now probably several thousand such bulletin boards in operation, serving various communities of users with common interests. Among the more attractive aspects of the emerging communication facilities are the prospects of their use to provide new avenues to the world and to other human beings for people who would otherwise be isolated because of illness or infirmity.

Wideband communication facilities, coupled with very high-resolution display technology, will make it increasingly feasible to tap human expertise at a distance. Indicative of the possibilities is a "telepathology" system, set up in March, 1990, to connect the pathology departments of Emory University Hospital and Grady Memorial Hospital in Atlanta, Georgia. The system includes a robotically controlled microscope at one of the facilities that can be manipulated (i.e., directed or focused) by a pathologist viewing an image on a high-definition television display at the other (Erickson, 1990a). Real-time images are transmitted by microwave in this case, but other wideband channels are also possibilities for the future.

The continuing expansion of communications facilities will make it possible to move very large amounts of information from almost any point to almost any other in a variety of forms, including text, speech, and still and moving pictures. Such ease of information transfer will have implications for publishers, movie houses, recording studios, and other industries that have traditionally been involved in the distribution of information in one form or another, but exactly what those implications are is difficult to say at this point. This ease of transfer also has unknown implications for the individual user of the information that will be available: "More information than we can ever conceivably want will be available to us within seconds. What do we do with it? How do we condense, correlate, and sort it so that humans can base decisions on it? This is one of the major challenges facing all science, all society" (Bromley, 1986, p. 628).

As more and more homes are linked to computer networks with two-way connections, the possibility of instantaneous polls and referenda becomes increasingly real. Exactly what effects such capabilities will have on government remains to be seen. The ability to gather opinions from large fractions of a population on almost any issue very quickly would seem to have the potential to make the process of governing more democratic and participatory. To the extent that the primary function of representatives is to represent the interests and preferences of constituents who, for practical reasons, are unable to represent their own, the need for them could diminish. Decision making could become more of a total-community activity because the will of the people on any given question could be ascertained directly and rapidly. In theory, this sounds just right, but one must wonder. Perhaps the ponderousness and inefficiency of the system of government that depends on representatives to express the will of the people they represent has some virtues that we fail to see, but that would be missed if they were no longer there. The fact that things move slowly through the legislative process gives time for issues to be debated, and the fact that effecting change through legislative action is difficult may contribute to a degree of stability and predictability in our daily lives that is more important than we realize.

Securing computer systems against theft, espionage, sabotage, vandalism, and other forms of mischief will become a matter of increasing concern as ever more functions become dependent on the reliable operation of computer systems, and these systems become accessible to more individuals because of their network connections. Already, the ways of working mischief on computer systems of all types are many, and some of them are extremely subtle and hard to prevent (System Security Study Committee, 1990). The well-publicized 2-day minor panic within the Internet community in 1988, caused by a relatively benign "worm," provided only a hint of the kind of havoc that could be caused by a truly destructive "virus" (Marshall, 1988). (A computer worm or virus is a bit of code that is capable of replicating itself and inserting itself into other programs where it can do various sorts of harm, either immediately or at future times. The distinction between worms and viruses is subtle and perhaps not worth maintaining; however, the term *worm* has usually connoted a somewhat less destructive type of pest than has *virus*.)

Tracing the origin of worms, viruses, and similar types of disruptive code is an extremely difficult, although not always impossible, task. Stoll (1989) has told the story of an intriguing bit of sleuthing that led to the identification and apprehension of one major perpetrator of this type of mischief. Denning (1990) warns that "we can expect steady increases in acts of crime, espionage, vandalism, and political terrorism by computer in the years ahead" (p. 10); unfortunately, the good guys have no monopoly on brains. One can only assume that the people who engage in computer crime or vandalism either for profit or for whatever pathetic satisfaction they get from complicating other people's lives will continue to find new and ever more devious ways to do harm. If individuals and organizations become more and more dependent on computer systems for occupational, business, and private functions, as they are very likely to do, their vulnerability to such activities is likely to increase, as well. In part as a consequence of this concern, there will be increasing tension between the need to safeguard information for purposes of national or corporate security or the protection of individual privacy and the desire for free and open information exchange. Communication security has been a long-standing concern of the government and, in particular, the Department of Defense; it is rapidly becoming a concern of the business community and individuals, as well.

Information has always been difficult to control. One cannot guard it against theft as one might guard the crown jewels of Great Britain or the Mona Lisa. Because it is so easy to copy and to disseminate, containment is next to impossible. An interesting aspect of modern communication technology is the fact that it makes the control of information more difficult than ever. Lucky (1989) points out that computer networks "cut not only across corporate hierarchies but between corporations and even countries. They help promote a new measure of nearness that has little to do with corporate ties. Information floats around these networks in a fairly uncontrolled manner" (p. 17). In general, the emergence and proliferation of computer networks, word-processing software, desktop publishing systems and other computer-based means of facilitating the production and distribution of electronic "documents" have complicated considerably the problem of establishing intellectual property rights (OTA, 1986c).

The growing dependence of society on information systems is sure to raise numerous conflict issues pertaining to privacy, accessibility, and constitutional rights for which no legal precedents exist (A. R. Miller, 1971; OTA, 1981, 1987a). As Branscomb (1991) points out, the legal system under which we operate "arose in a era of tangible things and relies on documentary evidence to validate transactions, incriminate miscreants and affirm contractual relations," and is now faced with the problem of dealing with a situation in which "letters, journals, photographs, conversations, videotapes, audiotapes and books merge into a single stream of undifferentiated electronic impulses" (p. 154). Kapor (1991) worries that failure of legal and social institutions to adapt effectively to the new computer-based modes of communication—especially in our concern for controlling computer crime—could result in access to the global electronic media being viewed as a privilege rather than as a right, and that the conditions of access could be antithetical to the ideal of free and open information exchange.

An activity that illustrates the kind of conflict of interests that can arise from the widespread use of information technology in the workplace is the electronic monitoring of office work (M. J. Smith, 1988; M. J. Smith, Carayon, & Maezio, 1986). As of 1987, it was estimated that the work evaluation of about 6 million office workers was based, at least in part, on computer-generated statistics. These are statistics produced from computer-based work-monitoring procedures that often involve the automatic recording of quantitative aspects of work performed with computer-based systems (key stroke counts, time spent on various activities, destination and duration of telephone calls). The use of computer-based work monitoring is expected to increase as the use of computer-based systems in office jobs becomes increasingly pervasive (OTA, 1987a): "Equipment and software for telephone-call accounting (tracking the time, destination, and cost of calls) make up the fastest growing segment of the telecommunication industry" (p. 5). Computer-based work monitoring clearly can produce some useful information relating to questions of efficiency and cost control (the same OTA report included an audit of telephone call accounting records, which found that about 33% of the off-network long-distance calls made by employees on the federal communications system were personal calls), but the practice also raises serious issues regarding privacy, fairness, and quality of life in the workplace.

The existence of ever more powerful information resources raises also questions of equity and distribution of power. Tesler (1991) points out that "the chasms between rich and poor could widen . . . if the latest computing paradigm [networked computing] creates still more opportunities for educated people and still fewer for the uneducated" (p. 93). This observation applies, it seems to me, both at the level of nations and at that of individuals. If information technology has the enormous potential that many observers believe it has, ensuring its availability to Third World nations as well as to socioeconomically disadvantaged individuals within the industrialized world will be essential if the gap between the haves and the have-nots is not to be widened.

SOFTWARE

Computers and computer-based systems are only as useful, of course, as the software that runs them permits them to be, and software development know-how is generally acknowledged to be progressing considerably more slowly than hardware know-how (although there is some question as to what such comparisons mean). In any case, software development, not hardware, is believed to be the factor that will most limit the rate at which computer resources are applied effectively to various enterprises in the future (Peled, 1987).

Programming

As of 1987, there were an estimated 2,000 software companies in the United States, offering about 30,000 software packages for sale (Forester, 1987). The U.S. software industry has been projected to grow in revenues at an annual rate of between 20% and 25% for the next few years, bringing it to $50 billion in revenues by 1993. The total worldwide software market by that time is expected to exceed $130 billion (U.S. Industrial Outlook, 1989). In terms of percentage increase, computer programming has been among the fastest growing occupations in the United States, although the rate of growth appears to have been declining recently (Kraft, 1987).

As a cognitive activity, programming is very demanding. It requires one to be able to think effectively about complex processes at various levels, from the minute details of algorithms to accomplish specific tasks, to the overall organization of many processes interacting in complicated ways. As an activity that requires the use of deductive and inductive reasoning, planning, designing, hypothesis testing, among other cognitive processes, it is proto-typical of many complex thinking tasks, and for that reason represents a rich focus of study for psychologists interested in thinking processes. Given the rate at which the population of professional programmers is growing, there are very practical reasons for studying programming as a process. How to tell who will be a productive programmer and who will not remains a challenge for personnel selection and training functions. Programmer training programs can always stand to be improved, and there is a continuing need for better programming tools and environments.

Programming is a much more diverse activity than it once was. It can be done, for example, using a great variety of languages and in varied environments that make quite different demands on the programmer. The difference between using a high-level language and using machine code or a low-level language is analogous in some respects to the difference between giving instructions to a skilled artisan and giving them to a neophyte. Imagine the difference in the instructions that one would give to a master carpenter regarding the building of a house and what one would have to give to a new apprentice regarding the same job. To the former, instructions can be brief and global, and can assume a wealth of knowledge about house building; to the latter, they must be very detailed, and consequently lengthy, and can assume very little. Terms of the trade that are full of meaning for the master carpenter will have to be defined for the novice, and so on. The analogy is not perfect, but it suffices to make a point if not pressed too hard. Higher level programming languages have the ability to translate instructions expressed in forms that are relatively easy for humans to use into machine code, which is more concrete and precise but less natural for a human programmer.

Beyond this general observation, it is difficult to say much about programming that applies universally, because the activity can differ greatly from situation to situation. Bobrow and Stefik (1986) distinguished among several styles of programming: procedure-oriented programming, object-oriented programming, access-oriented programming, logic programming, rule-based programming, and constraint-oriented programming. Not surprisingly, different programming styles are not equally well suited to all types of applications. What may be less intuitively clear is that even for a given application different styles structure the programming task in different ways, and consequently have different implications regarding how the programmer should conceptualize the problem.

Traditionally, programming has been thought of as telling the computer what to do. In the case of rule-based expert systems and spreadsheet packages, it may be described more accurately as telling the computer what to know. The developer of an expert system who wishes to use a preprogrammed structure sometimes referred to as a "shell," has the problem of specifying, usually in the form of if–then statements or production rules, the knowledge that constitutes expertise in the domain of interest. The user of a spreadsheet package has the task of specifying the functional relationships among the variables that are to be represented on the spreadsheets.

The individuality of programmers and the variability of programming styles is seen in the results of extended interviews with individuals who have distinguished themselves in this field (Lammers, 1986). This is not to suggest that there are no principles or practices that the novice programmer should learn, independently of the specific type of programming to be done, but it does appear that, excepting perhaps unusually well-structured situations, programming is a creative activity and, as such, lends itself to highly individualized approaches. This is, in part, what makes programming attractive to creative individuals; it also can present some perplexities to managers; as a colleague of mine once put it, managing programmers is a lot like herding cats.

On the other hand, some observers believe that programming, on balance, is becoming less of a creative activity than it once was. On the basis of an interview study of 667 software specialists—including programmers, systems analysts, software engineers, and managers—Kraft and Dubnoff (1986, reported in Kraft, 1987) concluded that, although some software specialists do some creative, challenging work, others have routine, repetitive, boring jobs and that, on the average, software work is becoming less skilled, not more. "What is remarkable about the software shop," Kraft observed, "is not how different it is from the machine shop, but how similar" (p. 109).

Software Evaluation

The problem of evaluating software remains essentially unsolved. At the highest level of analysis, we can distinguish between two types of software bugs. Most straightforward and easily detected are those that result from infractions by the programmer of the rules of syntax of the programming language used or from violations of other constraints of the program development system. Often, these types of bugs can be detected automatically by the program development system itself. The other major type of software bug derives from conceptual difficulties on the part of the programmer. Examples of this would be failure to initialize a program loop properly, use of an inappropriate formula to compute some mathematical function, or performance of major program elements in the wrong order.

Large complex programs provide countless opportunities for such errors at all levels of program organization. Usually, they do not make the program unrunnable; they just ensure that it will not do precisely what it was intended to do. In some cases, the consequences of such errors are sufficiently obvious that the errors are easy to track; in others, however, the consequences can be very subtle, or intermittent, and they can remain undiscovered for a very long time.

What makes software evaluation especially difficult is the fact that one can demonstrate the existence of bugs in a program, but, in principle, one can never demonstrate that a complex program is really bug-free. No matter how extensive the testing, there is always the possibility that there remain undetected bugs the effects of which will be seen only under low-probability conditions that the evaluators have failed to anticipate.

Programming Environments and Tools

Programming environments are intended to facilitate the composing, testing, debugging, running, and enhancing of programs. Such environments include tools for creating programs, for finding and fixing errors in them, and for tracing their execution. In thinking about computing environments, the appropriate model to have in mind is that of the individual at center stage, with the various available tools and resources positioned around him. The situation is satisfactory to the degree that the tools and resources the individual needs can be accessed with a minimum of effort. Because people typically work on a variety of projects more or less simultaneously, moving quickly from one to the other, the environment must support that style of activity. Tools and resources that take a long time to be brought into the workspace represent an impediment to efficiency.

Human factors researchers have given a lot of thought to the layout of physical workspaces. For the most part, computing environments are better thought of as virtual workspaces because most of the tools to which one wants access are pieces of software. There are organizational problems, however, that are analogous in some respects to the problems encountered in laying out physical workspaces. Some ways of organizing a virtual workspace are more efficient than others. This problem has not, however, been much studied.

Programmer Productivity

In view of the importance of computers and information technology more generally to the economy of the United States, and indeed to the world, and the dependence of these systems on the software that operates them, programmer productivity will be a continuing focus of attention. How to gauge the productivity of programmers is not entirely clear. There are as yet no widely agreed upon methods or metrics for quantifying the productivity of individual programmers. Lines of code produced per unit time is much too gross a measure, but nothing much more acceptable has yet been proposed. Any satisfactory measure of the productivity of the software industry as a whole must take into account such considerations as what the resulting software accomplishes, the amount of maintenance and repair the completed software packages require, and the degree to which program building blocks are reinvented many times rather than being widely shared.

There are countless tasks that are common to many computer systems. It would be an enormous waste of intellectual capital for every programmer who wants a computer system to be able to do x to have to write the program for doing x. To the degree that individual programmers write many programs that already exist, programming, as a profession, is less productive than it could be. The productivity of programmers in the aggregate could be increased by the development of procedures and policies to facilitate the sharing of code. This is not an easy problem to solve, however, because there are many impediments to code sharing. Usually, there is no easy way to discover whether a program that one is about to write already exists. Often developers of software wish to keep their programs proprietary, but even if one knows of an existing program that is suitable for one's purpose and proprietary rights are not an issue, program documentation may be poor, and programmers may feel it is just easier to produce their own code than to attempt to understand someone else's.

Although there is unquestionably a great deal of duplicated effort in programming, with people writing programs that have been written many times before, there are indications that programmers are increasingly making use of other programmers' codes, sometimes modifying them to make a better fit to specific applications. As Bobrow and Stefik (1986) put it, "today's programs are parts of larger complex systems, and the main activity of programming has moved from the origination of new programs to the modification of existing ones" (p. 956). This means that an important aspect of a program is the ease with which it can be used by a programmer other than the one who wrote it. Bobrow and Stefik point out that programs are increasingly judged on these characteristics, as well as how well and how quickly they accomplish their intended processing tasks. One of the main attractions of "object-oriented" programming is its emphasis on the incorporation of existing program code (objects) in emerging software systems and in the production of reusable code (new objects that will be useful in other systems). Objects are "black boxes" that take specific types of inputs and deliver specific types of outputs; in order to use them, it is not essential to know the details of their inner operations.

Programming of Parallel Processors

Almost all of the many thousands of programs that have been written to date were written to run on serial machines. With the rapidly increasing availability of parallel-architecture systems of very great capacity, there is more and more interest in producing programs that can exploit the potential of these machines. Parallelism can mean many things and can be implemented to varying degrees and in varying ways. However, attempting to apply parallel-processing hardware to the solutions of particular problems invariably requires decomposing the problems into components that can be tackled by individual processors. Some problems lend themselves to decomposition in obvious ways; others do not. Efforts are being made to develop methods to modify existing programs automatically so they can be run with increased efficiency on parallel systems. This means identifying the parts of these programs that are parallelizable, and then recompiling the programs so that the parallelizable components can be run on different processors simultaneously. Although there are systems that do such parallelizing for programs written in certain languages (Fortran, Pascal, Ada), it remains to be seen how far the automatic translation of sequential programs to parallel form can be pushed. I suspect that, in order for a program to exploit fully the processing power that exists in a multiprocessor system, the programmer must write it with parallel execution in mind. If this is true, there is a need for the development of new ideas, new tools, and new computing environments that facilitate thinking about parallel approaches to problem solving.

Knowledge of how to program parallel-architecture computers is considerably behind the efforts to develop such machines, and much more energy is going into hardware design than into the development of software (G. C. Fox & Messina, 1987). Because there are, as yet, few experts in writing programs for parallel processors, and automatic parallelization software is still very limited in what it can accomplish, the full potential power of parallel systems is seldom brought to bear on a specific problem. As Kuck, Davidson, Lawrie, and Sameh (1986) put it, "multiprocessor systems to date have not made parallel execution convenient, and few users have found it worth the large effort required to restructure their programs" (p. 968). As a result, programs that run on large multiprocessor systems seldom realize anything like the peak performance of these systems.

Gelernter (1989) uses the metaphor of the refinery to characterize some of the software systems that are currently being developed for parallel processing machines. An *information refinery,* in his terms, is a system that can "transform mere facts into knowledge on a vast scale" (p. 66). Two kinds of machines that hold promise for the realization of such systems, according to him, are information filters and smart databases. An information filter takes an incoming stream of data and transforms it into higher level knowledge; a smart database is then capable of finding interesting patterns by searching efficiently through large stores of existing data.

Gelernter describes an architecture for filtering that is especially well suited to implementation on parallel-processing machines. This architecture is referred to as the *trellis architecture* because it has a hierarchical organization in which the outputs of lower order components feed into a smaller number of higher order components at each level of the hierarchy. At the very bottom are modules connected to sensors in the real world. At the top are components that make classifications and decisions at a relatively high cognitive level. The trellis organization is inherently parallel. All of its modules run continuously and concurrently, the outputs of the lower level modules being passed to higher level modules as they are generated. Information can flow in both directions in the hierarchy, thus permitting feedback and control from higher level units to lower level ones. Gelernter notes too, that the trellis organization is *transparent* and *locally comprehensible,* which is to say that it is easy to figure out what each module does, and the way the modules are interconnected reflects the structure of the problem the machine is designed to solve. It follows directly from these properties that programmers are able to work effectively on individual modules without necessarily understanding the machine in its entirety, and this greatly facilitates the development of programs containing very large numbers of modules.

The trellis organization is reminiscent of some pioneering work done by Selfridge and Neisser on the problem of pattern recognition (Neisser, 1967; Selfridge, 1959; Selfridge & Neisser, 1960). These investigators developed a model in which the recognition process was assumed to be mediated by a hierarchy of processes as in the trellis organization described by Gelernter. Processes ran at a given level in parallel and the outputs at one level were inputs for those at the next higher level. The model was called Pandemonium because its operation resembled the uncoordinated activity of a collection of hierarchically organized demons. Data demons (feature analyzers, say) were the lowest order components. Each data demon inspected an input, looking for a specific feature. All demons at this level worked simultaneously and independently of each other, shouting their findings at the demons in the next level of the hierarchy. Each demon at that level (a character analyzer, say) listened for that particular combination of inputs that defined the pattern for which it watched. When a demon at this level recognized a character, it in turn shouted at demons at a still higher level that might be watching, for example, for specific words. Paradoxically, Pandemonium was programmed and tested on a serial-processing machine because its conception predated the existence of parallel-processing architectures; conceptually, however, it was a parallel process and well suited to implementation on a parallel-architecture system.

SOME EXPECTATIONS

As of 1988, purchase of information technology—computer and communication technology—accounted for about 40% of all new investment in plant and equipment in the United States; 10 years before that it accounted for about 20% (OTA, 1988c). By the end of the century the worldwide sales of the electronics industry could be $1 trillion (Chaudhari, 1986). This phenomenal growth in electronic circuit "consumption" is due, in part, to the explosive increase in the population of people who use computers daily in their jobs or for avocational purposes and, in part, also, to the increasing utilization of computing devices in automobiles, household appliances, wearable or pocket items, and children's toys. Nearly everyone is already a user of computing resources in one way or another, and in the future the number of ways in which people interact with computers, directly or indirectly, can only increase. Indeed, applications of computer and communication technology have proliferated so fast and are now so extensive that one hardly knows how to begin to consider them. Here I mention only a few of the directions the uses of information technology may be expected to take in the future.

More Software for the General Consumer

A great deal of practical software will be produced that is intended for use by the general consumer. Such software will provide information regarding nutrition, the environment, gardening (efficient use of available space, environmentally benign gardening practices), home finances, tax-return preparation, first aid and emergency medical treatment, retirement planning, will preparation, education planning, home repair, and countless other topics. Some such software packages are already available, and many others will be produced.

One expects that some of the many programs of these types that will be brought to market will prove to be very useful; it seems safe to assume, also, that many will not. Given the lack of a theory that could provide the basis for designing personal information systems, a lot of experimentation will be needed to evolve systems that provide users with the functionality they will want and the kind of interfaces they will find convenient to use. There are many opportunities here for human factors involvement. The market for personal information systems is undoubtedly enormous, but only those systems that individuals feel comfortable using and that prove to be useful will survive in this market.

Information Services

Computer-based information services addressed to a wide diversity of objectives—job posting, want ads, selected news, information searches—will proliferate. As a consequence of the capability for two-way communication via cable TV and home computers, there will be increasing opportunities for instant polling, including regional, national, or even international referenda. The potential for educational uses of computer and communication technology has barely begun to be realized, although 35 states are now delivering or plan to deliver some instruction via electronic communications systems (OTA, 1988a). Electronic desktop editing and publishing systems will continue to impact the publishing industry in major ways, and in conjunction with computer-based communication systems to which they can connect, they are likely to transform the way information is collected, represented, conveyed, and used.

There will exist extremely large collections of information in electronic form. There are many large collections in existence today that serve the interests of specific companies or governmental agencies. Major computer-based national information systems already in operation as of 1981 included the National Crime Information Center of the Federal Bureau of Investigation, the FEDWIRE electronic funds transfer network of the Federal Reserve System, various nationwide credit-card and check-authorization services, the air traffic control system operated by the FAA, the nationwide airline reservation systems operated by major air carriers, and the automatic quotation system for obtaining over-the-counter stock prices (OTA, 1981). A national information resource along the lines of that envisioned by Yourdon (1971) remains a possibility.

Some of the databases that are currently being developed for special purposes (e.g., those holding the results of efforts to decode segments of the human genome as well as DNA sequences from other organisms, and the astrophysical data that are being collected at numerous observatories and by space-probing satellites) are already large and are growing rapidly. Access to these databases is critical for scientists and technicians working in these areas. The problem of making the access easy becomes increasingly difficult as the amount of information in the databases grows, at exponential rates in some cases, and is stored in a variety of systems with different formats and query languages. As the amount of information that is "available" and relevant to users' interests continues to increase, the task of getting to individuals precisely the information they need (and not more), when they need it, and in a form in which they can use it, becomes ever more difficult. The complexity of the problem gets truly great when one thinks of providing access to extremely large general-purpose databases, such as the information stored in the Library of Congress.

Before many more decades, there will probably exist electronic books that can present, along with conventional prose, explanatory and elaborative information regarding specific parts of the text: definitions, explanations, maps, simulations, and answers to specific questions about the time, place, and circumstances of referenced events. "Virtual travel," in which one "travels" through electronic representations of spaces of interest (buildings, towns) may become widely used for both educational and recreational purposes. Experimental systems that permit virtual travel in a limited sense have already been built (Lipmann, 1984; K. S. Wilson, 1987). Electronic pocket fact-finders may be developed, containing the kinds of information now found in almanacs, fact books, yearbooks, and other repositories, that can provide answers to questions of fact immediately on request. Personal computers in the future will serve the dual role of being very powerful information-processing engines in their own right and providing an access port to a world of distributed information resources of an unimaginably rich variety.

The combination of greatly increased communication bandwidth and storage capacity resident in personal computers will open the way for far greater use of pictorial imagery for a large variety of purposes. When the bandwidth and storage capacity are sufficient to permit rapid transmission and temporary storage of large quantities of digitized visual images, uses of this technology will proliferate. One can imagine the time, not that many years hence, when, in order to view a movie of one's choice at the time of one's choice in one's home, it will not be necessary to acquire a videotape. All one will have to do is dial up the library, which will transmit the movie digitally to one's home computer where it will be stored for showing on one's home viewing system. There will be no need to maintain a home film library because it will be less expensive to have a copy transmitted from the library whenever one wishes to view it.

Teleoperator Systems and Telepresence

A *teleoperator system,* as the term is typically used, includes a machine that is operated from a distance and the human being who operates it. The person controls the machine through motions (grasping, pushing, pulling, turning) that he would make if he were in the position of the machine doing the work the machine is intended to do; and he receives visual and perhaps tactual information from the site of the machine that are similar in important respects to those he would receive if he were there. The human's motions may be amplified or given greater force by the machine. The idea is to couple the human operator's perceptual, cognitive, and motor capabilities to the machine's power and ability to operate in environments that would be hostile or inaccessible to people.

A particularly interesting aspect of teleoperator technology is the question of how to create for the human operator a sense of "remote presence" without being exposed to the risks of the situation in which the machine is functioning. One wants the human to experience sensory input that is similar in some ways but not identical to what he would experience if he were in the location of the machine. The experience has to be similar enough to create the feeling of presence, but not so similar as to reproduce the characteristics of the situation that would be harmful or dangerous. How to accomplish this is not yet known. Spatial and temporal perturbations in the visual feedback that the operator receives from the system have severe effects on performance (T. J. Smith & Stuart, 1990).

Some experimentation is being done by the U.S. Army with teleoperated vehicles. The remotely located operator sees a driver's view of the terrain through a helmet-mounted display that receives its input from twin television cameras positioned where the driver's head normally would be in the vehicle. An initial problem with this system is the fact that operators frequently experience a form of motion sickness resulting from the conflicting cues sent by the visual and vestibular systems to the brain (Uttal, 1989).

Most of the work to date on teleoperators has involved machines that are roughly comparable in size to their human operators. One can imagine a teleoperator system, however, in which the machine is orders of magnitude smaller or larger than the person who operates it from a distance. Ultrasmall teleoperator systems might be useful in medicine, for example, and ultralarge ones might be useful in heavy construction.

Artificial Intelligence (AI)

Work on machine intelligence seems likely to make steady progress. A computer program may yet win the world championship in chess. Although the original predictions of this event proved to be overly optimistic (from the point of view of the computer enthusiast), chess-playing programs have improved steadily. One such program recently attained a rating of about 2600, which indicates a playing strength in the bottom half of the grandmaster range; it is able to beat grandmasters about half the time under conditions of tournament play (Hsu, Anantharaman, Campbell, & Nowatzyk, 1990). (It is interesting to note that the program has a rating considerably higher than that of the highest rated player on its development team.) This program uses a selective search technique to explore possibilities several moves ahead, but it was not designed to emulate human thought. So far, experience with this program's search algorithm has shown playing strength to increase regularly with increases in processing speed.

Hsu et al. note that workers at IBM are developing a next-generation machine that will permit the examination of more than a billion positions per second and speculate that, if the relationship between processing speed and playing strength holds over this speed range, the program should attain a strength rating of 3400, 500 points above the rating of the current world champion, Gary Kasparov. Although the practical significance of a world-champion chess program is dubious, the symbolic importance of it, should it come to exist, is probably fairly substantial. Government funding for AI research and development, 90% of which comes from the Department of Defense, increased from $48.5 million in 1984 to $172.5 million in 1988 (U.S. Industrial Outlook, 1989).

More and more frequently, and in an increasing variety of contexts, all of us will find ourselves interacting with "smart" devices and devices that have the ability to learn. Almost nothing is known about how people will interact with machines that share many of their cognitive abilities, and it may not be possible to predict this very accurately until such devices become more common and some experience is gained in using them. The conceptual models that people develop vis-à-vis computer-based systems that have nontrivial cognitive abilities are likely to differ qualitatively from the models they develop of other types of systems. Of particular interest is the question of whether people who are not highly knowledgeable with respect to computer technology (and perhaps even those who are) will develop anthropomorphic models of system capabilities and impute to such systems more human-like abilities and characteristics than they really have. How to ensure that people's conceptions of these systems are reasonably consistent with the facts and that their attitudes and expectations relative to them are realistic are important questions for research.

Much has been written about the coming of smart homes in which lights, heating and cooling systems, security systems, phones and other communication equipment, and appliances of various sorts—including robotic house-cleaning devices—are connected to a control system that has enough intelligence to keep things running smoothly (e.g., to alert the fuel company if the furnace breaks down) and that can be accessed from anywhere in the house or even from remote locations. There has been considerable speculation regarding what capabilities one would want such a house to have, but which of the many possibilities that have been imagined will prove to be truly useful remains to be seen. The ability of a kitchen stove to turn off a burner that has been left on inadvertently in an empty house seems very useful; a bathtub that can fill itself with water of just the right temperature in response to a call from a car phone is of somewhat more doubtful value. Today's household appliances—washing machines, dryers, dishwashers, vacuum cleaners, food processors—do much of the work that was once done by hand. One thing that seems fairly certain is that the appliances of the future will have cognitive, as well as physical, capabilities.

If, as many anticipate, machines acquire the ability to carry on conversations with human beings (or with each other), we must wonder what the implications will be for people's perception of themselves, and, more generally, how people will adapt to the reality of interacting with machines with these types of capabilities. To date, speech has been a communication medium unique to human beings. We have shared it neither with other organisms nor with machines. We have very little idea of the psychological effects of having the world populated with machines that listen and talk. The experience gained with radios, television, and recording machines provides little insight into this question. One does not think of these devices as having the ability to use language. To be sure, they produce speech, but the speech they produce is perceived by the listener as coming from human beings, even if by an indirect route. Systems that have the ability to understand and generate language and to synthesize speech are in a different category altogether.

Expert Systems

More and more of the knowledge that is carried in people's heads, especially the kind of knowledge that constitutes expertise in the professions (medicine, law, engineering) will be codified and incorporated within expert systems. Such systems will be used for diagnosis, decision making, and design, and will be increasingly employed in business and industry. In some cases, expert systems are intended to function in place of human experts in specific contexts; in others their role is that of an assistant, associate, or advice giver. Sheridan (1980) makes the point that the degree to which decision making is, or can be, automated in computer-based systems should be thought of as a continuum. At one extreme are those systems in which control is vested 100% in the human operator: The human considers alternatives, and makes and implements the decision. At the other are those in which the computer is in complete control: The computer makes and implements a decision if it feels it should, and informs the human only if it feels this is warranted. Between these extremes are many intermediate cases, including some in which the actual decision option is chosen by the human (the computer offers a restricted set of alternatives and suggests one, but the human still makes and implements the final decision) and others in which it is selected by the computer (the computer makes and implements the decision, but must inform the human after the fact).

Expert systems have been developed within several specific application domains, notably computer systems configuration, locomotive repair, medical diagnosis, and oil exploration. Several systems are in operational use; many more are still considered experimental. The utility of such systems has been demonstrated, and much effort is going into both improving existing systems and developing others in new domains. Digital Equipment Corporation, for example, has, by its own count, over 50 knowledge-based systems supporting a variety of its activities "from designing custom integrated circuits to troubleshooting faulty systems" (Product Insight, 1990, p. 8). Digital estimates that XCON, its grandfather expert system, which has been used for several years to configure computer systems, now saves the company about $200 million a year. The system's knowledge base began in 1978 with 500 rules and grew to contain over 12,000 by 1990. According to a recent report from the U.S. Department of Commerce (cited in U.S. Industrial Outlook, 1989), about 80% of U.S. corporations have some involvement with expert systems at the present time and have, in total, implemented about 1500 of them. Use of these systems will probably be generally beneficial for the industries that employ them, but there may be some unpleasant surprises as a consequence of unexpected secondary effects in some cases (like the destabilizing effects of programmed trading on the stock market when the approach was adopted by a sufficient number of traders).

Motivation for the development of expert systems has several bases. First, human experts in specific domains are too few in number to be available whenever and wherever the expertise they have is needed. Second, expert systems can have certain abilities that human experts do not have (e.g., the ability to compute very rapidly). Third, there is the possibility of patterning an expert system after the best of the human experts in a given field. Finally, interest in expertise has also been motivated by educational concerns: Assuming that one of the objectives of education is to turn neophytes into experts in specific domains, it is important to know what constitutes expertise and how it can best be developed.

A common-sense definition of *expert* is "someone who knows a lot about a specific topic," although precisely how much one has to know to be perceived as knowing a lot is likely to depend on who is judging expertise. The attempt to build systems with problem-solving capability that compares favorably with that of human experts in specific domains has put new meaning and motivation into the question of what constitutes expertise. One of the major issues that has emerged from work in this area is the question of the relative importance to expert performance of domain-specific knowledge, on the one hand, and of domain-independent cognitive skills (e.g., in problem solving, reasoning, decision-making) on the other. Much of the debate on this question has been polarized and has promoted the importance of one factor at the expense of the other. It seems most likely that further research will reveal both domain-specific knowledge and domain-independent skills to be essential. Assuming this to be true, we will need to know how domain-specific knowledge and domain-independent skills interact in expert performance, and how to relate the teaching of domain-specific knowledge to the teaching of domain-independent skills in the development of expertise.

The building of an expert system requires, first, that the knowledge that constitutes expertise in the field of interest must be acquired, and then, that it be represented in a form that is usable by the computer. One might think that the way to acquire such knowledge is to consult the textbooks, manuals, and reference documents used by professionals in the field of interest. It appears, however, that the knowledge that is required is not readily obtainable from these sources; much of it seems to exist only in the heads of human experts. The first task of a builder of an expert system, therefore, is to get this knowledge from the experts. That is not an easy task, in part because experts do not know what they know, or at least they are unable to say explicitly what they know that constitutes their expertise when asked to do so. In that regard, experts are little different from the rest of us. None of us is able to say all we know on any specific subject unless the task is trivialized by the choice of an area in which we know next to nothing.

What experts and the rest of us can do, however, is draw on the relevant knowledge we have in order to answer specific questions, and we can use our knowledge in dealing with situations for which it is germane. To find out what experts know in their areas of expertise, one must ask them specific questions and observe them functioning in situations that provide an opportunity to exercise their expertise. A variety of approaches to knowledge acquisition have been developed as have several knowledge-acquisition tools (Boose, 1986, 1988; Garg-Janardan & Salvendy, 1988; Gruber, 1988; Shalin, Wisniewski, & Levi, 1988; Silvestro, 1988; Witten & MacDonald, 1988). Tools include AQUINAS (Boose & Bradshaw, 1987), ASTEK (Jacobson & Freiling, 1988), KNACK (Klinker, Genetet, & McDermott, 1988), KREME (Abrett & Burstein, 1987), MOLE (Eshelman, 1988), MORE (G. Kahn, Nolan, & McDermott, 1985a, 1985b), and SALT (Marcus, 1987).

An important aspect of knowledge acquisition that has not received a lot of attention to date is the problem of fitting newly acquired knowledge into an existing knowledge representation scheme (Lefkowitz & Lesser, 1988). Just as with human learners, so with electronic expert systems, newly acquired knowledge must usually be assimilitated within the context of an existing knowledge base. This means, at the least, relating the new knowledge to the knowledge that exists, and often it requires, also, the resolution of inconsistencies between what has just been found out and what was already "known."

Experience on efforts to build rule-based expert systems has shown that it is possible to represent a significant portion of the knowledge that appears to constitute expertise in certain areas with only a few hundred, or at most a few thousand, rules (R. Davis, 1986). It has shown also, however, that although a modest number of rules will suffice to handle the most common problems in specific areas, several times as many rules may be needed to handle the uncommon problems. The expertise represented in the expert systems that have been built to date tends to be very narrowly focused and sharply delimited; consequently, when faced with problems only slightly outside their domain, expert systems typically do not fail gracefully. Further, their "understanding" of their domain is somewhat free-floating and not anchored to a general knowledge base. Thus, a system that is able to diagnose specific illnesses may have considerable knowledge relative to the illnesses of interest, but lack any understanding of what a human being is.

Researchers have begun to look at some of the psychological factors involved in the development of expert systems (Boehm-Davis, 1988). These include how to acquire from experts the knowledge base a system must have, how to represent that knowledge in the system's software, how to make the system's knowledge available to the system's users, and how to give users an accurate understanding of the system's capabilities and limitations (Chignell & Peterson, 1988; Madni, 1988). The intense interest in building expert systems for practical purposes provides an opportunity for a synergistic coupling of systems building and research: Research on human expertise yields information system builders need; conversely, efforts to build systems force attention to problems relating to expertise that, apart from these efforts, might never be identified. Focusing research on these problems should deepen our understanding of expertise.

The prevailing assumption among developers of these systems seems to be that the more closely the reasoning process used by an expert system matches that used by a human being, the better the system will function. This is an assumption subject to challenge. What one wants from the computer in many situations (e.g., medical diagnosis, computer system configuration, manufacturing scheduling, anticipation of system failure) is the right answer. How it comes by the right answer is for practical purposes of secondary importance. A similar assumption regarding the desirability of a computer system's having reasoning capabilities like those of human beings is also made with respect to systems that are intended to provide an advisory service to human users. What is important in this case is not how closely these systems' abilities mirror those of human beings, but how effectively they augment human performance. Their value, again from a practical point of view, is not determined by their cleverness, but by their usefulness.

If the efforts to develop expert systems are highly successful, there may be some unanticipated social effects. An obvious possibility is the prospect of making specialized knowledge more readily available to potential users, including nonspecialists. There is also the possibility of changing the societal roles of at least some human experts. One consequence of the codification of expert knowledge that has been predicted is the erosion of the power of established professions, the idea being that professions derive their strength from their status as exclusive repositories and disseminators of specialized knowledge (C. Evans, 1979). In any case, to the degree that experts derive status and other rewards from the fact that they have knowledge that is not readily accessible to most other people, the value attached to human expertise could decrease or, perhaps more likely, the definition and functions of expertise could change. Research directed toward anticipating the psychological and social effects of the development and proliferation of expert systems could help to effect a graceful accommodation to that change.

Artificial Realities

An especially noteworthy current focus of research on display technology is the development of simulated environments that seem real and permit one to interact with the simulated entities much as one would interact with their actual counterparts in the real world, moving about, looking at objects from different viewing positions, picking things up, moving them from place to place, modifying them, and so forth. Foley (1987) notes the possibility of such "artificial realities" going beyond reality by modeling abstract entities like mathematical equations in concrete form, and changing scale, thereby permitting one to manipulate entities that are either too small or too large to be manipulated directly in real life.

Experimentation with artificial realities at the present time involves presenting visual simulations on helmet-mounted displays and sensing the user's interaction with "virtual" objects through electronically instrumented gloves. NASA, for example, is working on the development of artificial-reality techniques that will improve the effectiveness and efficiency with which human beings can interact with and control the operation of semiautonomous robots. Areas of research interest include the development and use of helmet-mounted three-dimensional visual displays, three-dimensional sound cueing, voice input–output, eye-movement and fixation tracking, gesture tracking, and tactile feedback from remote actuators.

Under a program called Super Cockpit, the U.S. Air Force is working on the development of a helmet-mounted display that will provide the pilot with out-of-cockpit information from the infrared as well as the visible part of the spectrum, thus providing nighttime, as well as daytime, vision. The helmet will sense the orientation and motion of the pilot's head, so the display can show the pilot what he would see if he were able to look out the aircraft in whatever direction his head is pointing. Head and eye direction can be sensed and used to provide target-heading information to the aircraft's weapons systems, which will be controllable by voice command. The display of actuatable virtual control switches is also a planned feature of the Super Cockpit (D. Williams, 1986).

In my view, there are few areas of application of computer technology that are more exciting or that hold greater promise for the future than the development and use of artificial realities. Although considerable work is being done in this area already, the systems that have been developed to date provide only vague hints of what should be possible in the foreseeable future. Eventually, if this technology advances as some observers expect it to, a medical student will be able to practice surgery on a virtual patient, a biology student will be able to roam around inside a virtual protein molecule, and a meteorologist will be able to fly, safely, into the center of a virtual hurricane and out again.

There are significant practical impediments to the realization of these fantasies soon, one of the most obvious being the enormous amounts of computing power that will be required to create and maintain a veridical simulation of any but the simplest realities. A high-resolution, three-dimensional, real-time visual representation of even a moderately complex dynamic situation requires more computing than even the world's largest current computers can supply. Providing the individual with the full-body experience of "being there" will also make very heavy computing demands, but developers have glimpsed the possibilities, and steady progress toward their realization seems inevitable.

An aspect of artificial realities and the closely related idea of telepresence that makes for entertaining and thought-provoking science fiction (e.g., Sanford, 1981) is the possibility that the technology could become sufficiently advanced that an individual would be unable to tell whether the "reality" of the moment was artificial or real. We are a long way from that point, but movement in that direction has been the cause of some concern. Hardison (1989) argues, for example, that "movies and television create an illusion of presence at the unfolding of events. Interactive environments like arcade games, training simulations, and artificial realities create illusions that are even more vivid. At their best, they come close to obliterating the difference between reality and illusion" (p. 321). Hardison sees all these technologies, in the aggregate, as contributing to a weakening to the human sense of what reality is. It is a thought that should not be too lightly dismissed.

* * *

The pervasiveness of computer technology means that we are all affected by it in many ways. It means also that most, if not all, of us are users of this technology directly or indirectly. We have much at stake in the way it continues to be developed and applied.

There are a host of human factors issues relating to information technology, only a few of which have been mentioned in this chapter. They include the development of tools and aids for computer designers; the design of network protocols and addressing standards; the design of terminals, personal computers, and workstations; the study of new forms of interperson communication made possible by computers and networks (e.g., electronic mail, electronic bulletin boards, teleconferencing); the development and application of new possibilities in visual imaging; the development of information services of many types and the fitting of them to the needs and preferences of their users; the study of programming; the development of programming aids and tools; the development of approaches to the evaluation of software and the measurement of programmer productivity; the further development of teleoperator technology and especially of more effective methods of coupling the human and machine components of teleoperator systems; and the integration of artificial intelligence and expert systems in the workplace in humane and cost-effective ways.

The challenges and opportunities for human factors research that information technology provides have not gone unnoticed. Human factors research activity has probably grown faster in this area since the 1970s than it

9. INFORMATION TECHNOLOGY

has in any other. Most of the work to date on the human factors of computer systems has focused on user-related issues: how to design user-friendly interfaces; how to make applications languages and software packages easy to use; how to ensure the adequacy of user documentation; how to design on-line "help" systems that are genuinely helpful; and so on. These issues fall under the general topic of person–computer interaction, which is discussed in the following chapter.

A specific problem area that has not yet received much attention relates to the needs of system developers, as opposed to those of end users. Many of the systems that are being developed today are sufficiently large and complex to involve teams of programmers and designers working concurrently on various aspects of the task. Seldom is it possible to design such systems in their entirety in advance of any development; more typically, major aspects of a system's design evolve during the process of its development. No single member of the design/development team is likely to understand the entire system in minute detail, because the complexity is too great. (The software system used in the United States to monitor and control commercial air traffic, which is the largest such system in existence, contains some 20 million lines of code [Bromley, 1986].) Ensuring the communication, coordination, and effective teamwork that are essential to such efforts involves a variety of human factors issues that have yet to be articulated.

Another research challenge that has received little attention in the past but that is likely to stimulate increasing interest in the future is the need to develop a better understanding of, and more effective approaches to, parallel problem solving. Until very recently all computers were sufficiently similar in design to be treated by programmers as more or less equivalent. All of them executed instructions in sequence, one at a time, which meant that programmers, and others who wished to use a computer to solve specific problems, were encouraged to structure any given problem solution as a sequence of steps that could be executed in a linear fashion. This is a natural approach for most of us anyway, because our own spontaneous problem-solving activity, to the extent that it is open to introspection, seems to proceed in a sequential, step-wise fashion.

With the appearance of parallel-processing architectures (of a great variety of designs), the situation has changed dramatically. When one has a computer at one's disposal that has dozens, hundreds, or possibly thousands of processors that can operate simultaneously, one has an incentive to structure one's problems so various aspects of them can be worked on at the same time. The fact is, however, that we do not yet know how to do this very well. Whether that is primarily because the development of parallel approaches is inherently more difficult than the development of serial ones, or simply because there have not been compelling practical reasons to attempt to develop parallel approaches in the past, or both, is not clear at this point.

What is clear is that technology has provided an incentive now to think along these lines, because the available hardware has advanced beyond our ability to exploit it as effectively as we would like. This is a problem area relating to computer use that could benefit from some focused research.

Electronic computers have been in existence for less than half a century and already—especially as coupled with communication technology—they have affected our lives profoundly in many different ways. The significance of the digital representation, rapid processing, and transmission of information can hardly be overstated. No one can say precisely what additional effects these technologies will have in the future, but it is hard to imagine that they will be any less pervasive or profound than they have been in the recent past. My own expectation is that they will change our lives in ways that we cannot now even imagine. Whether the changes, on balance, will be positive remains to be seen. It is not prudent to assume that they will naturally be so. We must work to ensure that they are.

10

Person–Computer Interaction

Human factors researchers have become increasingly interested in information technology as a focus of research since the mid-1960s. Turoff (1967) observed that although one might expect psychology to be able to contribute greatly to the construction of immediate-access computer systems, the evolution of systems to that date had not been much influenced by this field. My own search of several major human factors and applied psychology journals a couple of years later revealed very few articles relating to how people interact with computer-based systems (Nickerson, 1969). Things have changed very considerably during the ensuing years, and especially since about 1980. As one evidence of this change, consider the contents of *Human Factors* for the 3 years: 1965, 1975, and 1985.

I scanned each of these issues looking for articles focused on some aspect of the use of computers as an area of human factors research or engineering. In the 1965 volume, I found 1 such article, out of a total of 58 published that year. It was an article by Spesock and Lincoln entitled "Human factors aspects of digital computer programming for simulator control." The 1975 volume contained 2 articles, out of a total of 62, that focused on computer usage; both of these related to input techniques. One reported a study in which typists were taught to abbreviate frequently occurring words when entering documents via keyboard into a computer (Schoonard & Boies, 1975). The second compared three methods of positioning a cursor on an electronic display (Goodwin, 1975). A third article in 1975 involved the use of computer-generated displays in a driving simulator, but did not relate to computer use as such.

By 1985, the situation had changed considerably. Out of a total of 55

248

articles published in *Human Factors* that year, about one third had to do with human factors issues in the use of computers or computer-based systems. Topics included factory and office automation systems (Helander, 1985; Parsons, 1985; Sharit, 1985), artificial intelligence (Hillman, 1985), automation in training and education (Kearsley & Seidel, 1985), cockpit automation (Wiener, 1985), robotics (H. G. Shulman & Olex, 1985), VDT glare (Garcia & Wierwille, 1985), automatic speech recognition and generation (Damos, 1985; Schwab, Nusbaum, & Pisoni, 1985; Simpson, McCauley, Roland, Ruth, & Williges, 1985), touch-input devices (Beringer & Peterson, 1985), abbreviations for computer input (Ehrenreich, 1985), menu structuring (E. Lee & MacGregor, 1985; Seppala & Salvendy, 1985), the design of help facilities (Cohill & Williges, 1985), program documentation (Boehm-Davis & Fregly, 1985), and the use of simulation in the development of computer communication products (Francas, Goodman, & Dickinson, 1985).

Since about 1980, several new journals that focus on the human factors of computer-based systems and especially on human–computer interaction have been started. These include *Behavior and Information Technology, Human-Computer Interaction,* the *International Journal of Human–Computer Interaction,* and *Interacting with Computers.* Handbooks, compilations of readings, edited volumes, conference proceedings, and monographs on these topics have also been published. Examples are Card, Moran, and Newell (1983), Rupp (1984), Monk (1985), Nickerson (1986), D. A. Norman and Draper (1986), Baecker and Buxton (1987), Shneiderman (1987). A series entitled *People and Computers* is produced by the British Computer Society; another, *Human Factors in Computing System Series* (CHI Conference Proceedings) is published by the Association for Computing Machinery. Numerous other publications dealing with human factors in information technology and with person–computer interaction, in particular, have also appeared.

It is not surprising that computer technology had not yet become a primary focus of the human factors community by the mid-1960s, because the computer industry itself was barely getting off the ground. What is impressive is how dramatically things changed over a short time and how significant to human factors researchers, as well as to everyone else, computers and computer-based systems have become. There is perhaps no other focus of human factors work that is receiving more attention today than human–computer interaction, which seems appropriate given the pervasive influence of computer technology on our daily lives. An extensive review of research on this topic would be a major undertaking and is outside the scope of this book. I have reviewed some of this work elsewhere (Nickerson, 1986); all I wish to do here is point to some representative examples of work addressed to specific questions within this broad domain.

THE PERSON–COMPUTER INTERFACE

In terms of sheer number of studies, the most frequent subject of human factors research relating to computers and computer systems has been the person–computer interface. This should probably not be surprising, because the interface, by definition, is where the user and the computer come in contact. It is the space in which the interaction between user and machine occurs, the portal through which information is passed in both directions. Interface design increases in importance as the number of computers in existence grows ever larger, and as both the uses that are made of them and the community of users become more diverse. The rate of increase in the availability of computing power and the proliferation of uses of it have been phenomenal over the recent past. Every day new software becomes available to extend the range of tasks to which computer technology can be applied or to enhance the effectiveness of its application where it is already being used.

As computer-based systems become increasingly available and their range of potential usefulness continues to broaden, the population of users naturally grows and becomes more heterogeneous. No longer are these systems the province only of technologists; they are being used more and more in the workplace, in schools, and in the home, by farmers, sales people, managers, designers, investors, writers, students, and garden-variety folk. Proliferation and diversification of both uses and users seem highly likely to continue into the foreseeable future.

For these reasons, interface design has become a major concern of system developers; although computer technologists may be willing to suffer an interface that is difficult to master, provided a system has the desired functionality, other users are less tolerant in this regard. Issues of interface design and information representation will become increasingly important determinants of the effectiveness of interactive computer-based systems as the systems become used in the workplace and elsewhere by an ever-increasing percentage of the population. As Makhoul, Jelinek, Rabiner, Weinstein, and Zue (1990, p. 495) have put it: "Whoever controls the development of human–machine interfaces may own the key to controlling information technology going into the next century." Foley (1987) referred to the computer–user interface as perhaps the last frontier in computer design.

Interfaces may be described in terms of the physical entities that serve as the media for information exchange, which include visual displays, light pens, mice, keyboards, speakers, eyes, ears, and fingers. They also may be thought of in terms of the information that passes in both directions, and the ways in which information is represented for either human or computer use, which is appropriate to a view of person–computer interaction as an interaction between two information structures, one of which resides inside the computer and the other of which is in the user's head. Human factors issues that relate to the design of interfaces range from the characteristics of keyboards to the details of visual displays to a host of questions dealing with more cognitive aspects of person–computer interaction.

Some aspects of human–computer interface design are relatively well understood, especially those that are not unique to human–computer interfaces, but that pertain to more conventional systems as well. These include such features of visual displays as brightness, figure–ground contrast, character fonts (size, spacing, height-to-width ratios, stroke width), dot-matrix character design, label layouts, icon and pictorial symbol design, coding dimensions (size, color, location, shape, flash rate), dial and gauge design, scales (graduation markers, pointers), analog indicators, and multi-indicator layouts, among numerous others. They include, also, many aspects of the design of auditory displays, such as the relative effectiveness of various types of alarm or alerting signals, the detectability of signals with specified acoustic properties in different types and intensities of noise, signal distinc-tiveness and identifiability, direction indication, speed or rate indication, the relative merits of headsets versus loudspeakers in particular situations, and coding dimensions (intensity, frequency, repetition rate).

With the exception of the exploratory work that has been done using speech as an input–output medium, very little effort has been made to exploit sound to convey information to the computer user. The desire to improve the access that blind users have to computing facilities is a stimulus to such work, however, and a little effort in this direction is being made (A. D. N. Edwards, 1989). Although many visually impaired people use computers very effectively, the devices on which they depend for access—which typically use either large print, paperless Braille, or synthesized speech—are, in most cases, considerably less than ideal and could be improved upon from a human factors point of view (Aaronson & Gabias, 1987).

Less well understood than the aspects of interface design that are common to both computer systems and non-computer systems, are many that are more or less unique to computer-based systems. These have to do with such issues as the relative merits of menus and command languages (under what conditions is each preferred? how can the advantages of both be realized in a single system?), the design of menus (depth–breadth tradeoffs, naming of menu items) and command languages (naming of commands, syntax), the use of scrolling versus paging, the naming of variables (in programs, functions, and files), the use of speech for input or output, and the design of on-line help facilities and tutorial aids.

Many of the most challenging human factors issues relating to interface design arise because the technology provides such an enormous range of possibilities for representing information and for structuring user–computer dialogues. Moreover, the capabilities the technology provides have been changing so rapidly that design guidelines that are specific to particular display hardware and techniques can be obsolete almost before they are used. The computer provides very rich alternatives, for example, to the linear organization that characterizes the representation of information in books. Hypertext capabilities make it possible to organize information in multilevel structures that can be accessed in a variety of ways. Multimedia systems provide for the mixing of text, graphics, dynamic simulations, and sound—including speech—in the same document. Experimentation with such capabilities is just beginning, however, and much more of it will be necessary before we learn how to exploit them effectively or how to articulate guidelines for their design.

Video Display Terminals

Much attention has been given by researchers to the design of video display terminals (VDTs), especially those that make use of cathode ray tube (CRT) displays, as most do (Helander, Billingsley, & Schurick, 1984; National Research Council, 1983). Some of this attention has been motivated by complaints of stress or visual fatigue from people who use these devices more or less constantly in their work (Dainoff, Happ, & Crane, 1981; Muter, Latremouille, Treurniet, & Beam, 1982; Nordqvist, Ohlsson, & Nilsson, 1986; M. J. Smith, Cohen, Stammerjohn, & Happ, 1981). Results are difficult to interpret, however, in part because of lack of agreement across studies and in part because visual fatigue has proved to be very difficult to measure objectively (Miyao, Hacisalihzade, Allen, & Stark, 1989). The issue is likely to get increasing attention in the future because of the growing number of people who use VDTs on a regular basis.

As color monitors have become more and more common, investigators have attempted to determine whether some colors or color combinations make for more usable—less error-evoking or less stressful—displays than others. Although some differences have been reported, on balance the results seem to indicate that, providing it is not confounded with such factors as contrast and resolution, color is not a major determinant of performance or user comfort (Lovasik, Matthews, & Kergoat, 1989; Matthews, Lovasik, & Mertins, 1989; D. L. Post, 1985). The design of non-alphanumeric symbols for use on VDTs has also been the focus of some research. The performance measures that have been used to evaluate symbols and symbol sets typically have been some index of accuracy (error rate) or a measure of search time (Remington & Williams, 1986).

A number of studies have compared people performing specific tasks using VDTs and using hard-copy representations of the same information (W. H. Cushman, 1986; Gould & Grischkowsky, 1984; Waern & Rollenhagen, 1983); usually performance has been better with the hard-copy representations, but generalizations are risky because VDTs vary in quality over a considerable range. Not surprisingly, VDTs that have a relatively large number of high-resolution pixels compare more favorably with hard-copy print than do those that do not (Harpster, Freivalds, Shulman, & Leibowitz, 1989). Also as one might expect, the advantages of higher over lower resolution displays, in terms of readability, is more noticeable when the font that is used is relatively small (Miyao et al., 1989). Gould, Alfaro, et al. (1987) explored the effects of several variables that might be expected to account for the slower reading from CRT displays than from paper and concluded that the difference is probably due to a combination of variables rather than to any single one.

Although the fact has not received the attention it deserves, computer technology can be used to facilitate the reading task in ways that are not possible with conventional hard-copy text. This potential has not been developed much as yet, but some experimentation has been done with the use of on-line definitions that are callable by a reader of text on a CRT display (Lachman, 1989). Looking ahead, it seems inevitable that this technology will be used both to enhance the learning of reading by children and by adults who lack literacy and to enrich the reading experiences of competent readers. A system that can speak any word to which a reader points is well within the state of the art, although the quality of the synthesized speech leaves something to be desired. Systems that can give definitions of words on request, provide elucidating information (textual, pictorial) and answer questions posed by a reader about the material being read, run dynamic simulations of processes described or referenced by the text, and support comprehension in other ways are likely to become quite common.

Human–Computer Dialogues

Much of the work on the human factors of computer systems has focused on the objective of facilitating the interaction between the user and the system. The interaction typically has information flowing in both directions, from the user to the computer and from the computer to the user. In this respect, and in some other ways as well, the interaction is similar to conversational interaction between people and consequently is often referred to as a *dialogue*. Much of the work on human–computer interaction can be classified under the subject of dialogue design.

A widely held belief is that the appropriate objective for person–computer dialogue is that it be as "natural" as possible, where *natural* means similar to person-to-person dialogue. More generally, an intuition that many people who have thought about person–computer interaction seem to share is that the appropriate metaphor to represent the ideal person–computer relationship is that of a person–person partnership. Implicit in this metaphor—and explicit in much of the writing on person–computer interaction—is the idea that the computer part of this partnership should be as human-like as possible: It should think like a human, converse like a human, be friendly like a human, and so on. Although terms such as *conversational* and *user-friendly,* are used to describe existing systems, no one to my knowledge has claimed that these systems are really human-like in any very deep sense, but that is only because we do not know how to build such systems. Doing so still remains a goal.

I believe, however, that the partnership metaphor, with its attendant implicit acceptance of human-like qualities and behavior as the goal for computer systems, is probably wrong. Consider the specific idea that the appropriate goal for the developers of interactive systems is to produce systems in which the computer's end of a conversation will be very similar to the person's. It is at least a plausible hypothesis that the types of interactions between people and computers that will prove to be most effective and acceptable will differ considerably from our usual idea of a congenial interperson conversation (Nickerson, 1976). Thinking of the machine as a partner, colleague, or co-conversationalist may tend to limit one's imagination unduly about how an interaction might proceed. One may find it advantageous for one's interaction with a computer to have some asymmetries that would be inappropriate or objectionable for person-to-person conversation. One may not wish to feel the same obligation, for example, to be polite, patient, or tolerant of imprecision, ambiguity, and error when one is using a computer as one does when carrying on a conversation with another human being.

At the present time, human–computer dialogues usually involve the use of command languages, menus, or both. A *command language* is a vocabulary (a set of symbols) and a syntax (a set of rules for combining those symbols to make commands that the computer can interpret and execute). A *menu* is a list of options presented by the computer to the user on a display; the user can select any one option for the computer to execute. Menu-oriented systems minimize the user's learning task because they make the user's options explicit at any given point in the interaction, but they are relatively slow. Command-oriented systems tend to be faster in the hands of an experienced user, but they require more initial learning and information retention if they are to be used effectively. Many systems permit the same action to be identified either by pointing to a name or symbol on a menu or by typing a command.

How menus should be organized and the number of options they should contain are topics of continuing research (E. Lee & MacGregor, 1985; MacGregor, Lee, & Lam, 1986; Paap & Roske-Hofstrand, 1986). On the basis of a theoretical analysis of search times, Lee and MacGregor (1985) concluded that the optimal number of items to have on a menu is in the range of 4 to 8 under a wide variety of conditions. Paap and Roske-Hofstrand (1986) have presented analytic evidence, however, that if a menu is organized in such a way as to reduce the amount of search required to locate the option of interest, the optimal number may be much greater, possibly as great as 64.

Of course, hierarchically arranged menu systems need not have the same number of options on every menu; how many options should appear in any particular instance may depend on the way the subject matter of the data base is naturally, or conventionally, structured. K. L. Norman and Chin (1988) have shown that given a specific 256-item database and the constraint that a hierarchical selection tree be four levels deep, a "concave" structure ($8 \times 2 \times 2 \times 8$) was superior to other structures ($4 \times 4 \times 4 \times 4$, $8 \times 8 \times 2 \times 2$, $2 \times 2 \times 8 \times 8$, or $2 \times 8 \times 8 \times 2$) for some searches, whereas an "increasing" structure ($2 \times 2 \times 8 \times 8$) was superior for others. An extensive review of the work to date on the design and use of menus by K. L. Norman (1991) revealed the issues involved to be more numerous and complex than one who is unfamiliar with the problem is likely to assume.

Intuition in Interface Design

Although a variety of tools and systems are available that are intended to help interface designers (Farooq & Dominick, 1988), their development has been based, for the most part, on the intuitions of their developers and not on an understanding of the designer's task derived from careful empirical studies (Hartson & Hix, 1989). People who build computer-based systems are not insensitive to the importance of interface design; typically they try very hard to make their systems easy to use, and sometimes they succeed to a considerable degree. Inasmuch as many of the features of interfaces that contribute to a system's ease of use appear to be matters of common sense once they have been implemented, one might be tempted to assume that good intentions guided by intuition and common sense are enough to ensure good interface design. Yet, intuitively compelling features tend to be much easier to recognize as such after the fact than to come up with in the first place, and common-sense ideas as to what would constitute good design innovations often turn out to be wrong.

In general, intuition has not been a very reliable basis for predicting what people will like and what they will dislike, what they will buy and what they will not, or what they will find convenient to use and what they will prefer to do without. This should not be surprising, given that what is intuitively clear or commonsensical to one person can be intuitively obscure or nonsensical to another. Testimony to this fact is seen in the countless interface designs that have been implemented and then discarded or simply ignored. It is seen also in the diversity of ways in which the same information is represented, or the same action is identified, in existing systems. Every programming language, of which a very large number have been developed, reflects some developer's intuitions about the capabilities and characteristics a programming language should have, and in the aggregate they attest to the enormous range of these intuitions. Every text editor, of which there also are many, reflects the intuitions of a developer regarding how to expedite the process of composing and editing text. Every electronic mail system reveals the intuitions of a developer regarding the capabilities people need or want for processing messages. Other examples come readily to mind.

Differences in intuitions involve not only major issues of functionality and organization; they also relate to what might appear to be minor design decisions, such as how to identify individual commands. T. L. Roberts and Moran (1983), for example, compared nine text editors and discovered among them six different ways of invoking commands, including typing all or part of a verb, typing a one-letter mnemonic for the command name, holding down a control key while typing a one-letter mnemonic, typing a one-letter mnemonic on a chord set, pressing a special function key, and selecting a command from a displayed menu.

Intuitions differ even with respect to features for which one might expect to find a high degree of consensus. One might assume, for example, that people would tend to agree, on an intuitive basis, on what specific information-handling operations should be called. Some work by Landauer and his colleagues (Furnas, Landauer, Gomez, & Dumais, 1983; Landauer, Galotti, & Hartwell, 1983) calls this idea into question. Landauer et al. (1983) gave several typists a manuscript containing proofreaders' marks and asked them to write out instructions as to the changes that should be made. The probability of any two users using the same verb in the same situation was only about .08. In the vast majority of cases, the verb used most frequently by the subjects to describe a required operation was not the word used by the text editor against which their performance was compared (Ed of the UNIX system). The investigators concluded from this study that "intuitive guesses as to what is a 'common' or 'natural' name for a command are likely to be hazardous" (p. 502). The study illustrates quite convincingly that people's intuitions differ considerably as to what simple editing actions should be called, and it challenges the usefulness of such an intuitively compelling design principle as that of identifying commands by their natural names. In a study with a related finding, Streeter, Ackroff, and Taylor (1983) discovered that word abbreviations derived by rules involving vowel deletion and truncation were easier for subjects to learn and use than were abbreviations that people tended to use spontaneously.

System developers' intuitions about what potential users of a system will find convenient should not be trusted for the simple reason that it is difficult, if not impossible, for an individual who understands a complex system thoroughly to see that system through the eyes of a person who is encountering it for the first time. A common error made by people who have acquired some specialized knowledge or complex skill over a period of time is to underestimate how long it took them to learn what they now know, and how difficult it may be for a beginner to do what they are now able to do effortlessly and automatically. If substantiation of this point is needed, there is at least suggestive experimental evidence that people tend to overestimate the likelihood that other people know what they themselves know (Nickerson, Baddeley, & Freeman, 1987). This observation is similar to Gould and Lewis's (1983) suggestion that system designers typically underestimate the diversity of the prospective users of their systems and assume greater uniformity and ability than is warranted.

If the intuitions of system developers are not to be trusted, neither are those of potential users. The fact that an innovative feature is not intuitively convincing when first encountered by potential users is not compelling evidence that it is not a good feature from a human factors point of view; sometimes what prove to be useful and convenient features in the long run are counterintuitive at first look because they represent departures from traditional ways of doing things. More generally, although designers clearly should design to the capabilities and limitations of the intended users of whatever they are designing, users' perceptions of their needs are not necessarily the best point of departure for the development of products intended for their use.

Evolutionary Iterative Design

If intuition and common sense do not suffice to ensure good interface design from a human factors point of view, what will? The answer to this question is more complicated than one would like. The experimental psychology literature on perceptual, motor, and cognitive capabilities and limitations of human beings contains a wealth of information that is relevant to interface design. Unfortunately, this literature is not a practical source of guidance for the designer because the ratio of relevant information to irrelevant information (for any given designer's purposes) is not high, and the literature is not organized to make it easy to find the information one needs.

The human factors or engineering psychology literature is another source of relevant information. Much information that is appropriate to the design of such things as keyboards and visual display devices can be found in human engineering handbooks; design guidelines relating to the more cognitive issues of person–computer interaction are less thoroughly codified. Many issues of the latter type have yet to be articulated and explored, although research activity in this area is increasing rapidly, thanks to the increase in research articles dealing explicitly with interface issues.

Every major feature of present-day interfaces—windows, mice, pull-down and pop-up menus, icons—has a long history. Each, in its present form, is the result of several years of experimentation and evolution. The story of progress in the field is one of a small number of truly innovative developments scattered among countless incremental improvements on existing designs. Both the innovations and the incremental improvements have almost invariably come from people who are heavy users of existing systems. The experience of using a system serves as the catalyst to the generation of ideas regarding how to improve it or how to do things in a qualitatively different way.

Experience in using a system while it is under development can be important for end users as well as for developers. This is because potential users often acquire a clear idea of what they want or can expect from a system only by using it or an approximation to it. Frequently, a user's initial expectations are vague or inaccurate because of an unfamiliarity with the technology; equally often, these expectations may prove to be wrong because the use of a computer-based system changes the nature of the user's job in unanticipated ways. Nash (1989) makes the critical point that involving intended users of a system in the development process is not just a matter of asking them what they would like: People's thinking tends to be constrained by their habitual ways of doing things. What is needed is for users to have an opportunity to interact with the system, or a prototype thereof, as it is being developed and before its design features have been solidified. Only in this way can users get the understanding of a system's capabilities that will permit them to imagine realistically what using it would be like in the workaday world—and of how using it could change qualitatively what they do—and only with such an understanding is one likely to be able to make the kinds of design inputs that will help insure the system's usefulness and usability.

Recognition of these facts has led to the idea of *guided evolution*, or *iterative design*, as an approach to the development of complex person–computer systems. The fundamental assumption underlying this idea is that it is seldom, if ever, possible to design these systems independently of their implementation and use, and that the only effective approach is to use a closed-loop process that permits experience gained in using initial versions of an evolving system to influence the design of subsequent versions (Alavi, 1984; J. R. Cushman, 1983; Eason, 1982; Nickerson, 1986; Sheil, 1983). The assumption gets credence from the fact that attempts to design complex person–computer systems in advance of any implementation have not been successful. It has proved to be impossible to anticipate critical ways in which system components and capabilities would interact and how user requirements would change as a consequence of unanticipated capabilities such systems often provide.

The idea of guided evolution, or flexible evolving design, may have merit from an economic as well as human factors point of view. R. D. Reich (1989) has recently pointed out that in Japan, unlike in the United States, research and development activities are integrated with those of engineering and production. All the phases occur in parallel and in an interconnected way, with an innovation shuttling between development and production as trial uses yield insights regarding possible improvements in design. This is, according to Reich, one of the main reasons why Japanese manufacturers are winning the race to get electronics-based consumer products to market even while the United States continues to lead the world in the quantity and quality of research and development.

On the assumption that an iterative design process is the preferred, if not the only, way to ensure the development of interfaces that are seen as well-designed products, there are two types of activities that can greatly facilitate that process: task analysis and rapid prototyping (Nickerson & Pew, 1990). It is important to note, however, that although performing a task analysis is usually the most reasonable way to begin an interface design effort, task analyses can be helpful to the designer at points in the process other than at the beginning. This is because the availability of new technology can change, sometimes quite substantially and in completely unanticipated ways, the nature of the task one wishes to perform. When this happens, the results of an initial task analysis—although useful in getting the development moving in the desired direction—will be inadequate to guide its subsequent stages. Task analysis should itself be seen as an ongoing process that not only lays out in detail the task as it is initially understood but also tracks the way the task changes as a consequence of new capabilities that result from the emerging system.

If the desirability and effectiveness of specific interface features are best revealed as a consequence of users interacting with functioning systems, an effective way to facilitate progress is to make it easier for developers to bring prototype systems into existence. Rapid prototyping is becoming an accepted design methodology, and a variety of tools now exist to help the prototype builder.

In short, the process of designing interfaces should be a dynamic one in which, with the help of task analysis and rapid prototyping tools, one shapes a design by putting initial ideas to the test and converging on a set of features that are consistent with the demands placed on the user by the task of interest, as modified by the new technology in hand. Given this approach, the problem of evaluating designs must also be addressed in a dynamic way. A design should be evaluated more or less continuously as it is evolving and the results of the ongoing evaluation fed back into the continuing process of design.

Interface Evaluation

With the ability to build prototypes quickly, the challenge becomes that of evaluating their designs in a sufficiently cost-effective and timely way that implementors can factor the results into the designs of subsequent prototypes or operational systems. Although there is no generally agreed upon method for judging the quality of interface designs, there appears to be an emerging consensus as to some of the major criteria that should be used to assess effectiveness. These include functionality (what the interface, and the system more generally, permit the user to do), the speed with which benchmark tasks can be performed, error rates, error costs (some types of errors are more costly than others), user satisfaction, the time required to learn to use the system to some specified level of efficiency, and the robustness of the system under stress (a poorly designed interface may sometimes suffice under ideal working conditions but yield unacceptably high error rates under suboptimal or stressful conditions).

Interfaces that are intended to be used in crisis situations, such as a military engagement, a natural disaster, or a major industrial accident, pose a special challenge because their adequacy probably cannot be established with certainty by means of tests conducted under noncrisis conditions. The stress that an actual crisis causes, especially a life-threatening one, can have very large effects on human performance, and less-than-optimal design features of an interface that are inconsequential when the system is used under non-stressful conditions can become major liabilities in a crisis. This is a particularly sobering thought in view of the increasing dependence of military systems on computer-based technology. Nash (1989) describes this dependence this way: "By any measure, computers have infiltrated every aspect of modern armed forces. In weapon systems, computers and digital electronics are used for communications, navigation, weapon-path prediction and control, and for target identification, tracking and projection. Computer systems are at the heart of all activities in command, control, communications, and intelligence, including, increasingly, data management and interpretation" (p. 1). Nash expresses concern for what he refers to as "widespread confusion, among both military and civilian decision makers, about what tasks computers realistically can be expected to accomplish and on what time scale" (p. 3).

The problem of evaluating the effectiveness of systems that are intended to be used in specific crisis situations that we hope will seldom, if ever, arise was mentioned in chapter 5 as it relates to the use of simulation in training. It is a very difficult problem and one that is likely to increase in significance as the systems involved increase in complexity. Testing under simulated conditions can only go so far in providing evidence of how a system and its users will function when the stakes are real. Finding ways to address this problem effectively, short of creating real crises for purposes of performance evaluation, will continue to be a major challenge.

Needs of Novices and Experienced Users

One of the most enduring difficulties facing developers of computer-based systems has been the objective of designing interfaces that meet the needs of both novices and experienced users. Beginners need a great deal of structure, walk-through procedures, and on-line assistance; they tend to prefer to operate in a "react" mode, in which the computer explicitly presents the action alternatives that are available at any given moment, and the user's task is to select one from among them. Menu-driven systems fit these needs better than do systems one interacts with through keyboard commands. Experienced users need less structure and system-volunteered assistance; and they may often prefer, in the interest of time, to type commands rather than to select them from menus. Various approaches to the problem of satisfying the needs of both novices and experts have been tried, including having different interfaces for people with different levels of expertise.

One limitation of any approach that makes a sharp distinction between novices and experts is that users are not readily partitioned into two such groups. Expertise with respect to any system can vary continuously over a range from very low to very high, and any given user can be relatively expert with respect to some system capabilities and inexperienced with respect to others. What is needed is not an approach that meets only the requirements of users who are at one extreme of the expertise continuum or the other, but one that will accommodate all levels and mixes of know-how. Further, one wants interfaces that not only meet users' current needs, but ones that also make it easy for them to increase the degree and breadth of their expertise.

An example of an interface feature that accomplishes these goals to a considerable degree is now common to many menu-driven systems. Such systems are more likely than command-driven systems to provide novices with the structure and guidance regarding option availability that they need. The action options represented on the menus, however, may also be exercised via typed command (usually two or three key strokes), so experienced users can save the time of pulling down a menu and positioning the cursor if they know the command for the desired action. An especially helpful feature of some of these systems is the fact that the key combinations that code the commands are shown on the menus along with the corresponding English designations of the action alternatives.

If, for example, while typing these words using Microsoft Word on my Macintosh, I decide to put something in bold type, I can pull down the Format menu and click on **bold,** after highlighting the words I wish to emphasize. In looking at the menu, however, I will discover that I could get the same effect by simultaneously pressing the Shift and Control keys and the letter *B* before typing the words I wish to appear in bold. Thus, each time one selects a menu option, one has the opportunity to see the keystroke code that would yield the same action; so, in the process of doing what one does by preference as a beginner, one is shown constantly, but unobtrusively, the information one must assimilate in order to become a more expert user. The keystroke codes are relatively mnemonic: typically a Control character and the first letter of the name of the designated action. Although learning the codes is a convenience, there is no great pressure on users to remember them perfectly, because whenever they forget a code, they can resort to the use of the menu (whereupon, they again, see the code in the process of making the menu selection).

There is a need for more innovative thinking about unobtrusive ways to help novice users to become experts as an outgrowth of normal use. Unfortunately, it is still much too easy for users of most systems to continue to use those systems in far less than optimal ways, never discovering anything close to the full power of the systems they are using or the fact that the particular tasks they are accomplishing could be done more effectively and efficiently, if only they had a little more knowledge, which, at the moment, most systems are unable to volunteer.

The Invisible Interface

From the point of view of the user, there is a sense in which the interface should be invisible, or at least not a focus of attention. It should serve as a portal, an open door to the domain of interest. For most users, the computer is a means to an end. Its function is to help one perform some task, to solve some problem, to think about some topic, and one wants to focus one's attention on the task, the problem, or the topic, and not on the system one is using or on its interface. This idea has been promoted by several investigators who have suggested that we should begin to think about interfaces, and person–computer interaction more generally, in qualitatively different ways than we have thought about them in the past.

Draper (1986) captures this notion in his reference to a shift in conception of person–computer interaction from one that views input and output (I and O) as messages to and from the computer (or user) to one that views communication as taking place "by user and system alternately modifying and referring to the objects on the screen. Thus, whereas the implicit model of the overall user-plus-machine system used to be one in which there were two main modules (user and machine) communicating by messages (I/O), the new model consists of three modules—user, machine, and shared I/O" (p. 348). It sees user and computer writing on the same blackboard, as it were. Draper refers to this conceptualization as *inter-referential I/O*.

Laurel (1986) argues that both the tool metaphor and the idea of the computer as an intermediary put users in the position of having to persuade the system to do something that they would like to do themselves. She proposes an alternative metaphor that she calls *first-personness*. The idea is that the objective should be to give the user the experience of dealing directly with the constructs or variables of interest, much as the player of a video game such as Atari's Pole Position actually "drives" a simulated race car: "First-personness is enhanced by an interface that enables inputs and outputs that are more nearly like their real-world referents, in all relevant sensory modalities" (p. 77).

Laurel notes that simulators and games tend to be evolving in the direction of putting users in more life-like situations through incorporation of such capabilities as higher resolution graphics, faster animation, greater sound capabilities, motion platforms, and force feedback controllers. She suggests too, that in product-driven applications, indirect or symbolic representations and manipulations involving text and typing are being replaced with more direct and concrete ones involving spatial and graphic representation of data and physical pointing and speaking. The difference between the more conventional metaphors for interfaces and the metaphor proposed by Laurel is illustrated by the contrast between searching a database for information about, say, pyramids and doing something more analogous to "climbing them, looking around their musty innards, reading hieroglyphs, or reincarnating Pharaohs" (p. 85). Such a view of what one wants the experience to be like has implications for how we think about interfaces; in particular, it prompts us to focus on the kinds of experiences we want to have with interactive representations, rather than on what can be done within the constraints of current technology and conventions.

In a similar vein, Hutchins, Hollan, and Norman (1986) argue for what they refer to as *direct-manipulation interfaces*, interfaces that give one the feeling that one is directly manipulating the objects of interest: "Instead of describing the actions of interest, the user performs those actions. . . . The goal is to permit the user to act as if the representation is the thing itself" (p. 97). The intent is to have the interface present to the user "a world of action rather than a language of description," so that interacting with representations will be very similar, from the user's point of view, to interacting with whatever is being represented; what the user should experience is "a feeling of involvement directly with a world of objects, rather than of communicating with an intermediary" (p. 114). This is not to argue that all interactions with a computer system should be by means of direct manipulation of (virtual) objects—when direct manipulation is more appropriate and when it is not is a question for research—but it is to suggest that the objective should be conceptualized as that of helping people to interact more effectively, not with computers or computer-based systems, but with the information, the ideas, and the constructs to which these systems are intended to provide access.

Perhaps the most extreme statement of the idea of the invisible interface has come from the Computer Science Laboratory at Xerox's Palo Alto Research Center. Weiser (1991), the head of this group, argues that there is a need for new ways to think about computers that will bring the human world to the forefront and let the computers—that serve human goals—fade into the background. The computers should be invisible, just as the 20 or more motors that function in an automobile—starting the engine, cleaning the windshield, locking and unlocking doors—are invisible to most drivers. The very concept of an interface, in this view, gives way to the idea of an environment filled with a variety of displays, processes and other information resources, "a pleasant and effective 'place' to get things done" (p. 100).

The idea that is common to all these perspectives is that users of a computer system want to focus on the purpose for which they are using the system and not on the system itself nor on the characteristics of any display or other device that provides them access to it. In this sense, the system and its interface(s) should be invisible. However, as long as people are using computer-based systems, the design of the devices through which the use is mediated—the interfaces—whether much in the users' consciousness or not, will be a major determinant of the effectiveness of that use. Perhaps the best designed interfaces will be the least visible in this sense, but that does not diminish the importance or difficulty of the design problem.

Interface design will continue to be a major focus of human factors research relating to information technology and its use. The issues that will have to be addressed will increase in number and in scope. To date, the way information has passed between computer and user has been limited by almost exclusive dependence on small two-dimensional visual displays, keyboards, and such ancillary devices as lightpens and mice. Sound has been largely unexploited but will undoubtedly be used increasingly in the future. Three-dimensional and walk-around volumetric displays will become feasible. A wide range of input techniques will be explored in the context of work on the development of development of artificial realities. In short, the interfaces will become more varied and complex and will provide much richer environments within which people can work, learn, or play.

INFORMATION ACCESS AND USE

Interacting With Very Large Information Stores

In a prescient article published over 40 years ago in the *Atlantic Monthly,* Vannevar Bush (1945) imagined a device of the future, for which he coined the term *memex,* that would be a kind of mechanized private file and library: "A memex is a device in which an individual stores his books, records, and communications, and which is mechanized so that it may be consulted with exceeding speed and flexibility. It is an enlarged intimate supplement to his memory" (p. 106). Bush went on to describe how one might interact with such a device, which he imagined as having sufficient capacity that if the user were to insert 5,000 pages of material a day, it would take him hundreds of years to fill it up. Bush recognized that simply providing one with a device of such capacity did not assure it's usefulness, and he speculated a bit regarding the kinds of capabilities that might facilitate searching for information in such a database and for making the device function as a true memory aid.

Licklider (1965) coined the term *procognitive systems* to connote systems of the future that would facilitate interaction with large information stores:

> A basic part of the over-all aim for procognitive systems is to get the user of the fund of knowledge into something more nearly like an executive's or commander's position. He will still read and think and, hopefully, have insights and make discoveries, but he will not have to do all the searching himself nor all the transforming, nor all the testing for matching or compatibility that is involved in creative use of knowledge. He will say what operations he wants performed upon what parts of the body of knowledge, he will see whether the result makes sense, and then he will decide what to have done next. (p. 32)

Early seminal thinking about the potential that computer technology would offer for augmenting the human intellect came also from Engelbart and his colleagues (Engelbart, 1963; Engelbart & English, 1968). The visions of these and other pioneers in this field have not yet been realized fully, but they are not nearly as far-futuristic now as they were when they were first described.

In a small but visionary book, written nearly 30 years ago, Licklider (1965) considered what computer technology might mean for libraries of the future. Discounting pictures, Licklider estimated the total amount of information stored in all the libraries of the world to be roughly 10^{15} bits. The number of bits required to hold all of "solid" science and technology, he estimated to be about 10^{13} bits, which might be divided into 100 fields or 1,000 subfields with 10^{11} bits per field or 10^{10} bits per subfield, on the average. The total corpus of stored knowledge he estimated to be doubling every 15 to 20 years.

In reflecting on such estimates, it is important to distinguish between knowledge and data. Especially as a consequence of the space program, data are likely to accumulate at enormous rates in the future. The authors of a recent report from NASA's Information Systems Strategic Planning Project estimated that the annual volume of data from NASA research missions will rise from 0.5 terabits in 1989 to more than 2,500 terabits by the late 1990s (Rothenberg & Sander, 1990). (Low-earth reconnaissance satellites may be able to resolve objects as small as 6 inches in width or diameter [Richelson, 1991].) Twenty-five hundred terabits (2.5×10^{15} bits) is 2½ times the size of Licklider's estimate of the total amount of information stored in the world's libraries in 1965. How much the NASA data will add to the store of the world's knowledge after it has been processed and integrated with what is already known is impossible to say; of course, whatever the answer, such a volume of data poses great challenges for storage, transmission, processing, and interpretation.

Whether Licklider's estimates were accurate or not is not of great consequence for present purposes. What is important is the fact that the total body of stored knowledge is finite, and probably not greater than Licklider's estimate by many orders of magnitude. It is at least conceivable that at some point in the not greatly distant future, it will be possible to put all that information into a single, integrated, electronically accessible (or perhaps photoelectronically accessible) store. In the meantime, it is already possible to put very large bodies of information on electronic or photonic storage media, and the capacity of these media is continuing to increase at phenomenal rates. (According to Herskovitz, 1990, E-Systems of Dallas, TX, is developing a robotically controlled storage system with a capacity of more than 7×10^{15} bytes.) What is perhaps most important, in terms of implications for research, is the fact that our capacity to store information is already well ahead of our ability to interact with very large information stores in natural and productive ways.

One can think of information accessibility in the narrow sense and in the broad sense. By *narrow-sense accessibility*, I mean the kind of accessibility to focused databases that specialists often need to get their work done. A stockbroker or investment portfolio manager, for example, wants easy access to current equity prices and trading data, to the most recent financial analyses and vital statistics on specific companies, to customers' accounts, to current financial and economic news, and to a variety of investment data banks and information services. The automobile mechanic wants access to maintenance and repair information regarding all the makes of vehicles he may have to service, to parts and price lists, to addresses and phone numbers of suppliers. The attorney wants access to legal statutes and regulations, case histories, previous court decisions, sentences, and settlements. By *broad-sense accessibility*, I mean accessibility to information of all types that is available to the general public through libraries, news media, and information services to which anyone can subscribe.

The distinction is not a sharp one, but I believe it is helpful, nevertheless. Much more effort is likely to go into the provision of greater narrow-sense accessibility in the near-term future than into the expansion of broad-sense accessibility. In many businesses the rapidity and ease with which very specific information can be accessed will be a primary determinant of competitiveness. The investment portfolio manager, for example, who does not have easy access to current accurate financial information of a variety of sorts will be at a serious disadvantage.

Among the greatest challenges to developers of electronic information systems for the future, in my view, is the need for tools that will help us filter, search, and manage very large information stores. Most of us do not need access to more information in the raw, unorganized form in which we often receive it now; we already have more of that than we can deal with effectively. What we need are tools that will help us find the information we want when we want it and that will organize it and present it in forms conducive to our use. As Barker and Manji (1989) have put it, "technology has done much to accelerate the speed with which we can generate and disseminate information. However, it has achieved much less in terms of improving the speed with which we can assimilate and digest this information" (p. 345). Barker and his colleagues have been working on the development of pictorial methods to facilitate the exchange of information between computers and their users (Barker & Najah, 1985; Barker, Najah, & Manji, 1987). Browsing aids and structured search techniques need to be integrated within the same information system, because browsing often prompts interests and questions that are best answered by a structured search (R. H. Thompson & Croft, 1989).

In my limited experience, searches of computer databases based on key-word descriptors have not been very productive. I try one of these every few years, just to see if things have improved since the last time. My most recent experience was in connection with the writing of this book. I wanted to get some information on alternative energy sources, and, in particular, on geothermal, wind, and tidal sources. A search of several databases readily accessible to me, using the descriptor ([geothermal *or* wind *or* tidal] *and* [energy *or* power]) yielded the following titles:

A deep geoelectric survey of the Carnmenellis granite

A family of cometary globules around an infrared source near the Rosette nebula

Effects of nonstationarity on spectral analysis of mesoscale motions in the atmosphere

Application of the SP technique over Lagadas low enthalpy geothermal field, Greece

A fluid-inclusion investigation of the Tongonan geothermal field, Philippines

Modeling the heat extraction from the Rosemanowes HDR reservoir

Propagation of high-energy laser beams through the earth's atmosphere

On the dynamics and sound power levels of orchestral instruments

The influence of lateral mass flux on mixed convection over inclined surfaces in saturated porous media

Non-isothermal atmosphere, solar wind, shearing and pressing magnetic field and preflare loops

Role of the lifetime of ring current particles on the solar wind-magnetosphere power transfer during the intense geomagnetic storm of 28 August 1978

Propagation of planetary waves trapped by a zonal escarpment

Concentration distribution in the environment of nuclear installations

Environmental impacts of the proposed Severn and Mersey barrages

The librarian who conducted the search for me forwarded only these "hits" to me, from a larger set returned by the program, with the explanation that the items not forwarded were obviously irrelevant. I will probably try again, on another topic, in about 5 years. Key-word database searches are not the kind of tool I have in mind.

Denning (1987) who has pointed out that, with an estimated 825,000 scientists and engineers in the United States and some 1,500 journals that publish primary research, the production of scientific information is exceeding anyone's ability to assimilate and use it, has suggested that there are several computing technologies that can help to support the dissemination and use of the knowledge that science is producing:

> Databases can manage information, bibliographies, access to facts and data, and cross referencing. Algorithms can package procedures in forms that can be used reliably by those who do not know their details. Expert systems can store and apply rules for processes of inquiry. Local-area networks can link databases, algorithms, and expert systems with other processors such as simulators, statistical analyzers, and equation solvers. Concurrent processing can support "agents" that automatically seek out information while their owners are occupied with other tasks. And finally, heuristic searches can help sift through databases looking for new patterns and correlations. (p. 573)

Although the technology holds the potential to deliver these services, only beginnings on these possibilities have yet been realized.

There has been some discussion of the possibility of electronic journals among psychologists. In a recent issue of *Psychological Science* that focused on electronic publishing, Gardner (1990) gave the following description of what an electronic archive might be: "The electronic archive would retain the article as the fundamental unit of scientific communication. But it would publish articles, not journals, on demand for individual readers, in whatever format suits them best (and one option would be bound and printed serial issues, that is, a traditional journal). Instead of contracting with subscribers to purchase all that it publishes, it would communicate to them what it has, so they can acquire and read selectively" (p. 334).

A number of practical technical problems must be solved in order to make an electronic archive feasible. One such problem involves the handling of graphical material. Because different personal computers and workstations handle graphics differently, there will be a need for some standardization analogous to the use of ASCII code for text (Hunt, 1990). The idea of software "agents" that know something of the interests of a user of a very large database and that can act on the user's behalf in searching the database for information pertinent to those interests and organizing that information for the user's perusal has been part of this discussion, also (N. Yankelovich, 1990). Several experiments with electronic publishing are ongoing, notably by the Association for Computing Machinery, involving electronic libraries, electronic submissions of articles for publication, and videotape and CD ROM publishing (E. A. Fox, 1990).

Elsewhere (Nickerson, 1986), I have argued that the history of humanity could be told in terms of the techniques that have been developed for acquiring, storing, manipulating, and transmitting information, and that we may be in an early stage of a quantum leap in the accessibility of information to the individual, the implications of which could be as great as the invention of the printing press. Information is immeasurably more accessible to the average person today than it was before Gutenberg, and this greater accessibility has been directly responsible for the development of science and technology as we know them, and for the explosive production of knowledge over the last few centuries generally. It is easy now to imagine information being far more readily accessible than it currently is, and this within the bounds of what is technically feasible now or very likely to be so within the foreseeable future. The world is rapidly becoming "networked," and the access pathways to information repositories are becoming ever more capacious. One can foresee the time when, thanks to such networks and the existence of relatively inexpensive versatile terminals with which one can connect to them, when one will have immediate access to enormous information repositories of all types.

What is needed to make such connectivity really useful, and to increase access not only to the repositories but to the information within them, are powerful and sophisticated tools to help one get directly to the information one wants and to see it in a form or forms that are appropriate to one's intended uses of it. Sometimes the form will be text, sometimes a verbal answer to a specific question, sometimes a picture of a structure, sometimes a graph of a relationship, sometimes a dynamic simulation of a process. I find the prospect of greatly increased information accessibility to be an enormously exciting one, and I believe the effects of such an advance will be far-ranging and profound indeed. Many of the most intriguing questions facing human factors researchers in the future will relate to the human use of information and how information technology can facilitate, amplify, and enrich that use.

A concept that is much in the spirit of some of the visionary ideas of V. Bush (1945) and Licklider (1965) is that of a personal electronic agent or alter ego that each of us could have to help us interact effectively and efficiently with the world of information. Such an agent would know how to find its way around the various information resources that exist and would know enough about its owner to be able to identify items of genuine interest. One version of this idea is that of a knowledge robot, or *Knowbot,* a term coined and registered as a trademark by the Corporation for National Research Initiatives. Knowbots are "programs that move from machine to machine, possibly cloning themselves. . . . They communicate with one another, with various servers in the network and with users. In the future, much computer communication could consist of the interactions of Knowbots dispatched to do our bidding in a global landscape of networked computing and information resources" (V. G. Cerf, 1991, p. 74).

Information Representation and Presentation

The way information can be represented and presented for human consumption is constrained by the representational media available. The book, which has served so well both to store and convey information, will undoubtedly be a major medium for many years to come, if not indefinitely. It is a static medium, however, and severely limited in its ability to represent processes and dynamic concepts efficiently. Moving pictures, including those generated by time-lapsed photography, slow or fast motion, and animation, have some advantages over the book in this regard, but this medium has limitations of its own, and lacks some of the conveniences of paper media. Neither books nor motion pictures provide a convenient basis for development of interactive or adaptive representational techniques.

Computer-driven displays and videodisks provide an opportunity to explore new and qualitatively different ways of representing and presenting information. In particular, they make possible the support of dynamic, multimedia, interactive representation schemes. The feasibility of information presentation techniques that are much more flexible and versatile than what has been available in the past gives new meaning and urgency to the question of how information should be packaged for human use. It is easy to imagine interactive, dynamic, multimedia newspapers, maps, blueprints, circuit diagrams, instruction manuals, and informative documents of various types. The details of their design, however, are neither evident nor readily deducible from existing handbooks; realization of the potential that technology represents in this regard will require a better understanding than we now have of what aspects of a presentation method determine how readily information is assimilated and how well it is retained. When, for example, is animation appropriate? How can static and dynamic representations be used to advantage, both singly and in combination? How can the increased degrees of freedom in display design be used to enhance the attention-getting value or memorability of specific aspects of displays? How might eye fixations and eye movements be used as control inputs for displays?

Hypertext provides exciting new possibilities for representing and presenting information in nonlinear ways that are only just beginning to be exploited. I believe this will prove to be a powerful representational vehicle; it does not follow, however, that everything that traditionally has been represented as linear text should be recast in that form. Hardison (1989) expresses some skepticism about the effects of presenting Shakespeare, say, in hypertext form:

> What does hypertext do for—or to—*The Tempest?* . . . If you imagine a reader using hypertext, you have to imagine a constant movement from text to glossary to grammatical comment to classical dictionary to Bermuda map to textual variants to drawing of Ariel to text to *Sea Adventure* narrative to. . . . The process of "reading" hypertext is therefore different from what is normally meant by the phrase "I read the play". . . . When "read" in this way, the play tends to disappear into the hypertext like water in a sponge. . . . Computers make the disappearance of the text irreversible by plunging the reader into an information swamp from which there is no escape other than turning the computer off and pulling out a paperback copy of the play—or, better yet, attending a performance, since the performance does not have footnotes. (p. 263)

Still, Shakespeare can be read for different purposes, some of which might be facilitated by a hypertext representation and others of which might not, and some readers might enjoy reading a given work more than once, with different objectives on different occasions. In any case, there is much to be learned about how to use hypertext-type representations well.

The report from a National Research Council (1984) workshop on research needs relating to the interaction between information systems and their users concluded that "the most fundamental need of the field is a better and deeper understanding of the capabilities and limitations of human beings as seekers, creators, users, and transmitters of information, and of the implications of these capabilities and limitations for the design and use of information systems'' (p. 13). Topics suggested as foci of research included information representation, information seeking, information filtering and evaluation, and information organization and management. With respect to information representation, it was noted that computer technology provides an unprecedented opportunity to develop new ways of representing information, including the mixing, through multimedia technology, of different representational forms in whatever ways would be most instructive for human use. The research challenge comes from the fact that current understanding does not provide an adequate basis for judging what would constitute effective representational mixes in specific cases. How best to represent and present information for human use is a problem that manifests itself in countless contexts, and the search for answers has become considerably more complicated by virtue of the greatly extended range of possibilities that technology is making possible.

Navigating Information Spaces

Some of the problems of information presentation stem from the fact that only a limited amount of information can be displayed in a small space at one time, and when the display is a cathode ray tube, what is not displayed is really out of sight and in no specific location—not in a bookcase or on a shelf where it can be seen, albeit from a distance, with a change of gaze. The ephemeralness of the medium can make it difficult, sometimes, for users to maintain their bearings with respect to the information with which they are working and how it is, or should be, organized. An approach to this problem that has become familiar to many users patterns interfaces on a desktop metaphor, so they can display a number of partially overlaid objects—documents, files, file folders—bringing to the top of the pile whatever they want to attend at the moment. Other approaches have also been developed to help maintain one's orientation with, or navigate through, electronically stored information.

Researchers at Bell Communications Research (Bellcore) have developed a technique to help in navigating through textual material similar to what would be found in a conventional book (Egan et al., 1989; Furnas, 1982, 1986). An individual using a book focuses on only one section of a single page at any instant. However, the need to focus has different implications for computer-based systems than for books. In the latter case, one's orientation with respect to the whole database is less likely to be disrupted, because it is always there in its entirety; one's place with respect to the book as a whole is marked by the fact that one is looking at something on page 237, say, and all the other pages are physically there in the reader's hands. In the case of a computer-based system, one sees on the display a portion of a page of the underlying "book," as it were, but the part of the book that one is not looking at is not in evidence. This makes it much easier to lose one's place.

The technique developed by the Bellcore researchers provides a "fisheye" view of electronic text, thus permitting the user to focus on an area of interest without forgetting where that area fits within the overall organization of the document. On the right side of the display there appears the segment of text the user is looking at, while on the left is a partial Table of Contents (TOC). When the user clicks the mouse on a TOC item, the associated text appears in the text window and the TOC is modified to provide more detail regarding the structure of the selected segment. At any given time, the representation of the TOC provides detail about the immediate neighborhood of the user's attentional focus and only much coarser information about the more remote regions. This illustrates the principle of displaying information that is at the proper level of detail for the task at hand.

The problem of keeping one's place in an electronic text is analogous, in some respects, to that of maintaining one's orientation when viewing magnified regions of a very large scale integrated (VLSI) circuit design. When developing the details of a VLSI design, the designer will want to view different parts of the circuit at various levels of magnification. At one moment, the display may show one small portion of the circuit greatly enlarged, and at another moment another portion at a different scale. In moving from one display to another, the user can easily lose track of exactly where the section that is displayed is located within the overall circuit, especially because there may be many small regions that, when magnified, look very similar. One approach that has been used to help the user stay properly oriented in this situation is to overlay on the display a small inset map of the entire circuit. The map is a "minified" representation of the whole circuit, with the region that is currently on the display being indicated in some way, perhaps by shading.

Both of the navigation aids described here are conceptually simple and can be applied to relatively straightforward two-dimensional structures. (One might even argue that, for some purposes, text is better thought of as a one-dimensional structure.) Nevertheless, they help to keep users oriented with respect to the entire information structure with which they are working. The need for effective navigation aids is much greater when using multidimensional information structures that include hypertext and multimedia representations. Utting and Yankelovich (1988) described the problem very well: "The promise of hypermedia is the ability to produce complex, richly interconnected and cross-referenced bodies of multimedia information. The horror of hypermedia is the ability to produce complex, disorganized tangles of haphazardly connected documents. Hypermedia can either help to manage, visualize and internalize complexity or it can help make that complexity even more inscrutable" (p. 1). The development of aids that are effective with such structures will be a continuing human factors challenge as builders of information systems discover how to exploit more fully the potential the computer provides for representing information in multidimensional and dynamic ways.

Personal Information Systems

All of us have personal information that we have to manage in some fashion. We have information that we keep for tax purposes, we may keep a calendar, we probably have a file or collection of addresses and phone numbers, we may have a list of birthdays that are important to us. Well-organized people may keep a budget; even those of us who are not well organized are likely to have a variety of documents and other things the location of which we should know, and about which we probably should have some records. These include social security cards, credit cards, insurance policies, birth certificates, deeds, financial papers, receipts, warranties, academic records, military records, medical records, and many similar things. The list can easily get fairly long. One can imagine getting from a well-designed software package some welcome help in organizing one's personal information store. Of course, the kind of information that one might tend to keep in such a store would undoubtedly depend to a large degree on the ease with which that information could be organized, retained, and accessed as needed.

Designing personal information systems that will prove to be helpful to the average user over a long period of time, and will not become a burden, is probably not a trivially easy thing to do. For any information system to be useful it must be accurate, current, durable, and easy to use. For a computer-based information system to be accurate and current requires that it be easy for the user to enter new information and update or purge information that has outlived its relevance. The system also must be relatively fail-safe against inadvertent erasure and the consequent loss of information that may be difficult or impossible to replace.

When I was trying to finish my doctoral dissertation, I lived in fear of losing whatever was the current draft of that evolving document. I was sure that if I kept it in my car, the car would be stolen; if I left it in my office, the office building would burn down; if I kept it with me, I would inadvertently leave it someplace. The document was of little value to anyone else, but to me it represented not only the expectation of the attainment of a desired goal, but months of work that I could not stand the thought of having to repeat. My way of dealing with this phobia was to keep obsolete but almost-current copies of the manuscript in different places, so if I lost the current version I would have to reconstruct only the difference between it and an almost-current one. I did not, of course, invent the idea of keeping more than one copy of a valued document; presumably, that practice has been followed for as long as there have been documents that people did not wish to lose.

The possibility of the loss or destruction of documents is not unknown to users of electronic word-processing and document-preparation systems, and the conventional response is analogous to that of the users of paper. One makes back-up copies of the documents one wants to be sure not to lose. Computing centers typically have a formal back-up policy and periodically store on some secondary medium the contents of their primary systems. Many individual computer users have their own back-up discipline that they practice with varying degrees of regularity. Safeguarding electronic files has its own peculiar difficulties, however, and the individual who is prone to worry about the possibility of losing valued information that would be difficult or impossible to reconstruct will find ample opportunity to fret.

Electronically stored information is trivially easy to destroy inadvertently; and this can be an excessively discomfiting fact if one has several years' worth of work stored in one's computer files. Floppy discs and magnetic tapes can be lost as easily as paper. The faithful making of duplicate copies of files requires some discipline and is easily neglected—until one is reminded of its importance by the destruction of something of value. Unless one's file management practices include an effective policy for deleting obsolete files, one can find oneself with an absurd collection of many versions of documents representing various stages of their development. How often should one produce the duplicate of a file? Every time it is updated or otherwise modified ever so slightly, or only after major changes? How easy should one rest when all the copies of any given document are in electronic form? (People who have been accustomed to using paper for years may find it difficult to be completely at ease about valuable documents that do not exist in paper form.)

The use of computer technology to manage one's personal information store brings with it the possibility to keep on file many orders of magnitude more information than one would be likely to keep with a conventional paper-based system. Whether people will need, want, or be able to use such a capability is difficult to tell. If they are to use it, they will need information management tools that will make such use feasible. Having enormous amounts of information in one's filing system does little good unless that information is accurate, timely, and retrievable when needed. It is not clear how information management systems intended to be used for management of personal databases should be designed. It seems a little unlikely that information management systems developed for business use would be ideal for the management of personal databases. The question would benefit from some research.

Of course, software that is intended for personal use can be used only by people who have the computing hardware necessary to run it. Although the number of homes that have computers has been increasing rapidly and is likely to continue to do so, it seems likely that a great many people will not acquire home computers any time soon, either because they cannot afford them or because they have other preferred ways to spend whatever discretionary money they have. Still, it could be that in relatively few years very few homes will lack computing hardware because it will come automatically with the television set, which will be able to serve also as a video terminal.

On-Line Help and Advice

Most interactive systems today have some type of on-line help and advice capability. The variability across systems in this regard attests to the lack of consensus as to exactly what kinds of help or advice users need, how it should be packaged, and under what circumstances it should be delivered. The building of advice-giving capabilities into software is especially difficult because there are, as yet at least, no good ways to determine the receptivity of advisees to specific bits of advice that might be offered or to monitor the fluctuations in receptivity as a consequence of advice-giving activity.

Help facilities that can respond only to precisely formulated questions, the asking of which requires considerable system knowledge, are not really very helpful. O'Malley (1986), points out that a characteristic of the "local expert" in a community of computer users is an ability to use system documentation, and notes that a valuable service that experts provide is helping inexperienced users ask the right questions. This is important because novice users often have only a vague understanding of the kind of help they need.

On the other hand, providing unsolicited advice is always risky business. The advisee may consider it to be gratuitous, fail to understand it, be distracted by it, perceive it as patronizing, be offended by it in some other way, or be helped by it and sincerely appreciate the help. As Owen (1986) points out, the fact that an individual is using a particular facility is not a reliable indication of receptivity to information about the facility. The advice giver is well advised to be sensitive to the full range of possibilities and to give out only advice that can be offered judiciously and with a sensitivity to the fact that there are various ways in which it is likely to be received. Good tutors know when to give advice and when to refrain from doing so, and also how to shape the advice they do give so it will be not only understood but gratefully received.

In spite of these uncertainties, there are good reasons for attempting to build advice-giving capabilities into interactive systems, especially those that are sufficiently complex that useful capabilities are unlikely to be discovered spontaneously. As already noted, people can continue to use a system inefficiently for a long time, simply because they fail to discover system capabilities that they would be able to use to their advantage if they only knew of them. The idea that the system should be able to volunteer information about such capabilities, information that would enhance the user's expertise, is a very attractive one. The question is how to design advice-giving capabilities so as to have the advantages of advice-giving without the possible attendant drawbacks.

SPEECH AND NATURAL LANGUAGE

Speech production and recognition, natural language understanding, and automatic translation remain very difficult problems in computer science and are unlikely to be solved, in any very general sense, for many years to come. Significant progress is being made in these areas, however, and these technologies are becoming sufficiently mature to support useful applications (Makhoul et al., 1990; Rettig & Bates, 1988; Waldrop, 1988a; Waltz, 1983; R. C. Wood, 1987). Makhoul et al. note that, although they do not expect the challenges of spoken-language systems to be met fully soon, there are shorter-term goals that are both achievable and economically worthwhile. An example would be the development of "robust, operational voice-operated data entry and query systems with limited language-understanding capabilities in actual applications" (p. 482). Such systems could be especially useful in situations in which the user's hands and/or eyes were sufficiently occupied (or impaired) to preclude the use of a keyboard and/or visual display. Moreover, the technologies of natural-language understanding and automatic translation appear to be sufficiently advanced to support database query systems with limited capability, and automatic translation in situations in which posttranslation editing is possible.

Speech and natural language are often discussed as though they were one and the same problem, but they are not. It is possible to have a system with some speech capability, especially speech output capability, but no ability to understand or otherwise process natural language. Similarly, it is possible to have a system with a natural language capability that receives typed inputs but has no ability to deal with speech. Many people working on speech and natural language problems envision, ultimately, systems with both types of capabilities, but each area has its own set of research challenges.

Speech Production and Recognition

Speech as an output medium for computer systems has been in use for some time now in limited ways. Most of the human factors work that has been done in this area has focused on the intelligibility of computer-generated speech and how it depends on speech rate and a variety of segmental, prosodic, and semantic parameters (Huggins, 1978; Pisoni, 1982; Simpson & Marchionda-Frost, 1984; Simpson et al., 1985; Simpson & Navarro, 1984; Slowiaczek & Nusbaum, 1985). High intelligibility is, of course, the sine qua non of a useful speech production system, but qualitative characteristics of the speech that is produced, independently of its intelligibility, may also be an issue in determining the acceptability of a system to its users although this is not necessarily to suggest that speech generated by a computer should resemble as closely as possible that produced by human beings (Nickerson & Huggins, 1977; Simpson, 1981, 1983). The prevailing assumption that it would be desirable for computer-generated speech to sound as human as possible is one that needs empirical confirmation. The possibility that it might turn out to be advantageous and perhaps preferable to have talking computers sound like computers and not like human beings is at least conceivable.

Assuming for the moment that computers are provided with human-like speech, many questions then arise that cannot be answered from intuition and common sense alone. Human speech varies over a wide range along many dimensions. Which human speaker or speakers should computer speech emulate? How much emotion should it convey? How will the preferred characteristics of speech depend on the purpose the speech serves and the context in which it is used?

Speech recognition is a less mature technology than speech production, but progress is being made in this area as well. There are a number of computer-based systems that can recognize speech under certain conditions. Given the current state of the art, the problem is relatively simple if the vocabulary is limited to a few dozen words, the speaker speaks the words distinctly one at a time in a relatively quiet background, and the system has an opportunity to become accustomed to the speaker's voice. There are no systems yet that can do a good job of recognizing continuous speech, spoken in a noisy environment by random speakers with which the system has had no previous experience. There are many intermediate situations between these two extremes, however, and a few systems that can cope with some of them to varying degrees.

Speech understanding is a considerably more difficult problem. Recognition, as just described, does not imply understanding. A system may be able to recognize an utterance, in the sense of being able to produce a correct, printed, word-for-word representation of it without necessarily understanding its meaning. To understand speech is to get the meaning of what has been said, to be able, for example, to paraphrase an utterance or to act appropriately on it. Researchers are working on the development of systems that can understand speech as well as recognize it, but progress toward this very ambitious objective has been slow.

Language Understanding

The ultimate goal in natural language processing research is to develop procedures or systems that can "understand" natural language when it is presented in either printed or spoken form. A fundamental problem relating to this endeavor is the difficulty of determining when, whether, or in what sense, a bit of written or spoken language has been understood. This is a problem whether the understander is a machine or a person. In some instances the problem is not very severe: Salt-passing behavior seems like reasonably adequate evidence that one has understood the utterance, "Please pass the salt." But how does one tell whether a listener has understood, "Mathematics is the science of numbers and their operations"? Indeed, what does it mean to understand such an utterance?

For spoken-language systems one might say that the problem of language understanding picks up where the problem of speech recognition leaves off. Once the system has figured out the sequence of words that has been spoken, the problem becomes that of understanding what that sequence means. It is not really quite as straightforward as this, because information regarding meaning can often be combined with the results of analyses of the acoustical properties of a speech signal to help figure out what words were said. Most of the work that has been done on natural language, however, has used printed language as input and thus has been done independently of the speech-recognition problem.

Weischedel et al. (1990) argue that although it will be a long time—if ever—before computer systems have natural language capabilities that match those of human beings, systems with a useful degree of natural language understanding are possible now, and the capabilities of such systems should increase continuously as research on natural language processing proceeds. Bates, Meltzer, and Shea (1987) note that, for practical purposes, the question one must ask about the state of the art of natural-language understanding is not whether computers can understand English, but whether the current level of computer understanding of English is sufficient to make natural language systems practical. Their answer to this question is "yes." There are now several commercially available natural-language front-ends to databases that work well within certain constraints that more or less follow from the narrowness of the application (limited vocabulary size, limited number of meanings a word can have in the database domain) and especially with users who have adapted to them.

The pros and cons of the use of natural language for input to computer-based systems have been much discussed and debated, but relatively little empirical work has been done to evaluate the usefulness of the natural-language systems that exist. Such work probably would have been premature until quite recently, because the technology was not sufficiently advanced to justify it. However, given the commercial availability of a few natural-language systems that serve as interfaces to data management systems, research toward this end begins to be timely. One recent study of casual users of the natural-language system, Intellect, yielded generally positive results and led the investigators to conclude that, in spite of some limitations of the system, "natural language is an effective method of interaction for casual users with a good knowledge of the database, who perform question-answering tasks, in a restricted domain" (Capindale & Crawford, 1990).

Bates et al. (1987) suggest that a natural language interface should be preferred when the following criteria are met:

- Users don't have the time or motivation to endure more than two or three minutes of instruction, or use will be so infrequent users won't remember special commands.
- The system being interfaced to has a well-defined command language.
- Potential users understand most of the concepts involved in the domain (although they do not need to understand the underlying software).
- Flexible, ad hoc, not-absolutely-precise expression is needed.
- Simpler interfaces will not suffice. (p. 61)

On the basis of some fairly straightforward assumptions and the requirements of a state-of-the-art phoneme recognition algorithm, Makhoul et al. (1990) estimate that computing speeds of about 100 gigaflops (billion floating point operations per second) will be required to support spoken natural-language systems with real-time capability. While pointing out the advantages of natural language as a medium for person–computer interaction, Weischedel et al. (1990) argue that natural language should be seen "as a powerful addition to the repertoire of methods for human–machine interaction, and not as a replacement for those methods" (p. 437).

Semi-Automatic Translation

Fully automatic language translation was an active quest in the 1960s, but early optimism for a quick realization of this goal proved to have been based on a gross underestimation of the complexity of the problem. As the magnitude of the task became considerably more clear, work on the problem diminished almost to the point of nonexistence and remained so for several years. Interest in machine translation has been on the increase recently, however, and once again several companies in the United States and elsewhere are developing translation products. There is a difference between today's expectations and those associated with the pioneer efforts in this area, however; whereas the earlier workers hoped to develop fully automatic translation systems, those of today are attempting to build systems that are good enough to help translators perform their tasks. The belief is that a system that can translate at a moderately high level of accuracy, but not perfectly, could increase the productivity of translators considerably by producing draft translations that they could then edit.

* * *

Computer-based systems are unlikely to be able to use speech and natural language with anything like the facility with which they are used by human beings for many years to come, if ever. The problems encountered in trying to produce such systems are many and of surprising difficulty (Lucky, 1989; Nickerson, 1986). Steady progress is being made, however, and there is every reason to expect continuing progress in the future. As the research proceeds, so do efforts to bring to market products that apply this technology to practical ends. So far, these efforts have not met with enormous success, but in looking to the future, it seems inevitable that speech and natural language capabilities will find increasing use in commercial products. What one cannot be so sure about is the time course in which this will occur. Perhaps the technology will be infused into the marketplace over a period of many years as it is improved and as new applications are gradually identified. On the other hand, there is the possibility that the technology will cross a threshold that will lead to an explosion of applications. I believe the former scenario to be more likely than the latter, but would not want to bet much on it.

Of special interest in the present context is the fact that Makhoul et al. (1990) identify "developing human-factors methods for the design of user-friendly spoken-language systems, including the use of clarification dialogues and the efficient training of users" (p. 483) as a general category of "missing science," representing an important barrier to progress in the use of speech and natural language for person–computer interaction. The lack of an ergonomically sound design, they note, is often a major obstacle to the fielding of a spoken-language system. The need for clarification dialogues— including graceful means of resolving ambiguities and dealing with the occurrence of words not in the system's lexicon—and for effective user training procedures follows from the assumption that natural-language systems will be error-prone and limited in capability for some time to come.

Makhoul et al. note that the United States does not lead the world in the commercial use of spoken-language systems, in spite of the fact that it is generally in the lead in the basic requisite technologies. They speculate that one reason for this situation may be that the U.S. consumer demands a higher degree of convenience and ease of use than do consumers in other countries.

There is, they suggest, a substantial market for speech-recognition systems with small or medium-sized vocabularies in the near term, if only the human factors problems can be dealt with satisfactorily. Because user acceptance is the real measure of progress in spoken-language technology, the development of techniques that would do a better job of predicting user reaction to specific system features would be a significant contribution to the furthering of this technology.

The prospect of machines that listen and talk has many challenges for human factors research. Although we can try to imagine living in a world in which such machines are commonplace, we really do not know what that will be like. Some developers of this technology predict that the consequences of the existence of systems with natural language capability will be profound: "The impact of a breakthrough in computer use of natural languages will have as profound an effect on society as will breakthroughs in superconductors, inexpensive fusion, or genetic engineering. The impact of NLP [natural language processing] by machine will be even greater than the impact of microprocessor technology during the last 20 years. The rationale is simple: Natural language is fundamental to almost all business, military, and social activities; therefore, the applicability of NLP is almost limitless" (Weischedel et al., 1990, p. 451).

I, too, believe the consequences of having machines with natural-language capability are likely to be quite significant, but I suspect, also, that they will catch us by surprise in a variety of ways. On the whole, we have been exceedingly inept at predicting how emerging technologies will be applied and the kinds of consequences they will have. There is likely to be a considerable amount of trial-and-error experimentation with talking and listening artifacts before it becomes clear how this technology will affect our daily lives over the long term.

A primary focus of human factors research has been on problems relating to the ways in which people interact with machines. It is not at all clear what the full range of implications of machines having the ability to communicate in natural language will be with respect to this interaction. They seem likely to go beyond the simple fact of increasing the size of the set of input–output mode options, as, for example, did the invention of the mouse. Will machines that can listen and talk be perceived by their users as qualitatively different from machines that do not have these capabilities? Will users impute to machines with natural-language capability other properties or characteristics that have typically been associated with natural-language capability in the past? Will people's views of themselves change as a consequence of interacting with machines that have capabilities once thought to be uniquely human?

Such questions can be asked in a disinterested way as empirical questions, to be answered by empirical observation. In some cases, however, they may touch on issues about which we are not disinterested. We may believe it would be an unfortunate thing, for example, if the limited natural-language capability of a machine led people to impute to it other capabilities that natural-language users typically have, or if the existence of such machines proved to be detrimental to the self-image of people who do not understand the basis of their operation. These are human factors issues, perhaps of a different sort than usually arise in the context of person–machine interaction, but they are extremely important, in my view, and are likely to become increasingly so as progress on the development of natural-language capabilities for machines continues.

HUMAN UNDERSTANDING OF COMPUTER SYSTEMS

Tools differ in the degree to which their principles of operation are apparent to their users. There is nothing mysterious, at least on the surface, about how a hammer accomplishes its function. A manual typewriter, although considerably more complex than a hammer, is also not very mysterious: The user who is sufficiently interested to observe its operation closely will have no difficulty figuring out how the striking of a key eventuates in the appearance of a letter on a piece of paper. A telephone is a different matter. It is probably safe to assume that a very large percentage of the people who use the telephone daily have not the foggiest understanding of how the instrument and the system of which it is a part make conversation at great distances possible. It obviously works and it is simple to use. What more need one know? (It is worth noting in passing, however, that although we take the telephone very much for granted today, early users often found it intimidating and difficult to use [Lucky, 1989].)

To many people, computers are mysterious machines indeed. We can be sure that people who lack any technical understanding of computers invent their own theories of how they do what they do. In many cases, such theories, even if bizarre, cause no practical difficulties. They have essentially the same implications, for practical purposes, as the belief that the moon is made of green cheese. (Of course, one can make an argument that failure to have some basic appreciation of the operation of a device that is as ubiquitous in our society as the computer is a significant impediment to a reasonable understanding of our workaday world and an intellectual handicap of a sort.) In other cases, inaccurate theories of how computers do what they do can have practical implications. They can, for example, lead to ineffective behavior on the part of a user that produces results counter to the user's intentions, and they can contribute to certain types of emotional stress.

The latter possibility may be more important than we realize. Computers are not only mysterious to many people, but threatening in a variety of ways (Morrison, 1983; Ray & Minch, 1990; Zoltan & Chapanis, 1982). Sheridan

(1980) made this point and identified several factors relating to computers and computer control systems that can contribute to people's feeling of alienation. Paraphrased somewhat, Sheridan's factors are as follows:

- Worry about inferiority of one's own abilities (as compared to those of computer) and threat of obsolescence.
- Tendency for computer control to make human operators remote from their ultimate task.
- Devaluation of hard-earned skills by deskilling the tasks that have to be performed.
- The feeling of technological illiteracy by people not technically trained (which is most people).
- The tendency to attribute magical properties to the computer.
- The high stakes of decision making within large, complex, computer-controlled systems, such as nuclear power plants and air traffic control systems.
- The belief that intelligent machines are becoming more powerful than human beings.

It has been noted that people learning how to use word processing systems sometimes produce explanations of system performance that make the effects of errors seem reasonable, and thereby remove the motivation for correcting them (C. Lewis, 1986; C. Lewis & Mack, 1982; Mack, Lewis, & Carroll, 1983). Sometimes such spurious explanations appear to be based on a metaphorical understanding of a system's operation. The user likens the computer-based system to a more familiar tool or process and makes unwarranted assumptions about what will happen under specific circumstances. By analogy, one imputes to the computer-based system capabilities it does not have. Thus, if the system has displayed enough "intelligence" to handle certain commands as a human being would, the user may impute to the system the ability to understand other commands that would be readily understood by a human being.

Users of computer systems who have been given some conceptual training regarding how such systems are organized or how they work, perhaps with the help of suitable analogies or metaphors, typically do better than those whose training has focused only on use procedures (Rumelhart & Norman, 1981; J. P. Schwartz, Norman, & Shneiderman, 1985; Webb & Kramer, 1987). Metaphors can be very useful tools for instructing novices in the use of computer systems, but as K. L. Norman and Chin (1988) have pointed out, they can also sometimes lead a learner to expect a system to have capabilities or features that it lacks. We need to know how to exploit analogies and metaphors effectively to help people who are not technically trained to understand computer-based systems better, while still guarding against the misleading ideas that analogies and metaphors can sometimes foster.

Efforts to develop computer programs that can do things that when done by humans are considered evidence of intelligence have been going on for several decades. In the 1980s, this work captured the imagination of the business community and the general public, and the intensity of the effort to begin applying artificial intelligence to practical problems greatly increased. Where this will lead is still uncertain; however, it is highly likely that computer-based systems of the future will have more and more capabilities that might be described as cognitive or intellectual in character. What the limits of movement in this direction are we do not know, but it is clear that they have not yet been approached and that the set of things that people can do and machines cannot will continue to diminish for some time to come.

The emergence of computer-based systems that have nontrivial cognitive abilities raises the question of how these systems will be perceived, not only by people who interact with them but by those who do not. The question of how systems with some intelligence will be viewed by people who do not interact directly with them and, consequently, may have a very limited understanding of their real capabilities and limitations, is an especially interesting and important one. The probability of the emergence of grossly inaccurate conceptual models seems high in this case. The challenge to research in this area is both to determine how these systems are typically viewed by that portion of the public that is not technically oriented and to discover what might be done to preclude misconceptions that are dysfunctional or potentially destructive.

An aspect of this challenge that I believe deserves special attention is the need for a better understanding of how people's conceptions of machines with certain real or perceived humanlike capabilities affect their conceptions of themselves. This question has been the source of considerable concern and serious speculation (Mazlish, 1967; Roszak, 1986; Turkle, 1984), as well as a common theme in science fiction. It deserves careful study on a continuing basis. Particularly important, it seems to me, are efforts to prevent the proliferation of misconceptions about the capabilities of machines that have implications for people's views of what it means to be human.

INTERPERSON COMMUNICATION

Human factors researchers have focused a great deal on the question of how to improve person–computer interaction, and with good reason, given the exponentially increasing frequency with which such interaction occurs and how important it is, in many cases, that the interaction be smooth. But much more important, in my view, than issues relating to human–computer interaction are those that have to do with the use of computer technology to support communication between people, because communicating with each other is what we do, always and everywhere, and a lot depends on how well we do it.

Consider, for example, the office context. Independently of what classification scheme is used to study the way office workers spend their time—and there are numerous such classification schemes (Helander, 1985)—studies have found that the most common office activity, by far, is interperson communication. This communication takes a variety of forms, including face-to-face conversation, telephone conversation, written communication, and meetings (Bair, 1987). The emphasis on communication in office work seems unlikely to change as a consequence of the increasing use of computer-based technology, although the ways in which the communication occurs will certainly be affected.

Electronic mail and other forms of computer-supported interperson communication are very interesting phenomena from both technological and psychological points of view. The extent of this type of use of computer networks came as a complete surprise to network implementors. Movement of electronic mail is not among the reasons why the ARPANET was planned; nevertheless, within two years of the installation of the network, mail was its major source of traffic (Denning, 1989b). The exchange of electronic messages among users has remained a primary, if not the primary, use not only of the ARPANET and its successors but of other networks as well. We have not seen this fact publicized a lot. Perhaps it has been a tiny bit embarrassing to the computer industry to acknowledge that the main use we have found for such powerful technology is to facilitate such a mundane activity as the exchange of messages among people. One could argue, however, that facilitating communication among people is one of the most worthy goals that any technology could have.

Electronic bulletin boards represent a new form of communication that is likely, in my view, to prove to be an increasingly effective way of linking people with common interests and facilitating the sharing of information among them. This form of communication could be a fascinating focus of study by scientists interested in understanding how technology can impact interperson communication. Its use resembles the use of conventional bulletin boards in some respects, but differs considerably from it in others. It differs significantly also from other traditional modes—face-to-face, person-to-person, telephone, mail, formal and informal groups, lectures, radio, and television—and consequently, its study will probably involve the emergence of some new concepts and terminology.

An important feature that electronic bulletin boards share with conventional bulletin boards is their totally public nature. One does not, as a rule, use either type for passing private messages. Another commonality is the fact that an individual can participate as a bulletin board user in a variety of ways and to varying extents. One may read notices only, and at more or less frequent intervals, or one may post notices as well, again more or less frequently. Patterns of communication emerge with electronic bulletin boards, however, that are not typically seen with conventional ones. For example, multiperson "conversations" often occur. One person asks a question, or makes a comment; several others reply. The replies evoke further comments, perhaps from the source of the original question or comment, from other commentators, or from newcomers to the discussion.

The dynamic and fluid nature of the types of communication that occur through an electronic bulletin board is partially captured by the notion of many conversations, occurring simultaneously but asynchronously, starting at different times, extending over varying durations, and involving different people. Any given conversation can have any number of participants, and the number of participants in a conversation can change dynamically as some people drop out and others enter. Conversations that are started on a bulletin board sometimes split off and are carried on privately outside it. Also, individuals can participate in as many different conversations as they wish.

There are many questions relating to the use of electronic bulletin boards that could motivate some research. Answering some of these questions would require some attention to definitions and methodological conventions. How, for example, does one characterize a community of bulletin board users, given that individual users are likely to differ greatly in their frequency of reading and posting messages, and that many probably read only? Assuming that such definitional or methodological issues could be resolved, some simple statistical descriptions of bulletin board communities and use patterns would be of considerable interest.

What are the statistical characteristics of the messages that are posted on bulletin boards (distributions of lengths, of messages per unit time), of users, of messages per "conversation," of reply times (time between the appearance of a message and the appearance of the first reply to it)? How large can a bulletin board community be before it becomes unwieldy? What factors determine when a community of users of a bulletin board will reorganize itself into smaller, more focused, special-interest groups? What kinds of measures could be used to reflect the way the community of users of a bulletin board changes dynamically over time: users coming and going and changing their frequency of participation? Does communication via this medium lend itself more readily to emotional interaction (e.g., heated argumentation, "flaming") than do other means of communication? Do personal attacks via bulletin boards (or via electronic mail) evoke the same kinds of responses as do personal attacks in face-to-face interaction? Of what interest and use are bulletin boards to shut-ins and people who are unable to have the social contact they would like because of an inability to get around?

Harnad (1990) has noted the capability that electronic mail provides for researchers to distribute research findings or prepublication drafts of scientific articles to individuals, groups, or even an entire discipline for comment and interactive feedback, and has referred to this medium as *scholarly sky writing*. In Harnad's view, "sky writing offers the possibility of accelerating scholarly communication to something closer to the speed of thought while adding a globally active dimension that makes the medium radically different from any other" (p. 344). Levinson (1988) suggests teleconferencing technology could enable what he refers to as "the globalization of intellectual circles." The idea of such circles is contrasted with Marshall McLuhan's "global village," the inhabitants of which receive common information from television but usually are not able to create or exchange it: "The computer conferencing network establishes on a geographically irrelevant basis the possibilities for reception and initiation and exchange of information that heretofore have existed only in in-person, localized situations. The result is that members of these electronic communities know each other much as members of a real-life local intellectual community such as a university do, but as members of a global television audience do not" (Levinson, 1988, p. 205).

Complementary to the idea of sky writing is that of casting "does anybody know?" questions on a network. Readers of electronic bulletin boards find such questions posted on a variety of subjects addressed to the community at large in the hope that someone will have an answer and be willing to share it. Sproull and Kiesler (1991) described a study of information inquiries on a company-wide bulletin board in a company of several thousand employees. They found about 6 "does anybody know?" questions a day, each of which generated 8 replies, on the average, the vast majority of which came from people not acquainted with the questioner.

There is little question that computer networks, computer-based message technology, and information resources can provide people with access to other people and resources to whom, or to which, they did not have access before, and that they can facilitate interperson communication in a variety of ways. It is only prudent to ask whether it can also create communication problems or make communication more difficult in certain respects. There are two problems that I have heard users of this technology mention. One is the problem of inundation: Some people have found their greater accessibility to others to be burdensome because of the volume of electronic mail that they receive. Another problem that I have heard discussed is a direct consequence of how easy it is to fire off messages the instant after they have been written. When one uses conventional mail, one typically takes some time to compose a letter; after finishing it, one then has an opportunity to rethink whether to send it during the period between when it is put in an envelope and when it is actually posted. It is much easier to be impulsive with electronic mail.

The issue of increased access to people that modern communication provides, and promises to provide to an even greater degree in the future, is, in the eyes of some, a mixed blessing. Lucky (1989) undoubtedly expresses the sentiments of many in putting it this way: "I want to be able to reach other people whenever I wish, but I do not want other people to reach *me* whenever *they* wish" (p. 208). The popularity of unlisted telephone numbers attests to the fact that even without the increased accessibility that can be realized with the help of computer networks, wireless terminals, electronic message systems, and related developments, many people already considered themselves more accessible to the world at large than they wished to be. The tension between the opposing desires for access and for privacy is likely to manifest itself in a variety of ways as the technologies continue to develop.

How does computer-mediated communication differ psychologically from face-to-face communication? Is it, for example, emotionally cooler and less personal, as some have suggested (Kiesler, 1984; Short, Williams, & Christie, 1976)? Does it encourage some people to take more extreme positions and to vent anger more openly than they would in face-to-face situations (Sproull & Kiesler, 1991)? We know that people can project different personalities in different contexts. We know, too, that one may project different characteristics through different media. Some people come across quite differently over the phone, for example, than face-to-face. The driver's seat of an automobile is notorious for bringing out aspects of some people's personalities that seem to be evident in no other place. There is some anecdotal evidence that the personality traits that people convey through electronic mail systems are sometimes quite different from those they project in face-to-face encounters. It would be good to know whether this is true and, if it is, to understand it better.

When an individual projects different personalities through different media, the question naturally arises as to which of the projected personalities is the true one, but maybe the answer is that each is as valid as the other. Perhaps one's "real" personality is complex and is seldom fully revealed in a single context or through a single medium, and what the different media project are views of the same entity from different vantage points. So, if a person who appears to be docile in face-to-face situations becomes aggressive when using an electronic mail system, perhaps we should not ask whether he is "really" docile or "really" aggressive, but conclude that he is really docile in the one situation and really aggressive in the other.

One of the more thought-provoking findings to date about the difference between electronically mediated and face-to-face communication, in my view, is that social status seems to be a less important determinant of behavior in the former case than in the latter. The proportion of talk and influence of higher status individuals is less when communication is via electronic mail, and impediments to easy participation in meetings—a soft voice, a negative self-image—seem to be less important factors in electronically mediated communication (Sproull & Kiesler, 1991). This is not to suggest, of course, that electronically mediated communication is to be preferred to face-to-face communication in any general sense, but only to point out that there it may have advantages under certain circumstances.

A major hope for the future of electronic mail, electronic bulletin boards, and computer-based message technology more generally, is that they will further facilitate interperson communication by providing users with tools that will permit them more readily to share data, work spaces, and ideas through interconnected workstations and information repositories, using powerful capabilities for visual representation, modeling, data analysis, composition, and the exchanging of information at extremely high rates. Multimedia systems that can accommodate not only typed text, but facsimile, voice, and graphics will expand the existing capabilities appreciably. With such facilities as the base on which to build, the challenge now becomes that of adding functionality that will make possible in the future forms of communication that have not been possible in the past. Denning (1989b) has observed that the significance of the ARPANET and its derivatives is not in the networking technology so much as in the fundamental shifts in human practices that have resulted from their existence. This observation is likely to be as true in the 2020s as in the 1990s.

The possibility that information technology could significantly affect the ways in which people interact and communicate with each other was noted in a National Research Council (1984) report that focused on research needs relating to the interaction between information systems and their users:

> Information technology also will foster new modes of interaction among people, which in turn will stimulate the emergence of new social structures. Understanding the effects of these changes will require a better understanding than we now have of various aspects of human communication and interaction. We need to understand better, for example, the extent to which communication patterns in the workplace and elsewhere are determined by fundamental human characteristics as opposed to the capabilities and constraints of communication technology: does people's preference for face-to-face communication via speech (Chapanis, 1975) reflect a deep-seated human need or simply the limitations of currently available alternatives? Will computer-based information systems facilitate or inhibit communication? What will determine whether they help or hinder understanding across disciplines and cultures? Interpersonal communication clearly serves purposes other than that of verbal information exchange, although they may not always be recognized explicitly. It will be important to know to what extent these purposes can be met via electronic mail, shared electronic work spaces, electronic bulletin boards, and other new computer-based means of communication. (p. 8)

Personal computers and document preparation software have already changed dramatically the way some of the people who use these tools write. Because revising and reformatting a document is so much easier than it used to be, there is less need now to plan the structure of a document before starting to compose it. The composition process can also differ considerably from what it once was because editing is so trivially simple that there is much less pressure to say anything right the first time. There can be little doubt that these tools have made writers who use them more productive, in the sense that they get out more words per unit time; whether the quality of the finished products is better as a consequence of the use of electronic writing aids is much less clear. If there were a way to answer this question definitively, I would bet that the answer is "yes," but it would be a small wager. In one study, people made many more corrections and changes when composing on a word processor than when writing by hand, but did not, on balance, produce better documents (as evaluated by independent judges) as a consequence (Card, Robert, & Keenan, 1982).

I cannot resist wondering in print to what extent document preparation software—and I would not give mine up for the world—is responsible for the enormous quantity of junk mail that gets crammed into my mailbox daily, and now that I am on this digression, surely "personalized" letters, letters that have one's name sprinkled throughout to create the false impression of having been written specifically for the recipient, are among the more offensive instances of the depersonalization that has sometimes been associated with computerization, because they make a farce of the very idea of a personal touch. Although I mention this bit of banal deceitfulness to vent my sense of irritation as a recipient of such letters, there is a legitimate point to be made. Although computer technology has the potential to enhance interperson communication, it also has the potential to degrade it in a variety of ways. Figuring out how to keep the balance in the right direction is a human factors problem, at least in part.

There are a number of systems in one or another stage of development that are intended to provide writers with capabilities that will improve the quality of their products (Heidorn, Jensen, Miller, Byrd & Chodorow, 1982; MacDonald, Frase, Gingrich & Keenan, 1982). Whether they will have the desired effect remains to be seen. One wonders how Shakespeare's output might have differed had he had a word processor and sophisticated writing aids. But perhaps the question is irrelevant. Shakespeares come along infrequently and maybe they are above being influenced by their tools. Most of us lesser mortals can undoubtedly benefit from whatever help we can get from word processors, idea processors, or any other tools that offer some hope of improving our writing, whether at the level of spelling words correctly, putting the little marks where they belong, or helping us to figure out what we want to say and to say it somewhat more clearly than we otherwise might.

As computer-based systems are used more and more extensively to facilitate interperson communication, the forms of communication involved will increasingly include confidential transmissions and transactions that are intended to be legally binding. In many cases, it will be important that there be a way to determine, with a high degree of certainty, that given users are who they claim to be. To date, the most common approach to user authentication has been the use of passwords or other types of codes known only to individual users and the systems to which it provides access or to individuals to whom they wish to be identified with certainty, but codes have limitations and can be awkward in some circumstances. For some purposes, it would be desirable to have less formal authentication techniques, more analogous to the way we all use voice recognition to identify acquaintances on the phone and accept signatures that suffice for making legally binding commitments. The emergence of multimedia systems that include speech and facsimile capabilities will make both of these forms of authentication available in computer-based communication systems.

According to Makhoul et al. (1990), the state of the art in both speaker verification (verification that a speaker is who he claims to be) and speaker identification (identification of a speaker who does not necessarily claim a specific identity) now exceeds human performance on verification and identification tasks. Relative to speech recognition, speaker verification is considered a mature technology, and speaker verification products are commercially available. Makhoul et al. attribute the fact that these products are not yet in widespread use to human factors issues, at least in part. For the near-term future, much of the message traffic on computer networks will probably continue to be text that is entered via a keyboard terminal, so the development of effective approaches to user authentication will remain a challenge.

*   *   *

It has been over 30 years since Licklider (1960) wrote his seminal paper, "Man–Computer Symbiosis." During that time, many aspects of his vision of people interacting with computers to solve problems or to perform tasks more effectively than either people or computers could do alone have been realized. The challenge of finding synergistic ways to couple the capabilities of human beings with those of computers is as real today, however, as it was in 1960. Current uses of interactive systems to help design drugs, to explore the effects of stresses on materials, to help children learn physics, and to assist us in performing numerous other tasks have given us some tangible hints of the potential this technology represents. Human factors researchers, along with scientists and engineers from a variety of other fields, have much yet to contribute to the further realization of this potential.

Issues of interface design and information representation will become increasingly important determinants of the effectiveness of interactive computer-based systems as the systems become more and more versatile and are used in the workplace and elsewhere by an ever-increasing percentage of the population. There will be a continuing desire to develop more effective ways of getting information from the user to the computer. Typing and pointing, the two major input methods currently used, are effective but limited; speaking, grasping, and looking will be among the input options of many systems of the future. Eye fixation and eye movements may be monitored and used as input signals to make menu selections or to indicate desired actions in other ways, and speech will be used increasingly for both input and output.

There will always be a desire to invent new and better ways to represent information for human assimilation and use. Given the almost limitless possibilities for dynamic, multimodal, interactive representations, we are beginning now to be more constrained by our imaginations than by what the technology can deliver. Systems will be able to present information in a rich variety of representational forms, mixing text, speech and other sound with static and dynamic visual images, including process simulations. Hypertext, multimedia, and similar capabilities will make it possible for users to explore large information stores dynamically and interactively. Effective exploitation of such capabilities will be limited less by technology than by the speed with which the principles that can guide their utilization from a human factors perspective can be discovered and articulated.

Because both speech and pictures require more storage capacity and transmission bandwidth, by many orders of magnitude, than does text, the development of data compression procedures will be a continuing quest in both areas. Assessing the quality of compressed or synthesized speech and pictures will remain a matter of human judgment. The development of techniques that can predict listener and viewer reaction—and, in particular, perceived quality—on the basis of quantitative properties of the stimuli will continue to be an objective of people working in these fields, but the validation of such techniques, if they can be developed, will itself involve the use of judgmental procedures.

There are many other challenges and opportunities for research relating to person–computer interaction. How will the use of computer-based message systems and other communication tools affect the ways in which people communicate with each other? What kinds of tools will be needed to help people manage the enormous amounts of information they will have access to through computer networks and large information repositories? Under what conditions will speech be the preferred medium for person–computer interaction? Negroponte (1991) has made the thought-provoking observation that "independence of space and time is the single most valuable service and product we can provide humankind" (p. 108). How can we ensure that the further development and application of information technology will indeed decrease the space and time constraints under which people function and not rather further encumber their lives? The list of questions is easily extended.

Many of these questions are cognitive in nature, so much so that the distinction between applied engineering psychology and basic research on human cognition is likely to become increasingly blurred. Information systems are used to facilitate the performance of cognitive tasks, and more and more they are acquiring significant amounts of cognitive capabilities. Many of the questions that arise relate to the design of interfaces that will assure the usefulness of these systems, and they involve cognitive variables. The questions cannot yet be answered with much confidence because the necessary research on human cognition has not been done. The kind of research that is most needed is research that is addressed to relatively generic issues that are independent of specific systems, and to the discovery of principles that are applicable across a broad range of equipment. It must focus on basic questions of the capabilities and limitations of human beings as receivers, users, processors, and transmitters of information. That is not to claim that such research should always be done in abstract contexts and should never involve the use of real systems, but it is to suggest that it is possible to do research that is so narrowly focused on aspects of specific systems that it is not likely to have much general interest or applicability. We need better insights into some fairly fundamental issues, and it is research aimed at providing those insights that will do the most to evoke better interface designs and a more adequate understanding of how to use computer technology to the greatest advantage.

How should we think about the relationship between computer technology and people? What metaphors should we use to describe human computer interaction? What attitudes should condition our approach to this technology? Much use has been made of metaphors that cast the computer in the role of the person's partner, associate, or friend. I believe this to be unfortunate, misleading, and perhaps dangerous. I much prefer the view that sees computers and computer-based systems as tools to be used to serve human purposes. *Partner* and *associate* carry the notion of shared responsibility, which I see as inappropriate to the person–computer relationship on several counts; and the use of *friend* in this context demeans its traditional connotation. Computers are machines; people, in my view, are not.

Still, computers are extraordinarily powerful machines, like no others with which humankind has had very much experience. Information technology, more generally, has the potential to change life on this planet in ways that we have hardly begun to imagine. Some people believe that information technology will change things more dramatically than did the Industrial

Revolution; I believe it has the potential to do that. That potential encompasses many scenarios, some much more desirable than others. Within this technology are liberating, horizon-expanding and life-enriching possibilities. Like any powerful technology, this one has also the potential to be used for destructive and dehumanizing purposes. It contains the possibilities of concentrated power, manipulation, and repressive control. Learning to apply this technology effectively to desirable and humane ends while preventing misuses of it, and mitigating them when they occur, is surely one of the overarching challenges of our time.

11

Work

It is difficult to get a clear and unambiguous picture of what has been happening in the world of work in general, and in the American workplace in particular. Sometimes, observers are obviously describing different parts of the elephant, but even when this is not the case they sometimes use the same terms in different ways. Current-trend data lend themselves to various interpretations. A common reason for confusions with work data is uncertainty as to whether proportional or absolute numbers are being reported and, when proportional numbers are involved, lack of information as to how the proportions were derived. It may also be that the situation is sufficiently complex to leave all simple summary characterizations open to misinterpretation.

The picture gets even less clear as one tries to look ahead. It is very easy to find conflicting projections and guesses as to the future of work. Consider, for example, the top-level question of whether the demand for human labor—broadly defined—will grow or shrink. There can be little doubt that an increasing number of the tasks, including cognitive tasks, now performed by people will be done by machines, but there is no agreement on what this will mean with respect to the need for human participation in work. There is the fear that technology, especially in the form of automation, will make many jobs obsolete; counteracting this concern is the fact that the use of technology in the past has created more jobs than it has abolished and the hope that it will continue to do so, but there is no guarantee that what has been true in the past will continue to be so in the future; that, of course, is what makes the attempt to look ahead an interesting exercise.

OCCUPATIONAL TRENDS

As of 1988, the number of people employed in the United States was about 117 million, which was about 47.5% of the population, or about 62.5% of all adults, 16 years or older (Bureau of the Census, 1990). In 1950, the comparable figures were about 39.5% and 56.5%, respectively, so a larger percentage of the population is working today than was doing so then. This is evidence of more job opportunities per capita, but it probably also reflects a greater economic necessity for more wage earners per family and certainly an increase in the percentage of women who are working outside the home. Many people over 65 work part-time. Apparently, some alternative to the normal eight-hour workday would be of great interest to a large number of workers of retirement age (McConnel, Fleisher, Usher, & Kaplan, 1980).

Although the AFL-CIO (1983) has warned of the possibility of a permanent labor surplus in the United States, there appears to be some possibility that the country will actually experience a labor shortage in the near-term future. As of 1989, the growth rate of the U.S. workforce was less than 1.0%, down from about 2.5% in 1970 and continuing to fall (Vaughan & Berryman, 1989). A further slowdown in the rate of growth of the nation's labor force was expected over the decade following; in particular, the pool of young workers will shrink as a consequence of the "birth dearth" of the 1970s that followed the "baby boom" of the 1950s and 1960s. The National Center for Education Statistics (1990c) estimated a decline in the number of high school graduates in the United States from 2.8 million in 1988–1989 to about 2.5 million in 1993–1994.

During the 1970s, the labor force grew by 24.1 million; in the 1990s it is expected to grow by 15.6 million (Bureau of Labor Statistics, 1985). In fact, during the 1990s, the American workforce is expected to increase more slowly than at any other time during the previous 50 years (Rauch, 1989). This projected slowdown, coupled with disheartening statistics regarding school dropout rates and decreasing student achievement, has some economists worried about the possibility of an inadequate—and inadequately prepared—workforce to meet the country's needs around the turn of the century.

The mix of occupations has been changing in a fairly systematic way for the last century and a half. The percent of the U.S. labor force employed in agriculture, for example, has gone from over 70% in 1820 to about 3% in 1990. Farming has become a large-business enterprise as the number of active farms has decreased precipitously, and the average size of the farms that remain has increased commensurately. The mechanization of farming has made it also a capital-intensive business. Careful planning and managing of resources are critical components of this enterprise and can easily mean the difference between prosperity and financial failure. Information technology has been playing an increasingly important role in all aspects of farming for several years (Holt, 1985), and the importance of that role is unlikely to diminish in the future.

As the percentage of people engaged in farming decreased during the latter part of the 19th century and the early part of the 20th, the percentage employed in manufacturing increased until it accounted for the majority of the labor force. In more recent decades, the shift has been from the production of goods to the provision of services. Now the service sector accounts for about two thirds of the labor force, and the production of goods accounts for about one third. Ninety percent of the new jobs added to the United States economy from 1969 to 1976 were in the service sector (Ginzberg, 1982). Nearly all the new jobs created between now and the end of the century are expected to be there as well (Riche, 1988).

Until fairly recently, economists typically distinguished three major sectors of the economy: agriculture, manufacturing, and services. With the recent rapid increase in the number of jobs that involve some form of information processing, information is sometimes identified as a major sector as well. In fact, the information sector cuts across the other three, however, because information processing jobs exist in all of them. According to an OTA (1981) report, the information sector has had the largest number of workers in the United States since about 1955 and, as of 1980, accounted for more than 45% of the workforce. According to another estimate, the percent of the American workforce involved in information jobs increased from 17% in 1950 to 65% in the early 1980s (Naisbitt, 1984). This trend was noted some 30 years ago by Machlup (1962), who pointed out that the "knowledge industry" was then growing 2½ times as fast as other industries and that unemployment among people qualified to do only unskilled labor would be likely to become a serious problem in the future. Total capital investment per worker is now greater for workers in the information sector than for those in basic industrial activities (Quinn et al., 1987). Office work is becoming increasingly dependent on computer-based systems, and the trend is expected to continue; within a decade or so, computer terminals may be as common in offices as telephones (Chamot, 1987; OTA, 1987a).

There seems to be general agreement regarding the relative future demands for labor coming from the major economic sectors: The portion of the total workforce engaged directly in farming will remain very small (2% to 3% in the U.S.); the portion involved in manufacturing will continue to decrease; and the portion involved in service industries—already the majority—will continue to increase. The percent of workers involved in information work—whether in manufacturing or service industries—is also likely to grow during the near future, although the longer term effect of automation on these types of jobs is uncertain. Kraut (1987a) has argued that as information becomes easier to collect, process, and distribute electronically, more of us will become both information workers and information consumers:

> Writers reporting the latest fashion trends, receptionists announcing a telephone number, architects designing schoolhouses, insurance adjusters calculating the damage to our cars, clerks recording expense vouchers, teachers giving assignments, merchants calculating their income taxes, and spies assessing enemies' troop strength will all be using computers and telecommunications technologies to get their jobs done. Their fingers, eyes, and minds will be connected to silicon chips with wires of copper and glass. (p. 2)

JOB QUALITY

What about the quality of the new jobs that are being created? Again, the picture is not clear. According to the OTA (1988c), professional and managerial jobs accounted for about 45% of the job growth from 1980 to 1986. On the other hand, Thurow (1987) claims that the new jobs added to the economy during the 1980s have, by and large, been inferior in earning power to those that were being added a decade or so before. From 1973 to 1986, real average hourly wages of production and nonsupervisory workers decreased by 10% (Young, 1988).

It appears, also, that people move from job to job somewhat more quickly than they did in the past: The average time spent at the same job declined from 4.6 years in 1963 to 3.6 in 1980 (R. B. Reich, 1983). Reich notes that the average American holds 10 different jobs before retirement. The fast-food industry has an annual turnover in employment of greater than 1,000% (Bay, 1988). Undoubtedly, this is due in large part to the fact that this industry employs many young people while they are still in school, as well as many semi-retired people who want to work only on a part-time basis, but it is still an impressive statistic and one that has implications for the human factors of work. Industries with high turnover rates cannot afford to invest significant resources in employee training and must seek to minimize the skill requirements of their jobs.

Thurow (1987) notes an increasing inequality in the distribution of earnings in the American labor force, and attributes it, in part, to intense international competitive pressures and an increasing proportion of female workers in the United States. He also points out that during a period when male employment grew by 7.4 million jobs, 400,000 middle-income male jobs disappeared, and the new jobs that showed up were primarily in the low-earnings group. A solution to the problem, he suggests, is a higher rate of growth of productivity and enhanced international competitiveness. This, in turn, means a well-educated and highly skilled labor force, investment in capital equipment, and use of the most effective technologies for production. So far, Thurow claims, the United States is doing poorly on all of these counts.

Unionism appears to be playing a smaller role in the United States today than it did in the past. The percent of the American non-agricultural labor force that was unionized decreased from about 35% in 1945 to about 18% in 1985, and this downward trend is expected to continue into the 1990s ("Beyond unions," 1985). This change undoubtedly relates both to the shift of labor from manufacturing to the service sector and to a proliferation of small companies and highly specialized industries. In spite of the decline of unionism, people who have entered the U.S. workforce since World War II appear to have greater expectations regarding the quality of the workplace and respectful treatment by management than did those who entered before the war (D. Yankelovich, 1978). How important one considers one's job and company's purpose to be, relative to what one considers to be significant community or societal problems, appears to be an important determinant of job satisfaction for many workers (F. E. Emery & M. Emery, 1978; Levine, Taylor, & Davis, 1984).

Bjørn-Andersen and Kjærgaard (1987) make a distinction between people-driving and people-driven systems. People-driving systems, in their view, treat people as cogs in the machinery, and overly structure and program their operations. This approach may have been effective in the past, but it is unlikely to continue to be in the future. When, in order to survive in the competitive environments of today and tomorrow, corporations must be able to change quickly and creatively, they will need the type of employee who will not fit well in the people-driving model. They will be forced, if they wish to survive, to go to a people-driven model, in which skilled and motivated employees are given some scope and an opportunity to identify with a company's overall operation and products rather than only with a highly focused, specialized task.

TECHNOLOGY AND WORK

A major question of interest to anyone concerned about the future of work is that of how the increasing use of technology will change both the demand for human labor and the nature of the work that will be done. Among several related issues are concerns about the disruptive effects of worker displacement because of job obsolescence, about the need for retraining, and about the implications of future job opportunities for education. According to one assessment of the implications of continuing innovations in technology and production processes, as many as 75% of all workers now employed will need retraining by 2000 (Dentzer, 1989). The introduction of an innovative technology in the workplace often has the unintended and unanticipated effect of changing the nature and function of the organization (Coates, 1977). Originally, the technology is adopted in order to facilitate existing functions and activities. In time, however, the new technology is seen to permit the organization to do things it was not able to do before, and this may motivate a redirection of the organization's business or a redefinition of its role.

There has not yet been much very compelling evidence that automation or mechanization necessarily leads to a reduction in the need for human labor. The mechanization of farming resulted in an enormous decrease in the number of active farmers, but it also created many jobs involved in the production, sales, and servicing of farm equipment and machinery, the processing, packaging, transporting, and selling of farm products, and numerous derivative industries. In spite—or, perhaps in part, because—of increased mechanization in the banking industry, employment in banking grew by 50% between 1970 and 1980 (Ernst, 1982). There is also the fact that the introduction of technology can create goods and services and perhaps entire industries that did not exist before. The computer industry is a case in point. Of course, whether, on balance, technology creates more jobs than it abolishes, there is no doubt that, as a consequence of technological innovation, specific jobs become obsolete and individual workers are displaced in significant numbers (OTA, 1986d).

Aggregate statistics aside, the introduction of new technology in the workplace can have disruptive short-term effects on individuals, especially on those workers who lack the kind of broad-based skills that would facilitate retraining and new job placement. According to one set of data, about one third of workers laid off from a job either fail to find a new job within a reasonable time or have to accept one that pays at least 25% less than the one from which they were laid off (Osterman, 1988). Older workers typically find it more difficult to obtain new jobs when displaced from existing jobs than do younger workers (Sheppard, 1978). High unemployment rates are especially severe for people with less than high school education, and among high school dropouts, they are much worse for Blacks than for Whites and Hispanics, and somewhat worse for Hispanics than for Whites (Berlin, 1983).

On looking at the impact of office automation in the insurance industry, Feldberg and Glenn (1987) came to the conclusion that the net effect was an increase in the total number of jobs, in spite of a decrease in the number of traditional clerical jobs. They concluded, also, that the workers who were displaced by automation did not appear to benefit from the new jobs, because the latter were primarily technical-level positions that were filled largely by males and not by females, who were the ones that had held the clerical jobs that were discontinued. Applebaum (1985) has also suggested that, on balance, the new jobs, including clerical jobs, that are being created within the insurance industry are higher level jobs than those that are being eliminated. Feldberg and Glenn (1987) caution, however, that "a 'higher level' job can require more skills, even pay a higher salary, yet still be limited in autonomy, closely monitored, afford little discretion, and be tied to production quotas" (p. 96). They point out the need to learn more about the jobs that are being created in order to determine how they are or are not better than those they have replaced.

Finding humane and effective ways of dealing with displacement problems will continue to be a challenge to industry. The need for retraining of significant segments of the labor force could increase, as rapidly changing job opportunities make some skills obsolete and put others in greater demand. Goldstein and Gilliam (1990) have discussed a number of the implications that various expectations regarding the future workplace and composition of the workforce have for training needs. They note that, in spite of the fact that the need for new approaches to training is likely to be great, there does not yet appear to be a comprehensive federal government policy addressed to this need.

Considerable attention has been given to the effect of automation on white-collar work. One conclusion that seems safe to draw is that simply bringing computer technology into the office does not guarantee an improvement in office operations or an increase in productivity (Attewell, 1990; Bikson & Gutek, 1984; H. K. Bowen, 1986). Not everyone agrees, either, that bringing more and more technology into the office has generally had the effect of improving job quality (Braverman, 1974; Shaikin, 1988).

Regarding the question of whether the introduction of information technology has, on average, increased or decreased the skill requirements of white-collar workers, the results of research so far have been somewhat contradictory. Kraut (1987b) argues that this is a consequence, in part, of the fact that different studies have dealt with technology at different stages of maturity. The implications of information technology for a particular white-collar job may be quite different immediately on being introduced to the work situation from what they will be five years later. The job itself may change substantially over such a period, also. Scholars are debating, Kraut suggests, "whether new technology encourages job routinization and fragmentation, is a beneficiary of it, mitigates it, or is independent of it" (p. 15). Arguments can be marshalled for each of these possibilities, and that alone should make us cautious about drawing any sweepingly general conclusions about the effects of new technology on white-collar work.

Although much has been written about technology as an agent of change in the workplace, some observers have expressed a different view. Kraut's position is that technology has the potential for transforming work, but it does not now serve that function and may never do so. With respect to office automation in particular, J. C. Taylor (1987) has argued that, to date, it has been applied in outmoded ways that were appropriate to the industrial era, but not to the information age, and consequently the technology is not delivering the expected measurable improvements in effectiveness. In keeping with old values, managers are using the technology to mechanize old tasks rather than to develop new tasks that are more appropriate to the age.

Kraft (1987) predicts that, in the future, computer technology will be a conservative force that will be used to help maintain the status quo, especially with respect to patterns of authority and control:

> Although computer technology is dramatic, its application is not. The computer is a revolutionary tool that will have largely conventional effects in both the workplace and in society as a whole. In particular, the computer will help reproduce, not transform, traditional relations between manager and managed, between those who design and plan and those who carry out the designs and plans of others, between those who have jobs in which experience and judgment are valued and those with jobs in which experience and judgment are irrelevant. Computers will reproduce the divisions between people who are paid well and have reasonable job security and people who are paid badly and have little security. (p. 99)

"Whether people have interesting, challenging, rewarding, and skilled jobs," Kraft claims, "is primarily a political issue, not an engineering problem" (p. 111).

A common theme in these views is that new technologies that have applications in the workplace can be used in a variety of ways and to a variety of ends. There have been, and are continuing to be, a wide range of effects that are contingent not only on the specifics of the technology that is introduced, but on the organizational and social situation in which it is introduced, as well as on the details of the method of introduction and use (Bikson, 1987; Feldberg & Glenn, 1987). One can certainly find instances in which jobs were changed for the worse, from the workers' point of view, as well as those in which jobs were made more interesting and satisfying (Attewell & Rule, 1984; Kling, 1980). Technology lends itself to insensitive and dehumanizing applications in the white-collar workplace, as elsewhere, as well as to intelligent and humane ones.

Much has also been written about resistance to the introduction of new technology in the workplace, as a general phenomenon, and about the reluctance, especially of managers, to adapt to new computer technology, in particular. In a series of studies of 26 organizations, half in manufacturing and half in service provision, involving managerial, technical, and clerical workers, Bikson and his colleagues (summarized by Bikson, 1987) found little evidence of resistance to computer use. Workers in these studies frequently reported the need for more processing workstations and processing power than they currently had. Bikson also reported finding software functionality to be of greater concern to users than the details of interface design. A third noteworthy finding in these studies was less success in the sampled organizations in providing software support for higher management and professionals than for clerical, secretarial, and technical employees. Noting that these studies found no evidence that high-level employees are adverse to learning how to use a system, providing the functionality of the system makes it worth learning, Bikson took the position, not often encountered in the literature, that "questions of how to increase the power rather than the simplicity of user interfaces merit more attention" (p. 175).

The perceived ease of use of any system is dependent, to some degree, on the social environment in which the system is used (Blomberg, 1987). This is due, in part, to the fact that a major, and often preferred, way for people to obtain information about a system that they are beginning to use is by asking colleagues who are experienced users. Blomberg concluded from case studies of new users of photocopying equipment that a technology that appears to be difficult to use may be perceived favorably by employees if the support to make the technology understandable—especially in the form of experienced users—is readily available.

It seems safe to assume that the knowledge about the operation of any complex system that resides in the collective experience of a community of users is greater than the knowledge of any one of the community's members and certainly greater than that of the average member. A question of both practical and theoretical interest is how to make the aggregate knowledge available to all the members of the community. The most obvious possibility is to ensure the ease of communication among members of the group. In some cases, when the group functions operationally as a team, working together in a more or less face-to-face fashion, access is relatively straightforward. When the group does not function in this way, it can be a much more serious problem. Finding ways to share the corporate knowledge of the group, so as to make what the experts know easily available to the nonexperts, is a challenge.

One can find a variety of views among analysts regarding how the quality of office work may change in the future. Strassman (1985) and Uhlig, Farber, and Bair (1979) expect it to improve, from the worker's point of view, as a consequence of further advances in information technology and its use in office contexts. Bjorn-Andersen (1983) and Gregory and Nussbaum (1982) predict that the use of information technology in the office will benefit managers and professionals, but will work against the best interests of clerical workers by downgrading the skill requirements of their jobs and providing management with tighter control from the top. Feldberg and Glenn (1987) note the possibility that the reorganization of clerical work through the introduction of information technology into the office will sometimes make the work narrower, more fragmented, and more limited than it previously was. We may reasonably expect a considerable amount of variation in the way information technology is applied in offices in the future.

A challenge to human factors researchers will be to study these differences with a view to understanding better the variables—technological, psychological and social—that determine both productivity and worker satisfaction in technologically rich work settings. In evaluating the effects of new technology in the workplace (as in the home, school, or elsewhere), it will be important to observe situations long enough for the effects of novelty to dissipate. In some instances, it may take people some time to get used to the technology and to overcome their initial reluctance to use it. In other cases, people may use a system more frequently or more intensively initially, because of its novelty, than later, after it has become a standard fixture.

Many workers in the future, both in the office and elsewhere, will find themselves working with automated or semiautomated systems and with machines that have considerably more impressive capabilities—some of a cognitive nature—than were possessed by machines in the past. This fact probably has some implications that will become clear only as the situations involving such machines unfold. One aspect of the emergence of increasingly sophisticated machines and systems that can be anticipated, however, relates to the question of how to provide for their occasional failure. In the past, it has typically been assumed that if an automated process breaks down, it will be possible to fall back on a manual way of doing things, and the assumption has generally been a valid one. If a spot-welding robot ceases to function, a human welder can do the job; if the bar-code reader at the check-out counter refuses to read code, the cashier can punch in the prices manually; if the autopilot in an airplane malfunctions, the human pilot can take over the controls. Automated systems are beginning to emerge, however, that are sufficiently complex that the ability of human beings to operate them in the event of equipment failure is questionable; looking to the future, it seems inevitable that systems of even greater complexity than any that now exist will be developed. The complexity of some of these systems will be such that manual operation, in any meaningful sense, will not be an option; automation will be a necessary condition of their functioning at all. The design of adequate fall-back procedures that will ensure, in the case of various types of equipment malfunction, either the continuing operation of such systems or "soft," noncatastrophic shut-down poses new types of engineering challenges.

SKILL REQUIREMENTS OF FUTURE JOBS

There are differences of opinion regarding exactly how the skill requirements of whatever jobs there are in the future will change. There is the view that new jobs will increase most rapidly in professional, technical, and sales fields, and will demand much higher skill levels than do today's jobs; Johnston and Packer (1987), for example, predict that more than half of the new jobs that will be created between 1984 and 2000 will require education beyond high school, and that almost one third of them will be filled by college graduates. K. Miller (1989) has argued that the jobs that are created in the future are likely to require higher levels of education than those that will disappear. Projections from the National Science Foundation's econometric model show nonmanagement job opportunities in most areas in science and engineering increasing faster than in all other occupations combined until the end of the century. The largest increases are projected for computer and operations research analysts, electrical and electronics engineers, and biological scientists (Pool, 1990b).

The U.S. Department of Labor (1987) has claimed that the American labor force will need significant increases in the quality and quantity of skill training in the 1990s because the economy "will place a premium on highly-skilled workers who can work independently" (p. 45905). R. B. Reich (1983), also, has taken the position that the flexible-system production that the country must adopt if it is to remain a major economic power will put much greater demands on workers for skill, judgment, and initiative. Some researchers have argued that a general effect of the greater use of high technology in the workplace has increased the cognitive demands on the people who use it (W. C. Howell & Cooke, 1989). According to the results from a series of studies done at the National Center on Education and Employment, high technology industries differ from traditional industries in being more labor-intensive (and less capital-intensive), in employing people with more education, and in increasing their labor base at a greater rate (Bartel et al., 1989).

A special concern is the possibility of a coming shortage of scientists and engineers in the United States (Holden, 1989b). According to a National Research Council News Report (cited in Leeper, 1990), jobs requiring mathematical skills are increasing at nearly double the rate of overall employment, and if the needs of the workforce for people with such skills are to be met in 2000, it will be necessary for women and minority groups, who are currently greatly underrepresented among mathematically skilled workers, to get the education and training necessary to prepare for such jobs. Atkinson (1990) has warned that the supply of graduates at both the undergraduate and graduate levels in science and engineering is likely to be considerably below the demand for several decades to come. Although the percentage of 22-year-olds receiving a baccalaureate degree in science or engineering has held more or less constant ever since the 1960s, Atkinson referred to this statistic as "nothing less than a scandal," given the realities of global competition in today's world and the pervasiveness of science and technology in all aspects of our lives.

There is also the contrasting expectation that most of the new jobs—in absolute numbers—will have low to moderate skill requirements (Braverman, 1974). Roszak (1986) claims that two thirds of the new jobs created in the United States in the late 1970s and 1980s were low-skill, part-time service jobs. Among the 12 occupations that are expected to experience the largest absolute growth—that is, to produce the largest numbers of new jobs—9 have relatively low educational requirements: retail salesperson, janitor/cleaner, waiter/waitress, general office clerk, nursing aide/orderly/attendant, truck-driver, receptionist/information clerk, cashier, and guard (Bureau of the Census, 1990, Table 646).

Until recently, the fastest growing occupation in the country was that of computer programmer. It is still among the fastest growing, but is not at the top of the list, at least according to the Bureau of the Census (1987). It estimated the increase in the number of programmers from 1984 to 1995 at between 64% and 79% (from 341,000 in 1984 to between 559,000 and 609,000 in 1995). According to the most recent figures from the Bureau (1990), there were 519,000 programmers in 1988, and this number is now expected to increase to between 716,000 and 831,000 by 2000.

Some observers worry about the possible creation of a two-tiered society in the United States in which attractive job opportunities are available only to a small minority of the population (Cyert & Mowery, 1989; OTA, 1989a). Landauer (1988) assumes that the increasing abilities of machines to perform cognitively demanding tasks will mean a decreasing need for a large highly skilled labor force; he suggests that schools of the future should become less technical and more social in their orientation, and that by 2020 the humane, social, and cultural aspects of education may be more important to society than the technical ones. Levin and Rumberger (1987), also, have predicted a general decline in the educational requirements of jobs in the future as a consequence of an expected increase in the number of jobs with relatively low skill demands.

The figures cited in the preceding paragraphs illustrate how difficult it is to get a clear picture of how workforce demands and work opportunities are likely to change over the next few decades. The problem stems in part from the fact that growth or decline in specific jobs is sometimes reported in absolute numbers and sometimes in terms of rate of change. A slow rate of growth in a large population can easily result in a larger numerical increase than a faster rate of growth in smaller population: A 5% growth rate in a population of 10 million will produce a numerical increase of 500,000, for example, whereas a 10% growth rate in a population of 1 million will produce an increase of only 100,000. The picture is complicated by other factors as well.

One can be misled, for example, by statistics that relate jobs and worker education by the phenomenon of "credentialing." Suppose one notes that the average educational level attained by the employees in some industry has increased over a period of time. Does this mean that the educational requirements of the jobs in that industry have increased? That could be the case, but an alternative possibility is that the industry simply tends to hire the most highly educated people who apply for jobs, and that the average education level of the job applicants increased over the period of time of interest, perhaps because of a greater supply of highly educated people relative to the number of available jobs that really require workers with their credentials.

The situation is further obscured by the confusing, sometimes, of employment opportunities represented by replacement jobs (existing jobs vacated by workers because of retirement, death, or job changing) and those coming from newly created jobs. The tendency of workers in occupations with lower skill requirements to change jobs more frequently than those in occupations with higher skill requirements can create the impression of more job opportunities of the former type. There are more opportunities, in the sense that jobs become available more frequently, but not necessarily in the sense of the creation of new jobs to accommodate increased numbers of people in the workforce. Also, workforce skill requirements can change not only as a result of the creation of new jobs and the abolition of existing ones but also as a consequence of the changing requirements of persisting jobs (e.g., as the requirements of secretarial jobs have been changed by the introduction of word processing and related technologies).

Bailey (1990) argues that when such possibilities for confusion are properly discounted, the data support the conclusion that occupations requiring higher levels of education are now accounting for, and will continue to account for, more new job opportunities than occupations requiring lower levels of education. The greater educational requirements will come both from newly created jobs and from new demands of existing jobs, resulting from the introduction of new technology into the workplace, but the picture is still sufficiently obscure that we should be prepared for a range of possibilities regarding the kinds of demands jobs will place on workers in the years ahead.

THE CHANGING COMPOSITION OF THE WORKFORCE

The composition of the workforce and the culture and climate of organizations in which work is done have been changing for some time and are expected to continue to change over the near-term future (Offermann & Gowing, 1990). The composition of the workforce, for example, is expected to become older, more female, and more disadvantaged (Cascio & Zammuto, 1987; Vaydanoff, 1987). Somewhat over half of the new entries into the workforce during the remainder of this century will be women and roughly one third will be members of minority groups (Johnston & Packer, 1987). By the end of the century, about half of the entrants to the American labor force will be from minority groups that have traditionally received poorer education than the population as a whole (Baumol, 1989). Immigrants accounted for 22% of the United States's labor force growth between 1980 and 1987, and are expected to account for an even larger percentage of the growth over the decade following (OTA, 1990b). Given the educational deficiencies with which many individuals are entering the labor force, there is likely to be increasing interest on the part of industry to provide training programs for their employees, not only for job training, but for literacy, basic skills, and general education.

The anticipated "squaring" of the population (see Fig. 11.1) means not only that the labor force will be older on the average but that the number of workers at the upper end of the age range will increase much more than will the number at the lower end. This trend will put older workers in direct competition with newcomers to the job market in many instances, and given the rate at which job skill requirements are changing in some industries, it means also that the need for retraining to keep older workers competitive is likely to increase.

Unquestionably, there is a very substantial talent pool among women already in the labor force that is grossly under-utilized. Women are greatly underrepresented, for example, among top managers (Blau & Ferber, 1987); as of 1986, less than 2% of the officers of Fortune 500 companies were female (Offermann & Gowing, 1990). Gender discrimination is still a serious problem in the workplace at all levels. Kraft (1987) points out that when

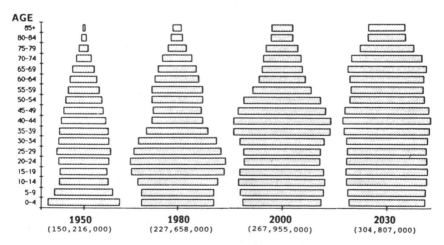

FIG. 11.1 Squaring the U.S. population pyramid, 1950–2030. (From the Transportation Research Board, 1988a.)

every reasonable variable is controlled for, women can expect to earn about 85% of what men with comparable credentials do for the same work. He notes also, that the higher one looks on the management ladder the greater the discrepancy one finds between male and female earnings: "To put this in the most dramatic terms, on average, the lowest paid male supervisor makes more than the highest paid female executive" (p. 105).

Between 1960 and 1980, the percent of U.S. women who were in the workforce went from 35% to 52%. Most of this increase was accounted for by married women, especially those with children. By 1982, women accounted for about 43% of the total workforce. Women currently account for almost two thirds of all new workers and are expected to represent about 47% of the workforce by 2000 (Riche, 1988). The distribution of this labor continues to be greatly skewed toward the lower paying positions, with women holding about 81% of clerical jobs and about 28% of managerial and administrative ones (Kraut, 1987b).

Murolo (1987) has given an historical account of how a two-tiered workforce came into existence in the insurance industry in the United States during the early part of the 20th century, and how routinized, repetitive jobs became the province of women while skilled work remained that of men. According to her analysis, the industry became even more gender-divided, with women being concentrated in the most routine jobs, between 1961 and 1980, while the industry was becoming rapidly computerized. The representation of women in clerical jobs went from 62% in 1950, to 74% in 1970, and then to 80% in 1982 (Gregory & Nussbaum, 1982).

As we attempt to look ahead, many aspects of the world of work are exceedingly unclear. We can be reasonably sure, however, that jobs and job requirements will change in significant ways, even if we cannot see precisely what the nature of those changes will be. We can also be quite certain that the composition of the workforce will be very different from what it was in the past; in this case, we can see relatively clearly what some of the major differences, at least in the near term, will be. Adapting to these changes, and doing so in a way that is consistent with the country's need for greater productivity from its workforce, with the goal of improving the quality of life in the workplace, will not be a trivially easy task.

WORK CONTEXTS

Human factors researchers and engineers are beginning to emphasize the importance not only of the design of technological systems for the workplace, but of the social and organizational contexts in which those systems are to be used. The term *macroergonomics* is sometimes used to connote this focus (Geirland, 1986, 1989; Hendrick, 1986):

The primary goal of macroergonomics is to develop design guidelines that would achieve a good fit among these [social and technological] subsystems. Inadequate attention to social and environmental aspects of the sociotechnical system may result in the development of a sophisticated technology that does not meet the needs of users, does not fit into the social and cultural milieus in which users operate, and does not support the larger mission of the group or organization in which it is introduced. By placing equal emphasis on the technical, social/organizational, and environmental aspects of design, macroergonomics offers an approach for avoiding these pitfalls. (Geirland, 1989, p. 2)

Hendrick (1986) has also argued that it is possible to do a good job of microergonomically designing parts and aspects of a system while failing to attend adequately to the macroergonomic design of the system as a whole and to the organizational context in which a system is to be used: "The notion here is that one cannot effectively design specific atomistic components of a sociotechnical system without first making scientific decisions about the overall organization, including how it is to be managed" (p. 1). The emergence of macroergonomics as the hallmark of the third generation of the discipline of human factors has been caused primarily, in his view, by two factors: (a) new technology and organizational automation, and (b) demographic and psychosocial changes in the workforce (e.g., the aging of the workforce and the greater cognitive complexity of job demands).

One of the macroergonomic effects of new technologies in the workplace that some observers anticipate is the decentralization, or distribution, of some work environments. In spite of some evidence of factors that will promote the growth of urban areas, Kasarda (1988) predicts, for example, that most employment growth during the next few decades will occur in the suburbs and exurbs. He foresees the continuing evolution of a new structure of business activities resembling a distributed network in important respects: "Within this new, more diffuse form of spatial organization, the traditional metropolitan CBD (Central Business District) has become just one specialized node in a multinodal, multiconnective urban field of interdependent activity centers extending outward as far as 100 miles from the metropolitan core" (p. 131).

Another anticipated trend is a tendency for many, if not most, new jobs to be with relatively small firms. Between 1978 and 1982, more than half of all new jobs were created by firms with fewer than 100 employees (Johnston & Packer, 1987). At the same time, many giant corporations have been downsizing their workforces in response to competitive pressures. Between 1980 and 1987, Fortune 500 companies reduced their workforces by almost 20%, dropping 3.1 million of 16.2 million employees (Christensen, 1990). As Fig. 11.2 shows, as of 1986 about 56% of all American workers worked for establishments with fewer than 100 employees; the comparable number in 1975 was about 54%. Perhaps of greater interest is the fact that about 27% of

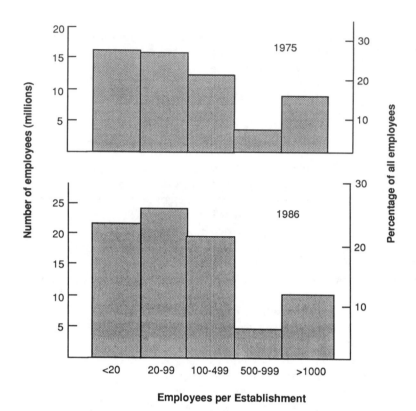

FIG. 11.2 Total number of employees (left scale) and percentage of all employees (right scale) working for U.S. establishments of specified sizes. (Bureau of the Census, 1990, Table 870).

all U.S. workers work for establishments with fewer than 20 employees. (*Establishment* is defined by the Bureau of the Census [1990, Table 872] as a "single physical location where business is conducted or where services or industrial operations are performed" (p. 528). The figures exclude government and railroad employees.) Another perspective on the situation is given by Fig. 11.3. Almost 99% of the approximately 5.8 million establishments in the United States have fewer than 100 employees, and about 87.5% have fewer than 20; for every establishment with more than 1,000 employees, there are about 1,000 with fewer than 20.

About 70% of the approximately 3.8 million enterprises in the United States have annual sales of less than $500,000; about 44% have sales between $100,000 and $500,000 (Bureau of the Census, 1990, Table 873). (An *enterprise* to the Bureau [1990, p. 519] is "a business organization under a single management and may include one or more establishments.") Even in Michigan, with its heavy concentration on the automotive industry, over

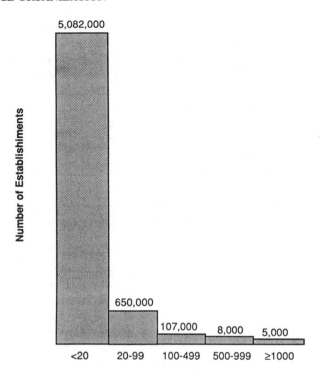

Number of Employees

FIG. 11.3 Number of United States establishments with the indicated numbers of employees, as of 1986. (Bureau of the Census, 1990, Table 872).

90% of the workers are with companies having fewer than 250 employees, and about half the value added by the automotive industry in that state comes from over 800 supplier companies with less than 250 employees each (Michigan Department of Commerce, 1987). In spite of the visibility of the giants, small businesses account for a large fraction of the American economy, and there is little reason to expect that fraction to decrease in the future. Some observers believe that among the effects of the increasing use of information technology to coordinate production and economic activities across firms will be a decreasing interest in vertical integration within individual companies, more buying in lieu of making, and a proliferation of smaller firms (Malone & Rockart, 1991).

WORKPLACE ORGANIZATION

A variety of factors are likely to affect the way workplaces are organized. The need for flexible-system production, for example, will cause significant emphasis to be put on integrated teams of workers, working collaboratively in a more participatory and egalitarian fashion than is typical in high-volume standardized production (R. B. Reich, 1983). Electronic mail and other computer-based communication facilities can change the way people communicate with and relate to each other, not only across great distances but within a single office complex. Some observers expect that the increasing use of networks, and networks of networks, in the workplace will cause organizations themselves to become less hierarchical and more network-like in structure (Dray, 1988; Long, 1987).

Two effects such an organizational change might have are greater accessibility of high-level managers to lower echelon workers and perhaps less of a need for some of the layers of management. By making people at all levels in an organization more accessible than they are either by phone or face-to-face, it can facilitate direct communication that need not go through nodes in an established hierarchy or chain of command. This may have other organizational implications, especially relative to the gatekeeping and information-filtering roles that middle managers often play in conventional hierarchical organizations (Crowston, Malone, & Lin, 1987); in particular, it could decrease the need for middle managers and first-line supervisors (Turnage, 1990). (Some researchers claim that the introduction of computer technology in the workplace in the past has often resulted in a disportionate increase in the size of the managerial staff [Helfgott, 1966; Hoos, 1961], but these studies predated the invasion of the office by the personal computer and the local area network; what was true then may not hold now or in the future, but then again . . .)

One view of the effects of information technology on corporate organization, and especially on communication and decision-making within the corporation, is that as corporate communications become more versatile and individuals within the corporation become more accessible through the communications system, irrespective of position, the flow of information changes from unidirectional to multidirectional, the organization becomes less hierarchical and rigid, and the decision making becomes more widely distributed. Another consequence of the emergence of such corporate communication networks is that the physical proximity of workers is less important than it used to be in giving an organization its identity and to hold it together.

Such expectations regarding organizational implications of information technology and its rapidly changing applications in the workplace raise numerous questions about their possible consequences. Sproull and Kiesler (1991) have expressed some of the questions: "Can people really work closely with one another when their only contact is through a computer? If employees interact through telecommuting, teleconferencing and electronic group discussions, what holds the organization together? Networking permits almost unlimited access to data and to other people. Where will management draw the line on freedom of access? What will the organization of the future look like?'' (p. 116).

For many years, the hierarchy has been the dominant organizational structure in the corporate world. (Essentially the same structure has also prevailed in military organizations.) Modern communications systems, and information technology more generally, appear to be invalidating many of the justifications for the ubiquity of this structure. Whether some other structure will emerge as the dominant one in the future is an interesting question. My hunch is that we will see a greater diversity of structures, with different companies organizing and reorganizing themselves in different ways to suit their own purposes—sometimes with good effect and sometimes not— because information technology makes such diversity feasible. Investigators of organizational aspects of work should have no scarcity of interesting phenomena to study in the years ahead.

TELECOMMUTING AND TELEWORK

Computer and telecommunications technology has the potential to make it possible for a significant fraction of the workforce to work at home or at near-home satellite offices at least part of the time (Galitz, 1984). Already there may be as many as 15 or 16 million "teleworkers" working at least part of the time at home (G. L. Martin, 1989). Among the first to write about the possibility of "telecommuting" were Nilles, Carlson, Gray, and Hanneman (1976). These investigators, writing shortly after the oil crisis of the early 1970s, argued that, for many purposes, the transmission of information could be substituted for the transportation of people to and from traditional places of work, and they stressed the potential importance of telecommuting for energy conservation and the saving of transportation costs. They estimated that a 1% replacement of urban commuting by telecommuting could reduce U.S. gasoline consumption by about 5 million barrels per year. Telecommuting, or telework, to use the more current term, could have other desirable consequences as well, such as a decrease in air pollution resulting from vehicle emissions, some relief of traffic congestion, and a reduction in office space requirements.

Huws, Korte, and Robinson (1990), point out that, although the concept of telework is a meaningful one and one that symbolizes to many people what the future of work may be, the term does not have a sufficiently stable definition to provide a basis for counting the number of teleworkers. Their own definition involves three variables: location of work, use of electronic equipment, and a communications link to the employer or contractor: ''We define telework as work the location of which is independent of the location of the employer or contractor and can be changed according to the wishes of the individual teleworker and/or the organization for which he or she is working. It is work which relies primarily or to a large extent on the use of electronic equipment, the results of which work are communicated remotely to the employer or contractor" (p. 10).

As to the symbolic significance of the idea of telework, they point out that it is positive for some people and negative for others: "For some it is a symbol of liberation from the daily grind of living in a suburb, commuting into the city, working from 9 to 5 and wearing stuffy suits, a chance to get to know one's family and pursue new leisure interests. For others, it has become a symbol of isolation, exploitation and the end of any possibility of temporary escape from the drudgery of housework and childcare" (p. 9). Telework need not be as liberating as it is sometimes pictured. It lends itself, as does any work with computer-based systems, to electronic monitoring, and employers have been known to drive hard bargains in pay-by-results situations (Chamot & Zalusky, 1985; Gregory, 1985). Again we are reminded that any technology can be used responsibly or irresponsibly, and in socially desirable or socially undesirable ways. And the more powerful the technology, the greater the potential in both directions.

Harkness (1977) has estimated that perhaps as many as 50% of all white-collar workers could work remotely at least part-time; however, to date, only a vanishingly small fraction of the nation's workforce works at home to any significant degree (Kraut, 1987b). People who do work at home tend to do it on a part-time basis, augmenting the work done at their conventional offices, not substituting for it. Some people who work at home do so because they have no option: Working at home is a condition of their employment. Others work at home by choice: Many of the people in this category work at home for some portion of the time and outside the home—perhaps at a traditional office—for the remainder. People who work at home by choice do so for a variety of reasons: to get greater flexibility of working hours, to be available for family needs, to have greater autonomy, or to avoid commuting. Organizations introduce telework for many reasons as well, including to cope with surplus work (work peaks), to recruit or retain workers with rare skills, to improve the motivation and productivity of employees, to reduce employee turnover and absenteeism, and to make it possible for employees to combine work and nonwork activities, like child care (Huws et al., 1990).

Employers also may look favorably on telework because by using it they can reduce overhead costs, by saving the cost of office space and perhaps even the cost of standard employee benefits (Kraut, 1987b; Olson, 1987). Another major reason for employers to be interested in telework is the possibility it provides of tapping a segment of the workforce that would otherwise not be available. The introduction of telework has been resisted by some corporations, perhaps because of a belief that employees will be less likely to identify with a company if they do not work on its premises (Olson, 1988) and certainly because many managers believe that telework is difficult to manage (DeSanctis, 1984). The skills required to manage telework effectively undoubtedly differ somewhat from those that have been seen as important in traditional organizations (G. E. Gordon & Kelly, 1986). Organized labor has been skeptical of telework, seeing in it the potential of some of the abuses of the past in some home-production work situations (Kraut, 1987b). This concern has been sufficiently strong to move the AFL-CIO and the National Association of Working Women to ask the U.S. Labor Department to ban home clerical work.

Why has it made sense, at least heretofore, for the white-collar workers of a given company to congregate daily at a common location—the office—to do their work? From the employer's point of view, having everyone together in one place has facilitated communication with the employees, and coordination and monitoring of the work. From the workers' vantage point, it has been necessary to go to the office because that is where the documents and data with which they work are kept; it is where their tools are. Often collaboration on tasks with colleagues is necessary, and that usually has required being in the same place at the same time. Undoubtedly there are other good business reasons for having a company's workers do their work under a single roof. The interesting question for the future is how many of these reasons will continue to be compelling ones, indeed how many of them are even now.

In thinking about this issue, it is important not to lose sight of the fact that there are, for many people, attractive aspects of working in an office with other people that have little to do with the tasks that must be performed: the routine of getting ready for work and going there (something people claim they miss when they retire); the social dimension of interacting with co-workers; even the commute can be a pleasant daily interlude, providing time to read a newspaper, listen to the radio, or talk with fellow commuters (which is not to deny that it can also be unpleasant and frustrating in some circumstances).

Computer and telecommunications technology has the potential to reduce greatly the amount of travel for the purpose of attending work-related out-of-town meetings. Whether this potential will be realized also is difficult to say. When asked why they attend meetings, people can readily give reasons. Among the most common are such professionally commendable purposes as seeking or giving information and problem solving (Hough & Penko, 1977). One must wonder, however, to what extent the usual methods for identifying reasons for attending meetings would uncover factors that were less commendable from a professional point of view, but possibly no less strong, such as the chance to visit a distant city, the opportunity to eat in interesting restaurants, meeting new people or renewing old acquaintances, or simply getting away from the daily routine and enjoying a change of scenery for a short time.

I do not wish to argue that teleconferencing will not be appreciated and widely used. What I do question is whether its use will have the effect of decreasing the amount of travel that is done for the purpose of having face-to-face meetings. We should not rule out the possibility that as teleconferencing technology becomes more effective and more widely available it will be used to advantage by many groups, but the frequency of face-to-face meetings and travel to and from them will not decrease as a consequence. If the potential that some people see in teleconferencing technology is realized, there may well be advantages to using the technology even in face-to-face meetings. This could be the case, for example, if teleconferencing systems include capabilities for aiding brainstorming, collaborative problem solving and group decision making. On the assumption that a group of experts has more knowledge and skills in the aggregate that are relevant to a problem in their area of expertise than does any member of the group individually, it would be useful to have an effective way to apply the "corporate" intellectual resources of the group to the problem at hand. One can imagine an interactive system with the ability to interrogate experts and, perhaps by means of many iterations, evolve a model of the knowledge of the experts as a group. Given such a knowledge base, the task would then be to develop and refine an approach to the problem that uses it to maximum advantage.

More generally, it is not necessarily the case that the widespread adoption of a technology that enables one to perform some function in a new way will make the old way of performing that function suddenly obsolete. Television has not put book publishers out of business, nor has television in combination with home video meant the closing of movie theaters. Electronic means for creating, storing, and transmitting "documents" has not yet produced the totally paperless office. (As for counterexamples, the telegraph and the railroad did put the pony express riders out of work, television did pretty much demolish radio theater, and the pocket calculator did make the analogue slide rule obsolete.) With respect to the effect of teleconferencing on travel, we shall simply have to wait and see; my (small) bet is that the amount of travel will not decrease.

*   *   *

Traditionally, human factors researchers have paid a lot of attention to work and the workplace. Unquestionably, there will continue to be a need for such a focus in the future. There will also be a need for some new approaches to the study of work to supplement the existing ones. Task analysis techniques have been used to great advantage in the design of work spaces and procedures and the organization of work teams. Task analysis will continue to be a useful tool, but, as conventionally done, it is limited when applied in anticipation of introducing new technology in a work setting. This is because new technology often changes qualitatively, sometimes dramatically and in unpredictable ways, the nature of the tasks that are performed. To be used effectively in these situations, task analysis must be done in a sufficiently flexible way as to take cognizance of the changing nature of the task and of what this means in terms of demands on the worker or work group.

Of the various ways in which the nature of the work that people do has been changing in the recent past, perhaps the most obvious and significant one at a general level is the shift from jobs involving the manufacturing and production of material goods to those involving the delivery of services and the manipulation of information. This trend, which is very likely to continue, suggests the need for new concepts and new techniques for studying work performance and work environments. The measurement of workload for primarily mental tasks illustrates this need. Measuring how hard one is working is relatively easy when the work is physical; it is much more difficult when the work is largely cognitive and not directly observable. Cognitive tasks, like physical ones, can vary greatly with respect to the demands they put on the individual's resources, and it is important to be able to determine how close the demands of any particular task come to exceeding the limit of what the individual can produce.

The measurement of mental workload has been seen as a significant challenge to human factors researchers for some time (Chiles & Alluisi, 1979; Kalsbeek, 1968; Moray, 1979; Sheridan & Simpson, 1979; Singleton, Fox, & Whitfield, 1971; R. C. Williges & Wierwille, 1979). Despite much attention to this subject and many research studies—Wierwille and Williges (1978) identified 28 different techniques that have been used—there is still no coherent theory of mental workload or a strong consensus among researchers about how best to measure it (Johannsen et al., 1979). Investigators have sometimes taken subjective indicators (subjects' ratings or opinions) and they have sometimes taken physiological measures that are assumed to correlate with effort, such as heart rate, sinus arrhythmia, electromyographic activity, pupil diameter, eye-blink frequency, and event-related brain potentials. Sometimes, both subjective and objective measures have been taken and correlations among them determined (Kramer, Sirevaag, & Broune, 1987; Vicente, Thornton, & Moray, 1987).

Another general approach to measuring workload is the indirect one of determining the effects of the demands of a primary task on the performance of a secondary task, the assumption being that the greater the workload imposed by the primary task, the less adequate will be the resources one has available to apply to the secondary task. One of the problems that has complicated workload measurement is the fact that subjective measures of workload have not always agreed with performance-based measures (Wickens & Yeh, 1983; Yeh & Wickens, 1988). In sum, in spite of the attention the problem has received, the measurement of mental workload remains a largely unsolved problem, and its practical significance is likely to increase as the cognitive demands of an increasing number of jobs become more important than the physical ones, and the consequences of exceeding a worker's capacity can, in some cases, be very serious. The complementary problem of boredom resulting from too little challenge in some job situations is also a significant one (R. A. Smith, 1981).

A much better understanding is also needed of the effects of aging, both physical and mental, on performance, especially as it relates to job requirements. The authors of a recent National Research Council report on Human Factors Research Needs for an Aging Population pointed out the existence of "large gaps in our knowledge base regarding the implications of age-related changes for the performance of everyday tasks and activities" (Czaja, 1990, p. 70). Apparently, many of the prevailing assumptions regarding decline do not have good empirical support (McEvoy & Casio, 1989); nevertheless, such assumptions get translated into policy regarding the hiring and retention of older workers. The same National Research Council report points out that relatively little research relevant to the needs of aging people has been done by human factors researchers—most research on aging has been done in the framework of medicine, gerontology, architectural safety, and engineering— but it notes numerous opportunities for human factors research relating to the needs of elderly people. In addition to work, it focuses on home activities, transportation, communication, safety and security, and leisure activities.

Some jobs of the future will require highly skilled workers because of the cognitive demands involved. There is little reason to believe, however, that boring, demeaning jobs will quickly become a thing of the past. Job quality and job satisfaction will remain legitimate human factors concerns for the foreseeable future. It is quite possible, of course, to increase workers' productivity and decrease their job satisfaction at the same time, and there are examples of this combination of effects from the introduction of new information technology in the workplace (J. A. Turner, 1984). It is also possible for workers to be pleased with the new technologies that they are using and at the same time be displeased with the conditions under which they are using them (Form & McMillan, 1983; Mumford, 1982). It would not be surprising to find this situation in the context of clerical work because, although clerical workers are expected to be among the most intensive users of many new technologies, they are least influential in determining the pace at which these technologies are introduced and the ways in which they are used (Iacono & Kling, 1987).

As just noted, it is important to distinguish between the effects of new technologies in the workplace and effects of the ways in which those technologies are introduced and applied. Job satisfaction seems to depend very much on issues other than the specific technology that one uses. According to one survey of 1,500 people nationwide, what makes jobs attractive are interesting work, good pay, adequate resources, sufficient authority, and congenial coworkers (Survey Research Center, University of Michigan, 1971). Organizational factors that have implications for inter-person communication also appear to be important. Some ways of organizing tend to isolate people performing a specific function from people performing other functions in the same company, and this can have a demoralizing effect. Routinization and consequent boredom are also significant issues. There is considerable evidence that a certain amount of mental challenge is an essential component of job satisfaction for many people (Locke, 1976). The challenge must be manageable, of course: If it exceeds the worker's ability to cope, it will produce failure and dissatisfaction, but if it is absent entirely, or trivial in degree, it will produce boredom and lower the sense of self-esteem and importance.

Unlike the problem of boredom, which has received relatively little attention, job-related stress has been a growing concern in recent years (Matteson & Ivancevich, 1987). The current cost to U.S. industry of stress-related symptoms resulting in absenteeism, increased medical expenses, and lost time has been estimated to be between $50 billion and $75 billion per year (OTA, 1987a). Stress has received a great deal of attention from researchers (Fidell, 1977; Lazarus & Monat, 1977); for the most part, however, the emphasis has been on the effects of relatively intense stress on performance over short periods of time as opposed to the effects of low-level stress on a continuing basis. Moreover, research results are not easily integrated into an encompassing theory or even into a cohesive body of knowledge, in part because of the wide variety of investigative techniques and performance measures that have been used and in part because of the sometimes conflicting findings from different studies (Bachrach, 1982; Theologus, Wheaton, Mirabella, Brahlek, & Fleishman, 1973). Even when attention is confined to a single stressor, findings have been sufficiently diverse and disparate to make concise summarization difficult (Rowell, 1978).

The stressfulness of many situations arises in part because of the individual's sense of a lack of control over the situation (Averill, 1973). A modicum of control, or at least perceived control, over one's personal situation and environment seems to be a very important aspect of one's degree of satisfaction with life in general (Barnes, 1980; Zimring, Weitzer, & Knight, 1982). Human reaction to, and performance under, stress more generally should continue to be a focus of human factors research. More needs to be learned about effects of heat, cold, sleep loss, perceived danger, hypoxia, noise, and other familiar stressors. There is also a need, however, for more attention to the cumulative and long-term effects of low levels of chronic stress that can result from unsatisfactory situations that persist in the workplace or in other contexts.

Some of the systems with which people interact in the workplace are extremely complex. Examples include high-performance aircraft, nuclear power plants, and automated assembly lines. As the complexity of these systems increases, their operation becomes ever more dependent on automated and semiautomated subsystems. The roles of humans in the operation of these systems tend to be supervisory, for the most part, and remedial on occasion. Although many, if not most, of the moment-to-moment functions are performed automatically without human intervention, people typically monitor the overall operation, make high-level control decisions, and attempt to do whatever is necessary to adjust the operation to unanticipated problems or subsystem failures.

One problem associated with many supervisory control situations is the fact that the demands on the human can swing between extremes. Long periods of "cognitive underload," a euphemism for boredom, may separate episodes of truly excessive demands, often in the context of great danger and the prospects of grave consequences following from less than optimal performance. We need a better understanding not only of the effects of long-term boredom in task situations that place too little demand on the individual, but also of how to ensure that people who function in such situations maintain the ability to perform adequately in the high-demand situations in which they may occasionally be placed abruptly after being underloaded for a long period of time.

There is also a need for greater attention to the "problem" of leisure. At least in developed countries, people may have more leisure or discretionary time in the future as a result of shorter work weeks, earlier retirement, and longer lives. How to spend leisure in satisfying ways could become an issue for increasing numbers of individuals. This seems like the right kind of problem to have, but there is evidence that for many people it is, in fact, a serious problem with which they do not know how to cope.

12

Decision Making and Policy Setting

Modern life presents us with some exceedingly complex problems; indeed, today's world appears to be much more complicated than the world of the past. Of course, this may be an illusion that derives from our vantage point. If so, it is a compelling one: International tensions and conflicts, energy utilization, environmental pollution, disruptive and dislocating effects of rapid technological change, and the interacting effects of national economies are but a few of the complexities of modern times. Such problems represent enormous challenges. Many of them involve tradeoffs: between the need for secrecy for purposes of national security and the need for accountability of government agencies, between the need for energy and the need to conserve natural resources and preserve the environment, between societal risks and the cost of decreasing them, between the purchasing power of the dollar and trade deficits, between personal freedom and the public good, and so on.

HIDDEN COSTS AND THE UBIQUITY OF TRADEOFFS

We have become aware in recent years that technological "advance" is often accomplished at a cost that is not apparent for some time. Industrial mass-production techniques have made consumer goods available on a much greater scale than would otherwise be possible, but it has also contributed to the pollution of the environment and to the depersonalization of work. The commercial nuclear reactor has provided a new source of energy, but at the expense of the prospects of nuclear accidents and the problem of the disposal of nuclear wastes. The use of chlorine in drinking water has been effective in

323

controlling organisms that cause typhoid and other infectious diseases. It has also left chloroform and other carcinogenic chlorinated hydrocarbons in the drinking water supply.

Commercially canned foods have a smaller danger of botulism than did foods preserved in the home, but the lead solder in some metal cans introduced a toxin in commercially canned foods not present in those put up at home. The use of pesticides and herbicides in agriculture have increased food yields greatly and decreased the cost of food to consumers, but it has exposed agricultural workers to hazardous chemicals and contributed to a variety of water pollution problems. Unpleasant surprises, often in the form of damage to the environment that is very hard to repair or reverse, have sensitized us to the need for better ways of anticipating the effects of technological developments, and of making informed decisions about possible technological innovations in the interest of improving the quality of life, or at least guarding against degrading it.

As a consequence of the energy crises of the early 1970s, the energy efficiency of business and residential buildings became a focus of national concern and activity. In addition to the financial incentive brought on by the high cost of heating fuel, the government stimulated efficiency-improvement projects by subsidizing their costs, at least in part, through tax policies. Homeowners replaced drafty windows, added insulation to their walls and tightened up their homes in numerous ways; owners of commercial buildings installed air circulating systems designed to minimize the loss to the outside of warm air in the winter and cool air in the summer.

Such steps improved the energy efficiency of buildings considerably, at least if efficiency is measured strictly in terms of what it costs to maintain a comfortable internal temperature. It now appears, however, that the benefits derived from making buildings more efficient may have been obtained, in some cases at least, at the cost of diminishing the quality of indoor air. According to an EPA report (1989a), indoor air pollution in the United States may be responsible, annually, for more than $1.2 billion in medical care and for between $4.7 billion and $5.3 billion in lost productivity. The problem stems in part from the fact that paints, synthetic carpeting, household chemicals, and numerous other materials commonly found in residences and commercial buildings emit, over time, a variety of toxic substances into the air, including formaldehyde, ammonia, sulphur dioxide, benzene, styrene, and other chemical agents. If a building does not "breathe," such agents can reach troublesome concentrations. Nothing, it would appear, is as simple as it seems.

The difficulty of getting a balanced understanding of the tradeoffs involved in the use of specific products can be seen in an exchange of letters to the editor of *Scientific American* regarding Frosch and Gallopoulos's (1989) discussion of the use of polyvinylchloride (PVC, or vinyl) in manufacturing.

C. N. Bush (1990) argues that the authors overstated the case against the use of PVC, whereas Borie (1990) contends that the dangers were understated. In their reply, Frosch and Gallopoulos (1990) point out that although the exclusive use of glass in the production of beverage containers has the advantage over plastics of being nontoxic, "making containers from glass consumes 20 times as much raw material, twice as much energy and 3.5 times as much water as making them from PVC. It also produces 18 times as much solid waste and twice the quantity of atmospheric emissions. The societal costs of lacerations and other serious injuries caused by broken glass containers should also be included in the accounting" (p. 13).

Making automobiles smaller and building them of lighter material is good both from the point of view of conserving energy and from that of protecting the environment, but occupants of small, light cars are at a considerable disadvantage relative to occupants of heavier automobiles when involved in an accident (Graham & Crandall, 1989; Partyka & Boehly, 1989); normalized to the number of registered cars of each type, fatalities are approximately 2.5 to 3 times greater for occupants of compacts and subcompacts than for full-sized cars (Partyka, 1989).

The generic decision problem involving the determination of acceptable tradeoffs between conflicting goals arises often in the context of energy production and environment protection. Energy producers and environmentalists find themselves frequently at loggerheads, for example, on questions pertaining to exploration and tapping of previously unknown or untapped fossil-fuel reserves. The dispute about the possibility of exploration and development of oilfields in the Arctic National Wildlife Refuge in Alaska (OTA, 1989b) is indicative of the kind of clash of values that occurs and will have to be resolved. What the general public considers an acceptable environmental price to pay for the availability of energy can vary as a function of how limited or threatened the energy flow appears to be; during or just following an energy crisis, concerns for environmental preservation are likely to diminish somewhat (Hummel, Levitt, & Loomis, 1978).

Tradeoffs can also be involved even when there is agreement as to goals. Methanol, which can be made from natural gas or coal, produces only about one tenth the hydrocarbon emissions per mile that gasoline does, but, although methanol is good news with respect to hydrocarbon emissions, it produces formaldehyde, which is a possible carcinogen (Wright, 1990a). Plastic composites are attractive materials for automotive construction because of their light weight and durability, but their durability can be a disadvantage from the point of view of waste management. As a consequence of public reaction against nuclear power following the Chernobyl accident, Sweden has committed itself to phasing out, by 2010, the 12 nuclear reactors that now provide over half its energy. This is a mix of good news and bad to the Lapplanders to the north of Sweden: On the one hand, their way of life was seriously threatened by the Chernobyl accident, but on the other, the most likely alternative to nuclear power as an energy supply for Sweden is the development of hydroelectric plants in areas Lapplanders use to herd reindeer (S. Stephens, 1987).

The reason I mention these examples of tradeoffs is to make the point that it is much easier to identify the serious problems that we face than to come up with effective approaches to solving them. The problems tend to be interconnected in complex, and often obscure, ways; simple, unidimensional "solutions" often make for appealing political rhetoric but are seldom effective from a long-range and global view. Narrowly conceived attacks on specific problems, however well intentioned, have all too often resulted in the trading of one problem for another; in many cases, the trade, in the long run, has been no bargain. Looking ahead, the opportunities for making undesirable trades that appear initially to be effective solutions to pressing problems are likely to be at least as plentiful as they have been in the past; a greater awareness on the part of decision makers of the importance of attempting to better understand the long-term effects of today's policy decisions would help, but without more adequate tools for making obscure tradeoffs explicit, greater awareness of the issue, by itself, will not suffice.

THE NEED FOR DECISION-MAKING TOOLS

Interactions among nations are extremely complex, disagreements and disputes are many-faceted, coalitions are fragile and tenuous, and long-term ramifications of policies and actions are often difficult or impossible to foresee. Solutions to problems can easily turn into problems that are worse than those that were solved. There is a great need for more effective tools to help people assess complex situations, to explore possible effects of alternative actions, to execute plans, to monitor the effects of programs, and to adapt to changing situations in rational and effective ways.

Needed tools include better and more accurate situation models, cost–benefit analyses, sensitivity analyses, and especially decision-making aids that can help policymakers and strategic planners to strike acceptable balances when tradeoffs are involved and, more generally, to cope rationally with complexities that may be beyond any individual's comprehension. In particular, there is a need for decision models of sufficient scope to provide a basis for relating local problems and proposed fixes to larger contexts so as to avoid what A. E. Kahn (1966) has referred to as the "tyranny of small decisions."

The most serious threats to a peaceful and stable future come less, I believe, from people or geopolitical entities with evil designs than from well-intentioned decision makers who oversimplify complex situations in their thinking, and are unable to appreciate the legitimacy of any view of the elephant but their own. This is not to discount the possibility of trouble coming from people who intend to cause trouble, which is real enough, but simply to argue that people with the best of intentions can wreak great havoc in a complex world if they are in positions of power and working on assumptions that are ill founded.

This concern applies both to policy decisions that are made in a careful, deliberative fashion, the effects of which are likely to become clear only after a long period of time, and those made hastily under the pressure of a crisis, with effects that are immediately disastrous. Blair and Kendall (1990) argue that the most plausible path to nuclear war is through the accidental, unauthorized, or inadvertent use of nuclear weapons. The possibility of an accident arising from an error of judgment is heightened by the complexity of the launch capabilities, the very short time that a decision maker has to respond in the event of a signaled attack, and the fallibility of early warning systems. In the United States, apparently an indication of a potential missile attack typically occurs several times a day: "Each year between 1979 and 1984, the only period for which official information is available, the North American Air Defence Command (NORAD) assessed about 2,600 unusual warning indications. One in twenty required further evaluation because it appeared to pose a threat" (Blair & Kendall, 1990, p. 55). Most false alarms are quickly recognized as such, but alarms persist long enough to trigger a nuclear alert on the order of once or twice a year.

The recent easing of tensions among the superpowers and what appears to be real progress toward nuclear arms reduction are encouraging developments relating to the possibility of the accidental triggering of a nuclear weapon; however, the possibility remains as long as such weapons exist and are poised for launch. Even as the superpowers appear to making some progress toward arms reduction, the number of countries with some nuclear weapons capability continues to grow. The probability of an inadvertent nuclear event seems unlikely to get comfortably close to zero as long as the number of fingers that could trigger one is on the increase.

Blair and Kendall point out that human factors play a key role in the risk of an accidental launch: "All those involved in the nuclear weapons chain, from top decision makers to launch officers, are subject to human frailities and instabilities. These frailities are aggravated by work conditions that are boring and isolated. (Duty on missile submarines adds the further stress of trying to adapt to an unnatural 18 hour 'day' for the 2 months of a typical patrol.) Such conditions can sometimes lead to severe behavioral problems, including drug or alcohol use and psychological or emotional instability" (p. 57). World leaders, they note, are not immune to stress, nor is it unknown for them to resort to alcohol or drugs to relieve it. Clearly, the challenge is to build the kinds of safeguards against accidental launch that are highly unlikely to be overridden by human error and eventually to do away all together with weapons that make it even remotely possible for large numbers of human beings to be obliterated "by mistake."

The importance of this challenge is illustrated by the occurrence of incidents like the downing of Iran Air Flight 655 by the USS *Vincennes* on July 3, 1988. Human factors issues were implicated in this disaster in several ways: Air traffic control at Bandar Abbas International Airport failed to warn the flight crew that a naval engagement between American and Iranian naval forces was in progress, the airliner was mistakenly identified by the *Vincennes* crew as an F-14 fighter, and the aircraft was erroneously reported by some officers in the Combat Information Center to be decreasing in altitude during the last few moments before the missile launch (Fogarty, 1988). To what extent these errors could legitimately be blamed on faulty human perception or judgment, as opposed to equipment limitations or procedural flaws, has been a matter of considerable debate (Defense Policy Panel, 1989; G. Klein, 1989).

One of the conclusions that came out of an investigation of the *Vincennes* tragedy was that "stress, task fixation, and unconscious distortion of data may have played a major role in this incident" (Fogarty, 1988, p. 45). It was hypothesized that after two of the numerous officers in the Combat Information Center became convinced that the approaching aircraft was an F-14, the subsequent data flow may have been distorted unconsciously to make the evidence fit a preconceived scenario. (The military recognizes a phenomenon called *scenario fulfillment*, which appears to be a variant of what has been referred to in a wide variety of contexts as *confirmation bias*.) The total time that elapsed from when the commanding officer of the USS *Vincennes* first became aware of the belief that his ship was threatened by an approaching unidentified aircraft until the command to fire was given was 3 minutes and 40 seconds, during which time his ship was engaged in a skirmish with hostile naval forces.

There are many practical reasons for wanting a much better understanding than we currently have of human judgment, decision making, and cognition more generally. Accidents on the highways, in the airways, in industry, and in the home often are the direct consequences of errors of judgment. No one knows what the total costs of such errors are, but can anyone doubt that they are extremely high and much greater than they need to be? Much has been learned about common ways in which judgment fails, but much still remains obscure, and the knowledge that has been gained through research has not, for the most part, led to effective remediation techniques.

Conflict Resolution

Nowhere is the need for more effective approaches to conflict resolution more apparent or more important than in the arena of disputes among nations that can lead to war. As of 1990, there were some 18 countries that have ballistic missiles in their military arsenal, or about double the number of a decade earlier (Nolan & Wheelon, 1990). Many of these missiles are capable of carrying not only conventional, but chemical or nuclear, warheads. At least a half dozen countries, in addition to the superpowers, are currently capable of producing chemical weapons. It seems inevitable that the number of countries possessing, or capable of making, a nuclear bomb (and other equally frightening instruments of destruction) will increase. The possibility of such weapons in the hands of Third World leaders bent on territorial expansion or engaged in adventures motivated by ideological fanaticism is worrisome indeed. As Nolan and Wheelon point out, "The third-world military buildup is perhaps more worrisome than its first-world prototype, for it is far more likely to find expression in war" (p. 34). Even more troublesome is the thought of such weapons in the possession of international terrorists. Nolan and Wheelon note also, that in addition to being weapons of destruction, missiles are symbols of technical prowess and political prestige, signs of great-power status; thus, some Third World nations are willing to acquire them at costs that may exceed their purely military value. Truly effective aids to conflict resolution should find many uses on the world scene.

Conflicts arise not only among nations, of course, but among political, ethnic, and religious factions within nations; among corporations; between labor and management; among rival organizations and special-interest groups; among street gangs; and between individuals. They differ in complexity and intensity, and in many particulars, but they share the property that the parties to the conflict have incompatible goals that cannot be simultaneously realized. The tragedy of many conflicts is that they escalate into situations with consequences that are much less desirable, from all parties' perspectives, than what could have been realized through negotiation and compromise during the conflicts' early stages. The need is for tools and techniques that will support such negotiation and compromise.

There is a great deal of interest at the moment in computer systems to support cooperative work. Such systems provide their users with shared access to databases and workspaces and with tools that facilitate communication and collaboration. Perhaps there is an opportunity to use this work as a point of departure for developing systems to facilitate the resolution of conflicts. One can imagine a system that parties to a conflict would use interactively in an effort to converge on a resolution that would be acceptable to both sides. The idea that such a system could be useful assumes, of course, that there are conflicts in which both parties desire an equitable resolution and the real problem is in coming to a common view as to what that should be. A system of the sort I have in mind would contain, or have access to, a database of facts relevant to the conflict, interactive capabilities to help the disputants make the parameters of the conflict explicit (e.g., assumptions, objectives, fears, tradeoffs), and modeling capabilities that would permit the testing by simulation of various compromises that the disputants wish to consider.

Developing such systems to the point of practical usefulness would be an ambitious undertaking. An interesting question from both a theoretical and a practical point of view is whether it is possible to develop a generic system that could be applied to a wide assortment of conflicts or whether it would be necessary to have special-purpose systems designed to apply to conflicts of specific types. It may be that the development of truly effective conflict-resolution systems would prove to be impossible. The problem of conflicts at all levels, from those involving individuals to those involving nations, is sufficiently serious, however, to warrant a major effort on the development of effective resolution techniques, even if the chances of success are deemed to be small. Moreover, even if the ultimate goal of developing systems that make the resolving of conflicts very much easier proved to be elusive, serious attempts to produce them should increase significantly our understanding of the nature of conflicts and provide some insights into the major impediments to their resolution.

Personal Decision Aids

All of us could use help from time to time in making decisions, especially decisions that are very important to us. There is some evidence that our ability to judge probabilities and to assess our own state of knowledge accurately may be less good in situations that are significant to us than in situations that are not (Sieber, 1974). There is evidence, also, that vested interests can influence the values of individuals and help determine the tradeoffs they are willing to make; for example, people are likely to be less bothered by air quality in their town of residence if they work for a major polluter than if they do not (Creer, Gray, & Treshow, 1970). We could use help in structuring problems (understanding problem situations and the action alternatives that are open to us), help in assessing our own beliefs and values as they pertain to problems of interest, and help in selecting among alternative courses of action (Jungermann, 1980).

Although numerous approaches to decision aiding have been developed and investigated in the laboratory, and a few decision aids have even made it to the marketplace, one suspects that the vast majority of people make decisions as they always have, depending much more on intuition than on analysis. Furthermore, the decision aids that have been used to date have not had a very impressive record of improving the quality of decisions made (Woods, 1986). These are not good reasons for abandoning the effort to develop effective aids; they do suggest, however, that laboratory work on decision making must be supplemented with more study of decision making in real-life situations and a greater effort to understand what is needed to improve decision quality in specific operational contexts.

Some of the most effective decision aiding in the future may come from facilities not thought of primarily as decision aids. Any technology that makes information more readily accessible to the average person should facilitate decision making generally. Especially helpful will be facilities that can provide selective information (e.g., answers to specific questions) on demand, and if computer networks become widely used for communication purposes by the general public, as many observers expect they will, it would not be surprising to see the use of "Does-anybody-know?" questions become a popular and effective means of acquiring information for the purpose of helping individuals arrive at decisions of just about any sort.

POLICY EVALUATION AND EFFECT PREDICTION

Evaluation of the consequences of government or corporate policies and the effectiveness of major programs and accurate representation of the findings to decision makers are recognized to be significant problems: "The quality of program evaluation information on some defense programs is so low that findings are misleading; in other cases, information is nonexistent; and in still others, data are incomplete (General Accounting Office, 1988, p. 12)." "The Medicare program's evaluation system today cannot support informed decisions" (p. 14). "EPA does not know whether many multibillion-dollar environmental programs are effective in achieving planned goals, because its evaluation information is misleading, inadequate, or incomplete" (p. 15). One must assume that as the number and complexity of programs increase, so will the difficulty of getting accurate assessments of their effects. As top-priority evaluation issues for the United States over the next few years, the General Accounting Office has identified: integrity of weapons systems, long-term medical care needs, cost-effectiveness of environmental systems, and excellence and competitiveness of American education.

The need for better techniques for predicting the effects of policy changes is acute in the area of nuclear power generation. There are 50 major facilities in the United States producing and processing nuclear material; many of these are now obsolete and represent risks to workers and the public of unknown magnitude involving "inadequate reactor safety systems to prevent or mitigate the consequences of a nuclear accident, poor management of hazardous and radioactive wastes, extensive soil and water contamination, and worker exposure to radiation" (General Accounting Office, 1987a, p. 13). How should these risks be balanced against the tens of billions of dollars that would be necessary to correct the problems? Holden (1990a) makes the point that the consequences of energy production for public health and safety are difficult to quantify with much confidence, because of different assumptions about many of the variables involved. The ranges of estimated hazards to public health from coal-fired and nuclear power plants are so wide as to provide little basis for preferring one of these energy sources over the other: "From both, the very size of the uncertainty is a significant liability" (p. 160).

A similar need is also apparent relative to the production and distribution of food. As Scrimshaw and Taylor (1980) put it, "almost super-human judgment seems to be called for in the area of food because it is just here that many promising initiatives have had unforeseen and unfavorable second-generation consequences, have not taken hold as widely as predicted, have failed disastrously or have simply been irrelevant" (p. 85).

In the area of environmental pollution (but not only here), the choice is often between making policy decisions on the basis of considerable uncertainty about future trends and inadequate understanding of the variables involved or postponing action at the risk of having to deal with more severe problems in the future. To be sure, environmental impact statements (EISs) are required by the National Environmental Policy Act for any major federal activity that can be expected to have an impact on the quality of the environment, and many states have similar impact study requirements for projects not involving federal funds. There are mixed reports, however, on how effective EISs have been in ensuring informed decision making (Suhrbier & Deakin, 1988).

E. P. Odum (1989) has argued strongly for a holistic top–down approach to environmental problems, in which an attempt is made to understand the inputs and outputs of major production systems or ecosystems as wholes, thus providing a basis for management policies that target long-term beneficial effects on the environment, as opposed to quick-fix approaches to the solution of local problems that can have negative effects from a longer term global perspective. Piecemeal approaches, he points out, can lead to the enhancement of one part of a life-support environment, such as agriculture, at the expense of degrading other vital components, such as soil and water supplies. The workability of such a holistic approach would require, however, a better understanding than now exists of the complex cause–effect relationships involved and how the critical variables interact. An attempt to implement models that make tradeoffs explicit could help to develop that understanding.

A continuing challenge for any entity that is attempting to allocate its limited resources to future development—whether it be an individual, a corporation or a nation—is the striking of a balance between focusing the available resources too narrowly and thereby missing out on growth that comes from possibilities not within that focus, and spreading the resources too thinly over many possibilities with insufficient commitment to any of them to realize much benefit from those that turn out to be winners.

Implicit in many of the foregoing comments about decision making and policy setting is a recognition that the consequences of decisions and policies often cannot be known with certainty. Further, even the current situation, the context in which a decision must be made or a policy set, may be known in only an imprecise or probabilistic way. Assessment of either the possible outcomes of decisions or of the "state of the world" in which decisions are made often involves the need to arrive at an opinion as to the magnitude or seriousness of one or more risks. Risk assessment is therefore an especially important aspect of decision making and policy setting, and how to do it is the topic to which we now turn.

DEALING WITH RISKS

I suspect that there is considerable agreement among policy makers and the general public alike on certain general principles relating to risk. No one with a rudimentary understanding of probability believes that zero-risk policies are attainable. Moreover, although people undoubtedly differ regarding what constitutes an acceptable degree of risk, there is probably broad agreement that we should work to make the actual levels as small as is practically feasible (although people may differ greatly with respect to what they do and do not consider to be practically feasible). One might expect also some consensus on the idea that, given limited resources, the greatest efforts at risk reduction should be directed where they are likely to accomplish the most good, which is to say that the highest priority should be given to the reduction of risks that are relatively high, that affect many people, and for which effective approaches to reduction can be designed.

In order to work effectively toward the reduction of risks that are relatively high, however, it is necessary to have a reasonable understanding of what the actual risks are. Without such an understanding, well-intentioned decisions can easily be made that commit large resources to the reduction of risks that are already very small, while ignoring, or perhaps unwittingly increasing, other risks that are considerably larger. And it is very hard to get the priorities straight. We seem to be able to get quite worked up about some problems while remaining relatively unconcerned about others, more or less independently of how serious they are. In 1988, for example, the equivalent of just two bags of medical waste that washed up on some shores of the northeast caused a big enough stir to result in the closing of dozens of beaches and the loss of $2 billion to $3 billion by local businesses, while the much greater danger stemming from nationwide substandard medical-waste incineration received scant notice (Hershkowitz, 1990). Part of the problem is that of limited and incomplete information. The appearance of used medical syringes on beaches was very well covered by the news media and therefore became part of everybody's consciousness; medical-waste incineration seldom gets much press and therefore escapes notice by most people most of the time.

The banning of Alar, the growth hormone used with apples, and the recall of 72 million bottles of Perrier water following the finding of samples with 19 parts per billion of benzene, have been pointed to as instances of overreactions to risks that were extremely small compared to others we accept daily with no complaint. W. Brookes (1990) has argued that the banning of Alar, which was motivated by a fear of cancer risk, may actually have raised the risk of getting cancer from apples because Alar had obviated the use of insecticides whose theoretical cancer risk is greater than that of the banned hormone. With respect to the Perrier water recall, Brookes cites an estimate by David Gaylor, head of biometry at the National Center for Toxicological Research in Jefferson, AR, that the additional cancer risk of drinking one bottle of the water every day for 70 years would be somewhere between 1 in 100,000 and 1 in 10 million. Brookes's point is not that efforts to reduce risks are not worth making, but that our priorities should be questioned: "The saving of lives is worth spending money on, but it doesn't make sense to spend vast amounts in the hope of saving a relative few when we cannot seem to afford to spend anything like those amounts on things that cause hundreds of times more deaths" (p. 47).

Actual Versus Perceived Risks

These examples point up the desirability of distinguishing between actual and perceived risk: *Actual risk* is the objective probability that some possible injurious or harmful event will occur; *perceived risk* is one's belief about the likelihood of the occurrence of such an event. Whenever one talks about objective probabilities, one risks stumbling into a philosophical quagmire, because what constitutes an objective probability, or even if there is such a thing, has been the focus of a long, and continuing, debate. Unfortunately, as desirable as such a distinction is, and as reasonable as it seems conceptually, it is not an easy one to make in any practically useful way.

Apostolakis (1990) states flatly that no probabilistic safety assessment (another name for probabilistic risk assessment) has ever been performed that does not use subjectivistic methods. This is because safety assessments typically deal with events that are sufficiently rare that relative-frequency-based statistics are not useful. Fischhoff (1989) argues that the distinction between actual and perceived risks is misconceived; although there are actual risks, nobody knows what they are. What are usually referred to as actual risks invariably represent some amount of judgment by the people who produced them: "In this light, what is commonly called the conflict between actual and perceived risk is better thought of as the conflict between two sets of risk perceptions: those of ranking scientists performing within their field of expertise and those of anybody else" (p. 270). He notes further that the quality of experts' assessments is itself ultimately a matter of judgment; and because expertise is narrowly distributed, the experts themselves are often called upon to judge the quality of their own judgments.

Slovic, Fischhoff, and Lichtenstein (1981) also point out that although risk assessments may appear to be objective, they are inherently subjective. Before one can assign a probability to a potential mishap, the potential mishap must have been identified, and identifying all potential mishaps is a very difficult thing to do, especially if few or none of them have ever occurred. Moreover, if potential mishaps have been omitted from an analysis because people have been unable to think of them, then their omission will obviously go undetected, and it seems almost unthinkable that some possibilities would not be overlooked when one is trying to generate a list of possible events with vanishingly small probabilities of occurrence.

Assuming—these difficulties notwithstanding—the distinction between actual and perceived risk to be a meaningful one, it is the perceived level of risk, not the actual one, that motivates behavior. When a small or modest risk is perceived to be very large, the anxiety produced by the misperception can result in a great deal of wasted effort and resources directed at the solution of fictitious or exaggerated problems. And when a really serious risk is perceived to be small, the problem is unlikely to get the attention it deserves. If a society is to deal effectively with the risks it faces, it must have, first, the means to assess those risks as accurately as possible and, second, the ability to communicate the results of those assessments to its members. The problem is complicated, of course, when there are advantages to be gained by special-interest groups in having specific risks either overstated or understated.

The problem of dealing rationally with risk is further complicated by the fact that decisions taken to mitigate a perceived risk can be costly even if the particular feared result does not come to pass (O'Hare, in press). Moreover, fear of a possible catastrophe is itself a psychological cost; if one believes that one's life is in imminent danger, the belief is unpleasant, whether or not it has any basis in fact. Also, fear-driven behavior can exact the tangible costs of preventive or protective measures as well as those of lost opportunities, and, to the extent that the fear was unfounded or grossly out of proportion to the actual risk, such costs would be wasted. O'Hare asks the interesting question of how we would behave if we believed that the cost of rare disasters to society was principally that of the anxiety they impose, rather than that of their actual occurrence.

Assessing Risks

Getting people to agree on the relative magnitudes of specific risks, or even on a process for assessing risks objectively, has proved to be very difficult. The evidence is quite compelling that as individuals we are not very good at assessing relative risks, at least insofar as real risks are reflected in incidence statistics. Scientists do not agree among themselves either about the basis of risk assessment when it comes to questions of probability, uncertainty, and long-term consequences (P. E. Gray, 1989). Especially difficult are situations for which there is no prior experience on which to base probability estimates. How, for example, does one assess the risks involved in experimentation with biogenetically altered organisms outside the laboratory? This has proved to be a difficult question on which to get a consensus among biologists and policymakers (H. I. Miller, Burris, Vidaver, & Wivel, 1990).

With respect to situations for which there are frequency data, research has shown that people often overestimate the frequency of infrequent risky events and underestimate the frequency of more frequent ones (Lichtenstein, Slovic, Fischhoff, Layman, & Coombs, 1978). For example, people tend to overestimate the frequency of death from sensational causes (homicide, tornadoes) while underestimating the frequency of death from mundane causes (asthma, diabetes). There is at least suggestive evidence, also, that estimates of the frequency of death from accidental causes may be inflated relative to estimates of death from natural causes because accidental deaths get more media coverage than do deaths from natural causes (Slovic, Fischhoff, & Lichtenstein, 1976). Deaths reported most frequently in the newspaper typically are not those that result from the most frequent causes (Coombs & Slovic, 1979).

The fact that the perception of risk can be affected by the occurrence of a single incident is demonstrated by the results of a longitudinal survey study conducted in the Netherlands. Dutch investigators who had sampled attitudes about nuclear energy two months before the Chernobyl accident were able to survey the same people at different times following the accident. Estimates of the probability of catastrophic accidents in Dutch nuclear power plants from the subjects in this study were much higher following the Chernobyl accident than before it (Verplanken, 1989). One may find other accounts of the effects of the Chernobyl accident on attitudes toward nuclear power generation and the perception of its riskiness by the general public in several countries in a special issue of *The Journal of Environmental Psychology* (Vol. 10, No. 2), which focused on "the psychological fallout" of the Chernobyl nuclear accident.

Fischhoff (1990) raises a cautionary note regarding the interpretation of the finding that people tend to overestimate low risks and underestimate high risks, pointing out that it has sometimes been distorted to justify viewing the public as ignorant of real risks and to make claims, in the context of discussions of risk management, that are unwarranted by the experimental outcomes. One study demonstrated that the accuracy of people's judgments and, in particular, whether people tend to over- or underestimate risks can depend on the details of the questions they are asked (Fischhoff & MacGregor, 1983). In any case, there are reasons to believe that decisions often get made on the basis of partial information and information whose interpretation is flawed.

This situation is exacerbated by the fact that the issues tend to be emotionally charged, and confirmation bias finds ideal conditions for expressing itself in numerous ways. It is also complicated by the fact that judgments of risk invariably involve assessment of benefits as well, at least implicitly (Fischhoff, 1989). In deciding whether to commit resources to the reduction of risk, for example, one must judge, at least implicitly, whether the benefit to be derived from the reduction is worth the cost of effecting it. It must be noted, too, that people do not always make decisions in such a way as to minimize risk; they may, in some cases, prefer a high-risk option to a low-risk one, because the benefits to be gained are considered worth the greater risk.

Covello, Sandman, and Slovic, cited by the National Research Council (1989c), identified several factors that, in their view, affect how the public perceives and evaluates risk. Their analysis is shown in Table 12.1. The claim is that people tend to be more concerned about unfamiliar risks that they cannot control, to which they are involuntarily exposed, and that have irreversible effects than about familiar, controllable risks that have reversible effects and to which they expose themselves voluntarily.

Unhappily, expertise is not a guarantee of great accuracy when it comes to quantifying risks. To be sure, experts are likely to have more objective data and are therefore less apt than others to produce extreme estimates of probabilities that relate to their areas of expertise, but experts and nonexperts alike tend to believe that their probability estimates are more accurate than they really are. When asked, for example, to set upper and lower bounds on estimates such that the chances would be 98 in 100 that the true value lies between them, people typically set the bounds such that a much smaller percentage, say 50% to 80%, of the true values fall between them (Lichtenstein, Fischhoff, & Phillips, 1977). Slovic et al. (1981) have listed a variety of ways in which experts may overlook or misjudge pathways to disaster: "failure to consider the ways in which human errors can affect technological systems . . . overconfidence in current scientific knowledge . . . failure to appreciate how a technological system functions as a whole . . . slowness in detecting chronic, cumulative environmental effects . . . [and] failure to anticipate human response to safety measures" (p. 159).

In spite of the thought-provoking results of much ingenious work on risk

TABLE 12.1 Qualitative Factors Affecting Risk Perception and Evaluation (from National Research Council, 1989c, p. 35)

Factor	Conditions Associated with Increased Public Concern	Conditions Associated with Decreased Public Concern
Catastrophic potential	Fatalities and injuries grouped in time and space	Fatalities and injuries scattered and random
Familiarity	Unfamiliar	Familiar
Understanding	Mechanisms or process not understood	Mechanisms or process understood
Controllability (personal)	Uncontrollable	Controllable
Voluntariness of exposure	Involuntary	Voluntary
Effects on children	Children specifically at risk	Children not specifically at risk
Effects manifestation	Delayed effects	Immediate effects
Effects on future	Risk to future generations	No risk to future generations
Victim identity	Identifiable victims	Statistical victims
Dread	Effects dreaded	Effects not dreaded
Trust in institutions	Lack of trust in responsible institutions	Trust in responsible institutions
Media attention	Much media attention	Little media attention
Accident history	Major and sometimes minor accidents	No major or minor accidents
Equity	Inequitable distribution of risks and benefits	Equitable distribution of risks and benefits
Benefits	Unclear benefits	Clear benefits
Reversibility	Effects irreversible	Effects reversible
Origin	Caused by human actions or failures	Caused by acts of nature or God

perception, we do not yet have a clear understanding of the accuracy of the opinions the general public holds with respect to significant risks to life and the quality thereof, or even of how best to determine that accuracy. This is an extremely important area for research. A democracy is predicated on the assumption that the public is sufficiently well informed to be the ultimate authority behind decisions regarding public policy. Research results that call that assumption into question deserve the closest scrutiny. If they stand up to such scrutiny, the challenge then becomes one of finding ways to correct whatever misperceptions the public may have.

A serious methodological difficulty associated with the study of risk is that of establishing cause–effect relationships involving remote causes. It is relatively straightforward to determine the number or percentage of deaths in a given population over a specified period of time that were due to lung cancer; however, it is another matter entirely to specify how many of those cancers resulted from exposure to x-rays or radon, or from aspiration of tobacco smoke, asbestos fibers, coal dust, or smog, or from some other cause or combination of causes. Estimates are derived statistically from studies in which the incidence of cancer among people who are believed to have had specific degrees of exposure to various cancer-causing agents is compared with its incidence among people who are believed to have had very different amounts of exposure, but these are estimates and are, thus, subject to considerable uncertainty. Relevant data also come from studies of the effects on laboratory animals of exposure to known doses of carcinogens, but extrapolating the results of these studies to humans is also less than straightforward, in part because the relationship between exposure and effect is sometimes known to be nonlinear even within a species (Morgan, 1981b).

Probabilistic risk assessment is a technique that has been developed for explicitly quantifying the riskiness of a process or operation. As applied to decisions in which the possibility of accidents is a consideration, it involves aggregating the costs of all possible accidents, each weighted by its estimated probability of occurrence. The technique can be applied iteratively so that if the risk is judged to be unacceptably high, the process or operation can be redesigned and the risk reassessed. The technique is now used widely in the nuclear power industry. Such assessments have been performed on about 40% of all the nuclear power reactors in the United States since the Three Mile Island incident, and many others are ongoing. It is used also in other industries in which system failures can be hazardous (Weinberg, 1989 1990).

There is some question, however, of the appropriateness of the technique when the probabilities are very small and the consequences are very large, which is to say, when the events of interest are considered highly improbable but their consequences catastrophic. Also, a limitation of any technique that requires experts to lay out the possible ways in which a system can fail or malfunction and assign probabilities to all the possibilities is that it is difficult to be certain that all the possibilities have been identified. It is extremely easy to fail to recall all the items in a set, even when dealing with a set with which one is quite familiar. Other formal techniques for risk assessment, such as fault tree analysis, can be helpful but also have serious limitations (Fischhoff, Lichtenstein, Slovic, Derby, & Keeney, 1981; McCormick, 1981): One significant risk associated with the use of these techniques is the risk of overconfidence in the results they yield.

Reducing and Managing Risks

Risk reduction typically, though not always, can be realized only at some cost. Sometimes the cost of reducing a risk below its current level is less than the value of the reduction; sometimes it is more. Often the cost of risk reduction and the expected cost of the risk itself are assumed to relate to level of risk in nonlinear and opposite ways: The cost of reducing risk by a fixed amount is much greater if the risk is already small than if it is large; similarly, the expected cost associated with an increase in level of risk depends on what the level of risk is before the increase occurs. From a strictly economic point of view, the trick is to set the level of risk at the point at which the total cost, which is the sum of the two cost components, is at its minimum. (The strictly economic view is not the only one that can be taken, of course, and sometimes it may be argued that a risk should be maintained below some threshold, somewhat independently of the cost.)

A special challenge for risk management is the problem of dealing with NIMBY (not-in-my-back-yard) conflicts. A NIMBY conflict is one in which there is general agreement that a particular facility (a hazardous waste processing plant, a prison, a halfway house for recovering substance abusers) is needed, but people are unwilling to have it in their own neighborhood. O'Hare (in press) points out that, at least in some cases, the NIMBY objection stems from the fear that something will go wrong, that is, the facility is acknowledged to pose no problem as long as it functions as intended, but there is concern that the plant will inadvertently release toxic wastes into the local environment, that a dangerous prisoner will escape, and so on. O'Hare discusses the paradoxical case in which neighbors of a proposed facility opt for not building it even when building it looks like the better option in terms of expected outcome—better for the neighbors as well as for everyone else. The resolution of the paradox, in O'Hare's view, lies in a recognition that anticipatory anxiety about a possible event, no matter how low the probability of the event may be, is a real social cost, and one that accrues over time, but the cost of anxiety is not captured by traditional analyses of such decision situations.

The practical problem of risk management is greatly complicated by the fact that the people who will bear the cost of specific risky practices are not always the same as those who will realize the benefits. In some cases, they do not even belong to the same generation. Certain energy policies, for example, could benefit this generation at considerable cost to future generations. Other policies could exact costs from the present generation to the benefit of future ones. Unfortunately, as human beings, we seem to share a fairly strong bias for situations that benefit us personally at someone else's expense. This is, of course, the moral of the "tragedy of the commons." There is some hope, perhaps with the help of negotiation techniques, of working out compromises between conflicting interests when the parties having those interests are able to communicate. It is much less clear that, as a species, we have whatever it takes to protect adequately the interests of those (e.g., future generations) who are unable to speak for themselves.

Recognition of the reality of anxiety costs raises issues of public policy in dealing with risks. When there is no feasible way to reduce a risk, for example, what should be the policy with respect to informing the public about the existence of that risk. Suppose a physician discovers that a patient has a 25% chance of becoming fatally ill in 20 years of an inherited disease for which there is no treatment, and the patient is blissfully ignorant of this risk. Does the physician have a moral obligation to inform the patient of what he knows about the situation? Obviously, arguments can be made either way, and perhaps one can find compelling analogies involving societal risks.

My own sense is that it would be difficult to justify withholding information from the public about risks that affect entire societies, so my assumption is that the goal should be to improve our corporate understanding of risks and to attempt to develop rational and effective ways to deal with them, remembering, as we do, that worry about risk can become a risk itself if carried to an extreme (Morgan, 1981a). There is no such thing as a risk-free existence, and the attempt to eliminate risk entirely can produce only disappointment and frustration. It is clear, however, that we sometimes accept risks that are higher than necessary without realizing we are doing so, and we sometimes believe risks to be much greater than they actually are. A better understanding of the true nature of the various risks we face should make it possible for us to take reasonable steps to reduce the larger ones to acceptable non-zero levels. Of course, the rub lies in the fuzziness of such words, in this context, as *reasonable* and *acceptable*. The problems of risk assessment and control are likely to represent interesting challenges to psychologists for a long time to come.

Risk Communication

Interest in risk communication is apparently quite high at the present time, but was not always so (National Research Council, 1989c; Plough & Krimsky, 1987). The right of members of the public to be informed about certain hazards and risks, whether or not they have any part in the decision-making process, is recognized in several federal laws, beginning with the Administrative Procedures Act of 1946 and continuing with several more recent pieces of legislation (National Research Council, 1989c). The Occupational Safety and Health Administration legislation and Title 3 of the Superfund Amendments and Reauthorization Act of 1986 require that certain individuals and communities be informed about specific hazards and risks involved in the handling of, or exposure to, certain materials and emissions. But what exactly does it mean to be "informed"? The National Research Council (1989c) perceived the problem of communicating risk to be a sufficiently important one that it commissioned a Committee on Risk Perception and Communication to "examine possibilities for improving social and personal choices on technological issues by improving risk communication" (p. ix).

In summarizing its work, the committee distinguished its conception of risk communication from the conventional idea of one-way messages from experts to nonexperts: "We construe risk communication to be successful to the extent that it raises the level of understanding of relevant issues or actions for those involved and satisfies them that they are adequately informed within the limits of available knowledge" (p. 2). The committee noted that one should not expect improved risk communication always to result in the reduction of conflict and improvement in risk management. Risk management decisions that benefit some people can harm others; moreover, inasmuch as people do not all share common interests and values, better understanding will not necessarily produce consensus on controversial issues.

The committee emphasized the interactive nature of risk communication, as it defined the process, and noted that it includes messages not strictly about risk: "Thus risk communication includes announcements, warnings, and instructions moving from expert sources to nonexpert audiences . . . but it also includes other kinds of messages—about risk information and information sources, about personal beliefs and feelings concerning risks and hazards, and about reactions to risk management actions and institutions. Not all these messages are strictly about risk, but all are material to risk management" (p. 22). The committee also noted a tradeoff between excessive thoroughness on the one hand and misleading simplicity on the other: "Preparing risk messages can involve choosing between a message that is so extensive and complex that only experts can understand it and a message that is more easily understood by nonexperts, but that is selective and thus subject to challenge as being inaccurate or manipulative" (p. 95).

The committee reviewed a variety of techniques for communicating about risks that can be used to influence people's thinking, techniques that go beyond the presentation of factual information. It recommended the establishment of guidelines for the content of risk messages; these can be used by responsible message designers to keep influence techniques under control so as not to bias the understanding of the messages' recipients. It noted, also, however, the impossibility of devising a complete guide to sources of potential bias on the basis of what is currently known. (What is currently in the literature that pertains to the problem of risk communication is reviewed in some detail by Fischhoff, 1989, in an appendix to the committee's report.) Moreover, the committee warned that "because the attendant choices are controversial, affecting important economic interests and strongly held values, participants in the decision process, including experts and their employers have incentives to appeal to emotions, distort facts, and otherwise use communication to influence the ultimate choice in the directions they desire. Thus there are no participants in debates on technological issues on whom non-experts and public officials can rely unquestioningly for unbiased information" (p. 23). Individuals must be prepared to make up their own minds on issues of risk, and the only way to ensure a broad perspective is to gather information from a variety of sources representing, insofar as one can manage it, the full spectrum of vested interests. The committee also pointed out that risks cannot be weighed against each other or even adequately understood without considering values. Consequently, values and interests should figure in debates regarding public decisions about risk.

* * *

Decision making and policy setting are inherently human activities. Even if we manage to turn such functions over to computers, to some degree, someone must write the programs that the computers will run and incorporate within them the rules by which the decisions will be made and the policies set. Much remains to be learned about these activities before we can be satisfied that either we or our electronic surrogates are well prepared to deal with the kinds of problems that will be encountered in the next few decades. We will need better techniques for identifying tradeoffs involved in important policy decisions and for predicting long-term consequences of actions aimed at correcting specific societal problems; more adequate models of complex systems of interacting variables that can be used to explore the implications of changes in one or more of those variables, and better ways of checking the validity of such models; more sensitive approaches for monitoring the consequences of policy decisions and evaluating outcomes relative to long-term objectives; more effective approaches to negotiation and tools that can be used to facilitate conflict resolution; better procedures for resource allocation and management; and fail-safe procedures for use in crisis situations in which errors of judgment can have disastrous effects.

There is also a need for improving our understanding of many aspects of human judgment and decision making: how they are affected by various types of stress, both acute and chronic; how to determine when an individual's capacity for effective decision making is lowered, or in danger of being lowered, as a consequence of fatigue, workload, boredom, chronic stress, or some other factor; and how people assimilate and integrate incoming information in time-critical situations in which the stakes riding on action choice are high.

Risk assessment and communication remain important challenges for research. How can people be helped to see the relative magnitudes of different risks—at least to the extent that they can be determined by objective means? What types of messages are understandable by the general public? What types tend to confuse or be misinterpreted? What approaches can be used to convey quantitative information about risks so that it will be understood by people without a quantitative bent? How can risk messages be constructed so as to distinguish clearly between uncertainties arising from incomplete scientific knowledge and disagreements about values and goals?

The problems of risk assessment and communication are exacerbated by the fact that the skill of estimation is largely ignored by formal education. The ability to estimate is one that all of us have constant opportunities to exercise, and anyone who lacks it is at a serious disadvantage in many contexts. In spite of its importance, it is seldom emphasized or even explicitly addressed in school, except perhaps in courses on statistics. Even here, however, the focus is on formulaic approaches to the estimation of population parameters on the basis of sample statistics; little attention is given to how one becomes adept at making the kinds of estimates of distances, weights, durations, quantities, and probabilities that must be made in countless situations in everyday life.

The report from the National Research Council's (1989c) Committee on Risk Perception and Communication makes it clear that making well-informed choices about risks is not a trivial matter; it requires a wide range of knowledge. As the committee put it:

> It depends on understanding of the physical, chemical, and biological mechanisms by which hazardous substances and activities cause harm; on knowledge about exposures to hazards or, where knowledge is incomplete, on analysis and modeling of exposures; on statistical expertise; on knowledge of the economic, social, aesthetic, ecological, and other costs and benefits of various options; an understanding of the social values reflected in differential reactions to the qualities of risks; on knowledge of the constraints on and responsibilities of risk managers; and on the ability to integrate these disparate kinds of knowledge, data, and analysis. (p. 38)

This is a tall order and one that is often impossible to fill. Given how difficult it is to understand in any very thorough sense many of the risks that are faced by individuals, societies, or the world as a whole, we should be suspicious of simplistic solutions to specific risky situations, even if we like the solutions or have proposed them ourselves.

I believe that significant progress on many real-world problems in the areas of complexity management, risk assessment, decision making, and the like will require an effective coupling of human and machine capabilities. The machines are likely to be complex, computer-based systems, and the people will often be functioning as teams rather than as individuals. This means there will be a need to attend to all the usual human factors issues that arise in the design of complex person–machine systems. I believe, also, however, that many of the problems are such that new approaches and new capabilities will have to be developed if real headway is to be made on them. Development of those approaches and capabilities will require some inventiveness and ingenuity. There will be many opportunities here for contributions by people with a sensitivity to human capabilities and limitations and some understanding of how to match them effectively with those of existing or future machines.

13

Quality of Life

In a sense, quality of life has been the underlying concern throughout this entire book. Enhancement of the quality of life—in the workplace, in the home, on the highways, in hospitals, in schools, wherever people spend time—is the appropriate ultimate objective of human factors as a discipline, or at least so it seems to me. Improving the design of workspaces and tools, enhancing the safety of appliances and vehicles, increasing the effectiveness of communication facilities, reducing the probability of human error and mitigating the harmful effects of errors that occur—these and many other foci of human factors research serve that objective.

I do not mean to argue that all work done under a human factors umbrella is necessarily motivated by an interest in improving the quality of life or even that it serves that purpose. (The fact that efforts to increase worker productivity have not always been sensitive to their effects on the quality of the life of the worker provides one notable counterexample, and one can find others.) However, I like to believe, and do believe, that, on balance, the raison d'etre of the discipline is to improve the human condition, especially insofar as it is affected by technology and the ways in which people relate to it.

In this penultimate chapter, I want to address several topics that pertain to quality-of-life issues, perhaps more obviously and explicitly than do many of the other subjects considered in the foregoing chapters. The topics I have chosen to include reflect personal interest to an extent; many others would have been equally appropriate for inclusion under a quality-of-life heading.

DESIGNING FOR SPECIAL POPULATIONS

Historically, in promulgating human factors design criteria, the objective has been to provide guidelines that would ensure the usability of equipment by people whose measurements did not deviate from the averages by more than a couple of standard deviations. The intent was not to design precisely for the average individual, but for a range of individuals from, say, the 5th to the 95th percentile. Theoretically, equipment designed to such specifications would be problematic only for truly extreme cases. This makes sense from a statistical point of view: If one can design a piece of equipment that is convenient for 90% of the population to use, that is no mean accomplishment, but extreme cases do exist. Some people do differ from the averages in one way or another by more than a couple of standard deviations, and the design problem cannot be considered completely solved as long as their needs are not met.

More generally, there are special populations of various types that represent special challenges to human factors researchers and others interested in the problem of designing equipment for human use. People with various types of disabilities constitute a number of such populations. I say "a number of such populations" advisedly, because I want to emphasize the fact that people with disabilities are not a homogeneous group with respect to their special needs for equipment design or in any other way. This is an obvious fact, yet much of what has been written about human factors and people with disabilities has not given it the emphasis it deserves. The most obvious way in which the collection of all people with disabilities differs from the collection of people without disabilities is that the range of individual differences is much greater in the former than in the latter. People with disabilities vary in all the ways in which people without disabilities vary and many others as well: The union of all disabled populations includes people who can see and people who are blind, people who can hear and people who are deaf, people who can walk and people who cannot; it includes people at all levels of intelligence and with all degrees of emotional stability, and so on. The population of people without disabilities is very homogeneous by contrast, containing, as it does, only people who can see, hear, walk, whose intelligence is above some specified level, and who are emotionally stable and "normal" in the ways that count in our society. The challenge of designing for people with disabilities is that of designing for diversity.

Meaningful incidence statistics regarding people with disabilities are very hard to get. An accurate count would be difficult even if it were clear whom should be counted, and that is not clear at all. Most types of impairments, like impaired vision or cognitive limitations, occur in varying degrees. How disabling must an impairment be in order to be considered a disability for the sake of our count? There is no easy answer to this question, because no matter where one sets the threshold, arguments can be made in favor of moving it up or down. Some physical conditions or illnesses—hay fever, allergies, high blood pressure—can be disabling under certain circumstances, but often are effectively controlled by behavior or medication. Should people with these conditions be counted as disabled or not? Many conditions that would be handicapping for some people are not so for others: I am missing the end of my right index finger; this could be a serious handicap if I were a concert pianist, but I am not, so I am usually blissfully unaware of its absence. How do we count people who are temporarily disabled, with a broken bone, say, or with a curable disease? What do we do about people who are seriously impaired but do not realize it: the alcoholic, for example, who will not admit to a problem, the individual who is having difficulty functioning effectively because of an undiagnosed brain tumor, the person who cannot hold a job because of a disabling neurosis that has been dismissed as obstinacy or a poor attitude?

In view of the difficulty of coming up with a definition of disability that would be universally viewed as appropriate and that could be applied unambiguously to determine who did and who did not satisfy it, it is not surprising to find that estimates of the number of disabled people in the United States vary from 10% or less to over 45% (Bowe, 1984a; OTA, 1982; C. S. Wilder, 1976, 1977; P. H. N. Wood, 1980). Because estimates are made by a variety of agencies and for a variety of purposes, it would be remarkable if large differences did not occur. In some cases, for example, when funding for a particular program is being sought, there is an incentive to make the estimate as high as possible, so we might expect to see some inflated numbers. There are reasons to suspect, however, that estimates often err on the low side, because they are based only on subsets of the general population. Pfeiffer (1985) points out that employment and vocational rehabilitation agencies usually focus on working-age people, educational agencies on school-age people, and so on. He notes, too, that estimates based on interview surveys can be flawed because people are sometimes unwilling to admit to a disability, fearing social stigmatization if they do so. Pfeiffer gives 30% as his own estimate of the size of the population that is disabled, including elderly disabled people, but notes it can be argued that this number is too low.

J. Elkind (1990), based on his effort to reconcile data from several sources, also arrived at an estimate that about one third of the population has some sort of disability. Using the categories used by the Bureau of the Census, excepting "emotionally disturbed" and "other health impairments," which he did not include, he derived the estimates shown in Table 13.1. Elkind noted that this estimate includes many people whose impairments are not serious or who get along well with simple prosthetic aids (e.g., glasses or hearing aids). On the other hand, the omission of people with emotional

TABLE 13.1 Incidence of Major Types of Disability (from Elkind, 1990)

Type of Disability	Percent of Population
Sensory	
Visually handicapped	3.4
Hearing handicapped	8.8
Motor	
Orthopedic impairments	9.6
Cognitive	
Specific learning disabilities	7.5
Speech impaired	0.9
Mentally retarded	1.2
Illiterate and semi-literate	5.0
Total	36.4

disturbances or other health impairments may mean overlooking sizable categories. Again consulting a number of sources, Elkind derived estimates of the number or people in each of the disability categories who have disabilities "severe enough to interfere with work or otherwise constitute a serious handicap" (p. 404). The estimate he arrived at with this criterion in mind was 21% of the total population.

As Elkind points out, the total shown in Table 13.1 is undoubtedly an overestimate (assuming the accuracy of the other numbers) because people with multiple disabilities would be represented in more than one category. (To illustrate the need to be sensitive to the possibility of double, or multiple, counting, Vanderheiden, 1990, gives examples of prevalence estimates in which the percentages in the individual disability categories add to considerably more than 100.) Citing Kraus and Stoddard (1989), Vanderheiden uses 12% to 20% of the total population as his estimate of the prevalence of disabilities or functional limitations in the United States.

Fortunately, for our purposes, an accurate estimate of the number of percentage of disabled people, by whatever criteria, is not essential. Even the most conservative estimates make it clear that the problem is large and worthy of considerable attention. Also, it seems inevitable that the number of people living with one or another type of physical or mental disability will increase, and possibly at a faster rate than the general population, as a consequence of ever more powerful life-saving and life-prolonging medical techniques. The challenge will be to ensure the quality of the prolonged lives, and that problem is a multidimensional one, involving not only physical or mental limitations and impairments, but attitudes, education, awareness, public policies, architectural and other barriers to mobility, employment practices, and equipment design.

For some time now there has been great interest in the United States and numerous other countries in increasing the opportunities for people with various types of disabilities to live independently, or in the least restrictive environments that are consistent with whatever needs they may have for daily assistance. The behavior of developmentally disabled adults appears to be determined to some degree by the characteristics of their living environment. Behavioral improvements have been observed, for example, when developmentally disabled adults have been moved from institutional barracks-like living quarters to environments that permitted greater privacy and personal control of their own space (Zimring et al., 1982).

Making independent living possible for an increasingly large proportion of the population poses many challenges for the development of a wide range of assistive devices, and the modifications of design features of buildings, vehicles, and appliances that limit the availability of the resources needed for independent living. A major problem at the present time is that of informing potential users of assistive technologies about what exists and how it can be acquired (Czaja, 1990). There are already several databases that contain information of this type. Casali and Williges (1990) describe six that provide information about hardware and software that is available for computer users with disabilities. They also discuss information not currently provided by these databases that would be useful to disabled individuals attempting to determine the relative suitability of available aids.

On July 19, 1990, President Bush signed into law the Americans with Disabilities Act, a landmark piece of legislation for people with disabilities and for the country as a whole. The law prohibits discrimination against people with disabilities in the workplace, in public accommodations, in transportation, and in communication services. Unlike the Rehabilitation Act of 1973, which prohibited discrimination by government agencies and companies that receive federal funds, the Americans with Disabilities Act prohibits discrimination against people with disabilities essentially everywhere. All public transit buses and passenger rail cars purchased in the future must be accessible, as must all new bus and train stations. Architectural barriers to accessibility in stores, restaurants, hotels, entertainment, and recreational facilities will have to be removed.

Employers will be required to make reasonable accommodations for job applicants with disabilities who are qualified for the work for which they are applying. The last provision is especially important in view of the fact that one of the most apparent effects of many types of disability is a decrease in job opportunities and, consequently, of earning power: Unemployment among disabled people who would like to work and could do so is several times higher than the national average, running as high as 30% or more in recent years (Kraus & Stoddard, 1989). About 28% of people with a work-related disability in the United States live below the poverty line, while the comparable figure for the total working-age population is 11% (Vachon, 1989–1990).

It has been the hope of some observers that there is, within modern technology and especially information technology, the potential to provide for many people with disabilities greater independence, more opportunities for employment, and easier access to the world in general. I believe this is a reasonable hope and one that is being realized to some degree, although the accomplishments to date have barely begun, in my view, to sample the full range of possibilities. Much of the work to date on applying technology, and especially personal computers, to the needs of people with disabilities has been done by individuals with disabilities who have seen opportunities to apply the technology to their own needs (Bowe, 1984b; Gergen & Hagen, 1985; McWilliam, 1984; A. Schwartz, 1984).

One of the more effective strategies that the human factors researcher, or anyone else with an interest in people with special needs, can use is that of looking for ways to tailor technology that is being developed for other purposes so that it will meet those special needs. Computer networks were not developed for the purpose of providing people with severe mobility constraints greater access to work opportunities and the world in general. The technology has the potential to do just that, however, although anything close to the full realization of that potential will require the application of some energy by people who want to see it realized. The personal computer was not developed in order to provide people who have severe communication disabilities, such as those sometimes associated with cerebral palsy or other motor impairments, with a tool that can mitigate those disabilities. Again, the technology has the potential to do that, as numerous users of it have amply demonstrated, but the potential has begun to be realized only as people with a special interest in the problem have done what was necessary to fit the technology to this end.

Robotic and teleoperator technologies have obvious potential for application to the design of prosthetic and orthotic devices. Some of the more difficult problems inhibiting the more rapid realization of this potential are less technological than cosmetic in nature (Sheridan, 1980): If a device is seen as too bizarre or unnatural, it will not be accepted. The ability to monitor low-intensity signals produced by muscles and to process these signals with computationally powerful microchips offers the possibility of enhancing the effectiveness and naturalness of assistive devices; however, sometimes effectiveness and naturalness will have to be traded off, to some degree, and how to make the tradeoff result in a device that is both useful and acceptable is a human factors problem.

The strategy of seeking, in technology that was developed for mainstream purposes, opportunities for application to special needs is a powerful one. To the extent that people who are interested in populations with special needs can anticipate developments intended to serve the societal mainstream, they can be in that much better a position to exploit those developments for their special interests when they come along. There may also be the possibility sometimes of influencing the mainstream development so as to facilitate its exploitation for the special needs.

I believe that helping to expand the opportunities for people with various types of disabilities to lead independent, productive lives is one of the most significant challenges to the human factors community for the future. The design of assistive devices, of accommodating workspaces, and of manageable living quarters has been of interest for some time and will become an even greater need as the number of people living with some type of functional impairment increases. Perhaps the greatest challenge is that of meeting the needs of people with low-incidence problems. Not surprisingly, disabilities that affect large numbers of people get more attention than do those that affect relatively few. Moreover, devices that can benefit large numbers of people are more likely to become commercially available because there is a greater chance of recovering the cost of the development of such devices through their sale. People with disabilities that occur in very small numbers typically have neither much economic leverage nor strong advocate lobbies, and consequently their special needs are likely to go unmet.

Vanderheiden (1990) has discussed the problem of designing things so as to be convenient for people with unusual needs as well as for those with more typical requirements. Although he acknowledges the impracticality—and perhaps the impossibility—of designing everything to be accessible to everyone regardless of their limitations, he argues that "for most types or degrees of impairment there are simple and low-cost (or no-cost) adaptations to product designs which can significantly increase their accessibility and usefulness to individuals with functional impairments" (p. 389). He also makes the important point that devices, such as household appliances, designed to be accessible to individuals with functional impairments should also be of interest to people who are not impaired because of the possibility that they someday will be. He notes too that designing devices so as to make them more accessible to people with physical limitations has often had the adventitious effect of making them easier to use by able-bodied people as well. Others have also argued that serious attempts to meet the needs of special populations, from a human factors point of view, will advance the field as a whole (Griffith, Gardner-Bonneau, Edwards, Elkind, & Williges, 1989).

AGING

Aging is hardly a new phenomenon. Individuals have been doing it for a rather long time. The aging of the population as a whole, however, is an event unique to our time. Dramatic increases in life expectancy during this century, in the United States and in many other parts of the world, coupled with changing patterns of birthrate, have resulted in a disproportionate increase in the number of older people and in other changes in the way population is, and is anticipated to be, distributed with respect to age. In the United States, these changes are sometimes referred to as the "squaring of the pyramid." As can be seen in Fig. 11.1 (from the Transportation Research Board, 1988a), in 1950 the number of people in each age group dropped off quite regularly for age groups beyond the 20s, but by 2030 each age group, except the few oldest, is expected to contain roughly the same number of people. (As a consequence of reductions in both fertility and mortality, populations are aging even more rapidly in some East Asian countries [e.g., Japan and Singapore] than in the United States [L. G. Martin, 1991].)

The percentage of the American population that is 65 or older increased from 4% in 1900 to 11% in 1980, and is expected to go to 20% probably before the middle of the 21st century (Hurd, 1989; Myers, 1985). The median age of the American population is expected to go from about 32 in 1985 to nearly 40 by 2020. The fastest growing age group in the country, in percentage terms, is the 85 and older group; it is expected to triple between 1980 and 2010 (Transportation Research Board, 1988a). This group is predominantly women, and is likely to continue to be for the near-term future.

Elderly men and women differ considerably with respect to marital status and living situation. About 70% of men over 75 have living spouses, whereas only about 30% of women over 75 do (Transportation Research Board, 1988a). More than one third of elderly people in the United States live alone (Bureau of the Census, 1985), and elderly women are about three times as likely as elderly men to do so. Among elderly people who live alone, women are about twice as likely as men to live below the poverty line (Hurd, 1989; OTA, 1985e). Persons over 75 years of age are three times as likely to be in an institution if they do not have a living spouse than if they do (Butler & Newacheck, 1981).

The continuing aging of the population will manifest itself in many ways: a significant increase in the number and percentage of elderly drivers, a much larger portion of the population in retirement, an increase in the number of people who could benefit from one or another sort of assistive device or service. In 1940, 1.7% of all drivers in the United States were over 65 years of age; in 1983, 11% were over 65 (Kostyniuk & Kitamura, 1987). That percentage will increase as the over-65 group continues to grow faster than the population as a whole. (The importance of the ability to drive to an elderly person's sense of well-being and general healthfulness may not be fully appreciated [Cutler, 1975].)

The fact that many elderly people have special requirements for housing is now fairly well recognized among designers and architects (Howell, 1980; Regnier & Pynoos, 1987). About 5 million elderly people with some functional limitations live in ordinary homes, and about 1.5 million live in nursing homes or other institutions (National Center for Health Statistics, 1987a, 1987b).

In attempting to understand the needs of elderly people, it is natural to focus on the limitations and impairments that often come with advancing age, and this seems appropriate, to an extent, because it is unquestionably true that the incidence of impairment is greater the older the age group and rises sharply for the very old. However, it is important not to equate old age with physical or mental disability. One often sees in the literature discussion of problems of "the disabled and elderly," almost as though the two groups were synonymous. They are not. The coupling of the groups might be defended on the grounds that many of the issues that must be considered in designing living environments and other facilities for the elderly, as a class, are the same as those that must be considered in designing facilities for people with disabilities, and this is so because of the high incidence of disabling conditions among the elderly and the relatively high probably that nondisabled elderly people will experience some such condition some time in the future as they continue to age. There is also a downside to this coupling, however, in that it tends to perpetuate the erroneous view that disability is an inevitable consequence of aging, and there are many elderly people who would not be considered disabled under any reasonable definition of the word.

The vast majority of people aged 65 and over do not fall into special categories, which is to say they require nothing more in the way of living aids than do people in other age brackets (Lawton, 1990). Less than 50% of people over 85 report any physical or mental limitations that affect their daily activities (Soldo & Longino, 1988). The majority of elderly people in the United States own their own homes (OTA, 1985e), and the desire for independence among elderly people is apparently very strong. Howell (1982) notes that among the few things that are clear about the preferences of aging Americans are "that they tend to: (1) reject nursing homes; (2) prefer not to live with adult children; and (3) predominantly stay in their *owned* homes even in the face of widowhood and health pressures" (p. 46). Elderly people represent an enormous resource for the nation and the world, with an excess of time and experience and a desire to contribute to family, community, and humankind generally. We need to be more innovative in finding ways to tap this resource.

We need also some new ways to think about aging and especially a greater recognition of the very great range of individual differences among elderly and aging people. Typically, when people in different age groups have been compared with respect to their ability to perform specific tasks, performance, on the average, has shown some decline with age for the groups beyond early adulthood; however, the variability within groups is large relative to the magnitude of any decline in the averages across groups, and the amount of variability in performance increases with age (Heron & Chown, 1976). A significant percentage of people in older groups do better at various tasks than the average level of performance of younger groups (Braune & Wickens, 1983). This kind of finding lies behind an interest in the concept of *functional age,* that how old one appears to be vis-à-vis one's ability to function in various capacities need not be the same as one's chronological age. This idea has led, in turn, to efforts to determine functional age relative to the demands of specific tasks, by means of tests designed to measure specific perceptual, motor, or cognitive capabilities (Borkan & Noris, 1980; Braune & Wickens, 1985; Salthouse, 1982).

I believe the work on functional aging and the underlying assumption that chronological age is not a justifiable basis for decisions regarding what individuals can and cannot do address an important issue that deserves more attention from researchers than it has yet received; however, functional age strikes me as not an altogether felicitous concept. Implicit within it is acceptance of the idea that decreasing competence is the natural and expected corollary of aging. To say that one's functional age is less (greater) than one's chronological age is to say that one can function better (more poorly) than would normally be expected of a person of that chronological age; to say that a person is functionally young (old) is to suggest that person is able to do what a young (old) person should be able to do. The underlying model that justifies this terminology is that the way things are and the way they ought to be is that function declines steadily with age. Might it not be better to talk simply of "functional ability" as it relates to the demands of specific tasks and, in making decisions regarding the suitability of individuals to perform those tasks, and to do so in terms of how they measure up on functional ability tests, but without reference, explicit or implicit, to age?

All this being said, it is undeniably the case that as people age the probability increases that they will experience one or more kinds of impairment and will require some accommodations in their living or working environments that were not necessary in their younger years. A growing challenge for the future will be the increasing population of very elderly people and the fact that, as a consequence of medical technology, many individuals will survive for years in a much poorer state of health than would have been possible in the past. One in four elderly persons is expected eventually to enter a nursing home where annual costs of care, as of 1987, exceeded $20,000 (General Accounting Office, 1987a). One simulation-based estimate of the annual cost of long-term care for the elderly puts the number at the equivalent of $120 billion (in 1987 dollars) per year by 2020 (Rivlin & Wiener, 1988). Many of these people could probably continue to live in their own homes with the help of appropriate assistive devices and other resources.

There are opportunities here for innovative exploitation of technology—

especially computer and communication technology—for assistive purposes. One might hope that what is learned about making restricted living and working quarters habitable in the space program might be used to advantage in designing spaces for people who, because of infirmity, are obliged to spend most of their time in very limited spaces. Especially important to people in such situations are "control centers" by means of which they can, in spite of limited mobility, acquire information and stimulus variety and achieve some social integration (Lawton, 1990).

The problem of providing long-term care for people with Alzheimer's disease and other dementias is of growing concern to the nation as the percentage of the population that lives to a very old age continues to increase faster than the population as a whole (OTA, 1987c). It is possible that in the future there will be many more people whose bodies have outlived their minds. Will we be better in the future than we have been in the past at dealing with the infirmities and disabilities that sometimes attend extreme old age? There is the unpleasant possibility that we will get better at prolonging life but not at improving its quality.

Perhaps the most difficult problems for many aging people to deal with are the loneliness and sense of uselessness that can come with the shedding of work responsibilities, the death of friends and acquaintances, restricted mobility, and declining sensory acuity and physical ability. The challenge to human factors researchers is to make it easier for elderly people to stay in touch with others and the world and to find ways to put their knowledge and talents to use. Computer networks, and communications technology more generally, have great potential, I believe, to be useful in this regard, but that potential has to be developed and explicitly applied to this purpose.

PERSONAL SAFETY AND HEALTH

The annual toll of preventable accidents in the United States is unknown, but undoubtedly enormous. According to Lescohier et al. (1990), the deaths due to accidental injuries are but the tip of the iceberg. They point out that in 1985, 57 million people—one person in four—were injured, and 2.3 million of these were hospitalized. Identifiable costs associated with these injuries—for medical expenses, lost work because of disability, and lost productivity because of premature death—has been estimated at $158 billion. According to McGinnis (1988–1989), more than 60% of all Americans who die each year die prematurely; that is, they did not have to die when they did. McGinnis estimates that between 40% and 70% of these deaths could have been prevented through better control of fewer than 10 risk factors, including poor diet, infrequent exercise, the use of tobacco and drugs, and the abuse of alcohol.

Table 13.2 shows the five most common causes of death in the United
States and risk factors associated with each cause. The cause that one is most
likely to associate immediately with human factors research is the one labeled
"Accidental injuries." Occupational hazards, in particular, have long been a
concern of the human factors community. Designing equipment and workspaces that are not only convenient but safe has been a major objective of the
profession and is likely to continue to be. Human factors researchers also have
something to offer with respect to the goal of mitigating some of the other
causes. Exercise physiology, the design of educational materials, and the
design of workspaces and procedures that provide sedentary workers with at
least the minimum of exercise that is consistent with good health are
examples of areas in which there are opportunities for human factors
involvement.

The major causes of accidental death in the United States are shown in
Table 13.3. As has already been noted, motor vehicle accidents account for
as many deaths annually as all other types of accident combined. The overall
decline in accidental death rate between 1970 and 1987 is unlikely to be a
consequence of the two years picked for comparison (these years were picked
only because they were the earliest and latest years reported in the source
cited), because the decline in almost all cases was gradual and regular over the

TABLE 13.2 The Five Leading Causes of Death in the United States and
Their Associated Risk Factors (from McGinnis, 1988–1989, Table 3; Source:
National Center for Health Statistics, 1987)

Cause of Death	Risk Factors
Cardiovascular disease	Tobacco use
	Elevated serum cholesterol
	High blood pressure
	Obesity
	Diabetes
	Sedentary lifestyle
Cancer	Tobacco use
	Improper diet
	Alcohol
	Occupational/environmental exposures
Cerebrovascular disease	High blood pressure
	Tobacco use
	Elevated serum cholesterol
Accidental injuries	Safety belt noncompliance
	Alcohol/substance abuse
	Reckless driving
	Occupational hazards
	Stress/fatigue
Chronic lung disease	Tobacco use
	Occupational/environmental exposures

TABLE 13.3 Number of Accidental Deaths per 100,000 population in the United States from Leading Causes in 1970 and 1987 (adapted from Hoffman, 1989)

	Year	
Cause	*1970*	*1987*
Motor vehicles	26.8	20.0
Falls	8.3	4.6
Drowning	3.9	2.2
Fire, burns	3.3	2.0
Poison (solid, liquid, gas)	2.6	2.2
Ingestion of food, object	1.4	1.3
Firearms	1.2	0.6
All other causes		6.9

intervening years. Everyday life in the United States has gotten safer since 1970 (at least as far as fatal accidents are concerned); people have become more careful, accident treatment has become more effective, or we have had a period of increasingly good luck. Whatever the cause or causes, the effect is a happy one. This is not to suggest, however, that accidental deaths should be no cause for concern: In spite of the favorable trend over the last couple of decades, the approximately 100,000 people who lost their lives accidentally in the United States in 1987, are too many, especially in view of the fact that many, if not most, accidents are preventable.

The psychology of accident prevention is very strange. There is some evidence, for example, that we spend about 1,000 times as much to prevent an accident that would claim 10 lives as we do to prevent one that would take a single life (O'Hare, in press). The amount of publicity given to accidents is not a simple multiplicative function of the number of people involved; a highway accident with 10 fatalities is much more than 10 times as likely to make front-page news than a highway accident that claims 1 life. Perhaps that is because we have become so accepting of the fact that over 100 people a day lose their lives on American highways that an accident that takes a life or two is not—except in a small town—really news. It is also true that the extent of news coverage that an accident receives is sometimes determined by variables that are independent of the number of victims involved. I have no data on the question, but am quite certain that a perusal of newspapers and other print media published over the several weeks immediately following the space shuttle *Challenger* disaster in 1986 would reveal that the coverage given to this event was many times greater than that given to several other accidents at about the same time that claimed as many or more lives.

The most common accidents suffered by elderly people are falls in bathtubs (Sterns, Barrett, & Alexander, 1985); approximately 30% to 50% of people 65 years of age and older who do not live in institutions fall during any given year. Falls account for more than half of all injuries to people over 75 and are the most common cause of accidental death for this age group (Czaja, 1990; Tideiksar & Fletcher, 1989). Many of the falls experienced by elderly people are preventable by relatively straightforward means. Better lighting, especially on stairs, is an obvious measure, because many elderly people have failing eyesight, especially in dim light, and the most serious falls often occur in stairwells. The installation of grab bars around toilets and bathtubs is another effective measure. Avoidance of appliances or furniture with sharp edges or corners is a commonsense dictum, and the careful placement of furniture can make it easier for people who are unsteady on their feet to move around within a room. Providing better information to elderly people who have mobility difficulties and to people who care for them regarding the causes of falls and preventive measures that can be taken could also help alleviate this problem.

Often, measures that are effective in preventing accidents and disease are known, but are not adopted voluntarily. We know, for example, that smoking increases the risk of cancer, yet people smoke. We know that ingesting alcohol before driving increases the risk of motor vehicle accidents, yet people drink and drive. We know that certain foods increase the risk of cardiovascular disease by increasing blood pressure or raising blood choles- terol levels, but many people make no effort to avoid these foods. How to motivate people to adopt simple preventive measures in the interest of protecting their health and quality of life is a question that deserves considerable attention.

Reasons to be interested in this question extend beyond a humanitarian wish to minimize suffering from unnecessary disease and accidents. There is also the fact that preventable diseases and accidents represent enormous costs to society in medical care, property damage, and lost productivity. Effective prevention is something from which we all gain. Unfortunately, both as individuals and as a society, we seem to be unaware of the magnitude of the stakes in this game. We need to understand better why measures that are known to be effective in preventing accidents or in mitigating their effects are often not acceptable to the general public. Why, for example, did 26 states rescind laws requiring the use of helmets by motorcycle riders in the 1970s at the cost of a 40% increase in motorcycle fatalities (Lescohier et al., 1990)?

Preventive measures have a major disadvantage as compared with therapies and treatments with respect to their ability to attract advocates and champi- ons. The problem is that the beneficiaries of effective preventive measures typically do not know who they are. Individuals whose lives have been saved by a sophisticated medical procedure or a wonder drug are keenly aware of their good fortune. People who live in good health but would have been dead had the smallpox vaccine never been invented have no inkling of their debt to this bit of preventive medicine. Ironically, it is perhaps those of us who have benefited most from effective preventive measures, those who have the most to be thankful for, who are least likely to be aware of the fact.

Studies aimed at providing a better understanding of factors and situations that cause, or increase the probability of, accidents continue to be a high priority for research. Another continuing challenge is that of effectively communicating causes of accidents and preventive measures that individuals can take. Although, as already noted, people often fail to take preventive measures when they are aware, or at least have been informed, of them and their effectiveness, it is undoubtedly also the case that people sometimes fail to take such measures because they are unaware of how to do so.

Preventing accidents through design has long been a goal of human factors research. Improving the safety of automobiles, household appliances, children's toys, and other consumer products has been a focus of considerable effort and should continue to be. A surprising aspect of some of this work is how simple the solutions to long standing problems sometimes appear to be once they have been found. For a long time, several infants died every year as a consequence of getting their necks stuck between crib slats that were spaced far enough apart to permit their bodies to fit between them, but too close together to permit their heads to pass through. The solution to this problem—requiring manufacturers to limit the interslat distance to a small enough amount that infants could not get their bodies through it—was simple in the extreme. Solutions to other problems have been equally obvious, once they have been pointed out by somebody. What is surprising is how long it sometimes takes for someone to see the obvious fix for the first time.

POPULATION GROWTH AND INCREASING URBANIZATION

I debated with myself whether to include a discussion of population growth in this book. It is a topic that, at least on the surface, has little to do with human factors. To my knowledge, human factors researchers have not focused on the issue, and it is not clear that the field has anything unique, or even particularly useful, to bring to it. On the other hand, in my view, any effort to look ahead that does not take population growth into account is ignoring what could be the single most significant determinant of the quality of life on this planet in the years ahead. Population growth acts as an amplifier on all the major problems, local and global, considered in this book.

This is very obvious, for example, with respect to the problems of energy and the environment. Energy is used to satisfy human needs and desires, and the demands for it can be expected to scale with the size of the population, although not in a simple linear way. The major cause of environmental change for some time has been human activity. The impact of that activity, also, should rise with population increases, again, probably in a nonlinear fashion, because relatively small increases in certain variables, such as greenhouse gases, can have very large effects. More generally, behavior that poses no threat to the environment when indulged in by a few million people, or a few hundred million, can become threatening indeed when practiced by a few billion. Whether human factors, as a discipline, has anything to contribute to a better understanding of the implications of a continuation of current trends in population growth or to a modification of those trends is a question that is worth raising. I believe the answer is not clear, at this point, but even if it turns out to be "no," a greater awareness of the problem is essential to a comprehension of the many other problems for which it is a forcing function.

The world population, currently at about 5 billion, has doubled since the middle of the 20th century and approximately quadrupled during the last 100 years. The rate of increase seems to be slowing somewhat in most of the world, and estimates of how long the next doubling will take vary considerably, but demographers typically see the 10 billion mark being reached perhaps before the middle of the 21st century and by the end of it at the latest. The Bureau of the Census (1987, Table 1435) projects that the world population will reach 6 billion by 2000.

At least since the time of Thomas Malthus, it has been recognized that population tends to increase exponentially; barring events that would cause major perturbations in the process, it tends to increase by a fixed percentage of its current size every year. In this respect, population growth is analogous to the growth of the value of an investment on which interest is compounded at a fixed rate. Any such process has a constant doubling time. Money invested at 7.5% interest, for example, will double in value approximately every 10 years. The simple fixed-interest model does not fit human population growth precisely, however, because the rate of increase has not been constant over time. At the present it is on the order of 2.0% per year, which gives a doubling time of approximately 35 years.

The fact that the world population is increasing exponentially has received very wide notice, not only in scientific publications but in the press and other popular media as well. Despite this, it is not at all clear that the nature of the process and its implications are very well understood, either by the general public or by the policymakers of the world, but the process has the potential to change life on this planet in unprecedented ways. One reason that we are not more anxious about the possibilities is the fact that processes that grow exponentially can produce trouble without much warning. Consider a pond that can support no more than 1 million fish; suppose it is stocked with 2 fish, and the fish population doubles every year. In 20 years, the pond will be saturated, but in the 19th year everything will seem fine, because the pond will be only half full.

Once, in a moment of philosophical musing, I wrote in my notebook: "Almost everyone who ever lived is dead." For some reason I can no longer remember, I thought this to be a profound observation. I had the view that the current generation could be thought of as the tip of an iceberg, the greatest portion of which, representing human beings who had come and gone, was submerged. I now realize that this probably is not a very accurate metaphor. It is more likely that the current generation represents a fairly sizable fraction of all the people who have ever lived. If, from the beginning, the population had doubled with each succeeding generation, the current generation would outnumber all previous generations combined, because any term in a doubling sequence is greater than the sum of all preceding terms. Although the population doubling time currently is not a lot longer than a generation (a generation being taken as the average age of parents when their children are born), it has not always been so; so, the simple doubling-every-generation model is not quite right. Nevertheless, given that the doubling time has been relatively short over the period during which the population has gone from a few hundred million to 5 billion, it would not be surprising to discover that a very sizable percentage of all the people who ever lived are alive today.

No one knows precisely the size of the human population the world could sustain. Estimates vary over a very large range. It seems highly unlikely, however, that a growth rate that yields a doubling every few decades can continue much longer without producing very serious problems. Nevertheless, expectations are that the population will continue to increase during the near-term future at something close to its current rate. The rate of population increase does seem to be slowing slightly in many parts of the world, but the situation varies greatly from country to country. Current birth rates are lowest in countries with the highest levels of economic development and the greatest availability of information about birth control and access to contraceptives. Over the next few decades, most of the increase in population is expected to occur in less developed countries (Keyfitz, 1989), specifically in Asia, Oceania, and Africa and, to a lesser extent, in Latin America; much less growth is anticipated in Europe, the (former) Soviet Union, the United States, and Canada (Friedlander, 1989).

The rate of population growth in the United States is expected to be slower over the next few decades than it has been in the recent past. The rate of increase is expected to differ for different segments of the population, proportional increases being greater for minorities (Hispanic, Black, and Asian residents) than for Whites, and greater for older than for younger age groups (Bureau of the Census, 1990; Kasarda, 1988). Changing immigration patterns are likely to affect the future population of the United States with respect to both ethnic composition and geopolitical distribution. Before the middle of this century, most immigrants came from Europe and many settled in the northeast. During the past two decades, an increasing proportion have come from countries to the south or west of the United States (Mexico, Latin America, and parts of Asia) and have tended to settle in the south and west of the country (Kasarda, 1988).

The aggregate effects of immigration and other factors that contribute to changes in relative population densities have been to foster a disproportionately large growth rate in the south and west, and this trend is expected to continue. Between 1980 and 1988, about 88% of the U.S. population increase of 19.3 million occurred in the south and west (Bureau of Census 1990, Table 27); according to the latest projections of the Bureau (1990, Table 29), about 96% of the country's population growth through 2010 (estimated at about 34.3 million) will take place in the same regions. More than half of this growth (about 18.6 million) is expected to occur in three states: California, Florida, and Texas. The Bureau projects an increase of about 2.1 million in the population of the northeast, a decrease of about 0.7 million in the midwest and increases of about 18.7 and 14.2 million in the south and the west, respectively.

If recent trends continue, the population in most countries will not only grow in size but will become increasingly concentrated in and around major cities (Vining, 1985); one prediction is for the world's population to go from about 17% urban in 1950 to over 50% urban by 2020 (Keyfitz, 1989). A trend toward increasing urbanization has also been seen in developed countries in the past: in the United States, for example, the portion of the population that lives in urban areas went from 40% in 1900 to 74% in 1980. As of 1980, 61% lived in areas that occupied less than 15% of the national domain (Lowry, 1988).

The rapidly expanding world population and the increasing density of people in urban areas represent challenges to any profession that concerns itself with the question of how to make efficient use of limited space. Crowding is not an unfamiliar condition, but it is one that we have not yet learned to deal with very effectively, and the fraction of the world population that lives and works under crowded conditions is almost certain to increase substantially over the foreseeable future. I do not mean to suggest that figuring out how to make people comfortable in ever smaller spaces is a long-term solution to the problem of population growth. Sooner or later the growth itself must be slowed, but even now, a better understanding of how to deal with crowding would be useful in many regions of the world and, under any plausible scenario for the future, the population is unlikely to be stabilized before it is considerably larger than it is at the present.

Considerable research has been done on the effects of crowding on human well-being. The results of this research are difficult to summarize succinctly because crowding has been measured in a variety of ways—persons per room in housing units, housing units per unit land area, persons per unit land area—and a variety of techniques have been used by investigators from several disciplines. On balance, the evidence seems to be that crowding of various types can have a deleterious effect on people, primarily by inducing psychological stress, which in turn can adversely affect general health (Cox et al., 1982; Epstein, 1981; Paulus, 1980). Apparently, the feeling of being crowded is not a simple linear function of density, however, but is determined in part by concomitant social and situational variables, such as the possibility for privacy and the degree of social stimulation (Pennebaker & Brittingham, 1982).

If the world's population does continue to increase at anything like its current rate, the problem will become ever more apparent through such effects as overcrowding in the cities, the scarcity of developable land in the suburbs, congested highways, and decreasing accessibility to recreational lands. In underdeveloped countries the effects of drought and other natural disasters will be more devastating as the degree of overpopulation increases. In spite of our ability to produce enough food to feed the world's current population and of the existence of food surpluses in parts of the world, hunger is already a very serious problem, largely because of distribution difficulties that are expected to continue (H. Kahn et al., 1976). Mahler (1980) has estimated that about one fourth of the people in the underdeveloped world have a food intake below the critical minimum level. Citing a World Bank estimate, Scrimshaw and Taylor (1980) put the number of undernourished people in the world at 1.1 billion. This problem is very likely to become increasingly severe as the world population continues to rise.

It is, of course, possible in principle for the population to decline, and such a decline could be forced by natural feedback mechanisms if the population gets to a size that cannot be sustained. It could also be reduced as a consequence of a deliberate slowing of the birthrate. There are social, psychological, and philosophical impediments to achieving negative population growth in the foreseeable future, even if it became clear that the size of the population had grown beyond the point of sustainability. Such considerations aside, achieving a zero growth rate, or even a positive rate much smaller than the current one, may pose some problems for world and regional economies that we do not yet know how to solve.

At least since the Industrial Revolution, the more developed countries of the world have looked to a continuously increasing population to provide demand for the production of ever more goods and services. Gross national product has been the primary indicator of economic well-being, and failure to accomplish some growth in excess of the rate of inflation from year to year has been taken as a sign of economic stagnation or worse. If businesses are to expand—and the prevailing assumption is that they must expand or fold—

they need a growing customer base; if government is to provide more and better services, it needs a growing tax base; and so on. The idea of a robust economy that is not growing is foreign to our thinking. Achieving such an economy, were that deemed desirable, would require some new perspectives and the solution of some problems that have not yet been articulated, including some that probably have not even been anticipated.

Does human factors, as a discipline, have anything to offer with respect to the objective of slowing the rate of population growth worldwide? Perhaps not, but then again, how much thought has been given to the question? Certainly not as much as it deserves. This is not the kind of problem on which human factors specialists have been trained to work, but it is a "human factors" problem in the most literal sense of the term. We have, I believe, an obligation to pay it some mind. I do not mean to argue that population growth should be of greater concern to human factors researchers than to scientists in other fields, but I do mean to suggest that it is rightfully as much of a concern for this field as for any other, and that it is a serious concern indeed.

Given the evidence that growth rates are lower in areas of the world where information regarding birth control is most readily available, one might argue that any professional group that concerns itself with communication and the technology of information representation and distribution has the potential to have an influence, at least indirectly, through the improvement of communication systems and techniques. There are also the general challenges of understanding more completely and communicating better the seriousness of the problem; these, too, are tasks that might represent opportunities for human factors researchers to make a contribution.

Why are people, in general, not more concerned about the problem of unconstrained population growth? Why are policymakers so complacent about this threat while they are able to get extremely concerned about problems with far less potentially devastating consequences? Does the apparent lack of concern stem from a lack of understanding of the nature of the problem, or is it simply a reflection of a general inability or unwillingness to attach a high priority to problems that will affect primarily generations other than our own? There is much suggestive evidence that, on the whole, we are willing to pay for our own comfort with our progeny's capital. We need to try to get better answers to these and similar questions than we now have.

Whether or not human factors researchers have anything to offer to the problem of controlling the world's population, it is quite clear that the fact of a rapidly growing population has implications for human factors research. Many of the problems on which human factors researchers have focused in the past will be exacerbated by having much larger numbers of people occupying the same space and tapping the same limited resources. Transportation safety, the effects of stress on human performance and well-being, the effects of human behavior on the environment, the need for effective complex-system modeling techniques and decision aids, as well as for better approaches to negotiation and conflict resolution are a few examples of problem areas that become more important and more difficult as the population grows, the density of people in already crowded areas of the world increases, and the competition for limited resources intensifies.

EQUITY AND INTERNATIONAL COLLABORATION

As I warned in the introductory chapter, I have focused throughout this book primarily, though not exclusively, on the United States and on problems that I believe will be of national concern in the future. I do not want to close, however, without stating explicitly that the major problems of the future—at least those that have been discussed here—are world problems and cannot be solved by any community, country, or region in isolation, without a sensitivity to the needs of the rest of the world and to the implications of local, national, or regional activity with respect those needs. Regarding the quality of life, the objective must be, as I see it, to improve it in every part of the globe, not only for altruistic reasons, but for quite self-serving ones as well. The world is much smaller than it used to be, and it is shrinking every day; enormous differences in the standards of living in different parts of the world are both troublesome on moral grounds and inherently destabilizing.

During the time it has taken the world population to increase by a little more than a factor of 3, the world economy has expanded by a factor of 20, the consumption of fossil fuels by a factor of 30, and industrial production by a factor of 50 (MacNeill, 1989). Economic and industrial expansion has not occurred uniformly around the globe, however, and has not outdistanced population expansion in all regions. In many of the less developed countries, the per capita income has decreased in recent years, and the gap between the poorest countries and the rest of the world has widened (Dadzie, 1980). The difference between the amounts of money spent on health care in affluent countries and in developing countries, for example, has been growing (Mahler, 1980). Even among industrialized countries, standards of living vary enormously. Americans have 55 square meters of living space per person, on the average; Western Europeans have 30; the Soviets and Eastern Europeans 20 or less, and the Chinese 6.5 (Chandler et al., 1990).

N. Rosenberg and Birdzell (1990) note that the growth of scientific knowledge and the rise of technology over the past 250 years have had the dual effect of dramatically increasing the material well-being of people in Western Europe and the United States and greatly magnifying economic inequality among nations. In the mid-1700s, they point out, the material welfare of an average Western European was not very different from that of someone in China or even in ancient Greece or Rome. Few people anywhere had more than the minimum required to sustain life. Beginning in the latter part of the 19th century the situation began to change:

> Between the mid-1700's and the present, per capita income increased ten-fold. The population of Europe grew five-fold and that of the U.S. eighty-fold. Infant mortality declined drastically and the average life span doubled. Famine was banished, and plagues disappeared. Food production, which in some countries had occupied as much as 90% of the working population, eventually came to occupy less than 5%. Nineteenth Century urbanization marched in step with developing technologies for improving sanitation, construction, communication, power distribution and other services. Urbanization and rising incomes led to changes in health and living standards, work patterns, values and other aspects of personal, family and community life. (p. 42)

These changes—usually associated, or equated, with the Industrial Revolution—were realized primarily in Western nations and had relatively little beneficial effect on large parts of the non-Western world. Some would argue that because of the advantages these changes gave to the countries that realized them, they made the rest of the world more vulnerable to exploitation, which indeed much of it suffered.

Whether the gap between have and have-not countries will close or increase in the future remains to be seen. Its failure to close would pose a threat to global stability and be a serious humanitarian concern. On the other hand, rapid and unconstrained industrialization and economic development of the less developed countries has the potential to exacerbate environmental problems out of proportion to anticipated population increases. To the extent that these countries not only increase greatly in population, but adopt the lifestyles of the developed countries, this will almost certainly be the case. Increasing the standard of living and reducing mass poverty worldwide are goals for the world community that are universally endorsed. Even partial realization of these goals, especially in view of the world's growing population, will require greatly increased economic activity around the globe, especially in developing countries. Managing that activity so that it does not impose an intolerably greater burden on the environment will be a major global challenge (MacNeill, 1989).

The world economy is changing rapidly at present. Further changes are likely to be spurred by the economic integration of Western Europe, by the further development of Third World countries as significant economic forces, and by the continuing movement of the Soviet Union, what, until recently, was Eastern Europe and perhaps China, in the direction of free enterprise, as well as by unanticipatable developments in various parts of the world. Possible realignments of international relations could have profound effects for world economics. Woodruff (1987) notes, for example, the major change that could result if Japan and China should become economic allies.

Dramatic changes—some of them totally unexpected—have been occurring in some parts of the world, notably the Soviet bloc and the Middle East, at a truly dizzying pace. Where these changes will lead is impossible to say, but clearly their effects, both military and economic, could be profound. Annual worldwide military expenditures reached $1 trillion in 1987 (Bureau of the Census, 1990, Table 542). One cannot help but wonder what it could mean to the world if, as a consequence of improved international relations, much of this money became available for the production of consumer goods and services, the development of new energy sources, the preservation of the environment, the modernization of education, the improvement of world health, and other objectives that are consistent with the general goal of raising the standard of living globally.

The economies of individual nations are becoming increasingly interdependent as the world economy becomes more integrated. International trade, currency exchange rates, and global economics more generally are expected to grow in importance as determinants of national practices, policies, and standards of living. A growing sensitivity to the finiteness of the earth's resources, of the interrelatedness of national problems, of the interdependence of national economies, and of the interacting effects of national policies is reflected in such currently fashionable concepts as "spaceship earth" and "the global village." The connectedness and interdependence of national economies is also reflected in the ease with which technology is transferred between countries and, conversely, in the difficulty that a country that develops a technological innovation has in maintaining a competitive edge in commercially exploiting it for any appreciable time.

World trade in goods and services has been rising faster than world GDP. From 1973 to 1984, GDP in the industrialized world grew at an average of 2.4% per year, whereas, during the same time, merchandise trade grew at a rate of 4.2% per year (Johnston & Packer, 1987). The increasing importance of world trade means that nations cannot expect indefinitely sustained growth in their local economies unless the world economy also grows. From this vantage point, the economic challenge for an individual nation is not so much to improve its trade balance or its competitiveness by capturing economic activity from other nations, but—because its economic success depends in part on that of other nations—to increase its own productivity while stimulating growth of the world economy at the same time.

An interesting and unanticipated consequence of the world-wide connectivity of computer-based communications systems has been the decoupling of the flow of money between countries from the international trade in goods and services: "Whereas world trade in goods and services amounts to only between $3 and $4 trillion per year, the financial transactions handled by just one intermediary, the Clearing House for International Payments, totalled $105 trillion in 1986, and transactions in early 1987 were proceeding at a rate of $200 trillion per year" (Quinn et al., 1987, p. 57). Instead of following goods, monies have been flowing toward high interest rates or stable economic and political conditions, thus making the value of a nation's currency less dependent on its position in international trade than on monetary policies, banking decisions, and events that affect foreign investment.

In short, the world economy of today differs from that of even decades ago in the fact that it truly is a *world* economy. International trade, multinational corporations, ease of international travel, worldwide communication networks, and numerous other factors ensure linkages and interdependencies among the economies of individual nations. More and more, producers must design products with a multinational consumer market in mind. Service providers, especially in urban areas, must be prepared to deal with people of many nations and many languages.

World economics is in a state of greater flux at present than at any other time in recent history. Long-standing ideologies and political systems are undergoing radical and, in many cases, completely unanticipated changes. International alliances, both political and economic, are shifting in dramatic ways. Controlled economies are giving way to market economies in major parts of the world. Pundits are offering up a panoply of scenarios with respect to how the world economy will continue to change in the near-term future. Predictions range from worldwide economic depression to an era of unprecedented global prosperity.

Do human factors researchers have any useful role to play in helping to shape the world economy of the future? To increase the chances that the actual scenario will resemble one of the more desirable possibilities? To help improve the quality of life everywhere? I believe the answer to that question is "yes." Ensuring that products are designed in such a way that they are safe and convenient to use has been a primary concern of human factors specialists from the beginning. Ensuring that products are designed for a world market rather than for, say, an American or European one is a natural extension of that interest and one that is particularly timely now.

One challenge that has been recognized for some time is for the development of international standards for such things as highway signs, danger and caution signals, and symbology that is used for important culture-free messages more generally. Considerable work has been done by the International Standards Organization on the problem of standardizing communication protocols so as to facilitate the transmission of information electronically over international computer networks. Much more work on standardization problems will undoubtedly be needed in the coming decades, and many of the issues that will have to be addressed should be of interest to the human factors community. Such work will be stimulated within Europe by the process that is already underway to integrate the European economic community within the 1990s. Inclusion of developing countries in efforts to define and promulgate standards will be an important aspect of helping to close the economic gap between the more developed and the less developed parts of the world.

Among the more pressing needs of many developing countries is the building and operating of an infrastructure that really works—communication systems and transportation systems, for example, that operate efficiently and on which one can depend. The field of human factors should have something important to offer to this problem, because rapid development of an infrastructure can mean embedding modern technology in a cultural context that is not well prepared to assimilate it. It is naive to assume that what works in an industrialized country that has evolved its technological infrastructure over several centuries will work equally well when suddenly introduced in a society that has been largely untouched by the Industrial Revolution. What is needed to strengthen the technological infrastructure of Third World countries is not a wholesale application of those systems and techniques that have proved to be effective in the industrialized world, but rather a better understanding of the situations in these countries, which could lead to a discovery of what is most likely to work there. Human factors know-how should be able to benefit Third World countries as they attempt to adapt various aspects of modern technology to their own cultural context and socioeconomic situations, but the human factors approaches must themselves be adapted to the contexts in which they are to be applied. Associations and societies representing human factors professions could play important roles in helping make available to Third World countries human factors resources that might be of use to them, and in facilitating collaborative efforts with local experts to understand and address local needs.

14

Epilogue

There can be no doubt that as we approach the threshold of the 21st century, the world faces some extremely serious problems. Many of these problems are the consequences, at least in part, of technology or the uses thereof. It is easy to generate a depressing list of examples:

- The proliferation of nightmarishly destructive power and the continuing threat of its intentional or accidental use.
- The growing pollution of the environment, disruption of ecological balances, and possible modification of world climate.
- The wasteful depletion of natural resources.
- The failure to deal effectively with regional problems of famine and hunger in spite of the overproduction of food at a global level.
- The possibility of increasing frequency and potency of terrorist activities by extremist groups around the world.
- Social unrest and instability generally, symptomized by crime, substance abuse, and malaise.
- The Orwellian specter of extensive electronic surveillance and other manipulative applications of technology.
- The possibility—with the shrinking of the globe as a consequence of the electronic and economic integration of its parts—of new opportunities for the misapplications of science and technology to the causing of worldwide mischief.
- The possibility that people who are disadvantaged in today's world will become even more so as the educational requirements of at least

the more influential, power-wielding and rewarding positions in both private and public sectors increase.

- The threat of overpopulation and its attendant complications.

If it is easy to list existing or potential problems, many of which are technology based at least in part, it is equally easy to list benefits that have resulted, or are likely to result, from technological advances as well:

- The excellent prospects of increasing life expectancy significantly in the future as the aging process becomes better understood. (People in developed countries are healthier and live longer now than ever before; in the United States, for example, the average life expectancy in 1985 was 74.7 years, as compared to 54.1 years in 1920; Bureau of the Census, 1987, Table 105).
- The likelihood of major breakthroughs in the understanding, treatment, and prevention of specific major diseases, such as cancer, diabetes, and schizophrenia.
- The possibility, as a consequence of further understanding of superconductivity and progress in the effective harnessing of alternative energy sources, of providing much cleaner and cheaper power in quantities that surpass the world's needs.
- The possibility of making information of all sorts much more readily accessible to people who wish to have it than it has been heretofore.
- The possibility of the development of tools, especially through the application of computer technology, that will greatly facilitate teaching and learning.
- The prospects, through emerging technologies, of creating new opportunities for work and independent living for people with various types of disabilities, and of expanding the life space of people with limited mobility.
- The prospects of greatly increased leisure time and many degrees of freedom regarding how to spend it.
- The prospects of breaking the chain that has confined us to this globe.
- The identification of new frontiers for exploration and discovery, resulting from the scientific advances in every domain of inquiry that are greatly increasing our understanding of the universe in which we live.

The most significant need for the future, in my view, is for a willingness to adopt a more reflective attitude toward the world and our places in it than we have tended to take in the past. We must be willing to rethink some of the tacit assumptions that we tend to make. We must be more explicit with ourselves regarding our personal values and more prepared to examine our deepest beliefs and live by those we find compelling. We cannot afford to let our attitudes and values be shaped by the image makers and hucksters of the world. We must learn to think for ourselves and to work for goals that reflection has convinced us are worthwhile and not only for things that someone else has persuaded us we ought to want. The world is rapidly becoming too small to permit us to survive unless we find a way to become more rational and better able to contain our destructive instincts, even as we find ever more powerful ways to express them.

Our adaptiveness as a species is at once a blessing and a curse. It is a blessing because without it we would not have survived. It is a curse to the degree that we adopt the attitude that whatever the future holds we will adapt to that, as we have always adapted in the past. As Botkin, Elmandjar, and Malitza (1979) have argued, we can no longer afford to think primarily in terms of adaptation. What has worked in the past may not work in the future. The changes in the world that we accommodated in the past were largely beyond our control; today, the most significant changes—those with the most profound implications for us—result largely from our own activities. Of course, we must try to adapt to things we cannot change, but we must also recognize that there is much we can change; only the past is inevitable. As long as the future is still the future, we have not only the opportunity, but the responsibility, to try to mold it in accordance with the values that we wish to preserve.

In the context of trying to look ahead, how should we think of human factors as a discipline? What should we see as its primary goals? Is it a service discipline whose raison d'etre is to help manufacturers do a better job at producing whatever they want to produce? Is its primary goal the development of better interfaces between people and machines, no matter what the ultimate purpose of the machines may be? Would it be naively idealistic to think of its highest purpose as that of helping to make the world a better place in which to live?

One of the assumptions that motivated the writing of this book is that an attempt, from time to time, to take a rather broad view of the kinds of problems that are likely to be faced by individuals, communities, nations, and the world in the near-term future should help provide a useful frame of reference in which to view the various objectives toward which one might work. In some of the problem areas discussed in this book, human factors work may have its greatest impact indirectly. When, for example, one enhances the usability of a tool that is used by people working on problems of, say, energy conservation, one has contributed, albeit indirectly, to the solution of the problem to which that tool is applied. By improving display designs, data management techniques, computer–user communication methods, tradeoff analyses, and decision-making techniques, one accomplishes something that could facilitate progress in many of these problem areas. However, reflection on what the problems are likely to be may reveal some that are not receiving the attention they deserve. It could also reveal that some planned project might be made more directly responsive to those problems if modified in certain ways, or it could convince one that the work that one plans to do is precisely what needs to be done. It is hard to see how a serious attempt to understand what the problems are, or are likely to be, can hurt.

Many of the challenges and opportunities for human factors research for the future are global, or at least international, in scope. Because of this, there should be many opportunities for collaboration among human factors researchers from countries around the world. Such collaboration should be sought and encouraged for at least two reasons. First, international and global problems will be more easily solved with international or global efforts. Second, greater communication among scientists around the world is not only in the interest of forging and maintaining world peace, it provides opportunities to identify common interests and to develop win–win approaches to problems of mutual concern.

The National Academy of Sciences has been supporting the idea of greater contact between U.S. and Eastern European scientists for some time, and has had agreements of scientific cooperation with Poland, Czechoslovakia, Romania, Yugoslavia, Hungary, Bulgaria, and East Germany for at least a decade, and considerably longer in some cases (McIntosh, 1990). How fitting an expansion of the connotation of *human factors* it would be to have the human factors community around the world take a leadership role in this regard, working toward the improvement of international communication and actively promoting collaborative efforts toward shared goals.

The industrialized world today provides us with creature comforts beyond the wildest dreams of our forebears, even those only two or three generations removed. Not only are there the comforts and conveniences that accrue from modern plumbing, heating, electrical appliances, communication facilities, and modes of transportation; the accumulated knowledge of humankind is available to the average person, as is its music and art. It is sobering to reflect on how much more immediately accessible the world's music is to today's ordinary consumer than it was to Mozart, Beethoven, or Brahms, and we only dimly see the countless ways in which technology could be used to make life longer, healthier, and fuller in the 21st century than it is today. It is also the case, however, that along with many benefits, technology and the way it has been used have brought problems of a type and magnitude not known to the world in preindustrial days. We are just beginning to understand the nature of some of these problems and the threat to the future that they pose if effective corrective action is not taken. It is probably pointless to debate whether, on balance, the pros outweigh the cons. I personally believe they do, and by a considerable margin. Others may see it differently. No matter; we have to take the world as we find it. How we find it is not our responsibility, but how we leave it is.

References

American Association of Colleges for Teacher Education (1989, February). [Resolution: Critical thinking].

Aaronson, D., & Gabias, P. (1987). Computer use by the visually impaired. *Behavior Research Methods, Instruments, and Computers, 19*, 275–282.

Abelson, P. H. (1987) Energy futures. *American Scientist, 75*, 584–593.

Abelson, P. H. (1990). Uncertainties about global warming [Editorial]. *Science, 247*, 1529.

Abrett, G., & Burstein, M. (1987). The KREME knowledge editing environment. *International Journal of Man–Machine Studies, 27*, 103–126.

Ackerman, R. K. (1990, December). Defense machinery gears up to fight environmental threat. *Signal*, pp. 35–38.

Adelman, M. A. (1987, September). *Are we heading towards another energy crisis?* Paper presented to the Oil Policy Seminar of the Petroleum Industry Research Foundation. Washington, DC.

Adult Performance Level Project. (1977). *Final report: The Adult Performance Level Study, University of Texas at Austin.* Austin, TX: Author.

AFL-CIO (1983). *The future of work: A report of the AFL-CIO committee on the evolution of work.* Washington, DC: Author.

Agnew, H. M. (1981). Gas cooled nuclear power reactors. *Scientific American, 244*(6), 55–63.

Ahead: A nation of illiterates? (1982). *U.S. News & World Report.*

Aharonowitz, Y., & Cohen, G. (1981). The microbiological production of pharmaceuticals. *Scientific American, 245*(3), 140–153.

Airport Network Study Panel. (1988). *Future development of the U.S. airport network: Preliminary report and recommended study plan.* Washington, DC: National Research Council, Transportation Research Board.

Alavi, M. (1984). An assessment of the prototyping approach to information systems development. *Communications of the ACM, 27*(6), 556–563.

Allgier, E. (1965). Accident involvements of senior drivers. *Traffic Digest and Review, 13*, 17–19.

Alvo, R. (1986, September). Lost loons of the northern lakes. *Natural History, 95*, 58–65.

American Association for the Advancement of Science. (1989). *Project 2061: Science for all Americans.* Washington, DC: Author.

American Public Transit Association. (1987). *1988 transit fact book.* Washington, DC: Author.

Ammons, J. C., Govindaraj, T., & Mitchell, C. M. (1988). Decision models for aiding flexible manufacturing system scheduling and control. *IEEE Transactions on Systems, Man and Cybernetics, SMC-18,* 744–756.

Apostolakis, G. (1990). The concept of probability in safety assessments of technological systems. *Science, 250,* 1359–1364.

Applebaum, E. (1985). *Technology and the redesign of work in the insurance industry.* Stanford, CA: Stanford University Institute for Research on Educational Finance and Governance.

Asimov, I. (1972). *Asimov's biographical encyclopedia of science and technology.* Garden City, NY: Doubleday.

Argote, L., & Epple, D. (1990). Learning curves in manufacturing. *Science, 247,* 920–924.

Arnold, J. R. (1980). The frontier in space. *American Scientist, 68,* 299–304.

Aronson, S. (1977). Bell's electrical toy: What's the use? The sociology of early telephone usage. In I. deSola Pool (Ed.), *The social impact of the telephone* (pp. 15–39). Cambridge, MA: MIT Press.

Arthur, W. B. (1990). Positive feedbacks in the economy. *Scientific American, 262*(2), 92–99.

Atkinson, R. C. (1990). Supply and demand for scientists and engineers: A national crisis in the making. *Science, 248,* 425–432.

Attewell, P. (1990). *Information technology and the productivity paradox.* Unpublished manuscript.

Attewell, P., & Rule, J. (1984). Computing and organizations: What we know and what we don't know. *Communications of the ACM, 27,* 1184–1192.

Ausubel, J. H. (1989). Regularities in technological development: An environmental view. In J. H. Ausubel & H. E. Sladovich (Eds.), p. 2 *Technology and environment* (pp. 70–91). Washington, DC: National Academy Press.

Averill, J. R. (1973). Personal control over aversive stimuli and its relationship to stress. *Psychological Bulletin, 80,* 286–303.

Ayres, R. U. (1989). Industrial metabolism. In J. H. Ausubel & H. E. Sladovich (Eds.), *Technology and environment* (pp. 23–69). Washington, DC: National Academy Press.

Bachrach, A. J. (1982). The human and extreme environments. In A. Baum & J. E. Singer (Eds.), *Advances in environmental psychology: Vol. 4. Environment and health* (pp. 210–236). Hillsdale, NJ: Lawrence Erlbaum Associates.

Baecker, R. M., & Buxton, W. A. S. (Eds.). (1987). *Readings in human–computer interaction: A multidisciplinary approach.* Los Altos, CA: Morgan Kaufman.

Baes, C. F., Goeller, H. E., Olson, J. S., & Rotty, R. M. (1977). Carbon dioxide and climate: The uncontrolled experiment. *American Scientist, 65,* 310–320.

Bailey, T. (1990). The changing occupational structure. *National Center on Education and Employment, 7,* 1–4.

Bair, J. H. (1987). User needs for office systems solutions. In R. E. Kraut (Ed.), *Technology and the transformation of white-collar work* (pp. 177–194). Hillsdale, NJ: Lawrence Erlbaum Associates.

Balzhiser, R. E. (1989). Meeting the near-term challenge for power plants. In J. H. Ausubel & H. E. Sladovich (Eds.), *Technology and environment* (pp. 95–113). Washington, DC: National Academy Press.

Balzhiser, R. E., & Yaeger, K. E. (1987). Coal-fired power plants for the future. *Scientific American, 257*(3), 100–107.

Banks, P. M., & Ride, S. K. (1989). Soviets in space. *Scientific American, 260*(2), 32–40.

Baran, P. (1964). On distributed communication networks. *IEEE Transactions on Communications Systems, CS-12,* 1–9.

Barker, P. G., & Najah, M. (1985). Pictorial interfaces to databases. *International Journal of Man–Machine Studies, 23,* 423–442.

Barker, P. G., Najah, M., & Manji, K. A. (1987). Pictorial communication with computers. *International Journal of Man–Machine Studies, 27,* 315–366.

Barker, P. G., & Manji, K. A. (1989). Pictorial dialogue methods. *Internatinoal Journal of Man-Machine Studies, 31,* 323–347.

Barnes, R. D. (1980). Perceived freedom and control and the built environment. In J. Harvey (Ed.), *Cognition, social behavior, and the designed environment.* Hillsdale, NJ: Lawrence Erlbaum Associates.

Barnola, J. M., Raynaud, D., Korotkevich, W. S., & Lorius, C. (1987). Vostik ice core provides 160,000 year record of atmospheric CO_2. *Nature, 329,* 408–414.

Baron, J., & Sternberg, R. J. (Eds.). (1986). *Teaching thinking skills: Theory and practice.* New York: W. H. Freeman.

Baron, S. (1988). Pilot control. In E. L. Wiener & D. C. Nagel (Eds.), *Human factors in aviation* (pp. 347–385). New York: Academic Press.

Baron, S., & Levison, W. H. (1980, October). The optimal control model: Status and future directions. *Proceedings of the International Conference on Cybernetics and Society* (pp. 90–101). New York: Institute of Electrical and Electronics Engineers.

Baron, S., Muralidharan, R., Lancraft, R., & Zacharias, G. (1980). *PROCRU: A model for analyzing crew procedures in approach to landing* (Contract Report No. CR-152397). Moffett Field, CA: NASA-Ames Research Center.

Bartel, A. P., Lichtenberg, F. R., & Vaughan, R. (1989) Technological change, trade, and the need for educated employees: Implications for economic policy. *National Center on Education and Employment NCEE Brief, 5,* 1–4.

Barton, J. K. (1988). What is biotechnology? In M. L. Good (Ed.), *Biotechnology and materials science: Chemistry for the future* (pp. 3–6). Washington, DC: American Chemical Society.

Bassen, H. I. (1986). From problem reporting to technological solutions. *Medical Instrumentation, 20*(1), 17–26.

Bate, R. T. (1988). The quantum-effect device: Tomorrow's transistor? *Scientific American, 258*(3), 96–100.

Bates, M., Meltzer, D., & Shea, S. (1987, May). Designing a practical interface. *AI Expert,* pp. 60–66.

Baum, A., & Singer, J. E. (Eds.). (1981). *Advances in environmental psychology: Vol. 3. Energy in psychological perspective.* Hillsdale, NJ: Lawrence Erlbaum Associates.

Baumol, W. J. (1989). Is there a U.S. productivity crisis? *Science, 243,* 611–615.

Bay, P. N. (1988). Commentary on personal mobility. *A look ahead: Year 2020* (Special Report No. 220). Washington, DC: Transportation Research Board.

Beardsley, T. (1987a). Science and the citizen. Two steps forward. . . . *Scientific American, 257*(3), 18.

Beardsley, T. (1987b). Science and the citizen: Computing cornucopia. *Scientific American, 257*(6), 22.

Beardsley, T. (1989a). Science and the citizen: Fuel for thought. *Scientific American, 261*(6), 20.

Beardsley, T. (1989b). Science and the citizen: Range war. *Scientific American, 261*(6), 20d.

Beardsley, T. (1990a). Science and the citizen: After shocks. California quake intensifies pressure for prediction. *Scientific American, 262*(1), 14.

Beardsley, T. (1990b). Science and the citizen: Slow boat to Mars. Can NASA get us to the red planet? *Scientific American, 262*(4), 14.

Beardsley, T. (1990c). Science and the citizen: Acid test. A mammoth assessment fails to find all the answers. *Scientific American, 262*(4), 18–20.

Beardsley, T. (1990d). Science and the citizen: Hubble's legacy. The space telescope launches a new era in astronomy. *Scientific American, 262*(6), 18–22.

Beardsley, T. (1990e). Science and the citizen: Boosting fusion. *Scientific American, 263*(3), 30–31.

Becker, L. J., & Seligman, C. (1978). Reducing air conditioning waste by signalling it is cool outside. *Personality and Social Psychology Bulletin, 4,* 412–415.

Beer, M., & Walton, E. (1990). Developing the competitive organization: Interventions and strategies. *American Psychologist, 45,* 154–161.

Bell, T. E. (1989b). Telecommunications. *IEEE Spectrum, 26,* 41–43.

Berger, S., Dertouzos, M. L., Lester, R. K., Solow, R. M., & Thurow, L. C. (1989). Toward a new industrial America. *Scientific American, 260*(6), 39–47.

Beringer, D. B., & Peterson, J. G. (1985). Underlying behavioral parameters of the operation of touch-input devices: Biases, models, and feedback. *Human Factors, 27,* 445–458.

Berlin, G. B. (1983). *Not working: Unskilled youth and displaced adults.* New York: Ford Foundation.

Bevington, R., & Rosenfeld, A. H. (1990). Energy for buildings and homes. *Scientific American, 263*(3), 76–86.

Beyond unions. (1985, July 8). *Business Week,* pp. 72–77.

Bickman, L. (1972). Environmental attitudes and actions. *Journal of Social Psychology, 87,* 323–324.

Bikson, T. K. (1987). Understanding the implementation of office technology. In R. E. Kraut (Ed.), *Technology and the transformation of white-collar work* (pp. 155–176). Hillsdale, NJ: Lawrence Erlbaum Associates.

Bikson, T. K., & Gutek, B. (1984). *Implementation of office automation.* Santa Monica, CA: The Rand Corporation.

Bissett, W., & Weaver, L. (1988). Working paper. In *Spills: The human–machine interface* (Great Lakes Science Advisory Board's Technological Committee Report to the International Joint Commission). Windsor, Ontario: Great Lakes Regional Office.

Bjørn-Andersen, N., & Kjærgaard, D. (1987). Choices on route to the office of tomorrow. In R. E. Kraut (Ed.), *Technology and the transformation of white-collar work* (pp. 237–251). Hillsdale, NJ: Lawrence Erlbaum Associates.

Blair, J. G., & Kendall, H. W. (1990). Accidental nuclear war. *Scientific American, 263*(6), 53–58.

Blau, F. D., & Ferber, M. A. (1987). Occupations and earnings of women workers. In K. S. Koziara, M. H. Moskow, & L. D. Tanner (Eds.), *Working women: Past, present, future* (pp. 37–68). Washington, DC: Bureau of National Affairs.

Bleviss, D. L., & Walzer, P. (1990). Energy for motor vehicles. *Scientific American, 263*(3), 102–109.

Bloch, E. (1984). Industry and universities: The case for a joint research effort in the semiconductor industry. In J. Botkin, D. Dimancescu, & R. Stata (Eds.), *Global stakes: The future of high technology in America* (pp. 180–186). New York: Penguin Books.

Blomberg, J. L. (1987). Social interaction and office communication: Effects on user evaluation of new technologies. In R. E. Kraut (Ed.), *Technology and the transformation of white-collar work* (pp. 195–210). Hillsdale, NJ: Lawrence Erlbaum Associates.

Bloom, J. (1990–1991, December/January). Byte-ing the bullet. *Parenting Magazine,* pp. 53–56.

Bobrow, D. G., & Stefik, M. J. (1986). Perspectives on artificial intelligence programming. *Science, 231,* 951–957.

Boden, T. A., Kanciruk, P., & Farrell, M. P. (1990). *TRENDS '90: A compendium of data on global change.* Oak Ridge, TN: Oak Ridge National Laboratory.

Boehm, B. W., & Mobley, R. L. (1969). Adaptive routing techniques for distributed communications systems. *IEEE Transactions on Communication Technology, COM-17,* 340–349.

Boehm-Davis, D. A. (Ed.). (1988). Special issue on expert systems. *Human Factors, 30,* 377–444.

Boehm-Davis, D. A., & Fregley, A. M. (1985). Documentation of concurrent programs. *Human Factors, 27,* 423–432.

Bolin, B., & Doos, B. R. (1986). *The greenhouse effect: Climatic change and ecosystems.* New York: Wiley.

Boorstin, D. (1984). *Books in our future: A report from the Librarian of Congress to the Congress.* Washington, DC: Congress of the United States, Joint Committee on the Library.

Boose, J. H. (1986). *Expertise transfer for expert system design.* New York: Elsevier.

Boose, J. H. (1988). Uses of repertory grid-centered knowledge acquisition tools for knowledge-based systems. *International Journal of Man–Machine Studies, 29,* 287–310.

Boose, J. H., & Bradshaw, J. (1987). Expertise transfer and complex problems: Using AQUINAS as a knowledge acquisition workbench for expert systems. *International Journal of Man–Machine Studies, 26,* 3–28.

Borchelt, R. (1989a). Brine seepage may delay disposal plans. *National Research Council News Report, 38*(5), 18–19.

Borchelt, R. (1989b). Finding the pollutants that cause forest damage. *National Research Council News Report, 39*(8), 14–15.

Borie, E. (1990). [Letter to the editor]. *Scientific American, 262*(5), 13.

Borkan, G. A., & Noris, A. H. (1980). Assessment of biological age using a profile of physical parameters. *Journal of Gerontology, 35,* 177–184.

Botkin, J. W., Dimancescu, D., & Stata, R. (Eds.). (1984). *Global stakes: The future of high technology in America.* New York: Penguin Books.

Botkin, J. W., Elmandjar, M., & Malitza, M. (1979). *No limits to learning: Bridging the human gap.* London: Pergamon Press.

Bowe, F. G. (1984a). *Disabled adults in America.* Washington, DC: President's Committee on Employment of the Handicapped.

Bowe, F. G. (1984b). *Personal computers and special needs.* Berkeley, CA: Sybex Computer Books.

Bowen, H. K. (1986). Advanced ceramics. *Scientific American, 255*(4), 168–176.

Branscomb, L. M. (1986). Science in 2006. *American Scientist, 74,* 650–658.

Branscomb, A. (1991). Common law for the electronic frontier. *Scientific American, 265*(3), 154–158.

Braune, R., & Wickens, C. D. (1983, October). *Individual differences and age-related performance assessment in aviators: Part I. Battery development and assessment* (Tech. Rep. No. EPL-83-4/NAMRL-83-1). Champaign, IL: University of Illinois, Engineering Psychology Lab.

Braune, R., & Wickens, C. D. (1985). The functional age profile: An objective decision criterion for the assessment of pilot performance capacities and capabilities. *Human Factors, 27,* 681–693.

Braverman, H. (1974). *Labor and monopoly capital: The degradation of work in the twentieth century.* New York: Monthly Review Press.

Brechner, K. C., & Linder, D. E. (1981). A social trap analysis of energy distribution systems. In A. Baum & J. E. Singer (Eds.), *Advances in environmental psychology: Vol. 3. Energy: Psychological perspectives* (pp. 27–51). Hillsdale, NJ: Lawrence Erlbaum Associates.

Broadbent, D. E. (1971). *Decision and stress.* London: Academic Press.

Brodsky, M. H. (1990). Progress in gallium arsenide semiconductors. *Scientific American, 262*(2), 68–75.

Broecker, W. S., & Denton, G. H. (1990). What drives carbon cycles? *Scientific American, 262*(1), 48–56.

Bromley, B. A. (1986). Physics: Natural philosophy and invention. *American Scientist, 74,* 622–639.

Bronfman, L. M., & Mattingly, T. J., Jr. (1976). Critical mass: Politics, technology, and the public interest. *Nuclear Safety, 17,* 539–549.

Brookes, W. (1990, Fall). Should companies pursue zero risk? *Best of Business Quarterly,* pp. 44–50.

Brookes, W. (1989, December). The global warming panic. *Forbes,* pp. 96–100.

Brown, E. L. (1961). *Newer dimensions of patient care.* New York: Russell Sage.

Brown, G. E., Jr. (1982). A congressional view of the coming information age. In R. A.

Kasschau, R. Lachman, & K. R. Laughery (Eds.), *Houston Symposium III: Information technology and psychology* (pp. 41–55). New York: Praeger.

Brunner, C., Hawkins, J., Mann, F., & Moeller, B. (1990). Designing inquiry. In B. Bowen (Ed.), *Design for learning: Research-based design of technology for learning* (pp. 27–34). Cupertino, CA: External Research, Apple Computer, Inc.

Buchanan, B. G. (1987). Expert systems: Applications in space. In T. B. Sheridan, D. S. Kruser, & S. Deutsch (Eds.), *Human factors in automated and robotic space systems: Proceedings of a symposium* (pp. 113–141). Washington, DC: National Research Council.

Budyko, M. I., Ronovn, A. B., & Yanshin, A. L. (1987). *History of the earth's atmosphere*. Berlin: Springer-Verlag.

Bureau of the Census. (1985). *Household and family characteristics: March 1984* (Current Population Reports, Series P-20, No. 398). Washington, DC: U.S. Department of Commerce.

Bureau of the Census. (1987). *Statistical abstract of the United States: 1987*. Washington DC: U.S. Department of Commerce.

Bureau of the Census. (1989). *Statistical abstract of the United States*. Washington, DC: U.S. Department of Commerce.

Bureau of the Census. (1990). *Statistical abstract of the United States*. Washington, DC: U.S. Department of Commerce.

Bureau of Labor Statistics. (1985). *Handbook of labor statistics*. Washington, DC: U.S. Department of Labor.

Bureau of Labor Statistics. (1987). *Consumer expenditure survey results from 1985*. Washington, DC: U.S. Department of Labor.

Bureau of Labor Statistics. (1988a). *Comparative real gross domestic product, real GDP per capita, and real GDP per employed person: Thirteen countries, 1950-1987*. Washington, DC: U.S. Department of Labor.

Bureau of Labor Statistics. (1988b). *International comparisons of manufacturing productivity and labor cost trends*. Washington, DC: U.S. Department of Labor.

Burner, R. A., & Lasaga, A. C. (1989). Modeling the geochemical carbon cycle. *Scientific American, 260*(3), 74–81.

Burnett, W. M., & Ban, S. D. (1989). Changing prospects for natural gas in the United States. *Science, 244*, 305–310.

Burns, J. O., Duric, N., Taylor, G. J., & Johnson, S. W. (1990). Observatories on the moon. *Scientific American, 262*(3), 42–49.

Burt, D. M. (1989). Mining the moon. *American Scientist, 77*, 574–579.

Bush, C. N. (1990). [Letter to the editor]. *Scientific American, 262*,(5), 13.

Bush, V. (1945, July). As we may think. *The Atlantic Monthly*, pp. 101–108.

Butler, L. H., & Newacheck, P. W. (1981). Health and social factors relevant to long-term care policy. In J. Meltzer, F. Farrow, & H. Richman (Eds.), *Policy options in long-term care*. Chicago: University of Chicago Press.

Byrne, G. (1988). Science achievement in schools called "distressingly low." *Science, 241*, 1751.

Byrne, G. (1989). U.S. students flunk math, science. *Science, 243*, 729.

Cantor, C. R. (1990). Orchestrating the Human Genome Project. *Science, 248*, 49–51.

Cantor, D., & Cantor, S. (Eds.). (1979). *Designing for therapeutic environments: A review of research*. New York: Wiley.

Capindale, R. A., & Crawford, R. G. (1990). Using a natural language interface with casual users. *International Journal of Man–Machine Studies, 32*, 341–362.

Card, S. K., Moran, T., & Newell, A. (1983). The keystroke-level model for user performance time with interactive systems. *Communications of the ACM, 23*, 396–410.

Card, S. K., Robert, J. M., & Keenan, L. N. (1982). *On-line composition of text*. Palo Alto, CA: Xerox Research Center.

Caro, P. W. (1988). Flight training and simulation. In E. L. Wiener & D. C. Nagel (Eds.), *Human factors in aviation* (pp. 229–261). New York: Academic Press.

Carson, R. (1962). *Silent spring.* Boston, MA: Houghton Mifflin.

Casali, S. P., & Williges, R. C. (1990). Data bases of accomodative aids for computer users with disabilities. *Human Factors, 32,* 407–422.

Cascio, W. F., & Zammuto, R. F. (1987). *Societal trends and staffing policies.* Denver: University of Colorado.

Cassidy, R. (1989, January). Research funding for 1989 won't even reach $131 billion. *Research and Development,* pp. 47–50.

Cassidy, R. (1990, April). Scientific visualization: A new computer research tool. Research and Development Magazine, pp. 50–60.

Cava, R. J. (1990a). Structural chemistry and the local charge picture of copper oxide superconductors. *Science, 247,* 656–602.

Cava, R. J. (1990b). Superconductors beyond 1-2-3. *Scientific American, 263*(2), 42–49.

Cerf, C., & Navasky, V. (1984). *The experts speak: The definitive compendium of authoritative misinformation.* New York: Pantheon Books.

Cerf, V. G. (1991). Networks. *Scientific American, 265*(3), 72–81.

Ceruzzi, P. (1986). An unforeseen revolution: Computers and expectations, 1935–1985. In J. J. Corn (Ed.), *Imagining tomorrow: History, technology, and the American future* (pp. 188–201). Cambridge, MA: MIT Press.

Chamot, D. (1987). Electronic work and the white-collar employee. In R. E. Kraut (Ed.), *Technology in the transformation of white-collar work* (pp. 23–33). Hillsdale, NJ: Lawrence Erlbaum Associates.

Chamot, D., & Zalusky, J. L. (1985). Use and misuse of workstations in the home. In *Office workstations in the home.* Washington, DC: National Academy Press.

Chandler, W. U., Makarov, A. A., & Dadi, C. (1990). Energy for the Soviet Union, Eastern Europe, and China. *Scientific American, 263*(3), 120–126.

Chapanis, A. (1975). Interactive human communication. *Scientific American, 232*(3), 36–42.

Chaudhari, P. (1986). Electronic and magnetic materials. *Scientific American, 255*(4), 136–144.

Cherfas, J. (1990). Europe: Betting heavily on fusion. *Science, 250,* 1500–1502.

Chignell, M. H., & Peterson, J. G. (1988). Strategic issues in knowledge engineering. *Human Factors, 30,* 381–394.

Childes, J. R. (1985, August). On land, at sea and in the air: Those polymer invaders are here. *Smithsonian,* pp. 76–87.

Childes, J. R. (1989, November). Flying cars were a dream that never got off the ground. *Smithsonian,* pp. 144–162.

Chiles, W. D., & Alluisi, E. A. (1979). On the specification of operator or occupational workload with performance-measurement methods. *Human Factors, 21,* 515–528.

Chipman, S. F., Segal, J. W., & Glaser, R. (Eds.). (1985). *Thinking and learning skills: Vol. 2. Research and open questions.* Hillsdale, NJ: Lawrence Erlbaum Associates.

Chou, T. W., McCullough, R. L., & Pipes, R. B. (1986). Composites. *Scientific American, 255*(4), 192–203.

Christensen, K. (1990). Externalized employment [Review of The Invisible Workforce]. *Science, 248,* 1428–1429.

Citro, C. F., & Kalton, G. (Eds.). (1989). *Surveying the nation's scientists and engineers: A data system for the 1990s.* Washington, DC: National Research Council, Committee on National Statistics.

Clarke, A. C. (1962). *Profiles of the future: A daring look at tomorrow's fantastic world.* New York: Bantam Books.

Clark, J. P., & Flemings, M. C. (1986). Advanced materials and the economy. *Scientific American, 255*(4), 50–57.

Clark, W. C. (1989a). The human ecology of global change. *International Social Science Journal, 41,* 315–345.

Clark, W. C. (1989b). Managing planet earth. *Scientific American, 261*(3), 47–54.

Clearwater, Y. A. (1985, July). A human place in outer space. *Psychology Today,* pp. 34–43.

Close, F. (1991). *Too hot to handle: The race for cold fusion.* London: Allen.

Coach, J. V., Garber, T., & Karpus, L. (1979). Response maintenance and paper recycling. *Journal of Environmental Systems, 8,* 127–137.

Coates, J. F. (1977). Aspects of innovation: Public policy issues in telecommunications development. *Telecommunications Policy, 1,* 11–23.

Cohen, D. K. (1988). Educational technology and school organization. In R. S. Nickerson & P. P. Zodhiates (Eds.), *Technology in education: Looking toward 2020* (pp. 231–264). Hillsdale, NJ: Lawrence Erlbaum Associates.

Cohen, S., & Weinstein, N. (1981). Non-auditory effects of noise on behavior and health. *Journal of Social Issues, 37*(1), 36–70.

Cohill, A. M., & Williges, R. C. (1985). Retrieval of HELP information for novice users of interactive computer systems. *Human Factors, 27,* 335–343.

Cole, R. E. (1982). Defusion of participatory work structures in Japan, Sweden, and the United States. In P. S. Goodman (Ed.), *Change in organizations* (pp. 166–225). San Francisco: Jossey-Bass.

Comer, J. P. (1988). Educating poor children. *Scientific American, 259*(5), 42–48.

Committee on Economic Development. (1987). *Children in need: Investment strategies for the educationally disadvantaged.* Author.

Committee to Evaluate Mass Balance Information for Facilities Handling Toxic Substances. (1990). *Tracking hazardous substances at manufacturing facilities: Engineering mass balance versus materials accounting.* Board on Environmental Studies and Toxicology, Commission on Geosciences, Environment, and Resources, National Research Council. Washington, DC: National Academy Press.

Committee on Production Technologies for Liquid Transportation Fuels. (1990). *Fuels to drive our future.* Washington, DC: National Academy Press.

Commoner, B. (1966). *Science and Survival.* New York: Viking Press. Compton, W. D., & Gjostein, N. A. (1986). Materials for ground transportation. *Scientific American, 255*(4), 92–100.

Compton, W. D., & Gjostein, N. A. (1986). Materials for ground transportation. *Scientific American, 255*(4), 92–100.

Computer Science and Technology Board. (1988). *The national challenge in computer science and technology.* Washington, DC: National Academy Press.

Confrey, J. (1990). Student conceptions, representations, and the design of software for mathematics education. In B. Bowen (Ed.), *Design for learning: Research-based design of technology for learning* (pp. 55–60). Cupertino, CA: External Research, Apple Computer, Inc.

Conner, J. (1988). Empty skies: Where have all the songbirds gone? *Harrowsmith,* pp. 35–45.

Cook, N. H. (1975). Computer-managed parts manufacture. *Scientific American, 232*(2), 22–29.

Cook, S. W., & Berrenberg, J. L. (1981). Approaches to encouraging conservation behavior: A review and conceptual framework. *Journal of Social Issues, 37*(2), 73–107.

Coombs, B., & Slovic, P. (1979). Newspaper coverage of causes of death. *Journalism Quarterly, 56,* 837–843, 849.

Corcoran, E. (1989). Science and business: A technological fix. *Scientific American, 261*(2), 60.

Corcoran, E. (1990a). Science and business: Milliken & Co. Managing the quality of a textile revolution. *Scientific American, 262*(4), 74–76.

Corcoran, E. (1990b). Science and business: Talking policy. The administration devises an industrial policy—sort of. *Scientific American, 262*(6), 82–84.

Corcoran, E. (1990c). Science and business: A thin line. *Scientific American, 263*(2), 98–100.

Corcoran, E. (1990d). Trends in materials: Diminishing dimensions. *Scientific American, 263*(5), 122–131.

Corcoran, E. (1991). Trends in computing: Calculating reality. *Scientific American, 264*(1), 100–109.

Corcoran, E., & Beardsley, T. (1990). The new space race. *Scientific American, 263*(1), 72–85.

Corcoran, E., & Wallich, P. (1990). The analytic economist: Green economists. *Scientific American, 262*(5), 86–88.

Corn, J. J. (1986). Epilogue. In J. J. Corn (Ed.), *Imagining tomorrow: History, technology, and the American future* (pp. 219–229). Cambridge, MA: MIT Press.

Cournoyer, D., & Caskey, C. T. (1990). Gene transfer into humans. *The New England Journal of Medicine, 323*, 601–602.

Covault, C. (1990, September 17). Exploration initiative work quickens as some lunar concepts avoid station. *Aviation Week & Space Technology*, pp. 36–37.

Cox, V. C., Paulus, P. D., McCain, G., & Karlovac, M. (1982). The relationship between crowding and health. In A. Baum & J. E. Singer (Eds.), *Advances in environmental psychology: Vol. 4. Environment and health* (pp. 271–294). Hillsdale, NJ: Lawrence Erlbaum Associates.

Craxton, R. S., McCrory, R. L., & Soures, J. M. (1986). Progress in laser fusion. *Scientific American, 255*(2), 68–79.

Creer, R., Gray, R., & Treshow, M. (1970). Differential responses to air pollution as an environmental health problem. *Journal of the Air Pollution Control Association, 2*, 814–818.

Crosson, P. R., & Rosenberg, N. J. (1989). Strategies for agriculture. *Scientific American, 261*(3), 128–135.

Crowston, K., Malone, T. W., & Lin, F. (1987). Cognitive science and organizational design: A case study in computer conferencing. *Human–Computer Interaction, 3*, 59–83.

Crump, J. H. (1979). A review of stress in air traffic control: Its measurement and effects. *Aviation, Space and Environmental Medicine, 50*, 243–248.

Crutzen, P. J., & Andreae, M. O. (1990). Biomass burning in the tropics: Impact on atmosphere chemistry and biogeochemical cycles. *Science, 250*, 1669–1678.

Cuban, L. (1986). *Teachers and machines*. New York: Teachers College.

Culliton, B. J. (1989). The dismal state of scientific literacy. *Science, 243*, 600.

Culotta, E. (1990). Can science education be saved? *Science, 250*, 1327–1330.

Curran, L. (1990, August). A small big picture. *Electronics*, pp. 42.

Curtis, K. K., & Bardon, M. (1983). *A national computing environment for academic research*. Washington, DC: National Science Foundation.

Cushman, J. R. (1983). Air–land battle mastery and C3 systems for the multi-national field commander. *Signal, 37*, 45–51.

Cushman, W. H. (1986). Reading from microfiche, a VDT, and the printed page: Subjective fatigue and performance. *Human Factors, 28*, 63–73.

Cutler, S. J. (1975). Transportation and changes in life's satisfaction. *Gerontologist, 15*, 155–159.

Cyert, R. M., & Mowery, D. C. (1989). Technology, employment and U.S. competitiveness. *Scientific American, 260*(5), 54–62.

Czaja, S. J. (Ed.). (1990). *Human factors research needs for an aging population*. Washington, DC: National Academy Press.

Dadzie, K. K. S. (1980). Economic development. *Scientific American, 243*(3), 58–65.

Dainoff, M. H., Happ, A., & Crane, P. (1981). Visual fatigue and occupational stress in VDT operators. *Human Factors, 23*, 421–438.

Dainoff, M. J., & Dainoff, M. H. (1987). *A manager's guide to ergonomics in the electronic office*. New York, NY: Wiley.

Dale, E. L., Jr. (1991). Is the U.S. really on the decline as an economic power? No. *Cosmos, 1*(1), 75–83.

Damos, D. (1985). The effect of asymmetric transfer and speech technology on dual-task performance. *Human Factors, 27*, 409–421.

Daniel, T. C. (1987). The legendary beauty of the Rocky Mountain region: Is it more than skin deep? *Journal of the History of the Behavioral Sciences, 24*, 18–23.

Daniel, T. C. (1990). Measuring the quality of the natural environment a psychological approach. *American Psychologist, 45*, 633–637.

Daniel, T. C., & Boster, R. S. (1976). *Measuring landscape aesthetics: The scenic beauty estimation method* (USDA Forest Service Research Paper No. 167) Fort Collins, CO: Rocky Mountain Forest and Range Experiment Station.

Daniel, T. C., & Vining, J. (1983). Methodological issues in the assessment of landscape quality. In I. Altman & J. Wohlwill (Eds.), *Human behavior and environment* (Vol. 4, pp. 39–84). New York: Plenum.

David, E. E., Jr. (1985). The federal support of mathematics. *Scientific American, 252*(5), 45–51.

Davies, D. W., & Barber, D. L. A. (1973). *Communications network for computers.* New York: Wiley.

Davis, B. D., & The Department of Microbiology and Molecular Genetics, Harvard Medical School. (1990). Perspective: The human genome and other initiatives. *Science, 249*, 342–343.

Davis, G. R. (1990). Energy for planet earth. *Scientific American, 263*(3), 54–62.

Davis, P. K., & Blumenthal, D. (1990). *The base of sand problem: A white paper on the state of military modeling.* Santa Monica, CA: The Rand Corporation.

Davis, R. (1986). Knowledge-based systems. *Science, 231*, 957–963.

Dean, L. M., Pugh, W. M., & Gunderson, N. E. (1975). Spatial and perceptual components of crowding: Effects on health and satisfaction. *Environment and Behavior, 7*, 225–236.

Dean, L. M., Pugh, W. M., & Gunderson, N. E. (1978). The behavioral effects of crowding: Definitions and methods. *Environment and Behavior, 10*, 417–431.

Debus, K. H. (1990). Mining with microbes. *Technology Review, 93*(6), 50–57.

de Chardin, P. T. (1959). *The phenomenon of man.* New York: Harper & Row.

Dede, C. J., Sullivan, T. R., & Scace, J. L. (1988). *Factors shaping the evolution of electronic documentation systems* (Tech. Rep. No. IM.4). Houston, TX: University of Houston, Research Institute for Computing and Information Systems.

Defense Policy Panel. (1989). *Iran Air flight 655 compensation: Hearings before the Defense Policy Panel of the Committee on Armed Services, House of Representatives, 100th Congress* (H.A.S.C. No. 100–119). Washington, DC: U.S. Government Printing Office.

Del Sesto, S. L. (1986). Wasn't the future of nuclear engineering wonderful? In J. J. Corn (Ed.), *Imagining tomorrow: History, technology, and the American future* (pp. 58–76). Cambridge, MA: MIT Press.

Demain, A. L., & Solomon, N. A. (1981). Industrial microbiology. *Scientific American, 245*(3), 66–75.

Demick, J., & Wapner, S. (1990). The role of psychological science in promoting environmental quality: Introduction. *American Psychologist, 45*, 631–632.

Denning, P. J. (1987). The science of computing: A new paradigm for science. *American Scientist, 75*, 572–573.

Denning, P. J. (1989a). The science of computing: Worldnet. *American Scientist, 77*, 432–434.

Denning, P. J. (1989b). The science of computing: The ARPANET after 20 years. *American Scientist, 77*, 530–534.

Denning, P. J. (1990). Science of computing: Stopping computer crimes. *American Scientist, 78*, 10–12.

Denning, P. J., & Tichy, W. F. (1990). Highly parallel computation. *Science, 250*, 1217–1222.

Dentzer, S. (1989, November). The Maypo culture. *Business Month*, pp. 26–34.

Department of Defense. (1990). *Total quality management guide: a two-volume guide for defense organizations.* Washington, DC: Author.

Dertouzos, M. L., Lester, R. K., Solow, R. M., & The MIT Commission on Industrial

Productivity. (1989). *Made in America: Regaining the productive edge*. Cambridge, MA: MIT Press.

Dervan, P. B. (1988). Synthetic tools for molecular biology. In M. L. Good (Ed.), *Biotechnology and materials science: Chemistry for the future* (pp. 21–29). Washington, DC: American Chemical Society.

DeSanctis, G. (1984). Attitudes towards telecommuting: Implications for work-at-home programs, *Information and Management, 7*.

Dickinson, R. E., & Cicerone, R. J. (1986). Future global warming from atmospheric trace gases. *Nature, 319*, 109–115.

Dickson, D., & Marshall, E. (1989). Europe recognizes the ozone threat. *Science, 243*, 1279.

Douglas, S. J. (1986). Amateur operators and American broadcasting: Shaping the future of radio. In J. J. Corn (Ed.) *Imagining tomorrow: History, technology, and the American future* (pp. 35–57). Cambridge, MA: MIT Press.

Draper, S. W. (1986). Display managers as the basis for user–machine communication. In D. A. Norman & S. W. Draper (Eds.), *User-centered system design: New perspectives on human–computer interaction* (pp. 339–352). Hillsdale, NJ: Lawrence Erlbaum Associates.

Dray, S. M. (1988). From tier to peer: Organizational adaptation to new computing architectures. *Ergonomics, 31*, 721–725.

Drexler, K. E. (1986). *Engines of creation: Challenges and choices of the last technological revolution*. New York: Anchor Press.

Duchemin, J. (1990a). *The "magnetic dream": A hard look at today's high speed rail options*. Unpublished manuscript, Commission of European Communities, Brussels, Belgium.

Duchemin, J. (1990b). *Very high speed trains: A new philosophy*. Unpublished manuscript, Commission of the European Communities, Brussels, Belgium.

Dunlap, R. E. (1987). Polls, pollution, and politics revisited: Public opinion on the environment in the Reagan era. *Environment, 29*, 34–37.

Dyson, F. (1989). *Infinite in all directions*. New York: Harper & Row.

Eason, K. D. (1982). The process of introducing information technology. *Behavior and Information Technology, 1*, 197–213.

Economy, J. (1988). High-strength composites. In M. L. Good (Ed.), *Biotechnology and materials science: Chemistry for the future* (pp. 117–128). Washington, DC: American Chemical Society.

Edmondson, B. (1988, February). Why adult education is hot. *American Demographics*, pp. 40–41.

Edwards, A. D. N. (1989). Modeling blind users' interactions with an auditory computer interface. *International Journal of Man–Machine Studies, 30*, 575–589.

Edwards, E. (1988). Introductory overview. In E. L. Wiener & D. C. Nagel (Eds.), *Human factors in aviation* (pp. 3–25). New York: Academic Press.

Egan, D. E., Remde, J. R., Gomez, L. M., Landauer, T. K., Eberhardt, J., & Lochbaum, C. C. (1989). Formative design-evaluation of superbook. *ACM Transactions on Information Systems, 7*, 30–57.

Ehrenreich, S. L. (1985). Computer abbreviations: Evidence and synthesis. *Human Factors, 27*, 143–155.

Ehrlich, P. R. (1968). *The population bomb*. New York: Ballantine books.

Ehrlich, P. R. (1987). Habitats in crises: Why we should care about the loss of species. *Wilderness*, pp. 12–15.

Eigler, D. D., & Schweizer, E. K. (1990). Positioning single atoms with a scanning tunnelling microscope. *Nature, 344*, 524–526.

Elkind, J. I. (1990). The incidence of disabilities in the United States. *Human Factors, 32*, 397–405.

Emery, F. E., & Emery, M. (1978). Searching. In J. W. Sutherland (Ed.), *Management handbook for public administrators* (pp. 245–263). New York: Van Nostrand Reinhold.

Emmanuel, W. R., Shugart, H. H., & Stevenson, M. P. (1985). *Climate Change, 7,* 29.

Engelbart, D. C. (1963). A conceptual framework for the augmentation of man's intellect. In P. W. Howerton & D. C. Weeks (Eds.), *Vistas in information handling* (Vol. I, pp. 1–29). Washington, DC: Spartan Books.

Engelbart, D. C., & English, W. K. (1968). A research center for augmenting human intellect. *Proceedings of the AFIPS Joint Computer Conference,* pp. 395–410.

Enting, I. G., & Mansbridge, J. V. (1987). The incompatibility of ice-core CO_2 data with reconstructions of biotic CO_2 sources. *Tellus, 39,* 318–325.

Environmental Protection Agency. (1974). *Information on levels of environmental noise requisite to protect public health and welfare with an adequate margin of safety* (Report No. 550/9–73–002). Washington, DC: Author.

Environmental Protection Agency (1987). *Unfinished business: A comparative assessment of environmental problems: Overview report.* Washington, DC: Environmental Protection Agency.

Environmental Protection Agency. (1989a). *EPA report to the Congress on indoor air.* Washington, DC: Author.

Environmental Protection Agency. (1989b). *Policy options for stabilizing global climate* [draft]. Washington, DC: Author.

Environmental Protection Agency (1989c). *EPA report to the congress on indoor air.* Washington, DC: Environmental Protection Agency.

Epstein, Y. M. (1981). Crowding stress and human behavior. *Journal of Social Issues, 37,* 126–144.

Erickson, D. (1990a). Do you see what I see? *Scientific American, 263*(1), 88–89.

Erickson, D. (1990b). Putting down roots: Genetically engineered plants head for the harvest. *Scientific American, 262*(5), 81–84.

Erickson, D. (1990c). Science and business: Down on the pharm. *Scientific American, 263*(2), 102–103.

Erikson, K. (1990, Fall). The fear you can't ignore. *Best of Business Quarterly,* pp. 52–59.

Erisman, A. M., & Neves, K. W. (1987). Advanced computing for manufacturing. *Scientific American, 257*(4), 162–169.

Ernst, M. L. (1982). The mechanization of commerce. *Scientific American, 247*(3), 132–145.

Ervine, R. D., & Chen, K. (1988–1989). Toward motoring smart. *Issues in Science and Technology, 5*(2), 92–97.

Eshelman, L. (1988). MOLE: A knowledge acquisition tool that buries certainty factors. *International Journal of Man–Machine Studies, 29,* 563–577.

Evans, C. (1979). *The micromillenium.* New York: Pocket Books.

Evans, G. W., & Jacobs, S. V. (1981). Air pollution and human behavior. *Journal of Social Issues, 37,* 95–125.

Evans, L., Frick, M. C., & Schwing, R. C. (1990). Is it safer to fly or drive? *Risk Analysis, 10,* 239–246.

Farman, J. C., Gardiner, G. B., & Shanklin, J. D. (1985). Large losses of total ozone in Antarctica reveal seasonal ClOx/NOx interaction. *Nature, 315,* 207–210.

Farooq, M. U., & Dominick, W. D. (1988). A survey of formal tools and models for developing user interfaces. *International Journal of Man–Machine Studies, 29,* 479–496.

Federal Highway Administration. (1987). *Urban and suburban highway congestion: Future national highway program, 1991 and beyond* (Working Paper No. 10). Washington, DC: U.S. Department of Transportation.

Feldberg, R. L., & Glenn, E. N. (1987). Technology and the transformation of clerical work. In R. E. Kraut (Ed.), *Technology and the transformation of white-collar work* (pp. 77–97). Hillsdale, NJ: Lawrence Erlbaum Associates.

Feurzeig, W. (1964, June). Conversational teaching machine. *Datamation,* pp. 38–42.

Feurzeig, W. (1987). Algebra slaves and agents in a Logo-based mathematics curriculum. In R.

W. Lawler & M. Yazdani (Eds.), *Artificial intelligence and education: Vol. 1. Learning environments and tutoring systems* (pp. 27–54). Norwood, NJ: Ablex.

Feurzeig, W., Munter, P., Swets, J., & Breen, M. (1964). Computer-aided teaching in medical diagnosis. *Journal of Medical Education, 39,* 746–754.

Fickett, A. P., Gellings, C. W., & Lovins, A. B. (1990). Efficient use of electricity. *Scientific American, 263*(3), 64–74.

Fidell, S. A. (1977). *Effects of stress on performance: The C3 system user. Vol. I: A review of research on human performance as it relates to the design and operation of command, control, and communication systems* (BBN Report No. 3459). Cambridge, MA: Bolt Beranek & Newman.

Fidell, S. A. (1978). Nationwide urban noise survey. *Journal of the Acoustical Society of America, 64,* 198–206.

Fidell, S. A., & Schultz, T. J. (1980). *A critical review of time-of-day weighting factors for cumulative measures of community noise exposure* (BBN Report No. 4216). Canoga Park, CA: BBN Laboratories, Inc.

Fine, A. (1986). Transplantation in the central nervous system. *Scientific American, 255*(2), 52–58.

Finkelman, J. M., & Kirschner, C. (1981). An information-processing interpretation of air traffic control stress. *Human Factors, 22,* 561–567.

Fischhoff, B. (1989). Risk: A guide to controversy. Appendix C. In *National Research Council: Improving risk communication* (pp. 211–319). Washington, DC: National Academy Press.

Fischhoff, B. (1990). Psychology and public policy: Tool or toolmaker? *American Psychologist, 45,* 647–653.

Fischhoff, B., Lichtenstein, S., Slovic, P., Derby, S. L., & Keeney, R. L. (1981). *Acceptable risk.* New York: Cambridge University Press.

Fischhoff, B., & MacGregor, D. (1983). Judged lethality: How much people seem to know depends upon how they are asked. *Risk Analysis, 3,* 229–236.

Fischhoff, B., Slovic, P., Lichtenstein, S., Read, S., & Combs, B. (1978). How safe is safe enough? A psychometric study of attitudes towards technological risks and benefits. *Policy Sciences, 9,* 127–152.

Fitzgerald, K. (1989). Medical electronics. *IEEE Spectrum, 26,* 67–69.

Flach, J. M. (1989, September). An ecological alternative to egg-sucking. *Human Factors Society Bulletin,* pp. 4–6.

Foerstel, K. (1989, September 25–October 1). How environmental legislation stands on Hill. *Roll Call: Environmental Policy Briefing,* p. 23.

Fogarty, W. M. (1988, July 12). *Formal investigation into the circumstances surrounding the downing of a commercial airliner by the USS* Vincennes *(CG 49) on 3 July 1988* [Memo to Commander in Chief, U.S. Central Command].

Foley, J. D. (1987). Interfaces for advanced computing. *Scientific American, 257*(4), 126–135.

Food and Drug Administration. (1984). *An overview of the medical device reporting regulation* (DHHS Publication No. FDA 85–4194). Washington, DC: U.S. Government Printing Office.

Forester, T. (1987). *High-tech society: The story of the information technology revolution.* Cambridge, MA: MIT press.

Form, W., & McMillan, D. (1983). Men, women, and machines. *Sociology of Work and Occupations, 10,* 147–178.

Foushee, H. C. (1990). *The national plan for aviation: Human factors* (Vols. I & II). Washington, DC: Federal Aviation Administration.

Foushee, H. C., & Helmreich, R. L. (1988). Group interaction and flight crew performance. In E. L. Wiener & D. C. Nagel (Eds.), *Human factors in aviation* (pp. 189–227). New York: Academic Press.

Foushee, H. C., Lauber, J. K., Baetge, M. M., & Acomb, D. B. (1986). *Crew performance as a function of exposure to high-density short-haul duty cycles* (NASA Tech. Memo. No. 88322). Moffett Field, CA: NASA-Ames Research Center.

Foushee, H. C., & Manos, K. L. (1981). Information transfer within the cockpit: Problems in intracockpit communications. In C. E. Billings & E. S. Cheaney (Ed.), *Information transfer problems in the aviation system* (NASA Tech. Paper No. 1875). Moffett Field, CA: NASA-Ames Research Center.

Fox, E. A. (1990). How to proceed toward electronic archives and publishing. *Psychological Science, 1,* 355–358.

Fox, G. C., & Messina, P. C. (1987). Advanced computer architectures. *Scientific American, 257*(4), 66–74.

Francas, M., Goodman, D., & Dickinson, J. (1985). A reliability study of task walk-through in the computer/communications industry. *Human Factors, 27,* 601–605.

French Embassy. (1990). Super-Concorde. *French Advances in Science and Technology, 4*(2), 6.

Friedlander, S. K. (1989). Environmental issues: Implications for engineering design and education. In J. H. Ausubel & H. E. Sladovich (Eds.), *Technology and environment* (pp. 167–181). Washington, DC: National Academy Press.

Friedli, H., Lütscher, H., Oeschger, H., Siegenthaler, U., & Stauffer, B. (1986). Ice core record of the 13C/12C ratio of atmospheric CO_2 in the past two centuries. *Nature, 324,* 237–238.

Friedmann, T. (1989). Progress toward human gene therapy. *Science, 244,* 1275–1281.

Frosch, R. A., & Gallopoulos, N. E. (1989). Strategies for manufacturing. *Scientific American, 261*(3), 144–152.

Frosch, R. A., & Gallopoulos, N. E. (1990). [Letter to the editor]. *Scientific American, 262*(5), p. 13.

Fulkerson, W., Judkins, R. R., & Sanghvi, M. J. (1990). Energy from fossil fuels. *Scientific American, 263*(3), 128–135.

Fulkerson, W., Reister, D. B., Perry, A. M., Crane, A. T., Cash, D. E., & Auerbach, S. I. (1989). Global warming: An energy technology R & D challenge. *Science, 246,* 868–869.

Fuller, R. B. (1969). *Utopia or oblivion: The prospects for humanity.* New York: Bantam Books.

Fuller, R. A., & Rosen, J. J. (1986). Materials for medicine. *Scientific American, 255*(4), 118–125.

Furnas, G. W. (1982). *The fisheye view: A new look at structured files* (Tech. Memo.). Murray Hill, NJ: Bell Laboratories.

Furnas, G. W. (1986). Generalized fisheye views. In M. Mantei & P. Orbeton (Eds.), *Human factors in computing systems: CHI'86 conference proceedings* (pp. 16–23). New York: Association of Computing Machinery.

Furnas, G. W., Landauer, T. K., Gomez, L. M., & Dumais, S. T. (1983). Statistical semantics: Analysis of the potential performance of keyword information systems. In J. C. Thomas & M. Schneider (Eds.), *Human factors and computer systems* (pp. xxx–xxx). Norwood, NJ: Ablex.

Gaden, E. L., Jr. (1981). Production methods in industrial microbiology. *Scientific American, 245*(3), 180–196.

Galitz, W. O. (1984). *The office environment: Automation's impact on tomorrow's workplace.* Willow Grove, PA: Administrative Management Society Foundation.

Garcia, K. D., & Wierwille, W. W. (1985). Effect of glare on performance of a VDT reading-comprehension task. *Human Factors 27,* 163–73.

Gardner, W. (1990). The electronic archive: Scientific publishing for the 1990s. *Psychological Science, 1,* 333–341.

Garg-Janardan, C., & Salvendy, G. (1988). A structured knowledge elicitation methodology for building expert systems. *International Journal of Man–Machine Studies, 29,* 377–406.

Garriott, O. K. (1974). Skylab report: Man's role in space research. *Science, 186,* 219–225.

Garriott, O. K., Parker, R. A. R., Lichtenberg, B. K., & Merbold, U. (1984). Payload crew members' view of Spacelab operations. *Science, 225,* 165–167.

Gasser, C. S., & Fraley, R. T. (1989). Genetically engineering plants for crop improvement. *Science, 244,* 1293–1299.

Geirland, J. (1986). Macroergonomics: The new wave. *Office Systems Ergonomics Report, 5,* pp. 3–11.

Geirland, J. (1989, September). Developing design guidelines for computer-supported cooperative work: A macro-ergonomic approach. *Human Factors Society Bulletin,* pp. 1–4.

Gelernter, D. (1989). The metamorphosis of information management. *Scientific American, 261*(2), 66–73.

Geller, E. S. (1981). Waste reduction and resource recovery: Strategy for energy conservation. In A. Baum & J. E. Singer (Eds.), *Advances in environmental psychology: Vol. 3. Energy: Psychological perspectives* (pp. 115–154). Hillsdale, NJ: Lawrence Erlbaum Associates.

Geller, E. S., Chaffee, J. L., & Ingram, R. E. (1975). Promoting paper-recycling on a university campus. *Journal of Environmental Systems, 5,* 39–57.

General Accounting Office. (1987a). *The Comptroller General's 1987 annual report.* Washington, DC: Author.

General Accounting Office. (1987b). *Education information: Changes in funds and priorities have affected production and quality.* Washington, DC: Author.

General Accounting Office. (1988). *Program evaluation issues.* Washington, DC: Author.

General Accounting Office. (1989). *Cancer treatment: National Cancer Institute's role in encouraging the use of breakthroughs.* Washington, DC: Author.

Gergen, M., & Hagen, D. (Eds.). (1985). Computer technology for the handicapped. *Proceedings of the 1984 Closing the Gap Conference.* Henderson, MN: Closing the Gap.

Getty, D. J., Pickett, R. M., D'Orsi, C. J., & Swets, J. A. (1988). Enhanced interpretation of diagnostic images. *Investigative Radiology, 23,* 240–252.

Gibbons, J. H., Blair, P. D., & Gwin, H. L. (1989). Strategies for energy use. *Scientific American, 261*(3), 136–143.

Gillan, D. G., Burns, M. J., Nicodemus, C. L., & Smith, R. L. (1987, November). The space station: Human factors and productivity. *Human Factors Society Bulletin,* pp. 1–3.

Gilleland, J. R., Nevins, W. M., & Kaiper, G. V. (1990). Moving ahead with fusion. *Issues in Science and Technology, 6*(4), 62–67.

Gilmartin, P. A. (1990, July). Nunn leads Democratic effort to shift defense resources to environmental research. *Aviation Week & Space Technology,* p. 22.

Ginzberg, E. (1982). The mechanization of work. *Scientific American, 247*(3), 66–75.

Giuliano, V. E. (1982). The mechanization of office work. *Scientific American, 247*(3), 148–164.

Glas, J. P. (1989). Protecting the ozone layer: A perspective from industry. In J. H. Ausubel & H. E. Sladovich (Eds.), *Technology and environment* (pp. 137–155). Washington, DC: National Academy Press.

Goddard, W. A., III (1988). Simulation of atoms and molecules: Chemical reaction in material science. In M. L. Good (Ed.), *Biotechnology and materials science: Chemistry for the future* (pp. 71–84). Washington, DC: American Chemical Society.

Golay, M. W., & Todreas, N. E. (1990). Advanced light-water reactors. *Scientific American, 262*(4), 82–89.

Goldstein, I. L., & Gilliam, P. (1990). Training system issues in the year 2000. *American Psychologist, 45,* 134–143.

Good, M. L. (1988). Preface. In M. L. Good (Ed.), *Biotechnology and materials science: Chemistry for the future* (pp. xiii–xvi). Washington, DC: American Chemical Society.

Goodwin, N. C. (1975). Cursor positioning on an electronic display using lightpen, lightgun, or keyboard for three basic tasks. *Human Factors, 17,* 289–295.

Gooler, D. D. (1986). *The education utility: The power to revitalize education and society.* Englewood Cliffs, NJ: Educational Technology Publications.

Gordon, J. (1990, October). Where the training goes. *Training*, pp. 51–69.

Gordon, G. E., & Kelly, M. M. (1986). *Telecommuting: How to make it work for you and your company*. Englewood Cliffs, NJ: Prentice-Hall.

Gore, A., Jr. (1989, September 25–October 1). A plan to meet the crisis. *Roll Call: Environmental Policy Briefing*, p. 1.

Gore, A. (1991). Infrastructure for the global village. *Scientific American, 265*(3), 150–153.

Goudie, A. (1982). *The human's impact: Man's role in environmental change*. Cambridge, MA: MIT Press.

Gould, J. D., Alfaro, L., Barnes, V., Finn, R., Grischkowsky, N., & Minuto, A. (1987). Reading is slower from CRT displays than from paper: Attempts to isolate a single-variable explanation. *Human Factors, 29*, 269–299.

Gould, J. D., & Grischkowsky, N. (1984). Doing the same work with hard copy and cathode ray tube (CRT) computer terminals. *Human Factors, 26*, 323–337.

Gould, J. D., & Lewis, C. (1983). Designing for usability: Key principles and what designers think. In A. Janda (Ed.), *Proceedings of the CHI'83 Conference on Human Factors in Computing Systems* (pp. 50–53). New York: Association for Computing Machinery.

Graeber, R. C. (1988). Air crew fatigue and circadian rhythmicity. In E. L. Wiener & D. C. Nagel (Eds.), *Human factors in aviation* (pp. 305–344). New York: Academic Press.

Graedel, T. E., & Crutzen, P. J. (1989). The changing atmosphere. *Scientific American, 261*(3), 58–68.

Graham, J. D., & Crandall, R. W. (1989). The effect of fuel economy standards on automobile safety. *Journal of Law and Economics, 32*, 97–118.

Gray, C. L., Jr., & Alson, J. A. (1989). The case for methanol. *Scientific American, 261*(5), 108–114.

Great Lakes Science Advisory Board's Technological Committee. (1988). Spills, the human machine interface. *Proceedings of the Workshops on Human Machine Interface*. Windsor, Ontario: Great Lakes Regional Office.

Green, K. C. (1989). A profile of undergraduates in the sciences. *American Scientist, 77*, 475–480.

Greene, D. L., Sperling, D., & McNutt, B. (1988). Transportation energy to the year 2020. In *A look ahead: Year 2020* (Special Report No. 220). Washington, DC: National Research Council, Transportation Research Board.

Greenwell, G. (1987). Science and the citizen: It's a dirty job. *Scientific American, 257*(4), 47–48.

Gregory, J. (1985). Clerical workers and new office technologies. In *Office workstations in the home*. Washington, DC: National Academy Press.

Gregory, J., & Nussbaum, K. (1982). Race against time: Automation of the office. *Office: Technology and People, 1*, 197–236.

Griffith, D., Gardner-Bonneau, D. J., Edwards, A. D. N., Elkind, J. I., & Williges, R. C. (1989). Human factors research with special populations will further advance the theory and practice of the human factors discipline. *Proceedings of the Human Factors Society 33rd Annual Meeting* (pp. 565–566). Santa Monica, CA: Human Factors Society.

Grotch, W. L. (1988). *Regional intercomparisons of general circulation model predictions and historical climate data* (Report #DOE/NBB 0084). Washington, DC: U.S. Department of Energy.

Grubb, M. J. (1990, July/August). The Cinderella options: A study of modernized renewable energy technologies. Part 1: A technical assessment. *Energy Policy*, pp. 525–542.

Gruber, T. R. (1988). Acquiring strategic knowledge from experts. *International Journal of Man–Machine Studies, 29*, 579–597.

Gulkis, S., Lubin, P. M., Meyer, S. S., & Silverberg, R. F. (1990). The cosmic background explorer. *Scientific American, 262*(1), 132–139.

Gunderson, E. K. E. (1963). Emotional symptoms in extremely isolated groups. *Archives of General Psychiatry, 9*, 74–80.

Gunderson, E. K. E. (Ed.). (1974). *Human adaptability to Antarctic conditions*. Washington, DC: American Geophysical Union.

Gunderson, E. K. E., & Nelson, P. D. (1963). Adaptation of small groups to extreme environments. *Aerospace Medicine, 34*, 1111–1115.

Gunn, G. (1982). The mechanization of design and manufacturing. *Scientific American, 247*(3), 114–130.

Häfele, W. (1990). Energy from nuclear power. *Scientific American, 263*(3), 137–144.

Hamakawa, Y. (1987). Photovoltaic power. *Scientific American, 255*(4), 86–92.

Hamilton, D. (1990). Briefings: Growth without new energy. *Science, 248*, 1486.

Hamilton, L. C. (1985). Self-reported and actual savings in a water conservation campaign. *Environment and Behavior, 17*, 315–326.

Hamilton, M. H. (1988, July 10). Employing new tools to recruit workers. *The Washington Post*, pp. H1, H3.

Hardin, G. (1968). The tragedy of the commons. *Science, 162*, 1243–1248.

Hardin, G. (1985). *Filters against folly: How to survive despite economists, ecologists and the merely eloquent*. New York: Penguin Books.

Hardison, O. B., Jr. (1989). *Disappearing through the skylight: Culture and technology in the twentieth century*. New York, NY: Penguin.

Harkness, R. C. (1977). *Technology assessment of telecommunications–transportation interactions*. Menlo Park, CA: Stanford Research Institute.

Harnad, S. (1990). Scholarly sky writing and the prepublication continuum of scientific inquiry. *Psychological Science, 1*, 342–344.

Harpster, J. L., Freivalds, A., Shulman, G. L., & Leibowitz, H. (1989). Visual performance on CRT screens and hard-copy displays. *Human Factors, 31*, 247–257.

Hart, S. G. (1988). Helicopter human factors. In E. L. Wiener & D. C. Nagel (Eds.), *Human factors in aviation* (pp. 591–638). New York: Academic Press.

Hartson, H. R., & Hix, D. (1989). Toward empirically derived methodologies and tools for human–computer interface development. *International Journal of Man–Machine Studies, 31*, 477–494.

Hatsopoulos, G. N., Krugman, P. R., & Summers, L. H. (1988). U.S. competitiveness: Beyond the trade deficit. *Science, 241*, 299–307.

Havera, S. P. & Belorose, F. C. (1985). *Wetlands, 4*, 19.

Heart, F. E. (1975). The ARPA network. In R. L. Grimsdale & F. F. Kuo (Eds.), *Computer communication networks: 1973 Proceedings of the NATO Advanced Study Institute* (pp. 19–33). Leyden, The Netherlands: Noordhoff International Publishing.

Heart, F. E., McKenzie, A., McQuillan, J., & Walden, D. (1978). *ARPANET completion report*. Cambridge, MA: Bolt Beranek & Newman.

Heidorn, G. E., Jensen, K., Miller, L. A., Byrd, R. J., & Chodorow, M. S. (1982). The EPISTLE text and critiquing system. *The IBM Systems Journal, 21*, 305–326.

Helander, M. G. (1985). Emerging office automation systems. *Human Factors, 27*, 3–20.

Helander, M. G. (Ed.). (1988). *Handbook of human computer interaction*. Amsterdam: North Holland.

Helander, M. G., Billingsley, P. A., & Schurick, J. M. (1984). An evaluation of human factors research on visual display terminals in the workplace. In F. A. Muckler (Ed.), *Human factors review: 1984* (pp. 55–129). Santa Monica, CA: Human Factors Society.

Helfgott, R. B. (1966). E.D.P. and the office workforce. *Industrial and Labor Relations Review, 19*, 503–516.

Helson, H. (1964). *Adaptation level theory*. New York: Harper & Row.

Hendrick, H. W. (1986). Macroergonomics: A conceptual model for integrating human factors with organizational design. In O. Brown, Jr. & H. W. Hendrick (Eds.), *Human factors in organizational design and management: Proceedings of the Second International Symposium on*

Human Factors in Organizational Design and Management (pp. 467–478). Amsterdam: Elsevier.

Herman, R., Ardekani, S. A., & Ausubel, J. H. (1989). Dematerialization. In J. H. Ausubel & H. E. Sladovich (Eds.), *Technology and environment* (pp. 50–69). Washington, DC: National Academy Press.

Heron, A., & Chown, S. (1976). *Age and function*. Boston, MA: Little, Brown.

Hershkowitz, A. (1990). Without a trace: Handling medical waste safely. *Technology Review, 93*(6), 35–40.

Herskovitz, S. B. (1990). The terabytes are coming: Storing and processing the data deluge. *Journal of Electronic Defense, 13,* 44–58.

Hibbard, W. R. (1986). Metals demand in the United States: An overview. *Materials and Society, 10,* 251–258.

Hill, P. G. (1977). *Power generation: Resources, hazards, technology, and costs*. Cambridge, MA: MIT Press.

Hillman, D. J. (1985). Artificial intelligence. *Human Factors, 27,* 21–31.

Hinman, C. W. (1986). Potential new crops. *Scientific American, 255*(1), 33–37.

Hively, W. (1989). Science observer: How bleak is the outlook for ozone? *American Scientist, 77,* 219–223.

Hoffman, M. S. (Ed.). (1989). *The world almanac and book of facts*. New York: Pharos Books, Scripps Howard.

Hofmann, D. J., Harder, J. W., Rolf, S. R., & Rosen, J. M. (1987). Balloon-borne observations of the development and vertical structure of the Antarctic ozone hole in 1986. *Nature, 326,* 59–62.

Hogan, D. L. (1987). Voice mail and office automation. *AFIPS Conference Proceedings, 56,* 43–48.

Holden, C. (1990a). Multidisciplinary look at a finite world. *Science, 249,* 18–19.

Holden, C. (1990b). Superconductivity: Japan versus U.S. science briefing. *Science, 248,* 681.

Holden, C. (1989a). Computers make slow progress in class. *Science, 244,* 906–909.

Holden, C. (1989b). Wanted: 675,000 future scientists and engineers. *Science, 244,* 1536–1537.

Holdren, J. P. (1990). Energy in transition. *Scientific American, 263*(3), 156–163.

Holt, D. A. (1985). Computers in production agriculture. *Science, 228,* 422–427.

Hoos, I. R. (1961). *Automation in the office*. Washington, DC: Public Affairs Press.

Hopkin, V. D. (1988a). Air traffic control. In E. L. Wiener & D. C. Nagel (Eds.), *Human factors in aviation* (pp. 639–663). New York: Academic Press.

Hopkin, V. D. (1988b). Boredom. *The controller,* pp. 1, 6–10.

Horgan, J. (1989a). Science and the citizen: Deep thought. *Scientific American, 260*(4), 28.

Horgan, J. (1989b). Science and the citizen: Big game forensics. *Scientific American, 261*(6), 27–30.

Hot rocks. (1990). *Technology Review, 93*(6), 9. MIT Reporter.

Hough, R. W., & Penko, R. R. (1977). *Teleconferencing systems: A state-of-the-art survey and preliminary analysis*. Palo Alto, CA: Stanford Research Institute.

Houghton, R. A., & Woodwell, G. M. (1989). Global climatic change. *Scientific American, 260*(4), 36–44.

Howe, R. T., Muller, R. S., Gabriel, K. J., & Trimmer, W. S. N. (1990). Silicon micromechanics: Sensors and actuators on a chip. *Spectrum, 27*(7), 29–35.

Howell, S. C. (1980). *Designing for aging: Patterns of use*. Cambridge, MA: MIT Press.

Howell, S. C. (1982). The mystery variable in health and aging. In A. Baum & J. E. Singer (Eds.), *Advances in environmental psychology: Vol. 4. Environment and health* (pp. 31–48). Hillsdale, NJ: Lawrence Erlbaum Associates.

Howell, W. C., & Cooke, M. J. (1989). Training the human information process: A look at cognitive models. In I. E. Goldstein (Ed.), *Training and development in work organiza-*

tions: Frontiers of industrial and organizational psychology (pp. 121–182). San Francisco: Jossey-Bass.

Hsu, F-H., Anantharaman, T., Campbell, M., & Nowatzyk, A. (1990). A grandmaster chess machine. *Scientific American, 263*(4), 44–50.

Hubbard, H. M. (1989). Photovoltaics today and tomorrow. *Science, 244*, 297–304.

Huggins, A. W. F. (1978). Speech timing and intelligibility. In J. Requin (Ed.), *Attention and performance VII* (pp. 278–298). Hillsdale, NJ: Lawrence Erlbaum Associates.

Hummel, C. F., Levitt, L., & Loomis, R. J. (1978). Perceptions of the energy crisis: Who is blamed and how do citizens react to environment–lifestyle trade-offs? *Environment and Behavior, 10*, 37–88.

Humphrey, C. R., Bord, R. J., Hammond, M. M., & Mann, S. H. (1977). Attitudes and conditions for cooperation in a paper recycling program. *Environment and Behavior, 9*, 107–124.

Hunt, E. (1990). People, pitfalls, and the electronic archive. *Psychological Science, 1*, 346–349.

Hurd, M. D. (1989). The economic status of the elderly. *Science, 244*, 659–664.

Hutchins, E. L., Hollan, J. D., & Norman, D. A. (1986). Direct manipulation interfaces. In D. A. Norman, & S. W. Draper (Eds.), *User centered design* (pp. 87–124). Hillsdale, NJ: Lawrence Erlbaum Associates.

Huws, U., Korte, W. B., & Robinson, S. (1990). *Telework: Towards the elusive office*. New York: Wiley.

Hwang, S-L., & Salvendy, G. (1984). Human supervisory performance in flexible manufacturing systems. *Proceedings of the Human Factors Society 28th Annual Meeting* (pp. 664–669). Santa Monica, CA: Human Factors Society.

Hwang, S-L., & Salvendy, G. (1988). Operator performance and subjective response in control of flexible manufacturing systems. *Work and Stress, 2*, 27–39.

Iacono, S., & Kling, R. (1987). Changing office technologies and transformations of clerical jobs: A historical perspective. In R. E. Kraut (Ed.), *Technology and the transformation of white-collar work* (pp. 53–75). Hillsdale, NJ: Lawrence Erlbaum Associates.

Ingersoll, A. P. (1983). The atmosphere. *Scientific American, 249*(3), 162–174.

Jackson, D. D. (1989, February). When 20 million tons of water flooded Johnstown. *Smithsonian*, pp. 50–61.

Jaco, E. G. (1972). Ecological aspects of patient care in hospital organization. In B. Georgopoulous (Ed.), *Organization research on health institutions* (pp. 223–254). Ann Arbor, MI: Institute for Social Research.

Jacobson, C., & Freiling, M. J. (1988). ASTEK: A multi-paradigm knowledge acquisition tool for complex structured knowledge. *International Journal of Man–Machine Studies, 29*, 311–327.

Jastrow, R. (1990). Global warming report [Letter to the Editor]. *Science, 247*, 14–15.

Jenkins, L. B., & MacDonald, W. B. (1989). Science teaching in the spirit of science. *Issues in Science and Technology, 5*(3), 60–65.

Jennings, D. M., Landweber, L. H., Fuchs, I. H., Farber, D. J., & Adrion, W. R. (1986). The next generation of personal computers. *Science, 231*, 943–950.

Johannsen, G., Moray, N., Pew, R., Rasmussen, J., Sanders, A., & Wickens, C. (1979). Final report of the experimental psychology group. In N. Moray (Ed.), *Mental workload: Its theory and measurement* (pp. 101–114). New York: Plenum.

Johnson, G. I., & Wilson, J. R. (Eds.). (1988). Ergonomics matters in advanced manufacturing technology. *Applied Ergonomics, 19*(1).

Johnson, L. R. (1990). Putting maglev on track. *Issues in Science and Technology, 6*(3), 71–76.

Johnston, W. B., & Packer, A. H. (1987). *Workforce 2000: Work and workers for the 21st century*. Indianapolis, IN: Hudson Institute.

Jones, D. M., & Chapman, A. J. (1984). *Noise and society*. London: Wiley.

Jones, P. D., & Wigley, T. M. L. (1990). Global warming trends. *Scientific American, 263*(2), 84–91.

Jones, P. D., Wigley, T. M. L., & Wright, P. (1986). Global temperature variation between 1861 and 1984. *Nature, 322,* 432–434.

Jungermann, H. (1980). Speculations about decision-theoretic aids for personal decision making. *Acta Psychologica, 45,* 7–34.

Kahane, C. J. (1989). *An evaluation of center high-mounted stop lamps based on 1987 data* (Report DOT-HS-807–442). Washington, DC: U.S. Department of Transportation.

Kahn, A. E. (1966). The tyranny of small decisions: Market failures, imperfections and the limits of economics. *Kyklos, 19,* 23–24.

Kahn, G., Nolan, S., & McDermott, J. (1985a). MORE: An intelligent knowledge acquisition tool. *Ninth Conference on Artificial Intelligence,* Los Angeles, CA.

Kahn, G., Nolan, S., & McDermott, J. (1985b). Strategies for knowledge acquisition. *IEEE Transactions on Pattern Analysis and Man-Machine Intelligence, 7*(5).

Kahn, H., Brown, W., & Martel, L. (1976). *The next two hundred years: A scenario for America and the world.* New York, NY: Quill.

Kahn, R. E. (Ed.). (1978). Packet networks [Special issue]. *Proceedings of the IEEE, 11,* 66.

Kahn, R. E. (1987). Networks for advanced computing. *Scientific American, 257*(4), 136–143.

Kaiser, E. T. (1988). Taming the chemistry of proteins. In M. L. Good (Ed.), *Biotechnology and materials science: Chemistry for the future* (pp. 43–52). Washington, DC: American Chemical Society.

Kalsbeek, J. W. H. (1968). Measurement of mental work and of acceptable work: Possible applications in industry. *International Journal of Production Research, 7,* 33–45.

Kanciruk, P. (1990, Spring). Information analysis centers supporting the global-change program. In *CDIAC Communications.* (pp. 1–7). Oak Ridge, TN: Oak Ridge National Laboratory, Carbon Dioxide Information Analysis Center.

Kapor, M. (1991). Civil liberties in cyperspace. *Scientific American, 265*(3), 158–164.

Kasarda, J. D. (1988). Population and employment change in the United States: Past, present, and future. In *A look ahead: Year 2020* (Special Report No. 220). Washington, DC: National Research Council, Transportation Research Board.

Kasl, S. V., White, M., Will, J., & Marcuse, P. (1982). Quality of the residential environment and mental health. In A. Baum & J. E. Singer (Eds.), *Advances in environmental psychology: Vol. 4 Environment and health* (pp. 1–30). Hillsdale, NJ: Lawrence Erlbaum Associates.

Kaufman, H. G. (1978). Continuing education and job performance: A longitudinal study. *Journal of Applied Psychology, 63,* 248–251.

Kay, A. C. (1984). Inventing the future. In P. H. Winston & K. A. Prendergast (Eds.), *The AI business* (pp. 103–112). Cambridge, MA: MIT Press.

Kay, A. C. (1991). Computers, networks and education. *Scientific American, 265*(3), 138–148.

Kearsley, G., & Seidel, R. J. (1985). Automation in training and education. *Human Factors, 27,* 61–74.

Keeling, C. D. (1986). *Atmospheric CO_2 concentrations: Mauna Loa Observatory, Hawaii: 1958–1986* (Report No. CDIAC NDP-001/R1). La Jolla, CA: Scripts Institute of Oceanography.

Kern, D. P., Kuech, T. F., Oprysko, M. M., Wagner, A., & Eastman, D. B. (1988). Future beam-controlled processing technologies for microelectronics. *Science, 241,* 936–944.

Kerr, R. A. (1988a). Is the greenhouse here? *Science, 239,* 559–561.

Kerr, R. A. (1988b). Report urges greenhouse action now. *Science, 241,* 23–24.

Kerr, R. A. (1989a). Research news: Hanson vs. the world on the greenhouse threat. *Science, 244,* 1041–1043.

Kerr, R. A. (1989b). Research news: Volcanoes can mottle the greenhouse. *Science, 245,* 127–128.

Kerr, R. A. (1989c). Research news: Oil and gas estimates plummet. *Science, 245,* 1330–1331.
Kerr, R. A. (1989d). Research news: Greenhouse skeptic out in the cold. *Science, 246,* 1118–1119.
Kerr, R. A. (1990). New greenhouse report puts down dissenters. *Science, 249,* 481–482.
Keyfitz, N. (1989). The growing human population. *Scientific American, 261*(3), 118–126.
Keyworth, G. A., II, & Abell, D. (1989). Policy forum: The third general of the space age. *Science, 245,* 16.
Kiernan, V. (1990). Reactor project hitches on to moon–Mars effort. *Science, 248,* 1482–1484.
Kiesler, S. (1984). Computer mediation of communication. *American Psychologist, 39,* 1123–1134.
Kiester, E., Jr. (1985, August). A deathly spell is hovering above the Black Forest. *Smithsonian,* 16 pp. 211–230.
Kinnaman, D. (1990). The next decade: What the future holds. *Technology and Learning, 11*(1), 42–49.
Kinnucan, P. (1981, September/October). How smart robots are becoming smarter. *High Technology,* pp. 32–40.
Kinoshita, J. (1989a). Science and the citizen: Misprints. *Scientific American, 261*(2), 12–12b.
Kinoshita, J. (1989b). Science in pictures: Neptune. *Scientific American, 261*(5), 82–91.
Klein, G. (1989, May). Do decision biases explain too much? *Human Factors Society Bulletin,* pp. 1–3.
Klein, L. R. (1988). Components of competitiveness. *Science, 241,* 308–313.
Kleinman, D. L., Baron, S., & Levison, W. H. (1971). A control theoretical approach to manned-vehicles systems analysis. *IEEE Transactions on Automated Control, AC-16,* 824–833.
Kling, R. (1980). Social analysis of computing: Theoretical orientations in recent empirical research. *Computing Surveys, 12,* 61–110.
Klinger, D., & Kuzmyak, R. (1986). *Personal travel in the United States, 1983–1984: Nationwide personal transportation study, 1,* Washington, DC: U.S. Department of Transportation, Federal Highway Administration.
Klinker, G., Genetet, S., & McDermott, J. (1988). Knowledge acquisition for evaluation systems. *International Journal of Man–Machine Studies, 29,* 715–731.
Koltnow, P. G. (1988). Advanced technology: Vehicle and automobile guidance. In *A look ahead: Year 2020* (Special Report No. 220). Washington, DC: National Research Council, Transportation Research Board.
Koshland, D. E. (1989). Sequences and consequences of the human genome [Editorial]. *Science, 246,* 189.
Kostyniuk, L. P., & Kitamura, R. (1987). Effects of aging and motorization on travel behavior: An exploration. *Transportation Research Record, 1135.* Washington, DC: National Research Council, Transportation Research Board.
Krafcik, J. F. (1989). A new diet for U.S. manufacturing. *Technology Review, 92*(1), 28–36.
Kraft, P. (1977). *Programmers and managers.* New York: Springer-Verlag.
Kraft, P. (1987). Computers and the automation of work. In R. E. Kraut (Ed.), *Technology and the transformation of white-collar work* (pp. 99–111). Hillsdale, NJ: Lawrence Erlbaum Associates.
Kramer, A. F., Sirevaag, E. J., & Broune, R. (1987). A psycho-physiological assessment of operator workload during simulated flight missions. *Human Factors, 29,* 145–160.
Kraus, L. E., & Stoddard, S. (1989). *Chartbook on disability in the United States: An InfoUse report.* Washington, DC: U.S. Department of Education, National Institute on Disability and Rehabilitation Research.
Krauskopf, K. B. (1988). *Radioactive waste disposal and geology space for a city.* New York: Chapman & Hall.

Krauskopf, K. B. (1990). Disposal of high-level nuclear waste: Is it possible? *Science, 249,* 1231–1232.

Kraut, R. E. (1987a). Social issues and white-collar technology: An overview. In R. E. Kraut (Ed.), *Technology and the transformation of white-collar work* (pp. 1–21). Hillsdale, NJ: Lawrence Erlbaum Associates.

Kraut, R. E. (1987b). Predicting the use of technology: The case of telework. In R. E. Kraut (Ed.), *Technology and the transformation of white-collar work* (pp. 113–133). Hillsdale, NJ: Lawrence Erlbaum Associates.

Kryder, M. H. (1987). Data-storage technologies for advanced computing. *Scientific American, 257*(4), 116–125.

Kryter, K. D. (1985). *The effects of noise on man.* New York: Academic Press.

Kuck, D. H., Davidson, E. S., Lawrie, D. H., & Sameh, A. H. (1986). Parallel supercomputing today and the cedar approach. *Science, 231,* 967–974.

La Brecque, M. (1989). Detecting climate change: I. Taking the world's shifting temperature. *Mosaic, 20*(4), 2–9.

La Brecque, M. (1990). Science observer: What's new in super computing? *American Scientist, 77,* 523–525.

Lachman, R. (1989). Comprehension aids for on-line reading of expository text. *Human Factors, 31,* 1–15.

Lammers, S. (1986). *Programmers at work.* Redmond, WA: Microsoft Press.

Landau, R. (1988). The U.S. economic growth. *Scientific American, 258*(6), 44–52.

Landauer, T. K. (1988). Education in a world of omnipotent and omniscient technology. In R. S. Nickerson & P. P. Zodhiates (Eds.), *Technology in education: Looking toward 2020* (pp. 11–24). Hillsdale, NJ: Lawrence Erlbaum Associates.

Landauer, T. K., Galotti, K. M., & Hartwell, S. (1983). Natural command names and initial learning: A study of text-editing terms. *Communications of the ACM, 26,* 495–503.

Landow, S. E., Panopoulos, N. J., & McFarland, B. L. (1989). Genetic engineering of bacteria from managed and natural habitats. *Science, 244,* 1300–1307.

Lang, L. (1988, July). The age of the supermap. *Computer Graphics World,* pp. 77–80.

la Riviere, J. W. M. (1989). Threats to the world's water. *Scientific American, 261*(3), 80–94.

Laub, L. (1986). What is CD ROM? In S. Lambert & S. Ropiequet (Eds.), *CD ROM: The new papyrus* (p. 47). Redmond, WA: Microsoft Press.

Laurel, B. K. (1986). Interface as mimesis. In D. A. Norman & S. W. Draper (Eds.), *User centered system design: New perspectives on human–computer interaction* (pp. 67–85). Hillsdale, NJ: Lawrence Erlbaum Associates.

Lauridsen, P. K. (1977). *Decreasing gasoline consumption in fleet owned automobiles through feedback and feedback-plus-lottery.* Unpublished thesis, Drake University, Des Moines, IA.

Lawn, R. M., & Vehar, G. A. (1986). The molecular genetics of hemophilia. *Scientific American, 254*(3), 48–54.

Lawrence, P. R., & Dyer, D. (1983). *Renewing American industry.* New York: The Free Press.

Lawton, M. P. (1990). Residential environment and self-directedness among older people. *American Psychologist, 45,* 638–640.

Lawton, M. P., & Nahemow, L. (1973). Ecology and the aging process. In C. Eisdorfer & M. P. Lawton (Eds.), *The psychology of adult development and aging.* Washington, DC: American Psychological Association.

Lax, P. D. (1982). *Report of the panel on large scale computing in science and engineering.* Washington, DC: National Science Foundation.

Lazarus, R. A., & Monat, A. (1977). Stress and coping: Some current issues and controversies. In A. Monat & R. S. Lazarus (Eds.), *Stress and coping* (pp. 1–11). New York: Columbia University Press.

Leahy, P. J. (1989, November 22). *Congressional Record,* #S16311.

Lee, C. (1990, October). Industry report 1990. *Training*, pp. 29–32.

Lee, E., & MacGregor, J. (1985). Minimizing user search time in menu retrieval systems. *Human Factors, 27,* 157–162.

Lee, T. H. (1989). Advanced fossil fuel systems and beyond. In J. H. Ausubel & H. E. Sladovich (Eds.), *Technology and environment* (pp. 114–136). Washington, DC: National Academy Press.

Leeper, P. (1990). More minorities in math is goal of new program. *National Research Council News Report, 30* (6), 2–4.

Lefkowitz, L. J., & Lesser, V. R. (1988). Knowledge acquisition as knowledge assimilation. *International Journal of Man–Machine Studies, 29,* 215–226.

Lehman, R. L., & Warren, H. E. (1978). Residential natural gas consumption: Evidence that conservation efforts to date have failed. *Science, 199,* 879–882.

Leibowitz, H. W. (1988). The human senses in flight. In E. L. Wiener & D. C. Nagel (Eds.), *Human factors in aviation* (pp. 83–110). New York: Academic Press.

Leibowitz, H. W., & Owens D. A. (1986, January). We drive by night. *Psychology Today,* pp. 54–58.

Lescohier, I., Gallagher, S. S., & Guyer, B. (1990). Not by accident. *Issues in Science and Technology, 6*(4), 35–42.

Lester, R. K. (1986). Rethinking nuclear power. *Scientific American, 254*(3), 31–39.

Leventhal, H., Nerenz, D. R., & Leventhal, E. (1982). Feelings of threat and private views of illness: Factors in dehumanization in the medical care system. In A. Baum & J. E. Singer (Eds.), *Advances in environmental psychology: Vol. 4. Environment and health* (pp. 85–114). Hillsdale, NJ: Lawrence Erlbaum Associates.

Levin, H., & Rumberger, R. (1987). Educational requirements for new technologies: Visions, possibilities, and current realities. *Educational Policy, 1,* 333–354.

Levine, M. F., Taylor, J. C., & Davis, L. E. (1984). Defining quality of working life. *Human Relations, 37,* 81–104.

Levinson, P. (1988). *Mind at large: Knowing in the technological age.* Greenwich, CT: JAI Press.

Lewin, K. (1935). *Dynamic theory of personality.* New York: McGraw-Hill.

Lewis, C. (1986). Understanding what's happening in system interactions. In D. A. Norman & S. W. Draper (Eds.), *User centered system design:* New perspectives on human–computer interaction (pp. 171–185). Hillsdale, NJ: Lawrence Erlbaum Associates.

Lewis, C. H., & Mack, R. L. (1982). Learning to use a text processing system: Evidence from "thinking aloud" protocols. In *Proceedings of the Conference on Human Factors in Computer Systems* (pp. 387–392). New York: Association of Computing Machinery.

Lewis, D., Hara, D., & Revis, J. (1988). The role of public infrastructure in the 21st century. In *A look ahead: Year 2020* (Special Report No. 220). Washington, DC: National Research Council, Transportation Research Board.

Lewis, J. S., & Lewis, R. A. (1987). *Space resources: Breaking the bonds of earth.* New York: Columbia University Press.

Ley, P. (1977). Psychological studies of doctor–patient communication. In S. Rachman (Ed.), *Contributions to medical psychology* (Vol. 1, pp. 9–42). New York: Pergamon Press.

Lichtenstein, S., Fischhoff, B., & Phillips, L. D. (1977). Calibration of probabilities: The state of the art. In H. Jungermann & G. deZeeuw (Eds.), *Decision making and change in human affairs.* Amsterdam: T. Reidel.

Lichtenstein, S., Slovic, P., Fischhoff, B. Layman, M., & Coombs, B. (1978). Judged frequency of lethal events. *Journal of Experimental Psychology: Human Learning and Memory, 4,* 551–578.

Licklider, J. C. R. (1960). Man–computer symbiosis. *Institute of Radio Engineers Transactions on Human Factors Electronics, HFE-1,* 4–11.

Licklider, J. C. R. (1965). *Libraries of the future.* Cambridge, MA: MIT Press.

Liedl, G. L. (1986). The science of materials. *Scientific American, 255*(4), 126–134.

Likens, G. E., Wright, R. F., Galloway, J. N., & Butler, T. J. (1979). Acid rain. *Scientific American, 241*(4), 43–51.

Lindzen, R. S. (1990). Global warming report [Letter to the Editor]. *Science, 247,* 14.

Lipmann, A. (1984). Imaging and interactivity. *Proceedings of the 15th Joint Conference on Image Technology.* New York: Association for Computing Machinery.

Lippard, S. J. (1988). Molecular basis of drug design. In M. L. Good (Ed.), *Biotechnology and materials science: Chemistry for the future* (pp. 31–42). Washington, DC: American Chemical Society.

Lipske, M. (1990, April). How much is enough? *National Wildlife,* pp. 18–22.

Locke, E. (1976). The nature and causes of job satisfaction. In M. Dunnette (Ed.), *Handbook of industrial and organizational psychology* (pp. 1297–1349). Chicago: Rand McNally College Publishing.

Loeb, M. (1986). *Noise and human efficiency.* Chichester, England: Wiley.

Loehr, R., & Lash, J. (1990). *Reducing risk: Setting priorities and strategies for environmental protection* (Report of the Science Advisory Board, Relative Risk Reduction Strategies Committee). Washington, DC: Environmental Protection Agency.

Loftus, J. P., Bond, R., & Patton, R. M. (1982). Astronaut activity. Workshop proceedings: Space human factors, 1, Leesburg, VA.

Logsdon, J. M., & Williamson, R. A. (1989). U.S. access to space. *Scientific American, 260*(3), 34–40.

Long, R. J. (1987). *New office information technology: Human and managerial implications.* London: Croom Helm.

Lovasik, J. V., Matthews, M. L., & Kergoat, H. (1989). Neural, optical, and search performance in prolonged viewing of chromatic displays. *Human Factors, 31,* 273–289.

Lowry, I. S. (1988). Planning for urban sprawl. In *A look ahead: Year 2020* (Special Report No. 220). Washington, DC: National Research Council, Transportation Research Board.

Lucky, R. W. (1989). *Silicon dreams: Information, man, and machine.* New York: St. Martin's Press.

MacDonald, G. J. (1985). *Climate change and acid rain,* (Report No. MP8600010). McLean, VA: Mitre Corporation.

MacDonald, G. J. (1989). Scientific bases for the greenhouse effect. *Mitre Journal, 1,* 205–222.

MacGregor, D. G., & Slovic, P. (1989). Perception of risk in automotive systems. *Human Factors, 31,* 377–389.

MacGregor, J., Lee, E., & Lam, N. (1986). Optimizing the structure of database menu indexes: A decision model of menu search. *Human Factors, 28,* 387–400.

Machlup, F. (1962). *The production and distribution of knowledge in the United States.* Princeton, NJ: Princeton University Press.

Mack, R. L., Lewis, C. H., & Carroll, J. M. (1983). Learning to use word processors: Problems and prospects. *ACM Transactions on Office Information Systems, 1,* 254–271.

Mackenzie, J. D., & Claridge, R. C. (1979). Glass and ceramics from lunar materials. *Space manufacturing facilities, III,* 135–140. American Institute of Aeronautics and Astronotics.

MacNeill, J. (1989). Strategies for sustainable economic development. *Scientific American, 261*(3), 154–165.

Madni, A. M. (1988). The role of human factors in expert systems design and acceptance. *Human Factors, 30,* 395–414.

Mahler, H. (1980). People. *Scientific American, 243*(3), 66–77.

Makhoul, J., Jelinek, F., Rabiner, L., Weinstein, C., & Zue, V. (1990). Spoken language systems. *Annual Review of Computer Science, 4,* 481–501.

Malone, T. W., & Rockart, J. F. (1991). Computers, networks and the corporation. *Scientific American, 265*(3), 128–136.

Manzer, L. E. (1990). The CFC–ozone issue: Progress on the development of alternatives to CFCs. *Science, 249,* 31–35.

Marcus, M. (1987). Taking backtracking with a grain of SALT. *International Journal of Man–Machine Studies, 26,* 383–398.

Marovelli, R. L., & Karhnak, J. M. (1982). The mechanization of mining. *Scientific American, 247*(3), 90–102.

Marrill, T., & Roberts, L. A. (1966). Cooperative network of timesharing computers. *Proceedings of the AFIPS 1966 Spring Joint Computer Conference, 28,* 425–431.

Marshall, E. (1988). Worm invades computer networks. *Science, 242,* 855–856.

Marshall, E. (1989a). EPA's plan for cooling the global greenhouse. *Science, 243,* 1544–1545.

Marshall, E. (1989b). Space station science: Up in the air. *Science, 246,* 110–112.

Marshall, E., & Crawford, M. (1988). Space station: At the brink. *Science, 241,* 22.

Martin, G. L. (1989). The utility of speech input in user–computer interfaces. *International Journal of Man–Machine Studies, 30,* 355–375.

Martin, L. G. (1991). Population aging policies in east Asia and the United States. *Science, 251,* 527–531.

Marvin, C. (1988). *When old technologies were new: Thinking about electric communication in the late nineteenth century.* New York: Oxford University Press.

Massey, D. (1988, February). Airports spin the wheel of fortune. *American Demographics,* pp. 42–49.

Massey, W. E. (1989). Science education in the United States: What the scientific community can do. *Science, 245,* 915–921.

Matteson, M. T., & Ivancevich, J. M. (1987). *Controlling work stress.* San Francisco: Jossey-Bass.

Matthews, M. L., Lovasik, J. V., & Mertins, K. (1989). Visual performance and subjective discomfort in prolonged viewing of chromatic displays. *Human Factors, 31,* 259 271.

Mayo, J. S. (1986). Materials for information and communication. *Scientific American, 255*(4), 58–65.

Mazlish, B. (1967). The fourth discontinuity. *Technology and Culture, 8,* 1–15.

McCain, G., Cox, V. C., & Paulus, P. B. (1976). The relationship between illness complaints and degree of crowding in a prison environment. *Environment and Behavior, 8,* 283–290.

MacDonald, N. H., Frase, L. T., Gingrich, P. S., & Keenan, S. A. (1982). The writer's workbench: Computer aids for text analysis. *IEEE Transactions on Communications, 30,* 105–110.

McClelland, L. (1980). *Encouraging energy conservation in multifamily housing: RUBS and other methods of allocating energy costs to residents* Boulder: University of Colorado, (Report to the U.S. Department of Energy). Institute of Behavioral Science.

McClelland, L., & Cantor, R. J. (1981). Psychological research on energy conservation: Context, approaches, methods. In A. Baum & J. E. Singer (Eds.), *Advances in environmental psychology: Vol. 3. Energy: Psychological perspectives* (pp. 1–25). Hillsdale, NJ: Lawrence Erlbaum Associates.

McClelland, L., & Cook, S. W. (1979). Energy conservation effects of continuous in-home feedback in all electric homes. *Journal of Environmental Systems, 9,* 169–173.

McClelland, L., & Cook, S. W. (1980). Promoting energy conservation in waste-metered apartments through financial incentives. *Journal of Applied Social Psychology, 10,* 19–31.

McConnel, S. R., Fleisher, D., Usher, C. E., & Kaplan, B. H. (1980). *Alternative work options for older workers: A feasibility study.* Los Angeles: Ethel Percy Andrus Gerontology Center.

McCormick, N. J. (1981). *Reliability and risk analysis.* New York: Academic Press.

McEvoy, G. M., & Cascio, W. F. (1989). Cumulative evidence of the relationship between employee age and job performance. *Journal of Applied Psychology, 74,* 1–5.

McGinnis, J. M. (1988–1989). National priorities in disease prevention. *Issues in Science and Technology, 5*(2), 46–52.

McHale, J. (1976). *The changing information environment.* Boulder, CO: Westview Press.

McIntosh, H. (1990). U.S.–East European workshops address mutual problems. *National Research Council News Report, 40*(7), 9–11.

McWilliams, P. A. (1984). *Personal computers and the disabled.* New York: Doubleday.

Medawar, P. (1984). *Pluto's republic.* New York: Oxford University Press.

Meier, M. F. (1984). Contribution of small glaciers to global sea level. *Science, 226,* 1418–1421.

Mekjavic, I., Banister, E., & Morrison, J. (1988). *Environmental ergonomics: Sustaining human performance in harsh environments.* Bristol, PA: Taylor & Francis.

Merland, G. (1989). Global reforestation could play a significant role in addressing the CO_2 problem. In D. E. Reichle (Eds.), *Environmental Sciences Division Annual Progress Report for Period Ending September 30, 1988* (ORNL-6521). Oak Ridge, TN: Oak Ridge National Laboratory.

Metz, C. E. (1986). ROC methodology in radiologic imaging. *Investigative Radiology, 21,* 720–733.

Michigan Department of Commerce. (1987). *Automobiles in Michigan.* Lansing, MI: Author.

Miller, A. R. (1971). *The assault on privacy: Computers, data banks, and dossiers.* Ann Arbor, MI: University of Michigan Press.

Miller, C. O. (1988). Systems safety. In E. L. Wiener & D. C. Nagel (Eds.), *Human factors in aviation* (pp. 53–80). New York: Academic Press.

Miller, D. C. (1986). Finally it works: Now it must "play in Peoria." In S. Lambert & S. Ropiequet (Eds.), *CD ROM: The new papyrus.* Redmond, WA: Microsoft Press.

Miller, G. A. (1988). The challenge of universal literacy. *Science, 241,* 1293–1299.

Miller, H. I., Burris, R. H., Vidaver, A. K., & Wivel, N. A. (1990). Risk-based oversight of experiments in the environment. *Science, 250,* 490–491.

Miller, J. G. (1965). Living systems: Basic concepts. *Behavioral Science, 10,* 193–237.

Miller, K. (1989). *Retraining the American workforce.* Reading, MA: Addison-Wesley.

Milstein, J. S. (1976). *Attitudes, knowledge and behavior of American consumers regarding energy conservation with some implications for governmental action.* Washington, DC: Federal Energy Administration.

Milstein, J. S. (1977). *How consumers feel about energy: Attitudes and behavior during the winter and spring of 1976–1977.* Washington, DC: Federal Energy Administration.

Mincer, J. (1989). *Labor market effects of human capital and of its adjustment to technological change.* New York: Columbia University Press.

Mishel, L. (1988). *Manufacturing numbers: How inaccurate statistics conceal U.S. industrial decline.* Washington, DC: Economic Policy Institute.

Mitchell, J. F. B. (1989). The greenhouse effect and climate change. *Reviews of Geophysics, 27,* 115–129.

Mitchell, T. M. (1987). AI systems in the space station. In T. B. Sheridan, D. S. Kruser, & S. Deutsch (Eds.), *Human factors in automated and robotic space systems: Proceedings of a symposium* (pp. 91–112). Washington, DC: National Research Council.

Miyao, M., Hacisalihzade, S. S., Allen, J. S., & Stark, L. W. (1989). Effects of VDT resolution on visual fatigue and readability: An eye movement approach. *Ergonomics, 32,* 603–614.

Moffat, A. S. (1991). Methanol powered cars get ready to hit the road. *Science, 251,* 514–515.

Mohnen, V. A. (1988). The challenge of acid rain. *Scientific American, 259*(2), 30–38.

Monk, A. (Ed.). (1985). *Fundamentals of human–computer interaction.* Orlando, FL: Academic Press.

Monmaney, T. (1985, April). Complex window on life's most basic molecules. *Smithsonian,* pp. 114–119.

Moore, T. (1990, June). Excellent forecast or wind. *EPRI Journal,* pp. 14–25.

Moray, N. P. (1979). Preface. In N. P. Moray (Ed.), *Mental workload: Its theory and measurement* (pp. v–viii). New York: Plenum.

Moray, N. P., & Huey, B. M. (Eds.). (1988). *Human factors research and nuclear safety.* Washington, DC: National Academy Press.

Moretti, P. M., & Divone, L. V. (1986). Modern windmills. *Scientific American, 254*(6), 110–118.

Morgan, M. G. (1981a). Choosing and managing technology-induced risk. *IEEE Spectrum, 18*(12), 53–60.

Morgan, M. G. (1981b). Probing the question of technology-induced risk. *IEEE Spectrum, 18*(11), 58–64.

Morgan, M. G. (1989). Space policy: Getting there from here. *Issues in Science and Technology, 5*(3), 72–77.

Morrison, P. R. (1983). A survey of attitudes towards computers. *Communications of the ACM, 26,* 1051–1057.

Mullis, I. V. S., & Jenkins, L. B. (1988). *The science report card: Elements of risk and recovery.* Princeton, NJ: Educational Testing Service.

Mumford, E. (1982). *Designing secretaries.* Manchester, England: University of Manchester Business School Press.

Murolo, P. (1987). White-collar women and the rationalization of clerical work. In R. E. Kraut (Ed.), *Technology and the transformation of white-collar work* (pp. 35–51). Hillsdale, NJ: Lawrence Erlbaum Associates.

Murphy, E. E. (1989). Transportation. *IEEE Spectrum, 26*(1), 62–63.

Murray, H. A. (1938). *Explorations in personality.* New York: Oxford.

Muter, P., Latremouille, S. A., Treurniet, W. C., & Beam, P. (1982). Extended reading of continuous text on television screens. *Human Factors, 24,* 501–508.

Myers, G. C. (1985). Aging and world-wide population changes. In R. H. Binstock & E. Shanas (Eds.), *Handbook of aging and the social sciences.* New York: Van Nostrand Reinhold.

Nadis, S. (1990). Hydrogen dreams. *Technology Review, 93*(6), 20–21.

Nagel, D. C. (1988). Human error in aviation operations. In E. L. Wiener & D. C. Nagel (Eds.), *Human factors in aviation* (pp. 263–303). New York: Academic Press.

Naisbitt, J., (1984). *Megatrends.* New York: Warner Books.

NASA/Johnson Space Center. (1979). *Space operations center: A concept analysis* (Report No. JSC-16277). Houston: Author.

NASA Task Force. (1989). *Report of the 90-day study on human exploration of the moon and Mars.* Washington, DC: NASA.

Nash, T. (1989). *Human–computer systems in the military context.* Batavia, IL: Fermi National Accelerator Laboratory.

Nathan, R. R. (1991). Is the U.S. really on the decline as an economic power? Yes. *Cosmos, 1*(1), 75–78.

National Academy of Engineering. (1989). *Technology and competitiveness: A statement of the Council of the National Academy of Engineering.* Washington, DC: Author.

National Academy of Science. (1983). *Changing climate: Report of the Carbon Dioxide Assessment Committee.* Washington, DC: National Academy Press.

National Assessment of Educational Progress. (1981). *Reading, thinking, writing: A report on the 1979–1980 assessment.* Princeton, NJ: Author.

National Center for Education Statistics. (1990a). *Announcement: Federal funds for education, fiscal years 1980–1989, now available* (Report No. NCES 90–662a). Washington, DC: U.S. Department of Education, Office of Educational Research and Improvement.

National Center for Education Statistics. (1990b). *Targeted forecast: Public elementary and secondary current expenditures: 1987–88 to 1993–94* (Report No. NCES 90–686). Washington, DC: U.S. Department of Education, Office of Educational Research and Improvement.

National Center for Education Statistics. (1990c). *Targeted forecast: High school graduates: 1987–88 to 1993–94*. (Report No. NCES 90–691). Washington, DC: U.S. Department of Education, Office of Educational Research and Improvement.

National Center for Health Statistics. (1987a). *Aging in the eighties: Functional limitations of individuals age 65 years and over* (Advance Data No. 133). Hyattsville, MD: Public Health Service.

National Commission on Excellence in Education. (1983). *A nation at risk: The imperative for educational reform*. Washington, DC: U.S. Department of Education.

National Commission on Space. (1986). *Pioneering the space frontier*. New York: Bantam Books.

National Center for Health Statistics. (1987b). *Vital statistics of the United States: Vol. 2. Mortality, Part A*. Hyattsville, MD: Public Health Service.

National Highway Traffic Safety Administration. (1989). *Fatal accident reporting system 1988*. Report DOT HS-807-507. Washington, DC: U.S. Department of Transportation.

National Information and Education Utilities Corporation. (1990). *The education utility*. Unpublished manuscript, National Information and Education Utilities Corporation, Vienna, VA.

National Research Council. (1977). *Drinking water and health*. Washington, DC: National Academy of Sciences.

National Research Council. (1983). *Video display, work, and vision*. Washington, DC: National Academy Press.

National Research Council. (1984). *Research needs on the interaction between information systems and their users: Report of a workshop*. Washington, DC: National Academy Press.

National Research Council. (1988) *Toward a national research network*. Washington, DC: National Academy Press.

National Research Council (1987). *Current issues in atmospheric change*. Washington, DC: National Academy Press.

National Research Council. (1989a). Advice for President Bush. *National Research Council Newsreport, 34*(3), 9–12.

National Research Council. (1989b). *Alternative agriculture*. Washington, DC: National Academy Press.

National Research Council. (1989c). *Improving risk communication*. Committee on Risk Perception and Communication, National Research Council. Washington, DC: National Academy Press.

National Science Board. (1986). *Undergraduate science, mathematics and engineering education*. Washington, DC: Author.

National Science Board. (1987). *Science and engineering indicators—1987*. Washington DC: Author.

National Science Board. (1989). *Loss of biological diversity: A global crisis requiring international solutions*. Washington, DC: Author.

National Science Foundation. (1988). *International science and technology data update: 1988*. (Special Report No. NSF 89-307). Washington, DC: Author.

Negroponte, N. P. (1991). Products and services for computer networks. *Scientific American, 265*(3), 106–113.

Neisser, U. (1967). *Cognitive psychology*. New York: Appleton-Century-Crofts.

Newell, A. (1987). Keynote address: Human factors research for the NASA space station. In T. B. Sheridan, D. S. Kruser, & S. Deutsch (Eds.), *Human factors in automated and robotic space systems: Proceedings of a symposium* (pp. 17–27). Washington, DC: National Research Council.

Newell, R. E., Reichle, H. G., Jr., & Seiler, W. (1989). Carbon monoxide and the burning earth. *Scientific American, 261*(4), 82–88.

Newman, D. (1987). Local and long distance computer networking for science classrooms. *Educational Technology, 27*(6), 20–23.

Newman, D. (1990a). Cognitive and technical issues in the design of educational computer networking. In L. Harasion (Ed.), *Online education: Perspectives on a new medium*. New York: Praeger.

Newman, D. (1990b). Why a school with Apple II computers should get a Macintosh network. In B. Bowen (Ed.) *Design for learning: Research-based design of technology for learning* (pp. 89–96). Cupertino, CA: External Research, Apple Computer, Inc.

Nickerson, R. S. (1969). Man–computer interaction: A challenge for human factors research. *Ergonomics, 12*, 501–517. (Reprinted in *IEEE Transactions: Man–Machine Systems, MMS-10*, 164–180.)

Nickerson, R. S. (1976). On conversational interaction with computers. In S. Treu (Ed.), *User-oriented design of interactive graphics systems: Proceedings of the ACM/SIGGRAPH Workshop* (pp. 101–113). Pittsburgh, PA.

Nickerson, R. S. (1986). *Using computers: Human factors in information technology*. Cambridge, MA: MIT Press.

Nickerson, R. S. (1987). Productivity in the space station. In T. B. Sheridan, D. S. Kruger, & S. Deutsch (Eds.) *Human factors in automated and robotic space systems: Proceedings of a symposium* (pp. 31–81). Washington, DC: National Research Council, Committee on Human Factors.

Nickerson, R. S. (1991). Understanding and controlling environmental change: Challenges and opportunities for information technology. In N. Moray, W. R. Ferrell, & W. B. Rouse (Eds.), *Robotics, control and society: Essays in honor of Thomas B. Sheridan* (pp. 225–252). New York: Taylor & Francis.

Nickerson, R. S. (1988). Technology in education: Possible influences on context, purposes, content and methods. In R. S. Nickerson & P. P. Zodhiates (Eds.), *Technology in education: Looking toward 2020* (pp. 285–317). Hillsdale, NJ: Lawrence Erlbaum Associates.

Nickerson, R. S., Baddeley, A., & Freeman, B. (1987). Are people's estimates of what other people know influenced by what they themselves know? *Acta Psychological, 64*, 245–259.

Nickerson, R. S., & Huggins, A. W. F. (1977). *The assessment of speech quality* (BBN Report No. 3486). Cambridge, MA: Bolt, Beranek & Newman.

Nickerson, R. S., Perkins, D., & Smith, E. E. (1985). *The teaching of thinking*. Hillsdale, NJ: Lawrence Erlbaum Associates.

Nickerson, R. S., & Pew, R. W. (1990, July). Toward more compatible human–computer interfaces. *Spectrum*, pp. 40–43.

Nickerson, R. S., & Zodhiates, P. P. (Eds.). (1988). *Technology in education: Looking toward 2020*. Hillsdale, NJ: Lawrence Erlbaum Associates.

Nierenberg, W. A. (1990). Global warming report [Letter to the Editor]. *Science, 247*, 14.

Nilles, J. M., Carlson, F. R., Gray, P., & Hanneman, G. J. (1976). *The telecommunications–transportation tradeoff*. New York: Wiley.

Nitzel, M. T., & Winett, R. A. (1977). Demographics, attitudes, and behavioral responses to important environmental events. *American Journal of Community Psychology, 5*, 195–206.

Nolan, J. B., & Wheelon, A. D. (1990). The Third World ballistic missiles. *Scientific American, 263*(2), 34–40.

Nordqvist, T., Ohlsson, K., & Nilsson, L-G. (1986). Fatigue and reading of text on videotext. *Human Factors, 28*, 353–363.

Norman, C. (1989). HDTV: The technology duJour. *Science, 244*, 761–764.

Norman, D. A., & Draper, S. W. (1986). *User centered system design: New perspectives on human–computer interaction*. Hillsdale, NJ: Lawrence Erlbaum Associates.

Norman, K. L. (1991). *The psychology of menu selection: Designing cognitive control at the human/computer interface*. Norwood, NJ: Ablex.

Norman, K. L., & Chin, J. P. (1988). The effect of tree structure on search in a hierarchical menu selection system. *Behaviour and Information Technology, 7*, 51–65.

Oakley, L. C., & Oakley, R. C. (1990, June/July). Long Island Education Utility, Inc.: A revolution in education, *Personal Investing News*, pp. 6–7.

Odum, E. P. (1989). Input management of production systems. *Science, 243,* 177–182.

Odum, H. T. (1971). *Environment, power, and society.* New York: Wiley Interscience.

Offermann, L. R., & Gowing, M. K. (1990). Organizations of the future: Changes and challenges. *American Psychologist, 45,* 95–108.

Office of Educational Research and Improvement. (1990, Summer). A quarter of the nation. *OERI Bulletin,* pp. 1–2.

Office of Science and Technology Policy. (1989). *The federal high performance computing program.* Washington, DC: Executive Office of the President.

Office of Technology Assessment. (1981). *Computer-based national information systems: Technology and public policy issues.* Washington, DC: Author.

Office of Technology Assessment. (1982). *Technology and handicapped people.* Washington, DC: Author.

Office of Technology Assessment. (1984a). *Acid rain and transported air pollutants: Implications for public policy.* Washington, DC: Author.

Office of Technology Assessment. (1984b). *Protecting the nation's groundwater from contamination: Vols. 1 & 2.* Washington, DC: Author.

Office of Technology Assessment. (1985a). *Federal government information technology: Electronic surveillance and civil liberties.* Washington, DC: Author.

Office of Technology Assessment. (1985b). *Information technology R & D: Critical trends and issues.* Washington, DC: Author.

Office of Technology Assessment. (1985c). *International cooperation and competition in civilian space activities.* Washington, DC: Author.

Office of Technology Assessment. (1985d). *New electric power technologies.* Washington, DC: Author.

Office of Technology Assessment. (1985e). *Technology and aging in America.* Washington, DC: U.S. Government Printing Office.

Office of Technology Assessment. (1986a). *Assessing biological diversity in the United States: Data considerations. Background paper #2.* Washington, DC: Author.

Office of Technology Assessment. (1986b). *Federal government information technology: Electronic record systems and individual privacy.* Washington, DC: Author.

Office of Technology Assessment. (1986c). *Intellectual property rights in an age of electronics and information.* Washington, DC: Author.

Office of Technology Assessment. (1986d). *Technology and structural unemployment: Re-employing displaced adults.* Washington, DC: Author.

Office of Technology Assessment. (1987a). *The electronic supervisor: New technology, new tensions.* Washington, DC: Author.

Office of Technology Assessment. (1987b). *Life-sustaining technologies and the elderly.* Washington, DC: Author.

Office of Technology Assessment. (1987c). *Losing a million minds: Confronting the tragedy of Alzheimer's disease and other dementias.* Washington, DC: Author.

Office of Technology Assessment. (1987d). *Ownership of human tissues and cells: Special report.* Washington, DC: Author.

Office of Technology Assessment. (1988a). *Power on! New tools for teaching and learning.* Washington, DC: Author.

Office of Technology Assessment. (1988b). *Safe skies for tomorrow: Aviation safety in a competitive environment.* Washington, DC: Author.

Office of Technology Assessment. (1988c). *Technology and the American economic transition.* Washington, DC: Author.

Office of Technology Assessment. (1988d). *Urban ozone and the Clean Air Act: Problems and proposals for change.* Washington, DC: Author.

Office of Technology Assessment. (1989a). *Annual report to the Congress for fiscal year 1988.* Washington, DC: Author.

Office of Technology Assessment. (1989b). *Oil production in the Arctic National Wildlife Refuge: The technology and the Alaskan oil context* (OTA Report Brief). Washington, DC: Author.

Office of Technology Assessment. (1990a). *Making things better: Competing in manufacturing*. Washington, DC: Author.

Office of Technology Assessment. (1990b). *Worker training: Competing in the new international economy* (Report No. OTA-ITE-457). Washington, DC: U.S. Government Printing Office.

O'Hare, M. (in press). Risk anticipation as a social cost. In C. Haar & J. Kayden (Eds.), *The NIMBY problem* Cambridge, MA: Lincoln Institute of Land Policy.

O'Leary, P. R., Walsh, P. W., & Ham, R. K. (1988). Managing solid waste. *Scientific American, 259*(6), 36–42.

Olsen, M. E. (1981). Consumers' attitudes toward energy conservation. *Journal of Social Issues, 37,* 108–131.

Olsen, R. (1978). *The effect of the hospital environment*. Unpublished doctoral dissertation, The City University of New York, New York.

Olson, M. H. (1987). Telework: Practical experience and future prospects. In R. E. Kraut (Ed.), *Technology and the transformation of white-collar work* (pp. 135–152). Hillsdale, NJ: Lawrence Erlbaum Associates.

Olson, M. H. (1988). Organizational barriers to telework. In W. B. Korte, W. J. Steinle, & S. Robinson (Eds.), *Telework: Present situation and future development of a new form of work*. Amsterdam: North Holland.

O'Malley, C. E. (1986). Helping users help themselves. In D. A. Norman & S. W. Draper (Eds.), *User centered system design: New perspectives on human–computer interaction* (pp. 361–398). Hillsdale, NJ: Lawrence Erlbaum N Associates.

Osterman, P. (1988). *Employment futures: Reorganization, dislocation, and public policy*. New York: Oxford University Press.

Owens, D. (1986). Naive theories of computation. In D. A. Norman & S. W. Draper (Eds.), *User centered system design: New perspectives on human–computer interaction* (pp. 187–200). Hillsdale, NJ: Lawrence Erlbaum Associates.

Owens, M. R. (1988). *Preliminary staff report on educational research, development, and dissemination: Reclaiming a vision of the federal role for the 1990s and beyond*. Majority staff, Subcommittee on Select Education of the Committee on Education and Labor, U.S. House of Representatives.

Paap, K. R., & Roske-Hofstrand, R. J. (1986). The optimal number of many options per panel. *Human Factors, 28,* 377–386.

Panel on the Improvement of Tropical and Subtropical Rangelands. (1990). *The improvement of tropical and subtropical rangelands*. Board on Science and Technology for International Development, Office of International Affairs, National Research Council, Washington, DC: National Academy Press.

Parker, J. F., Duffy J. W., & Christensen, D. G. (1981). *A flight investigation of simulated data link communications during single pilot IFR flight*. Falls Church, VA: BioTechnology Inc.

Parsons, H. M. (1985). Automation and the individual: Comprehensive and comparative views. *Human Factors, 27,* 99–112.

Parthenopoulos, D. A., & Rentzepis, P. M. (1989). Three-dimensional optical storage memory. *Science, 245,* 843–845.

Partridge, C. (1990). A faster data delivery. *Unix Review, 8*(3), 43–48.

Partyka, S. (1983). *Comparison by age of drivers in two car fatal crashes*. In Transportation in an aging society: Improving mobility and safety for older persons (Vol. 1, p. 84). Washington, DC: National Academy Press.

Partyka, S. C. (1989). Registration-based fatality rates by car size from 1978 through 1987. In *Papers on car size: Safety and trends* (DOT Report HS-807-44, pp. 45–72). Washington, DC: U.S. Department of Transportation, National Highway Traffic Safety Administration.

Partyka, S. C., & Boehly, W. A. (1989). Passenger car weight and injury severity in single vehicle nonrollover crashes. In *Papers on car size: Safety and trends* (DOT Report HS-807–44, pp. 73–107). Washington, DC: U.S. Department of Transportation, National Highway Traffic Safety Administration.

Passell, P. (1989, November 19). Cure for greenhouse effect: The costs will be staggering. The *New York Times*.

Paulus, P. B. (1980). Crowding. In P. B. Paulus (Ed.), *Psychology of group influence*. Hillsdale, NJ: Lawrence Erlbaum Associates.

Pea, R., Boyle, E., & de Vogel, R. (1990). Design spaces for multimedia composing tools. In B. Bowen (Ed.), *Design for learning: Research-based design of technology for learning* (pp. 37–42). Cupertino, CA: External Research, Apple Computer, Inc.

Pelca, J. (1990). Bitnet headed for new frontiers. *Science, 247,* 520.

Peled, A. (1987). The next computer revolution. *Scientific American, 257*(4), 56–64.

Peltier, W. R., & Tushingham, A. M. (1989). Global sea level rise and the greenhouse effect: Might they be connected? *Science, 244,* 806–810.

Pennebaker, J. W., & Brittingham, J. L. (1982). Environmental and sensory cues affecting the perception of physical symptoms. In A. Baum & J. E. Singer (Eds.), *Advances in environmental psychology: Vol. 4. Environment and health* (pp. 115–135). Hillsdale, NJ: Lawrence Erlbaum Associates.

Perchonok, K. (1972). *Accident cause analysis*. Ithaca, NY: Cornell Aeronautical Laboratory.

Perry, T. S. (1981). Engineering education: Coping with the crisis. *IEEE Spectrum, 18*(11), 65–71.

Perry, T. S. (1989). Science observer: A new world of viewer controlled video. *American Scientist, 77,* 130–132.

Peterson, W. L. (1984, June). The rage to automate: An interview with Captain Mel Hoagland. *Airline Pilot,* pp. 15–17.

Petsko, G. A. (1988). Protein engineering. In M. L. Good (Ed.), *Biotechnology and materials science: Chemistry for the future* (pp. 53–60). Washington, DC: American Chemical Society.

Pew, R. W., Baron, S., Feehrer, C. E., & Miller, D. C. (1977). *Critical review and analysis of performance models applicable to man-machines systems evaluation* (Tech. Rep. No. 3446). Cambridge, MA: Bolt, Beranek & Newman.

Pfeiffer, D. (1985). *The number of disabled persons in the U.S. and its policy implications*. Paper based on presentation at the 1985 meeting of the Association on Handicapped Student Service programs in Post-Secondary Education, Atlanta, GA.

Phaff, H. J. (1981). Industrial microorganisms. *Scientific American, 245*(3), 76–89.

Pickett, R. M., & Triggs, T. T. (1975). Human factors in health care. Lexington, MA: DC Heath and Company.

Pillsbury, A. F. (1981). The salinity of rivers. *Scientific American, 245*(1), 54–65.

Pines, M. (1987). *Mapping the human genome*. Bethesda, MD: Howard Hughes Medical Institute.

Pisoni, D. B. (1982). Perception of speech: The human listener as a cognitive interface. *Speech Technology, 1,* 10–23.

Platt, J. (1973). Social traps. *American Psychologist, 28,* 641–651.

Plough, A., & Krimsky, S. (1987). The emergence of risk communication studies: Social and political context. *Science, Technology and Human Values, 12*(34), 4–10.

Pool, R. (1989a). Fusion breakthrough? *Science, 243,* 1661–1662.

Pool, R. (1989b). IBM wins a patent for thallium superconduction. *Science, 246,* 320.

Pool, R. (1989c). Zero resistance at 250K? *Science, 246,* 320.

Pool, R. (1990a). A small revolution gets underway. *Science, 247,* 26–27.

Pool, R. (1990b). Who will do science in the 1990s? *Science, 248,* 433–435.

Port, O. (1989, March 13). Quantum effect chips. *Business Week,* pp. 70–74.

Post, D. L. (1985). Effects of color on CRT symbol legibility. *SID Digest, 16,* 196–199.

Post, W. M., Peng, T-H., Emmanuel, W. R., King, A. W., Dale, V. H., & DeAngelis, D. L. (1990). The global carbon cycle. *American Scientist, 78,* 310–326.

Postel, S. (1985, June). Thirsty in a water-rich world. *International Wildlife,* pp. 32–37.

Powell, C. S. (1990). Science and business: Plastic goes green. *Scientific American, 263*(2), 101.

Powledge, T. M. (1989). What shall we do about science education? *The AAAS Observer, 5,* pp. 1, 6, 7.

President's Commission on Industrial Competitiveness. (1985). *Global competition: The new reality, Vol. 2.* Washington, DC: U.S. Government Printing Office.

Press, F. (1981). *Report on the computational needs for physics.* Washington, DC: National Academy of Sciences.

Product Insight (1990, October). AI comes of age: Builds on Digital's knowledge base. Digital Equipment Corporation, pp. 8–11.

Proffitt, M. H., Fahey, D. W., Kelly, K. K., & Tuck, A. F. (1989). High latitude ozone loss outside the Antarctic ozone hole. *Nature, 342,* 233–237.

Psotka, J., Massey, L. D., & Mutter, S. A. (Eds.). (1988). *Intelligent tutoring systems.* Hillsdale, NJ: Lawrence Erlbaum Associates.

Pursel, V. G., Pinkert, C. A., Miller, K. F., Bolt, D. J., Campbell, R. G., Palmiter, R. D., Brinster, R. L., & Hammer, R. E. (1989). Genetic engineering of livestock. *Science, 244,* 1281–1288.

Quinn, J. B., Baruch, J. J., & Paquette, P. C. (1987). Technology in services. *Scientific American, 257*(6), 50–58.

Rafelski, J., & Jones, S. E. (1987). Cold nuclear fusion. *Scientific American, 257*(4), 84–89.

Ramanathan, V. (1988). The greenhouse theory of climate change: A test by an inadvertent global experiment. *Science, 240,* 293–299.

Ramanathan, V., Cess, R. D., Harrison, E. F., Minnis, P., Barkstrom, B. R., Ahmad, E., & Hartmann, D. (1989). Cloud-radiative forcing and climate: Results from the earth radiation budget experiment. *Science, 243,* 57–63.

Rasmussen, W. D. (1982). The mechanization of agriculture. *Scientific American, 247*(3), 76–89.

Rathje, W. L. (1989, December). Rubbish! *The Atlantic Monthly,* pp. 99–109.

Rauch, J. (1989, August). Kids as capital. *The Atlantic Monthly,* pp. 56–61.

Ray, N. M., & Minch, R. P. (1990). Computer anxiety and alienation: Toward a definitive and parsimonious measure. *Human Factors, 32,* 477–491.

Reddy, K. N., & Goldemberg, J. (1990). Energy for the developing world. *Scientific American, 263*(3), 110–118.

Reganold, J. P., Papendick, R. I., & Parr, J. F. (1990). Sustainable agriculture. *Scientific American, 262*(6), 112–120.

Regnier, V., & Pynoos, J. (Eds.). (1987). *Housing the aged.* New York: Elsevier.

Reich, R. B. (1983). *The next American frontier.* New York: Times Books.

Reich, R. B. (1987, May, June). Entrepreneurship reconsidered: The team as hero. *Harvard Business Review,* pp. 77–83.

Reich, R. B. (1989). The quiet path to technological preeminence. *Scientific American, 261*(4), 41–47.

Reichel, D. A., & Geller, E. S. (1981). Applications of behavioral analysis for conserving transportation energy. In A. Baum & J. E. Singer (Eds.), *Advances in environmental psychology: Vol. 3. Energy: Psychological perspectives* (pp. 53–91). Hillsdale, NJ: Lawrence Erlbaum Associates.

Reisner, M. (1988–1989). The next water war: Cities versus agriculture. *Issues in Science and Technology, 5*(2), 98–102.

Reizenstein, J. E. (1982). Hospital design and human behavior: A review of the recent literature.

In A. Baum & J. E. Singer (Eds.), *Advances in environmental psychology: Vol. 4. Environment and health* (pp. 137–169). Hillsdale, NJ: Lawrence Erlbaum Associates.

Remington, R., & Williams, D. (1986). On the selection and evaluation of visual display symbology: Factors influencing search and identification times. *Human Factors, 28,* 407–420.

Reno, A. T. (1988). Personal mobility in the United States. In *A look ahead: Year 2020* (Special Report No. 220). Washington, DC: National Research Council, Transportation Research Board.

Repetto, R. (1990). Deforestation in the tropics. *Scientific American, 262*(4), 36–42.

Resnick, L. B., & Johnson, A. (1988). Intelligent people: Cognitive theory and the future of computer-assisted learning. In R. S. Nickerson & P. P. Zodhiates (Eds.), *Technology in education: Looking toward 2020* (pp. 231–264). Hillsdale, NJ: Lawrence Erlbaum Associates.

Rettig, M., & Bates, M. (1988, July). How to choose natural language software. *AI Expert,* pp. 41–49.

Revelle, R. (1982). Carbon dioxide and world climate. *Scientific American, 247*(2), 35–43.

Revelle, R., & Suess, H. (1957). Carbon dioxide exchange between atmosphere and ocean and the question of an increase of atmospheric CO_2 during the past decades. *Tellus, 9,* 18–27.

Richards, J., Barowy, W., & Levin, D. (1992). Computer simulations in the science classroom. *Journal of Science Education and Technology, 1,* 67–79.

Riche, M. F. (1988, February). America's new workers. *American Demographics,* pp. 34–41.

Richelson, J. T. (1991). The future of space reconnaisance. *Scientific American, 264*(1), 38–44.

Ritchie, M. L. (1988). General aviation. In E. L. Wiener & D. C. Nagel (Eds.), *Human factors in aviation* (pp. 561–589). New York: Academic Press.

Rivlin, A. M., & Wiener, J. M., with Hanley, R. J., & Spence, D. A. (1988). *Caring for the disabled elderly: Who will pay?* Washington, DC: Brookings Institution.

Roberts, L. (1988). A sequencing reality check. *Science, 242,* 1245.

Roberts, L. (1989a). Ethical questions haunt new genetic technologies. *Science, 243,* 1134–1136.

Roberts, L. (1989b). Genome project underway, at last. *Science, 243,* 167–168.

Roberts, L. (1989c). Global warming: Blaming the sun. *Science, 246,* 992.

Roberts, L. (1989d). New chip may speed genome analysis. *Science, 244,* 655–656.

Roberts, L. (1989e). New game plan for genome mapping. *Science, 245,* 1438–1440.

Roberts, L. (1990a). Genome backlash going full force. *Science, 248,* 804.

Roberts, L. (1990b). News and comment: Down to the wire for the NF gene. *Science, 249,* 236–238.

Roberts, L. (1990c). News and comment: Counting on science at EPA. *Science, 249,* 616–618.

Roberts, N., Carter, R., Davis, F., & Feurzeig, W. (1989). Power tools for algebra problem solving. *Journal of Mathematical Behavior, 8,* 251–265.

Roberts, P. O., & Fauth, G. R. (1988). The outlook for commercial freight. In *A look ahead: Year 2020* (Special Report No. 220). Washington, DC: National Research Council, Transportation Research Board.

Roberts, T. L., & Moran, T. P. (1983). The evaluation of text editors: Methodology and empirical results. *Communications of the ACM, 26,* 265–283.

Rodhe, H. (1990). A comparison of the contribution of various gases to the greenhouse effect. *Science, 248,* 1217–1219.

Rose, A. H. (1981). Microbiological production of food and drinks. *Scientific American, 245*(3), 126–138.

Rosenberg, N., & Birdzell, L. E., Jr. (1990). Science, technology and the western miracle. *Scientific American, 263*(5), 42–54.

Rosenberg, N. J., Easterling, W. E., III, Crosson, P. R., & Darmstadter, J. (Eds.). (1989). *Greenhouse warming: Abatement and adaptation.* Washington, DC: Resources for the Future.

Rosenberg, S. A., Aebersold, P., Cornetta, K., Kasid, A., Morgan, R. A., Moen, R., Karson, E. M., Lotze, M. T., Yang, J. C., Topalian, S. L., Merino, M. J., Culver, K., Miller, A. D., Blaese, R. M., & Anderson, W. F. (1990). Gene transfer into humans: Immunotherapy of patients with advanced melanoma, using tumor-infiltrating lymphocytes modified by retroviral gene transduction. *The New England Journal of Medicine, 323,* 570–578.

Rosenfeld, A. H., & Hafemeister, D. (1988). Energy efficient buildings. *Scientific American, 258*(4), 78–85.

Ross, M. H., & Steinmeyer, D. (1990). Energy for industry. *Scientific American, 263*(3), 88–98.

Rossi, P. H. (1989). *Down and out in America.* Chicago, IL: University of Chicago Press.

Roszak, T. (1986). *The cult of information: The folklore of computers and the true art of thinking.* New York: Pantheon Books.

Rothenberg, J., Sander, M. (1990). *Report of the information systems strategic planning project.* Washington, DC: NASA.

Rouse, W. B. (1980). *Systems engineering models of human–machine interaction.* New York: North Holland.

Rouse, W. B. (1988). Adaptive aiding for human/computer control. *Human Factors, 30,* 431–443.

Rowell, L. B. (1978). Human adjustments and adaptation to heat stress: Where and how? In L. J. Folinsbee, J. A. Wagner, J. F. Borgia, B. L. Drinkwater, J. A. Gliner, & J. F. Bedi (Eds.), *Environmental stress: Individual adaptations.* New York: Academic Press.

Rowles, G. D. (1978). *Prisoners of space? Exploring the geographical experiences of older people.* Boulder, CO: Westview Press.

Rumelhart, D. E., & Norman, D. A. (1981). Analogical processes in learning. In J. R. Anderson (Ed.), *Cognitive skills and their acquisition* (pp. 255–359). Hillsdale, NJ: Lawrence Erlbaum Associates.

Runnion, A., Watson, J. D., & McWahorter, J. (1978). Energy savings in interstate transportation through feedback and reinforcement. *Journal of Organizational Behavior Management, 1,* 180–191.

Rupp, B. (1984). *Human factors of work stations with visual displays.* Yorktown Heights, NY: IBM Corporation.

Russell, M. (1988). Ozone pollution: The hard choices. *Science, 241,* 1275–1276.

Salt Institute (1980). *Survey of salt, calcium chloride, and abrasive use in the United States and Canada.* Alexandria, VA: Author.

Salthouse, T. A. (1982). *Adult cognition: An experimental psychology of human aging.* New York: Springer-Verlag.

Samdahl, D. M., & Robertson, R. (1989). Social determinants of environmental concern: Specification and test of the model. *Environment and Behavior, 21,* 57–81.

Sanderson, P. M. (1989). The human planning and scheduling role in advanced manufacturing systems: An emerging human factors domain. *Human Factors, 31,* 635–666.

Sanford, D. W. (1981). Where was I? In D. R. Hofstadter & D. C. Dennett (Eds.), *The mind's I: Fantasies and reflections on self and soul* (pp. 232–240). New York: Basic Books.

Sassin, W. (1980). Energy. *Scientific American, 243*(3), 118–132.

Sayer, M., & Sreenivas, K. (1990). Ceramic thin films: Fabrication and applications. *Science, 247,* 1056–1060.

Scardamalia, M., & Bereiter, C. (1990a) Computer supported intentional learning environments (CSILE). In B. Bowen (Ed.), *Design for learning: Research-based design of technology for learning* (pp. 5–14). Cupertino, CA: External Research, Apple Computer, Inc.

Scardamalia, M., & Bereiter, C. (1990b). Schools as knowledge-building communities. In S. Strauss (Ed.), *Human development* (Vol. 5). Norwood, NJ: Ablex.

Scardamalia, M., Bereiter, C., McLean, R., Swallow, J., & Woodruff, E. (1989). Computer-supported intentional learning environments. *Journal of Educational Computing Research, 5,* 51–68.

Schahn, J., & Holzer, E. (1990). Studies of individual environmental concern: The role of knowledge, gender, and background variables. *Environment and Behavior, 22,* 767–786.

Schein, E. H. (1990). Organizational culture. *American Psychologist, 45,* 109–119.

Schlesinger, L. A., & Oshry, B. (1984). Quality of worklife and the supervisor: Model in the middle. *Organizational Dynamics, 13,* 4–20.

Schlesinger, M. E., & Mitchell, J. F. B. (1987). Climate model simulations of the equilibrium climatic to increased carbon dioxide. *Review of Geophysics, 25,* 760–798.

Schlesinger, W. H., Reynolds, J. F., Cunningham, G. L., Huenneke, L. F., Gerrell, W. M., Virginia, R. A., & Whitford, W. G. (1990). Biological feedbacks in global desertification. *Science, 247,* 1043–1048.

Schneider, C. (1989, November 17). *Global warming prevention act* [Testimony before the House of Representatives]. Congressional Record, E3889–E3891.

Schneider, S. H. (1989). The greenhouse effect: Science and policy. *Science, 243,* 771–781.

Schoolland, K. (1990). *Shogun's ghost.* New York: Bergin & Garvey.

Schoonard, J. W., & Boies, S. J. (1975). Short-type: A behavioral analysis of typing and text entry. *Human Factors, 17,* 203–214.

Schulze, E. D. (1989). Air pollution and forest decline in a spruce *(Picea abies)* forest. *Science, 244,* 776–783.

Schurr, S. H. (1963). Energy. *Scientific American, 209*(3), 110–127.

Schwab, E. C., Nusbaum, H. C., & Pisoni, D. B. (1985). Some effects of training on the perception of synthetic speech. *Human Factors, 27* 395–408.

Schwartz, A. (Ed.). (1984). *Handbook of microcomputer applications in communication disorders.* San Diego: College-Hill Press.

Schwartz, J. P., Norman, K. L., & Shneiderman, B. (1985). *Performance on content-free menus as a function of study method* (Report No. CAR-TR-110). Baltimore, MD: University of Maryland, Center for Automation Research.

Schwartz, J. (in press). *The Geometric Supposer: What is it a case of?* Hillsdale, NJ: Lawrence Erlbaum Associates.

Schwartz, S. E. (1989). Acid deposition: Unraveling a regional phenomenon. *Science, 243,* 753–763.

Science and the citizen (1987). *Scientific American, 256*(6), 22–24.

Scrimshaw, N. S., & Taylor, L. (1980). Food. *Scientific American, 243*(3), 78–88.

Secretary of State for Transport. (1989, November). *Investigation into the Clapham Junction railway accident.* London: Her Majesty's Stationary Office.

Segal, H. P. (1986). The technological utopians. In J. J. Corn (Ed.), *Imagining tomorrow: History, technology, and the American future* (pp. 119–136). Cambridge, MA: MIT Press.

Seinfeld, J. H. (1989). Urban air pollution: State of the science. *Science, 243,* 752–754.

Seitz, F., Jastrow, R., & Nierenberg, W. A. (1989). *Scientific perspectives on the greenhouse problem.* Washington, DC: George C. Marshall Institute.

Selfridge, O. G. (1959). Pandemonium: A paradigm for learning. In *The mechanization of thought processes.* London: Her Majesty's Stationary Office.

Selfridge, O. G., & Neisser, U. (1960). Pattern recognition by machine. *Scientific American, 203*(2), 60–68.

Seligman, C., Becker, L. J., & Darley, J. M. (1981). Encouraging residential energy conservation through feedback. In A. Baum & J. E. Singer (Eds.), *Advances in environmental psychology: Vol. 3. Energy: Psychological perspectives* (pp. 93–113). Hillsdale, NJ: Lawrence Erlbaum Associates.

Seligman, C., & Darley, J. M. (1977). Feedback as a means of decreasing residential energy consumption. *Journal of Applied Psychology, 62,* 363–368.

Seligman, M. P. (1975). *Helplessness: On depression development and death.* San Francisco: W. H. Freeman.

Seppala, P., & Salvendy, G. (1985). Impact of depth of menu hierarchy on performance effectiveness in a supervisory task: Computerized flexible manufacturing system. *Human Factors, 27,* 713–722.

Shaikin, H. (1988). *Work transformed: Automation and labor in the computer age.* Lexington, MA: Lexington Books.

Shalin, V. L., Wisniewski, E. J., & Levi, K. R. (1988). A formal analysis of machine learning systems for knowledge acquisition. *International Journal of Man–Machine Studies, 29,* 429–446.

Sharit, J. (1985). Supervisory control of a flexible manufacturing system. *Human Factors, 27,* 47–59.

Sharit, J., Chang, T. C., & Salvendy, G. (1987). Technical and human aspects of computer-aided manufacturing. In G. Salvendy (Ed.), *Handbook of human factors* (pp. 1694–1723). New York: Wiley.

Shaw, R. W. (1987). Air pollution biparticles. *Scientific American, 257*(2), 96–103.

Sheil, B. (1983). Power tools for programmers. *Datamation, 29* pp. 131–144.

Sheldrick, M. G. (1990). Driving while automated. *Scientific American, 263*(1), 86–88.

Sheppard, H. L. (1978). *Research and development strategy on employee related problems of older workers.* Washington, DC: U.S. Department of Labor, Manpower Administration.

Sheridan, T. B. (1980). Computer control and human alienation. *Technology Review, 83*(1), 60–67, 71–73.

Sheridan, T. B. (1987). Supervisory control. In G. Salvendy (Ed.), *Handbook of human factors: Ergonomics.* New York: Wiley.

Sheridan, T. B. (1988). The system perspective. In E. L. Wiener & D. C. Nagel (Eds.), *Human factors in aviation* (pp. 27–51). New York: Academic Press.

Sheridan, T. B., & Simpson, R. W. (1979). *Toward the definition and measurement of the mental workload of transport pilots* (Tech. Rep. No. DOT-OS-70055) Cambridge, MA: MIT, Flight Transportation Man–Machine Laboratory.

Shiba, S. (1989a, April). Managers of quality. *Look Japan,* pp. 30–31.

Shiba, S. (1989b, May). Quality knows no bounds. *Look Japan,* pp. 30–31.

Shiba, S. (1989c, July). Lessons in quality. *Look Japan,* pp. 32–33.

Shiba, S. (1989d, August). Moving in the right circles. *Look Japan,* pp. 28–29

Shiba, S. (1989e, September). Quality, from top to bottom. *Look Japan,* pp. 36–37.

Shiba, S. (1989f, October). Universal quality. *Look Japan,* pp. 32–33.

Shinar, D. (1978). *Psychology on the road.* New York: Wiley.

Shleifer, A., & Vishny, R. W. (1990). The takeover wave of the 1980s. *Science, 249,* 745–749.

Shneiderman, B. (1987). *Designing the user interface: Strategies for effective human–computer interaction.* Reading, MA: Addison-Wesley.

Short, J., Williams, E., & Christie, B. (1976). *A social psychology of telecommunications.* New York: Wiley.

Shukla, J., Nobre, C., & Sellers, P. (1990). Amazon deforestation and climate change. *Science, 247,* 1322–1325.

Shulman, H. G., & Olex, M. B. (1985). Designing the user-friendly robot: A case history. *Human Factors, 27,* 91–98.

Shulman, S. (1989, July). When a nuclear reactor dies, 98 million dollars is a cheap funeral. *Smithsonian,* pp. 56–69.

Shumate, P. W., Jr. (1989). Optical fibers reach into homes. *IEEE Spectrum, 26,*(2), 43–47.

Sieber, J. E. (1974). Effects of decision importance on ability to generate warranted subjective uncertainty. *Journal of Personality and Social Psychology, 30,* 688–694.

Silvestro, K. (1988). Using explanations for knowledge-base acquisition. *International Journal of Man–Machine Studies, 29,* 159–169.

Simmons, H. E. (1988). Biotechnology: A new marriage of chemistry and biology. In M. L. Good (Ed.), *Biotechnology and materials science: Chemistry for the future* (pp. 7–19). Washington, DC: American Chemical Society.

Simpson, C. A. (1981). Evaluation of synthesized voice approach callouts (SYNCALL). In J. Moraal & K. F. Kraiss (Eds.), *Manned systems design: Methods, equipment, and applications* (pp. 375–393). New York: Plenum.

Simpson, C. A. (1983). Evaluating computer speech devices for your application. In J. C. Warren (Ed.), *Proceedings of the Seventh West Coast Computer Faire* (pp. 395–401). Woodside, CA: West Coast Computer Faire.

Simpson, C. A., & Marchionda-Frost, K. (1984). Synthesized speech rate and pitch effects on intelligibility of warning messages for pilots. *Human Factors, 26,* 509–517.

Simpson, C. A., McCauley, M. E., Roland, E. F., Ruth, J. C., & Williges, B. H. (1985). System design for speech recognition and generation. *Human Factors, 27,* 115–141.

Simpson, C. A., & Navarro, T. N. (1984). Intelligibility of computer-generated speech as a function of multiple factors. In *Proceedings of the National Aerospace and Electronics Conference* (pp. 932–940). New York: IEEE.

Simutis, Z. M., Ward, J. S., Harman, J., Farr, B. J., & Kern, R. P. (1988). *ARI research in basic skills education: An overview* (Research Report 1486). Alexandria VA: U. S. Army Research Institute for the Behavioral and Social Sciences.

Sind, P. M. (1990). Human factors in medical equipment design: An emerging and expanding frontier. *Human Factors Society Bulletin,* pp. 1–4.

Singleton, W. T., Fox, J. G., & Whitfield, D. (Eds.). (1971). *Measurement of man at work.* London: Taylor & Francis.

Sleeman, D., & Brown, J. S. (Eds.). (1982). *Intelligent tutoring systems.* New York: Academic Press.

Sleight, A. W. (1988). Chemistry of high-temperature superconductors. *Science, 242,* 1519–1527.

Slichter, W. P. (1988). Chemical reaction in material science. In M. L. Good (Ed.), *Biotechnology and materials science: Chemistry for the future* (pp. 63–69). Washington, DC: American Chemical Society.

Slovic, P., Fischhoff, B., & Lichtenstein, S. (1976). Cognitive processes and societal risk taking. In J. S. Carroll & J. W. Payne (Eds.), *Cognition and social behavior* (pp. 165–184). Hillsdale, NJ: Lawrence Erlbaum Associates.

Slovic, P., Fischhoff, B., & Lichtenstein, S. (1981). Perception and acceptability of risk from energy systems. In A. Baum & J. E. Singer (Eds.), *Advances in environmental psychology: Vol. 3. Energy: Psychological perspectives* (pp. 155–169). Hillsdale, NJ: Lawrence Erlbaum Associates.

Slowiaczek, L. M., & Nusbaum, H. C. (1985). Effects of speech rate and pitch contour on the perception of synthetic speech. *Human Factors, 27,* 701–712.

Smith, M. J. (1988, Feb xxx). Electronic performance monitoring at the workplace: Part of a new industrial revolution. *Human Factors Society Bulletin,* pp. 1–3.

Smith, M. J., Carayon, P., & Maezio, K. (1986). *Motivational, behavioral, and psychological implications of electronic monitoring of worker performance.* Washington, DC: Office of Technology Assessment.

Smith, M. J., Cohen, B. G. F., Stammerjohn, L. W., & Happ, A. (1981). An investigation of health complaints and job stress in video display operations. *Human Factors, 23,* 387–400.

Smith, R. A., Alexander, R. B., & Wolman, M. G. (1987). Water-quality trends in the nation's rivers. *Science, 235,* 1607.

Smith, R. P. (1981). Boredom: A review. *Human Factors, 23,* 329–340.

Smith, T. J., & Stuart, M. A. (1990). Human factors of teleoperation in space. *Proceedings of the Human Factors Society 34th Annual Meeting* (Vol. I., pp. 116–120).

Sobey, A. J. (1988). Technology and the future of transportation: An industrial view. In *A look ahead: Year 2020* (Special Report No. 220). Washington, DC: Transportation Research Board, National Research Council.

Soldo, B., & Longino, C. (1988). Social and physical environments for the vulnerable aged. In *The social and built environment in an older society.* Washington, DC: National Academy Press.

Sommer, R., & Dewar, R. (1963). The physical environment of the ward. In E. Freidson (Ed.), *The hospital in modern society* (pp. 319–342). New York: The Free Press.

Southwest Research Institute. (1982). *Final report on biotechnology research requirements for aeronautical systems through the year 2000* (Vol. I). San Antonio, TX: Author.

Spesock, G. J., & Lincoln, R. S. (1965). Human factors aspects of digital computer programming for simulator control. *Human Factors, 5,* 473–482.

Spencer, R. W., & Christy, J. R. (1990). Precise monitoring of global temperature trends from satellites. *Science, 247,* 1558–1562.

Sproull, L., & Kiesler, S. (1991). Computers, networks and work. *Scientific American, 265(3).* 116–123.

Stahl, F. W. (1987). Genetic recombination. *Scientific American, 256(2),* 90–101.

Stanislaw, H. (1987). Methodological considerations for the assessment of traffic safety trends and interventions. *Human Factors, 29,* 361–366.

Steinberg, M. A. (1986). Material for aerospace. *Scientific American, 255(4),* 66–72.

Steinhart, P. (1990, July). No net loss. *Audubon,* pp. 18–21.

Stephens, J. C., Cavanaugh, M. L., Gradie, M. I., Mador, M. L., & Kidd, K. K. (1990). Mapping the human genome: Current status. *Science, 250,* 237–244.

Stephens, S. (1987, December). Lapplife after Chernobyl. *Natural History,* pp. 32–41.

Stern, P. C., & Kirkpatrick, E. M. (1977). Energy behavior. *Environment, 19,* 10–15.

Sterns, H. L., Barrett, G. V., & Alexander, R. A. (1985). Accidents and the aging individual. In J. E. Birren & K. W. Schaie (Eds.), *Handbook of the psychology of aging* (2nd ed.) New York: Van Nostrand Reinhold.

Stever, H. G., & Bodde, D. L. (1989). Space policy: Deciding where to go. *Issues in Science and Technology, 5(3),* 66–71.

Stewart, D. (1989, July). Doing the dishes: A TV bonanza is up there in orbit. *Smithsonian,* pp. 156–169.

Stewart, D. (1990, August). New machines are smaller than a hair, and do real work. *Smithsonian,* pp. 84–96.

Stokes, A. F., & Wickens, C. D. (1988). Aviation displays. In E. L. Wiener & D. C. Nagel (Eds.), *Human factors in aviation* (pp. 387–431). New York: Academic Press.

Stokols, D. (1990). Instrumental and spiritual views of people–environment relations. *American Psychologist, 45,* 641–646.

Stokols, D., & Altman, I. (1987). *Handbook of environmental psychology* (Vols. 1 & 2). New York: Wiley.

Stolarksi, R. S. (1988). The Antarctic ozone hole. *Scientific American, 258(1),* 30–36.

Stoll, C. (1989). *The cuckoo's egg.* New York: Doubleday.

Strassman, P. A. (1985). *Information payoff: The transformation of work in the electronic age.* New York: The Free Press.

Streeter, L. A., Ackroff, J. M., & Taylor, G. A. (1983). On abbreviating command names. *Bell System Technical Journal, 62,* 1807–1826.

Streeter, L. A., & Vitello, D. (1986). A profile of drivers' map-reading abilities. *Human Factors, 28,* 223–239.

Streff, F. M., & Molnar, L. J. (1990). *Estimating costs of traffic crashes and index crimes: Tools for improved decision making.* Ann Arbor, MI: The University of Michigan Transportation Research Institute.

Studt, T. (1990, March). Biodegradable plastics: New technologies for waste management. *R & D Magazine,* pp. 50–56.

Suhrbier, J. H., & Deakin, E. A. (1988). Environmental considerations in a 2020 transportation plan: Constraints or opportunities? In *A look ahead: Year 2020* (Special Report No. 220). Washington, DC: National Research Council, Transportation Research Board.

Sun, M. (1989). Emissions trading goes global. *Science, 247,* 520–521.

Sundstrom, E., DeMeuse, K. P., & Futrell, D. (1990). Workteams: Applications and effectiveness. *American Psychologist, 45,* 120–133.

Survey Research Center, University of Michigan. (1971). *Survey of working conditions.* Washington, DC: U.S. Government Printing Office.

Swets, J. A. (1979). ROC analysis applied to the evaluation of medical imaging techniques. *Investigative Radiology, 14,* 109–121.

Swets, J. A. (1988). Measuring the accuracy of diagnostic systems. *Science, 240,* 1285–1293.

Swets, J. A., Getty, D. J., Pickett, R. M., D'Orsi, C. J., Seltzer, S. E., & McNeil, B. J. (1991). Enhancing and evaluating diagnostic accuracy. *Medical Decision Making, 11,* 9–18.

System Security Study Committee. (1990). *Computers at risk: Safe computing in the information age.* National Research Council Washington, DC: National Academy Press.

Szekely, J. (1987). Can advanced technology save the U.S. steel industry? *Scientific American, 257*(1), 34–41.

Tans, P. P., Fung, I. Y., & Takahashi, T. (1990). Observational constraints on the global CO_2 budget. *Science, 247,* 1431–1438.

Tapping the treasures of the rainforest. (1990, November/December). *Ecosource,* pp. 23–24.

Taylor, F. W. (1913). *Principles of scientific management.* New York: Harper & Brothers.

Taylor, J. C. (1987). Job design and quality of working life. In R. E. Kraut (Ed.), *Technology and the transformation of white-collar work* (pp. 211–235). Hillsdale, NJ: Lawrence Erlbaum Associates.

Taylor, J. J. (1989). Improved and safer nuclear power. *Science, 244,* 318–325.

Taylor, K. (1990, February 5). G.I.S. merges health, geographic data. *Federal Computer Week,* p. 8.

Tchobanoglous, G., Theisen, G. H., & Eliassen, R. E. (1977). *Solid wastes: Engineering principles and management issues.* New York: McGraw-Hill.

Technology Administration. (1990). *Emerging technologies: A survey of technical and economic opportunities.* Washington, DC: U.S. Department of Commerce.

Tesler, L. G. (1991). Networked computing in the 1990s. *Scientific American, 265*(3), 86–93.

Theologus, G. C., Wheaton, G. R., Mirabella, A., Brahlek, R. E., & Fleishman, E. A. (1973). *Development of a standardized battery of performance tests for the assessment of noise stress effects* (Report No. CR-2149). Washington, DC: NASA.

Thompson, C. B. (Ed.). (1914). *Scientific management: A collection of the more significant articles describing the Taylor system of management.* Cambridge, MA: Harvard University Press.

Thompson, R. H., & Croft, W. B. (1989). Support for browsing in an intelligent text retrieval system. *International Journal of Man–Machine Studies, 30,* 639–668.

Thurow, L. C. (1987). A surge in inequality. *Scientific American, 256*(5), 30–37.

Tideiksar, R., & Fletcher, D. (1989). Keeping the elderly on their feet. *Issues in Science and Technology, 5*(3), 78–81.

Tiegdje, J. M., Colwell, R. K., Grossman, Y. L., Hodson, R. E., Lenski, R. E., Mack, R. M., & Regal, P. J. (1989, xxx). The planned introduction of genetically engineered organisms: Ecological considerations and recommendations in ecology. *Ecology, 7,* 298–315.

Toch, T. (1984, September). America's quest for universal literacy. *Education Week,* pp. L3–5.

Toffler, A. (1971). *Future shock.* New York: Bantam Books.

Toynbee, A. J. (1946). *A study of history* (abridged by D. C. Somerrell). New York: Oxford University Press.

Toffler, A. (1980). *The third wave.* New York: Phantom Books.

Transportation Research Board. (1988a). *Transportation in an aging society: Improving mobility and safety for older persons, Vol. 1* (Special Report No. 218). Washington, DC: National Academy Press.

Transportation Research Board. (1988b). Session summaries. In *A look ahead: Year 2020* (Special Report No. 220). Washington, DC: National Academy Press.

Transportation Research Board (1988c). Transportation in an aging society: Improving mobility and safety for older persons (Vol. 1, p. 2). Washington, DC: National Academy Press.

Transportation Research Board. (1989). *Safety belts, air bags and child restraints: Research to address emerging policy questions* (Special Report No. 224). Washington, DC: National Academy Press.

Transportation Research Board. (1990a). *Airport system capacity: Strategic choices* (Special Report No. 226). Washington, DC: National Academy Press.

Transportation Research Board. (1990b). *Safety research for a changing highway environment* (Special Report No. 229). Washington, DC: National Academy Press.

Tugend, A. (1984, March 6). Half of Chicago students drop out, study finds. *Education Week,* p. 10.

Turkle, S. (1984). *The second self: Computers and the human spirit.* New York: Simon & Schuster.

Turnage, J. T. (1990). The challenge of new workplace technology for psychology. *American Psychologist, 45,* 171–178.

Turner, J. A. (1984). Computer mediated work: The interplay between technology and job claims representatives in the social security administration. *Communications of the ACM, 27,* 1210–1217.

Turner, L. (1989). Three plants, three futures. *Technology Review, 92*(1), 38–45.

Turoff, M. (1967, xxx). *Knowns and unknowns in immediate access time-shared systems.* Paper presented at the American Management Association Conference on The Computer Utility, New York.

Udall, J. R. (1986, January). Losing our liquid assets. *National Wildlife,* pp. 50–55.

U.S. Congress. (1989). *The status of the nation's highways and bridges.* Committee on Public Works and Transportation, Print 101–2. Washington, DC: Government Printing Office.

Uhlig, R. P., Farber, D. J., & Bair, J. H. (1979). *The office of the future: Computers and communications.* New York: North Holland.

United Nations Environment Program. (1987). *Montreal protocol on substances that deplete the ozone layer.* Montreal: Author.

U.S. Air Force. (undated). *Project Forecast II: The Air Force Tomorrow Team* [Executive Summary]. U.S. Air Force.

U.S. Department of the Army. (1990). *Army technology base master plan: Vol. I.* Washington, DC: Author.

U.S. Department of Defense. (1990). *Total quality management guide: A two volume guide for defense organizations.* Washington, DC: Author.

U.S. Department of Education. (1982). *On adult learning.* Washington, DC: U.S. Government. Printing Office.

U.S. Department of Energy. (1988a). *An assessment of the natural gas resource base of the United States* (Report No. DOE/W/31109/HL). Washington, DC: Author.

U.S. Department of Energy. (1988b). *Energy conservation multi-year plan: 1990–1994.* Washington, DC: National Academy of Sciences Press.

U.S. Department of Energy. (1990). *Building the advanced climate model: Draft plan for CHAMMP.* Unpublished manuscript.

U.S. Department of Labor. (1987). Apprenticeship 2000. *Federal Register, 52*(231), 45905–45908.

U.S. Industrial Outlook. (1989). *1989 U.S. Industrial Outlook.* Washington DC: U.S. Department of Commerce, International Trade Administration.

Uttal, Ql, W. R. (1989). Teleoperators. *Scientific American, 261*(6), 124–129.

Utting, K., & Yankelovich, N. (1988). *Context and orientation in hypermedia networks.* Unpublished manuscript. Providence, RI: Brown University, Institute for Research in Information and Scholarship.

Vachon, R. A. (1989–1990). Employing the disabled. *Issues in Science and Technology, 6*(2), 44–50.

Van Allen, J. A. (1986). Space science, space technology and the space station. *Scientific American, 254*(1), 32–39.

Vanderheiden, G. C. (1990). Thirty-something million: Should they be exceptions? *Human Factors, 32,* 383–396.

Vaughan, R. J. (1989). *Education, training and labor markets: Summary and policy implications of recent research by Jacob Mincer.* New York: Columbia University Press.

Vaughan, R. J., & Berryman, S. E. (1989). Employer-sponsored training: Current status, future possibilities. *National Center on Education and Employment Brief, 4,* 1–4.

Vaydanoff, P. (1987). Women's work, family, and health. In K. S. Koziara, M. H. Moskow, & L. D. Tanner (Eds.), *Working women: Past, present, future.* Washington, DC: Bureau of National Affairs.

Verma, I. M. (1990). Gene therapy. *Scientific American, 263*(5), 68–84.

Verplanken, B. (1989). Beliefs, attitudes, and intentions toward nuclear energy before and after Chernobyl in a longitudinal within-subjects design. *Environment and Behavior, 21,* 371–392.

Vicente, K. J. (1990, November). A few implications of an ecological approach to human factors. *Human Factors Society Bulletin,* pp. 1–4.

Vicente, K. J., Thornton, D. C., & Moray, N. (1987). Spectral analysis of sinus arythmia: A measure of mental effort. *Human Factors, 29,* 183–194.

Vining, D. R., Jr. (1985). The growth of core regions in the Third World. *Scientific American, 252*(4), 42–49.

Vining, J., & Ereo, A. (1990). What makes a recycler? A comparison of recyclers and nonrecyclers. *Environment and Behavior, 22,* 55–73.

Waern, Y., & Rollenhagen, C. (1983). Reading text from visual display units (VDUs). *International Journal of Man–Machine Studies, 18,* 441–465.

Waldrop, M. M. (1988a). A landmark in speech recognition. *Science, 240,* 1615.

Waldrop, M. M. (1988b). National academy looks at computing's future. *Science, 241,* 1436.

Wall, T. T., Clegg, C. W., & Kemp, N. J. (1987). *The human side of advanced manufacturing.* New York: Wiley.

Wallace, D. R. (1985). Wetlands in America: Labyrinth and temple. *Wilderness,* pp. 12–27.

Wallich, P. (1989). Science and business: Manufacturing intelligence. *Scientific American, 261*(6), 100–102.

Wallich, P. (1990a). The analytical economist: Sixteen tons of coal. *Scientific American, 262*(6), 94–95.

Wallich, P. (1990b). Murky water: Just what role do oceans play in absorbing greenhouse gases? *Scientific American, 262*(5), 25–26.

Waltz, D. L. (1983, November). Helping computers understand natural language. *IEEE Spectrum,* pp. 81–84.

Warriner, G. K., McDougall, G. H. G., & Claxton, J. D. (1984). Any data or none at all? Living with inaccuracies in self-reports of residential energy consumption. *Environment and Behavior, 16,* 503–526.

Watson, J. D. (1990). The Human Genome Project: Past, present, and future. *Science, 248,* 44–49.

Watson, R. T., Prather, M. J., & Kurylo, M. J. (1988). *Present state of knowledge of the upper atmosphere 1988: An assessment report* (Reference Publication No. 1208). Washington, DC: NASA.

Webb, J. W., & Kramer, A. F. (1987). Learning hierarchical menu systems: A comparative investigation of analogical and pictorial formats. *Proceedings of the Human Factors Society, 31st Annual Meeting* (pp. 978–982). Santa Monica, CA: Human Factors Society.

Weinberg, A. M. (1988–1989). Energy policy in an age of uncertainty. *Issues in Science and Technology, 5*(2), 81–85.

Weinberg, A. M. (1989–1990). Engineering in an age of anxiety. *Issues in Science and Technology*, 6(2), 37–43.

Weinberg, C. J., & Williams, R. H. (1990). Energy from the sun. *Scientific American, 263*(3), 146–155.

Weinberg, R. A. (1985). The molecules of life. *Scientific American, 253*(4), 48–57.

Weiner, J. (1989, February). Glacial bubbles are telling us what was in Ice Age air. *Smithsonian*, pp. 78–87.

Weischedel, R., Carbonell, J., Grosz, B., Lehnert, W., Marcus, M., Perrault, R., & Wilensky, R. (1990). Natural language processing. *Annual Review of Computer Science, 4*, 435–452.

Weiser, M. (1991). The computer for the 21st century. *Scientific American, 265*(3), 94–104.

Welter, T. R. (1990, August 20). A farewell to arms. *Industry Week*, pp. 36–42.

Weybrew, B. B. (1961). Human factors and work environment: II. The impact of isolation upon personnel. *Journal of Occupational Medicine, 3*, 290–294.

Weybrew, B. B. (1963). Psychological problems of prolonged marine submergence. In N. Burms, R. Chambers, & E. Hendler (Eds.), *Unusual environments and human behavior* (pp. 87–125). New York: Macmillan.

White, R. M. (1990). The great climate debate. *Scientific American, 263*(1), 36–43.

White, R. W. (1959). Motivation reconsidered: The concept of confidence. *Psychological Review, 66*, 297–333.

Whiteside, G. M. (1988). Materials for advanced electronic devices: Chemical reaction in material science. In M. L. Good (Ed.), *Biotechnology and materials science: Chemistry for the future* (pp. 85–99). Washington, DC: American Chemical Society.

Wickens, C. D., & Flach, J. M. (1988). Information processing. In E. L. Wiener & D. C. Nagel (Eds.), *Human factors in aviation* (pp. 111–155). New York: Academic Press.

Wickens, C. D., & Yeh, Y-Y. (1983). The disassociation between subjective workload and performance: A multiple resource approach. *Proceedings of the Human Factors Society 27th Annual Meeting* (pp. 238–244). Santa Monica, CA: Human Factors Society.

Wiener, E. L. (1983). Computers in the cockpit: But what about the pilots? *Proceedings of the Second Aerospace Behavioral Engineering Technology Conference Proceedings* (pp. 453–457). Warrendale, PA: Society for Automotive Engineers.

Wiener, E. L. (1985). Beyond the sterile cockpit. *Human Factors, 27*, 75–90.

Wiener, E. L. (1989). *Human factors of advanced technology ("Glass Cockpit") transport aircraft* (Contractor Report No. 177528). NASA.

Wiener, E. L. (1990, August). The golden era of human factors in commercial aviation. *Human Factors Bulletin*, pp. 1–2.

Wiener, E. L., & Currey, R. E. (1980). Flight deck automation: Promises and problems. *Ergonomics, 23*, 955–1011.

Wiener, N. (1950). *The human use of human beings: Cybernetics and society*. Boston, MA: Houghton Mifflin.

Wiener, N. (1966). *God and Golem, Inc.: A comment on certain points where cybernetics impinges on religion*. Cambridge, MA: MIT Press.

Wierwille, W. W., & Williges, R. C. (1978). *Survey and analysis of operator workload assessment techniques* (Tech. Rep. No. S-78-101). Blacksburg, VA: Systemetrics, Inc.

Wigley, T. M. L., & Schlesinger, M. E. (1985). Analytic solution for the effect of increasing CO_2 on global mean temperature. *Nature, 315*, 649–652.

Wilder, C. S. (1967). *Health characteristics of persons with chronic activity limitation, United States, 1974* Health Resources Administration (No. 36) Washington, DC: U.S. Department of Health, Education and Welfare.

Wilder, C. S. (1977). *Limitation of activity due to chronic conditions, United States, 1974* Health Resources Administration. Washington, DC: U.S. Department of Health, Education and Welfare, National Center for Health Statistics.

Wilfong, G. T. (1989). Robotics and automation. *Computer, 22*(3), 6–7.

Williams, D. (1986). Development of "Super Cockpit" plan will allow fliers to be quicker on the draw. *Defense News,* pp. 12, 32.

Williams, S., Fenn, S., & Clausen, T. (1990). Renewing renewable energy. *Issues in Science and Technology, 6*(3), 64–70.

Williges, R. C., & Wierwille, W. W. (1979). Behavioral measures of aircrew mental workload. *Human Factors, 21,* 549–574.

Wilson, E. O. (1989). Threats to biodiversity. *Scientific American, 261*(3), 108–116.

Wilson, E. O., & Peter, F. M. (1988). *Biodiversity.* Washington, DC: National Academy Press.

Wilson, H. J., MacCready, P. B., & Kyle, C. R. (1989). Lessons of (Sunraycer). *Scientific American, 260*(3), 90–97.

Wilson, J. W. (1986, March). An ill wind. *Natural History,* pp. 48–51.

Wilson, K. S. (1987). *The Palenque optical disk prototype* (Tech. Rep. No. 44). New York: Bank Street College of Education.

Wilson, W. J. (1987). *The truly disadvantaged: The inner city, the underclass, and public policy.* Chicago, IL: University of Chicago Press.

Wise, J. A. (1986, May). The space station: Human factors and habitability. *Human Factors Society Bulletin,* pp. 1–3.

Witten, I. H., & MacDonald, B. A. (1988). Using concept learning for knowledge acquisition. *International Journal of Man–Machine Studies, 29,* 171–196.

Wolkomir, R. (1986, January). A high-tech attack on traffic jams helps motorists go with the flow. *Smithsonian,* pp. 42–51.

Wollard, K., & Zorpette, G. (1989). Power and energy. *IEEE Spectrum, 26*(1), 56–58.

Wood, P. H. N. (1980). *International classification or impairments, disabilities, and handicaps: A manual of classifications relating to the consequences of disease.* Geneva, Switzerland: World Health Organization.

Wood, R. C. (1987, November). The language advantage: Japan's machine translators rule the market. *High Technology Business,* p. 17.

Woodruff, W. (1987). *The world at bay: 1500 to the present.* Unpublished manuscript.

Woods, D. (1986). Paradigms for intelligent decision support. In E. Hollnagel, J. Francini, & D. Woods (Eds.), *Intelligent decision support in process environment.* Berlin: Springer-Verlag.

World Commission on Environment and Development (1987). *Our common future.* New York: Oxford University Press.

Woteki, T. (1977). *The Princeton Omnibus Experiment: Some effects of retrofits on space heating requirements* (Report No. 43). Princeton, NJ: Princeton University, Center for Energy and Environmental Studies.

Wright, K. (1990a). The shape of things to go. *Scientific American, 262*(5), 92–101.

Wright, K. (1990b). Trends in communication: The road to the global village. *Scientific American, 262*(3), 83–94.

Wrighton, M. S. (1988). Chemistry of materials for energy production, conversion, and storage. In M. L. Good (Ed.), *Biotechnology and materials science: Chemistry for the future* (pp. 101–115). Washington, DC: American Chemical Society.

Yam, P. (1991). Atomic turn-on. *Scientific American, 265*(5), 20.

Yankelovich, D. (1978, May). The new psychological contracts at work. *Psychology Today,* pp. 46–50.

Yankelovich, N. (1990). Three pieces of the puzzle: Wide-area hypermedia, information agents and on-line reference works. *Psychological Science, 1,* 350–352.

Yankelovich, N., Haan, B. J., Meyrowitz, N. K., & Drucker, S. M. (1988, January). Intermedia: The concept and the construction of a seamless information environment. *Computer,* pp. 81–96.

Yeh, Y-Y., & Wickens, C. D. (1988). Disassociation of performance and subjective measures of workload. *Human Factors, 30,* 111–120.

Young, J. A. (1984). An agenda for the electronics industry. In J. Botkin, D. Dimancescu, & R. Stata (Eds.) *Global stakes: The future of high technology in America* (pp. 172–177). New York: Penguin Books.

Young, J. A. (1988). Technology and competitiveness: A key to the economic future of the United States. *Science, 241,* 313–316.

Yourdon, E. (1971). Maybe the computers can save us after all. *Computers and Automation, 20,* 21–26.

Zimring, C., Weitzer, R., & Knight, R. C. (1982). Opportunity for control and the designed environment: The case of an institution for the developmentally disabled. In A. Baum & J. E. Singer (Eds.), *Advances in environmental psychology: Vol. 4. Environment and health* (pp. 171–210). Hillsdale, NJ: Lawrence Erlbaum Associates.

Zoltan, E., & Chapanis, A. (1982). What do professional persons think about computers? *Behavior and Information Technology, 1,* 55–68.

Zwally, H. J. (in press). Growth of Greenland ice sheet. In N. Moray, W. R. Ferrell, & W. Rouse (Eds.), *Robots, control and society.* New York: Taylor & Francis.

Author Index

A

Aaronson, D., 251, **375**
Abell, D., 68, 191, **395**
Abelson, P. H., 50, 51, 88, 125, **375**
Abrett, G., 242, **375**
Ackerman, R. K., 127, 375
Ackroff, J. M., 256, **413**
Acomb, D. B., 164, **387**
Adelman, M. A., 50, 375
Adrion, W. R., 220, 221, **393**
Aebersold, P., 200, **408**
Agnew, H. M., 67, 375
Aharonowitz, Y., 198, **375**
Ahmad, E., 83, **407**
Alavi, M., 258, 375
Alexander, R. A., 357, **412**
Alexander, R. B., 97, **413**
Alfaro, L., 253, **390**
Allen, J. S., 252, 253, **400**
Allgier, E., 174, **375**
Alluisi, E. A., 319, **381**
Alson, J. A., 69, 92, 168, **390**
Altman, I., 129, 130, **413**
Alvo, R., 91, **375**
Ammons, J. C., 17, **376**
Anantharaman, T., 238, **393**
Anderson, W. F., 200, **408**
Andreae, M. O., 101, **383**
Apostolakis, G., 334, **376**

Applebaum, E., 301, **376**
Ardekani, S. A., 106, 111, 116, 180, **392**
Argote, L., 34, **376**
Arnold, J. R., 186, **376**
Aronson, S., 4, **376**
Arthur, W. B., 21, **376**
Asimov, I., 78, **376**
Atkinson, R. C., 141, 306, **376**
Attlewell, P., 32, 41, 302, 303, **376**
Auerbach, S. I., **388**
Ausubel, J. H., 3, 5, 106, 111, 116, 180, **376**, **392**
Averill, J. R., 321, **376**
Ayres, R. U., 81, 105, **376**

B

Bachrach, A. J., 321, **376**
Baddeley, A., 257, **403**
Baecker, R. M., 249, **376**
Baes, C. F., 79, **376**
Baetge, M. M., 164, **387**
Bailey, T., 308, **376**
Bair, J. H., 34, 286, 304, **376**, **414**
Balzhiser, R. E., 52, 68, 125, **376**
Ban, S. D., 48, **380**
Banister, E., 131, **399**
Banks, P. M., 190, **376**
Baran, P., 219, **376**
Barber, D. L. A., 219, **384**

421

Bardon, M., 220, **383**
Barker, P. G., 267, **376**, 377
Barkstrom, B. R., 83, **407**
Barnes, R. D., 321, 377
Barnes, V., 253, **390**
Barnola, J. M., 81, 377
Baron, J., 145, 377
Baron, S., 182, 377, **395**, **406**
Barowy, W., **408**
Barrett, G. V., 357, **413**
Bartel, A. P., 139, 141, 306, 377
Barton, J. K., 195, 377
Baruch, J. J., 24, 28, 298, 368, **407**
Bassen, H. I., 377
Bate, R. T., 214, 377
Bates, M., 277, 279, 280, 377, **408**
Baum, A., 71, 113, 377
Baumol, W. J., 26, 27, 29, 309, 377
Bay, P. N., 170, 299, 377
Beam, P., 252, **401**
Beardsley, T., 19, 66, 68, 70, 103, 118, 185, 190, 191, 215, 377, **382**
Becker, L. J., 47, 72, 377, **410**
Beer, M., 35, **378**
Bell, T. E., 191, 222, **378**
Belorose, F. C., 97, **391**
Bereiter, C., 149, **409**
Berger, S., 30, 35, 152, **378**
Beringer, D. B., 249, **378**
Berlin, G. B., 301, **378**
Berrenberg, J. L., 59, **382**
Berryman, S. E., 152, 154, 297, **416**
Bevington, R., 55, 56, 71, **378**
Bickman, L., 108, **378**
Bikson, T. K., 302, 303, **378**
Billingsley, P. A., 252, **391**
Birdzell, L. E., Jr., 365, **408**
Bissett, W., 120, **378**
Bjorn-Andersen, N., 300, 304, **378**
Blaese, R. M., 408, **408**
Blair, J. G., 327, **378**
Blair, P. D., 47, 48, 54, 55, **389**
Blau, F. D., 309, **378**
Bleviss, D. L., 168, **378**
Bloch, E., 139, **378**
Blomberg, J. L., 304, **378**
Bloom, J., 148, **378**
Blumenthal, D., 156, **384**
Bobrow, D. G., 229, 232, **378**
Bodde, D. L., 190, 192, **413**
Boden, T. A., 120, **378**
Boehly, W. A., 325, **405**

Boehm, B. W., 219, **378**
Boehm-Davis, D. A., 242, 249, **378**
Boies, S. J., 248, **410**
Bolin, B., 78, **378**
Bolt, D. J., 207, **407**
Bond, R., 192, 193, **398**
Boorstin, D., 142, **379**
Boose, J. H., 242, **379**
Borchelt, R., 101, 113, **379**
Bord, R. J., 108, **393**
Borie, E., 325, **379**
Borkan, G. A., 354, **379**
Boster, R. S., 132, **383**
Botkin, J. W., 24, 139, 372, **379**
Bowe, F. G., 347, 350, **379**
Bowen, H. K., 302, **379**
Boyle, E., 150, **406**
Bradshaw, J., 242, **379**
Brahlek, R. E., 321, **414**
Branscomb, A., 226, **379**
Branscomb, L. M., 201, **379**
Braune, R., 354, **379**
Braverman, H., 302, 307, **379**
Brechner, K. C., 129, **379**
Breen, M., 148, **386**
Brinster, R. L., 207, **407**
Brittingham, J. L., 198, 363, **406**
Broadbent, D. E., 132, **379**
Brodsky, M. H., 214, **379**
Broecker, W. S., 80, 83, **379**
Bromley, B. A., 213, 225, 246, **379**
Bronfman, L. M., 67, **379**
Brookes, W., 86, 334, **379**
Broune, R., 319, **395**
Brown, E. L., 199, **379**
Brown, G. E., Jr., 148, **379**
Brown, J. S., 148, **412**
Brown, W., 50, 162, 363, **394**
Brunner, C., 150, **380**
Buchanan, B. G., 189, **380**
Budyko, M. I., 78, **380**
Burner, R. A., 78, **380**
Burnett, W. M., 48, **380**
Burns, J. O., 186, 191, **380**
Burns, M. J., 187, **389**
Burris, R. H., 336, **400**
Burstein, M., 242, **375**
Burt, D. M., 186, **380**
Bush, C. N., 325, **380**
Bush, V., 265, 270, **380**
Butler, L. H., 352, **380**
Butler, T. J., 91, **398**

Buxton, W. A. S., 249, 376
Byrd, R. J., 291, 391
Byrne, G., 141, 380

C

Campbell, M., 238, 393
Campbell, R. G., 207, 407
Cantor, C. R., 202, 380
Cantor, D., 199, 380
Cantor, R. J., 71, 72, 399
Cantor, S., 199, 380
Capindale, R. A., 279, 280, 282, 380
Carayon, P., 227, 412
Carbonell, J., 279, 280, 282, 417
Card, S. K., 249, 291, 380
Carlson, F. R., 403
Caro, P. W., 155, 381
Carroll, J. M., 284, 398
Carson, R., 76, 127, 130, 381
Carter, R., 150, 408
Casali, S. P., 349, 381
Cascio, W. F., 308, 320, 381, 399
Cash, D. E., 89, 91, 388
Caskey, C. T., 200, 382
Cassidy, R., 143, 224, 381
Cava, R. J., 61, 381
Cavanaugh, M. L., 204, 413
Cerf, C., 4, 5, 381
Cerf, V. G., 219, 270, 381
Ceruzzi, P., 5, 381
Cess, R. D., 83, 407
Chaffee, J. L., 107, 388
Chamlot, D.,298, 316, 381
Chandler, W. U., 53, 92, 134, 170, 365, 381
Chang, T. C., 17, 411
Chapanis, A., 283, 290, 381, 419
Chapman, A. J., 132, 393
Chaudhari, P., 214, 234, 381
Chen, K., 170, 172, 386
Cherfas, J., 66, 381
Chignell, M. H., 242, 381
Childes, J. R., 18, 19, 172, 381
Chiles, W. D., 319, 381
Chin, J. P., 255, 284, 403
Chipman, S. F., 145, 381
Chodorow, M. S., 291, 391
Chou, T. W., 18, 381
Chown, S., 354, 392
Christensen, D. G., 163, 405
Christensen, K., 311, 381

Christie, B., 289, 411
Christy, J. R., 85, 413
Cicerone, R. J., 78, 385
Citro, C. F., 7, 381
Claridge, R. C., 186, 398
Clark, A. C., 4, 381
Clark, J. P., 20, 179, 180, 381
Clark, W. C., 76, 120, 382
Clausen, T., 62, 74, 127, 418
Claxton, J. D., 60, 416
Clearwater, Y. A., 187, 382
Clegg, C. W., 416
Close, F., 66, 382
Coach, J. V., 107, 382
Coates, J. F., 301, 382
Cohen, B. G. F., 252, 412
Cohen, D. K., 147, 150, 382
Cohen, G., 207, 375
Cohen, S., 133, 382
Cohill, A. M., 249, 382
Cole, R. E., 40, 382
Colwell, R. K., 208, 414
Combs, B., 67, 387
Comer, J. P., 142, 382
Commoner, B., 119, 382
Compton, W. D., 168, 170, 172, 382
Confrey, J., 150, 382
Conner, J., 104, 382
Cook, N. H., 15, 382
Cook, S. W., 59, 72, 382, 399
Cooke, M. J., 306, 392
Coombs, B., 336, 382, 397
Corcoran, E., 19, 20, 21, 23, 31, 68, 133, 190, 215, 217, 382, 383
Corn, J. J., 6, 383
Cornetta, K., 200, 408
Cournoyer, D., 200, 383
Covault, C., 186, 187, 383
Cox, V. C., 188, 189, 363, 383, 399
Crandall, R. W., 325, 390
Crane, A. T., 89, 91, 388
Crane, P., 252, 383
Crawford, M., 185, 398
Crawford, R. G., 280, 380
Craxton, R. S., 66, 383
Creer, R., 330, 383
Croft, W. B., 267, 414
Crosson, P. R., 78, 97, 102, 383, 408
Crowston, K., 314, 383
Crump, J. H., 166, 383
Crutzen, P. J., 94, 101, 383, 390
Cuban, L., 150, 383

Culliton, B. J., 142, **383**
Culotta, E., 144, **383**
Culver, K., 200, **408**
Cunningham, G. L., 102, **410**
Curran, L., 224, **383**
Currey, R. E., 167, **417**
Curtis, K. K., 220, **383**
Cushman, J. R., 258, **383**
Cushman, W. H., 252, **383**
Cutler, S. J., 352, **383**
Cyert, R. M., 15, 20, 26, 30, 307, **383**
Czaja, S. J., 320, 349, 358, **383**

D

Dadi, C., 53, 92, 134, 170, 365, **381**
Dadzie, K. K. S., 365, **383**
Dainoff, M. H., 252, **383**
Dainoff, M. J., **383**
Dale, E. L., Jr., 28, 44, **383**
Dale, V. H., 79, 80, 81, **407**
Damos, D., 249, **383**
Daniel, T. C., 132, **384**
Darley, J. M., 47, 72, **410**
Darmstadter, J., 78, **408**
David, E. E., Jr., 141, **384**
Davidson, E. S., 233, **396**
Davies, D. W., 219, **384**
Davis, B. D., 205, **384**
Davis, F., 150, **408**
Davis, G. R., 48, **384**
Davis, L. E., 300, **397**
Davis, P. K., 156, **384**
Davis, R., 242, **384**
Deakin, E. A., 92, 332, **413**
Dean, L. M., 188, **384**
DeAngelis, D. L., 79, 80, 81, **407**
Debus, K. H., 208, **384**
de Chardin, P. T., 115, **384**
Dede, C. J., 223, **384**
Del Sesto, S. L., 6, **384**
Demain, A. L., 207, **384**
DeMeuse, K. P., 40, **414**
Demick, J., 131, **384**
Denning, P. J., 215, 222, 226, 268, 286, 290, **384**
Denton, G. H., 80, 83, **379**
Dentzer, S., 10, 13, 30, 142, 153, 196, 300, **384**
Derby, S. L., 339, **387**
Dertouzos, M. L., 14, 15, 27, 28, 30, 35, 152, **378**, **384**

Dervan, P. B., 202, 204, **385**
DeSanctis, G., 317, **385**
de Vogel, R., 150, **406**
Dewar, R., 199, **413**
Dickinson, J., 78, 249, **387**
Dickirson, R. E., 78, **385**
Dickson, D., 124, **385**
Dimancescu, D., 24, 139, 141, **379**
Divone, L. V., 62, **401**
Dominick, W. D., 255, **386**
Doos, B. R., 78, **378**
D'Orsi, C. J., 197, **389**, **414**
Douglas, S. J., 5, **385**
Draper, S. W., 249, 262, **385**, **403**
Dray, S. M., 314, **385**
Drexler, K. E., 23, **385**
Drucker, S. M., 150, **418**
Duchemin, J., 177, **385**
Duffy, J. W., 163, **405**
Dumais, S. T., 256, **388**
Dunlap, R. E., 59, **385**
Duric, N., 186, 191, **380**
Dyer, D., 15, **396**
Dyson, F., 192, **385**

E

Eason, K. D., 258, **385**
Easterling, W. E., III, 78, **408**
Eastman, D. B., 214, **394**
Eberhardt, J., 272, **385**
Economy, J., 19, **385**
Edmondson, B., 139, **385**
Edwards, A. D. N., 251, 351, **385**, **390**
Edwards, E., 168, **385**
Egan, D. E., 272, **385**
Ehrenreich, S. L., 249, **385**
Ehrlich, P. R., 104, 130, **385**
Eigler, D. D., 23, **385**
Eliassen, R. E., 105, **414**
Elkind, J. I., 347, 348, 351, **385**, **390**
Elmandjar, M., 372, **379**
Emery, F. E., 300, **385**
Emery, M., 300, **385**
Emmanuel, W. R., 79, 80, 81, 102 **386**, **407**
Engelbart, D. C., 265, **386**
English, W. K., 265, **386**
Enting, I. G., 81, **386**
Epple, D., 34, **376**
Epstein, Y. M., 363, **386**
Ereo, A., 108, **416**

Erickson, D., 195, 208, 224, **386**
Erikson, K., 117, **386**
Erisman, A. M., 215, **386**
Ernst, M. L., 160, 301, **386**
Ervine, R. D., 170, 172, **386**
Eshelman, L., 242, **386**
Evans, C., 7, 243, **386**
Evans, G. W., 93, **386**
Evans, L., 175, **386**

F

Fahey, D. W., 94, **407**
Farber, D. J., 220, 221, 304, 393, **414**
Farman, J. C., 94, **386**
Farooq, M. U., 255, **386**
Farr, B. J., 155, **412**
Farrell, M. P., 120, **378**
Fauth, G. R., 11, 168, **408**
Feehrer, C. E., 182, **406**
Feldberg, R. L., 301, 302, 303, 304, **386**
Fenn, S., 62, 74, 127, **417**
Ferber, M. A., 309, **378**
Feurzeig, W., 148, 150, **387**, **408**
Fickett, A. P., 47, 54, 56, **387**
Fidell, S. A., 129, 132, 321, **387**
Fine, A., 197, **387**
Finkelman, J. M., 166, **387**
Finn, R., 253, **390**
Fischhoff, B., 67, 334, 335, 336, 337, 339, 342, 387, 397, **412**
Fitzgerald, K., 196, **387**
Flach, J. M., 131, 183, 387, **416**
Fleisher, D., 297, **399**
Fleishman, E. A., 321, **414**
Flemings, M. C., 20, 179, 180, **381**
Fletcher, D., 358, **414**
Foerstel, K., 126, **387**
Fogarty, W. M., 328, **387**
Foley, J. D., 244, 250, **387**
Forester, T., 218, 222, 228, **387**
Form, W., 320, **387**
Foushee, H. C., 163, 164, 167, **388**
Fox, E. A., 269, **388**
Fox, G. C., 233, **388**
Fox, J. G., 319, **412**
Fraley, R. T., 207, **388**
Francas, M., 249, **388**
Frase, L. T., 291, **399**
Freeman, B., 257, **403**
Fregley, A. M., 249, **378**
Freiling, M. J., 242, **393**

Freivalds, A., 253, **391**
Frick, M. C., 175, **386**
Friedlander, S. K., 107, 136, 361, **388**
Friedli, H., 81, **388**
Friedmann, T., 200, **388**
Frosch, R. A., 134, 324, 325, **388**
Fuchs, I. H., 220, 221, **393**
Fulkerson, W., 51, 89, 91, **388**
Fuller, R. A., 197, **388**
Fuller, R. B., 115, **388**
Fung, I. Y., 80, **414**
Furnas, G. W., 256, 272, **388**
Futrell, D., 40, **414**

G

Gabias, P., 251, **375**
Gabriel, K. J., 23, **392**
Gaden, E. L., Jr., 207, **388**
Galitz, W. O., 315, **388**
Gallagher, S. S., 174, 355, 358, **397**
Gallopoulos, N. E., 134, 324, 325, **388**
Galloway, J. N., 91, **398**
Galotti, K. M., 256, **396**
Garber, T., 107, **381**
Garcia, K. D., 249, **388**
Gardiner, G. B., **386**
Gardner, W., 110, 268, **388**
Gardner-Bonneau, D. J., 351, **390**
Garg-Janardan, C., 242, **388**
Garriott, O. K., 193, **388**
Gasser, C. S., 207, **389**
Geirland, J., 310, 311, **389**
Gelernter, D., 233, **389**
Geller, E. S., 59, 72, 73, 107, 108, 389, **407**
Gellings, C. W., 47, 54, 56, **386**
Genetet, S., 242, **395**
Gergen, M., 350, **389**
Gerrell, W. M., 102, **410**
Getty, D. J., 197, 389, **414**
Gibbons, J. H., 47, 48, 54, 55, **389**
Gillan, D. G., 187, **389**
Gilleland, J. R., 66, **389**
Gilliam, P., 302, **389**
Gilmartin, P. A., 127, **389**
Gingrich, P. S., 291, **399**
Ginzberg, E., 140, 298, **389**
Giuliano, V. E., 32, 218, **389**
Gjostein, N. A., 168, 170, 172, **382**
Glas, J. P., 94, 95, **389**
Glaser, R., 145, **381**

Glenn, E. N., 301, 302, 303, 304, **386**
Goddard, W. A., III, 19, 22, **389**
Goeller, H. E., 79, **376**
Golay, M. W., 67, **389**
Goldemberg, J., 58, 74, 101, **407**
Goldstein, I. L., 302, **389**
Gomez, L. M., 256, 272, **385, 388**
Good, M. L., 207, **389**
Goodman, D., 249, **388**
Goodwin, N. C., 248, **389**
Gooler, D. D., 150, **390**
Gordon, G. E., 317, **390**
Gore, A., 221, **390**
Gore, A., Jr., 126, **390**
Goudie, A., 76, **390**
Gould, J. D., 252, 253, 257, **390**
Govindaraj, T., 17, **376**
Gowing, M. K., 308, 309, **404**
Gradie, M. I., 204, **413**
Graeber, R. C., 164, **390**
Graedel, T. E., 94, **390**
Graham, J. D., 325, **390**
Gray, C. L., Jr., 69, 92, 168, **390**
Gray, P., 67, 136, 315, 336, **403**
Gray, R., 330, **383**
Green, K. C., 141, **390**
Greene, D. L., 54, 87, 179, 180, **390**
Greenwell, G., 207, **390**
Gregory, J., 304, 310, 316, **390**
Griffith, D., 351, **390**
Grischkowsky, N., 252, 253, **390**
Grossman, Y. L., 208, **414**
Grosz, B., 279, 280, 282, **417**
Grotch, W. L., 85, **390**
Grubb, M. J., 62, 64, 65, **390**
Gruber, T. R., 242, **390**
Gulkis, S., 191, **390**
Gunderson, E. K. E., 188, **390, 391**
Gunderson, N. E., 188, **384**
Gunn, G., 31, **391**
Gutek, B., 302, **378**
Guyer, B., 174, 355, 358, **397**
Gwin, H. L., 47, 48, 54, 55, **389**

H

Haan, B. J., 150, **418**
Hacisalihzade, S. S., 252, 253, **400**
Hafele, W., 65, 67, 68, **391**
Hafemeister, D., 54, **409**
Hagen, D., 350, **389**
Ham, R. K., 105, **404**

Hamakawa, Y., 64, **391**
Hamilton, D., 54, 99, **391**
Hamilton, L. C., **391**
Hamilton, M. H., 153, **391**
Hammer, R. E., 207, **407**
Hammond, M. M., 108, **393**
Hanley, R. J., **408**
Hanneman, G. J., 315, **403**
Happ, A., 252, **383, 412**
Hara, D., 9, 10, 26, **397**
Harder, J. W., 94, **392**
Hardin, G., 4, 128, 130, **391**
Hardison, O. B., Jr., 245, 271, **391**
Harkness, R. C., 316, **391**
Harman, J., 155, **412**
Harnad, S., 288, **391**
Harpster, J. L., 253, **391**
Harrison, E. F., 83, **407**
Hart, S. G., 165, **391**
Hartmann, D., 83, **407**
Hartson, H. R., 255, **391**
Hartwell, S., 256, **396**
Hatsopoulos, G. N., 10, 12, 13, 26, 27, **391**
Havera, S. P., 97, **391**
Hawkins, J., 150, **380**
Heart, F. E., 219, **391**
Heirdorn, G. E., 291, **391**
Helander, M. G., 249, 252, 286, **391**
Helfgott, R. B., 314, **391**
Helmreich, R. L., 164, **387**
Helson, H., 130, **391**
Hendrick, H. W., 310, 311, **391**
Herman, R., 106, 111, 116, 180, **392**
Heron, A., 354, **392**
Hershkowitz, A., 112, 333, **392**
Herskovitz, S. B., 191, 217, 266, **392**
Hibbard, W. R., 97, **392**
Hill, P. G., 47, 56, 62, **392**
Hillman, D. J., 249, **392**
Hinman, C. W., 102, **392**
Hively, W., 86, 95, **392**
Hix, D., 255, **391**
Hodson, R. E., 208 **414**
Hoffman, D. J., 94, **392**
Hoffman, M. S., 49, 357, **392**
Hogan, D. L., 224, **392**
Holden, C., 87, 134, 146, 306, 331, **392**
Holdren, J. P., 89, 93, **392**
Hollan, J. D., 263, **393**
Holt, D. A., 298, **392**
Holzer, E., 132, **409**

Hoos, I. R., 314, **392**
Hopkin, V. D., 167, **392**
Horgan, J., 104, **392**
Hough, R. W., 317, **392**
Houghton, R. A., 78, 82, **392**
Howe, R. T., 23, **392**
Howell, S. C., 352, 353, **392**
Howell, W. C., 306, **392**
Hsu, F. H., 238, **393**
Hubbard, H. M., 64, **393**
Huenneke, L. F., 102, **410**
Huey, B. M., 52, 117, **400**
Huggins, A. W. F., 278, **393**, **403**
Hummel, C. F., 325, **393**
Humphrey, C. R., 108, **393**
Hunt, E., 269, **392**
Hurd, M. D., 352, **392**
Hutchins, E. L., 263, **392**
Huws, U., 315, 316, **393**
Hwang, S. L., 17, **393**

I

Iacono, S., 320, **393**
Ingersoll, A. P., 81, 84, **393**
Ingram, R. E., **388**
Ivancevich, J. M., 321, **399**

J

Jackson, D. D., 63, **393**
Jaco, E. G., 199, **393**
Jacobs, S. V., 93, **386**
Jacobson, C., 242, **393**
Jastrow, R., 79, 83, 86, **393**, **410**
Jelinek, F., 250, 277, 280, 281, 292, **398**
Jenkins, L. B., 140, 144, **393**, **401**
Jennings, D. M., 220, 221, **393**
Jensen, K., 291, **391**
Johannsen, G., 319, **393**
Johnson, A., 148, **408**
Johnson, G. I., 17, **393**
Johnson, L. R., 161, 177, **393**
Johnson, S. W., 186, 191, **380**
Johnston, W. B., 306, 309, 311, 10, 26, 27, 367, **393**
Jones, D. M., 132, **393**
Jones, P. D., 78, 79, **394**
Jones, S. E., 66, **407**
Judkins, R. R., 51, **388**
Jungermann, H., 330, **394**

K

Kahane, C. J., 173, **394**
Kahn, A. E., 326, **394**
Kahn, G., 242, **394**
Kahn, H., 50, 162, 363, **394**
Kahn, R. E., 220, 221, **394**
Kaiper, G. V., 66, **389**
Kaiser, E. T., 195, **394**
Kalsbeek, J. W. H., 319, **394**
Kalton, G., 7, **381**
Kanciruk, P., 120, 121, **378**, **394**
Kaplan, B. H., 297, **399**
Kapor, M., 227, **394**
Karhnak, J. M., 31, **399**
Karlovac, M., 188, 189, 363, **383**
Karpus, L., 107, **381**
Karson, E. M., 200, **408**
Kasarda, J. D., 138, 161, 169, 311, 361, 362, **394**
Kasid, A., 200, **408**
Kasl, S. V., 129, 131, **394**
Kaufman, H. G., 153, **394**
Kay, A. C., 2, 146, **394**
Kearsley, G., 249, **394**
Keeling, C. D., 81, **394**
Keenan, L. N., 291, **380**
Keenan, S. A., 291, **399**
Keeney, R. L., 339, **387**
Kelly, K. K., 94, **407**
Kelly, M. M., 317, **389**
Kemp, N. J., 17, **416**
Kendall, H. W., 327, **378**
Kergoat, H., 252, **398**
Kern, D. P., 214, **394**
Kern, R. P., 155, **412**
Kerr, R. A., 50, 78, 79, 87, **395**
Keyfitz, N., 100, 170, 361, 362, **395**
Keyworth, G. A., II, 68, 191, **395**
Kidd, K. K., 204, **413**
Kierman, V., 68, **395**
Kiesler, S., 288, 289, 290, 314, **395**, **413**
Kiester, E., Jr., 101, **395**
King, A. W., 79, 80, 81, **407**
Kinnaman, D., 146, **395**
Kinnucan, P., 17, **395**
Kinoshita, J., 191, **395**
Kirkpatrick, E. M., 129, **413**
Kirschner, C., 166, **387**
Kitamura, R., 170, 352, **395**
Kjaergaard, D., 300, **378**
Klein, G., 328, **395**

Klein, L. R., 12, 24, 26, 30, **395**
Kleinman, D. L., 182, **395**
Kling, R., 303, 320, 393, **395**
Klinger, D., 169, **395**
Klinker, G., 242, **395**
Knight, R. C., 321, 349, **419**
Koltnow, P. G., 171, 172, **395**
Korotkevich, W. S., 81, **377**
Korte, W. B., 315, 316, **393**
Koshland, D. E., 205, **395**
Kostyniuk, L. P., 170, 352, **395**
Krafcik, J. F., 35, **395**
Kraft, P., 228, 230, 303, 309, **395**
Kramer, A. F., 284, 319, **395, 416**
Kraus, L. E., 348, 349, **395**
Krauskopf, K. B., 113, 114, **396**
Kraut, R. E., 298, 302, 310, 316, 317, **396**
Krimsky, S., 341, **406**
Krugman, P. R., 10, 12, 13, 26, 27, **391**
Kryder, M. H., 217, **396**
Kryter, K. D., 132, **396**
Kuck, D. H., 233, **396**
Kuech, T. F., 214, **394**
Kurylo, M. J., 94, **416**
Kuzmyak, R., 169, **395**
Kyle, C. R., 65, **418**

L

La Brecque, M., 80, 215, 220, **396**
Lachman, R., 253, **396**
Lam, N., 254, **398**
Lammers, S., 229, **396**
Lancraft, R., 182, **377**
Landau, R. 10, 15, **396**
Landauer, T. K., 256, 272, 307, **385, 388,**
  396
Landow, S. E., 207, **396**
Landweber, L. H., 220, **393**
Lang, L., 121, **396**
la Riviere, J. W. M., 95, 97, 98, 99, **396**
Lasaga, A. C., 78, **380**
Lash, J., 122, **397**
Latremouille, S. A., 252, **401**
Laub, L., 216, **396**
Lauber, J. K., 164, **387**
Laurel, B. K., 263, **396**
Lauridsen, P. K., 73, **396**
Lawn, R. M., 201, **396**
Lawrence, P. R., 15, **396**
Lawrie, D. H., 233, **396**
Lawton, M. P., 129, 130, 353, 355, **396**

Lax, P. D., 220, **396**
Layman, M., 336, **397**
Lazarus, R. A., 321, **396**
Leahy, P. J., 101, **396**
Lee, C., 152, 153, **397**
Lee, E., 249, 254, **397, 398**
Lee, T. H., 49, 50, **397**
Leeper, P., 306, **397**
Lefkowitz, L. J., 242, **397**
Lehman, R. L., 74, **397**
Lehnert, W., 279, 280, 282, **417**
Leibowitz, H., 253, **391**
Leibowitz, H. W., 174, 183, **397**
Lenski, R. E., 208, **414**
Lescohier, I., 174, 355, 358, **397**
Lesser, V. R., 242, **397**
Lester, R. K., 14, 15, 27, 28, 30, 35, 67,
  152, **378, 384, 397**
Levanthal, H., 198, 210, **397**
Leventhal, E., 198, 210, **397**
Levi, K. R., 242, **411**
Levin, D., **408**
Levin, H., 307, **397**
Levine, M. F., 300, **397**
Levinson, P., 288, **397**
Levison, W. H., 182, **377, 395**
Levitt, L., 325, **393**
Lewin, K., 130, **397**
Lewis, C., 257, 284, **397, 390**
Lewis, C. H., 284, **397, 398**
Lewis, D., 9, 10, 26, **397**
Lewis, J. D., 257, **389**
Lewis, J. S., 186, **397**
Lewis, R. A., 186, **397**
Ley, P., 198, **397**
Lichtenberg, B. K., 193, **388**
Lichtenberg, F. R., 139, 193, 306, **377**
Lichtenstein, S., 67, 335, 336, 337, 339,
  387, **397, 412**
Licklider, J. C. R., 265, 270, 292, **397**
Liedl, G. L., 18, **398**
Likens, G. E., 91, **398**
Lin, F., 314, **383**
Lincoln, R. S., **413**
Linder, D. E., 129, **379**
Lindzen, R. S., 79, **398**
Lippard, S. J., **398**
Lippman, A., 236, **398**
Lipske, M., 103, 104, **398**
Lochbaum, C. C., 272, **385**
Locke, E., 321, **398**
Loeb, M., 132, **398**

Loehr, R., 122, **398**
Loftus, J. P., 192, 193, **398**
Logsdon, J. M., 190, **398**
Long, R. J., 314, **398**
Longino, C., 353, **412**
Loomis, R. J., 325, **393**
Lorius, C., 81, **377**
Lotze, M. T., 200, **408**
Lovasik, J. V., 252, **398, 399**
Lovins, A. B., 47, 54, 56, **386**
Lowry, I. S., 169, 170, 362, **398**
Lubin, P. M., 191, **390**
Lucky, R. W., 226, 281, 283, 289, **398**
Lütscher, H., 81, **388**

M

MacCready, P. B., 65, **417**
MacDonald, B. A., 242, **418**
MacDonald, G. J., 83, 90, **398**
MacDonald, N. H., 291, **399**
MacDonald, W. B., 144, **393**
MacGregor, D., 337, **387**
MacGregor, D. G., 176, **398**
MacGregor, J., 249, 254, 397, **398**
Machlup, F., 298, **398**
Mack, R. L., 284, 397, **398**
Mack, R. M., 208, **414**
Mackenzie, J. D., 186, **398**
MacNeill, J., 102, 133, 365, 366, **398**
Madni, A. M., 242, **398**
Mador, M. L., 204, **413**
Maezio, K., 227, **412**
Mahler, H., 363, 365, **398**
Makarov, A. A., 53, 92, 134, 170, 365, **381**
Makhoul, J., 250, 277, 280, 281, 292, **398**
Malitza, M., 372, **379**
Malone, T. W., 17, 32, 313, 314, 383, **398**
Manji, K. A., 267, **376, 377**
Mann, F., 150, **380**
Mann, S. H., 108, **393**
Manos, K. L., 164, **388**
Mansbridge, J. V., 81, **385**
Manzer, L. E., 86, **399**
Marchionda-Frost, K., 278, **412**
Marcus, M., 242, 279, 280, 282, 399, **417**
Marcuse, P., 129, 131, **394**
Marovelli, R. L., 31, **399**
Marrill, T., 219, **399**
Marshall, E., 82, 92, 124, 185, 188, 225, **384, 399**

Martel, L., 50, 162, 363, **394**
Martin, G. L., 315, **399**
Martin, L. G., 352, **399**
Marvin, C., 4, **399**
Massey, D., 161, **399**
Massey, L. D., 148, 155, **407**
Massey, W. E., 12, **399**
Matteson, M. T., 321, **399**
Matthews, M. L., 252, 398, **399**
Mattingly, T. J., Jr., 67, **379**
Mayo, J. S., 217, **399**
Mazlish, B., 285, **399**
McCain, G., 188, 363, 383, **399**
McCauley, M. E., 249, 278, **412**
McClelland, L., 71, 72, **399**
McConnel, S. R., 297, **399**
McCormick, N. J., 339, **399**
McCrory, R. L., 66, **383**
McCullough, R. L., 18, **381**
McDermott, J., 242, 394, **395**
McDougall, G. H. G., 60, **416**
McEvoy, G. M., 320, **399**
McFarland, B. L., 207, **396**
McGinnis, J. M., 355, 356, **400**
McHale, J., **400**
McIntosh, H., 373, **400**
McKenzie, A., 219, **391**
McLean, R., 149, **409**
McMillan, D., 320, **387**
McNeil, B. J., 197, **414**
McNutt, B., 54, 87, 179, 180, **390**
McQuillan, J., 219, **391**
McWahorter, J., 73, **408**
McWilliams, P. A., 350, **400**
Medawar, P., 4, **400**
Meier, M. F., 83, **400**
Mekjavic, I., 131, **400**
Meltzer, D., 279, 280, **377**
Merbold, U., 193, **388**
Merino, M. J., 200, **408**
Merland, G., 101, **400**
Mertins, K., 252, **399**
Messina, P. C., 233, **387**
Metz, C. E., 196, **400**
Meyer, S. S., 191, **390**
Meyrowitz, N. K., 150, **418**
Miller, A. D., 200, **408**
Miller, A. R., 226, **400**
Miller, C. O., 164, **400**
Miller, D. C., 182, 216, 400, **406**
Miller, G. A., 140, **400**
Miller, H. I., 336, **400**

Miller, J. G., 115, **400**
Miller, K., 306, **400**
Miller, K. F., 207, **407**
Miller, L. A., 291, **391**
Milstein, J. S., 58, 59, **400**
Mincer, J., 139, **400**
Minch, R. P., 283, **407**
Minnis, P., 83, **407**
Minuto, A., 253, **389**
Mirabella, A., 321, **414**
Mishel, L., 28, **400**
Mitchell, C. M., **376**
Mitchell, J. F. B., 82, 85, **400**, **410**
Mitchell, T. M., 17, 93, 189, **400**
Miyao, M., 252, 253, **400**
Mobley, R. L., 219, **378**
Moeller, B., 150, **380**
Moen, R., 200, **408**
Moffat, A. S., 70, **400**
Mohnen, V. A., 91, **400**
Molnar, L. J., 173, **413**
Monat, A., 321, **396**
Monk, A., 249, **400**
Monmaney, T., 209, **400**
Moore, T., 63, **400**
Moran, T., 249, **380**
Moran, T. P., 256, **408**
Moray, N., 319, 393, **416**
Moray, N. P., 52, 117, **401**
Moretti, P. M., 62, **401**
Morgan, M. G., 190, 339, 341, **401**
Morgan, R. A., 200, **408**
Morrison, J., 131, **399**
Morrison, P. R., 283, **401**
Mowery, D. C., 15, 20, 26, 30, 139, 307, **383**
Muller, R. S., 23, **392**
Mullis, I. V. S., 140, **401**
Mumford, E., 320, **401**
Munter, P., 148, **386**
Muralidharan, R., 182, **377**
Murolo, P., 310, **401**
Murphy, E. E., 177, **401**
Murray, H. A., 130, **401**
Muter, P., 252, **401**
Mutter, S. A., 148, 155, **407**
Myers, G. C., 352, **401**

N

Nadis, S., 70, **401**
Nagel, D. C., 164, 165, **401**

Nahemow, L., 130, **396**
Naisbitt, J., 298, **401**
Najah, M., 267, **376**
Nash, T., 258, 260, **401**
Nathan, R. R., 10, 44, **401**
Navarro, T. N., 278, **412**
Navasky, V., 4, **381**
Negroponte, N. P., 148, 293, **402**
Neisser, U., 202, 234, **402**, **410**
Nelson, P. D., 188, **391**
Nerenz, D. R., 198, 210, **397**
Neves, K. W., 215, **386**
Nevins, W. M., 66, **389**
Newacheck, P. W., 352, **380**
Newell, A., 188, 189, 249, **380**, **402**
Newell, R. E., 93, **402**
Newman, D., 150, 151, **403**
Nickerson, R. S., 77, 145, 146, 187, 189, 193, 248, 249, 254, 257, 258, 259, 269, 278, 281, **403**
Nicodemus, C. L., 187, **389**
Nierenberg, W. A., 79, 83, **403**, **410**
Nilles, J. M., 315, **403**
Nilsson, L. G., 252, **403**
Nitzel, M. T., 59, **403**
Nobre, C., 99, **411**
Nolan, J. B., 329, **403**
Nolan, S., 242, **393**
Nordqvist, T., 252, **403**
Noris, A. H., 354, **379**
Norman, C., 20, **404**
Norman, D. A., 249, 263, 284, **393**, **403**, **409**
Norman, K. L., 255, 284, **403**, **410**
Nowatzyk, A., 238, **393**
Nusbaum, H. C., 249, 278, **409**, **412**
Nussbaum, K., 304, 310, **390**

O

Oakley, L. C., 151, **403**
Oakley, R. C., 151, **403**
Odum, E. P., 332, **404**
Odum, H. T., **404**
Oeschger, H., 81, **388**
Offermann, L. R., 308, 309, **404**
O'Hare, M., 335, 340, 357, **404**
Ohlsson, K., 252, **403**
O'Leary, P. R., 105, **404**
Olex, M. B., 249, **411**
Olsen, M. E., 57, **404**

Olsen, R., 199, **404**
Olson, J. S., 79, **376**
Olson, M. H., 316, 317, **404**, **405**
O'Malley, C. E., 276, **405**
Oprysko, M. M., 214, **394**
Oshry, B., 35, **410**
Osterman, P., 301, **405**
Owens, D., 276, **405**
Owens, D. A., 174, **396**
Owens, M. R., 142, **405**

P

Paap, K. R., 254, 255, **405**
Packer, A. H., 10, 26, 27, 306, 309, 311, 367, **393**
Palmiter, R. D., 207, **407**
Panopoulos, N. J., 207, **396**
Papendick, R. I., 102, 103, **407**
Paquette, P. C., 24, 28, 298, 368, **407**
Parker, J. F., 163, **405**
Parker, R. A. R., 193, **388**
Parr, J. F., 102, 103, **407**
Parsons, H. M., 249, **405**
Parthenopoulos, D. A., 217, **405**
Partridge, C., 221, **405**
Partyka, S., 174, **405**
Partyka, S. C., 325, **405**, **406**
Passell, P., 87, **406**
Patton, R. M., 192, **398**
Paulus, P. B., 188, 189, 363, **399**, **406**
Paulus, P. D., 188, 189, 363, **383**
Pea, R., 150, **406**
Pelca, J., 222, **406**
Peled, A., 228, **406**
Peltier, W. R., 87, **406**
Peng, T. H., 79, 80, 81, **407**
Penko, R. R., 317, **392**
Pennebaker, J. W., 198, 363, **406**
Perchonok, K., 174, **406**
Perkins, D., 145, **403**
Perrault, R., 279, 280, 282, **417**
Perry, A. M., 89, 91, **388**
Perry, T. S., 141, **406**
Peterson, J. G., 242, 378, **381**
Peterson, W. L., 167, 249, **406**
Petsko, G. A., 195, 197, **406**
Pew, R., 319, **393**
Pew, R. W., 182, 259, **403**, **406**
Pfeiffer, D., 347, **406**
Phaff, H. J., 207, **406**
Phillips, L. D., 337, **397**

Pickett, R. M., 196, 197, **389**, **407**, **413**, **414**
Pillsbury, A. F., 101, **406**
Pines, M., 201, 205, **406**
Pinkert, C. A., 207, **407**
Pipes, R. B., 18, **381**
Pisoni, D. B., 249, 278, **406**, **409**
Platt, J., 128, **406**
Plough, A., 341, **406**
Pool, R., 22, 61, 66, 306, **406**
Port, O., 214, **406**
Post, D. L., 252, **407**
Post, W. M., 79, 80, 81, **407**
Postel, S., 95, 97, **407**
Powell, C. S., 107, **407**
Powledge, T. M., 143, **407**
Prather, M. J., 94, **416**
Press, F., 220, **406**
Proffitt, M. H., 94, **407**
Psotka, J., 148, 155, **407**
Pugh, W. M., 188, **384**
Pursel, V. G., 207, **407**
Pynoos, J., 352, **407**

Q

Quinn, J. B., 24, 28, 298, 368, **407**

R

Rabiner, L., 250, 277, 280, 281, 292, **398**
Rafelski, J., 66, **407**
Ramanathan, V., 78, **83**, **407**
Rasmussen, J., 319, **393**
Rasmussen, W. D., 31, **407**
Rathje, W. L., 105, 106, **407**
Rauch, J., 297, **407**
Ray, N. M., 283, **407**
Raynaud, D., 81, **377**
Read, S., 67, **387**
Reddy, K. N., 58, 74, 101, **407**
Regal, P. J., 208, **414**
Reganold, J. P., 102, **407**
Regnier, V., 352, **407**
Reich, R. B., 13, 14, 40, 259, 299, 306, 314, **407**
Reichel, D. A., 59, 72, 73, **407**
Reichle, H. G., Jr., 93, **402**
Reisner, M., 97, **407**
Reister, D. B., 89, 91, **388**
Reizenstein, J. E., 199, **407**

Remde, J. R., 272, **385**
Remington, R., 252, **408**
Reno, A. T., 169, 170, **408**
Rentzepis, P. M., 217, **405**
Repetto, R., 99, 100, **408**
Resnick, L. B., 148, **408**
Rettig, M., 277, **408**
Revelle, R., 78, 101, **408**
Revis, J., 9, 10, 26, **397**
Reynolds, J. F., 102, **410**
Richards, J., **408**
Riche, M. F., 298, 310, **408**
Richelson, J. T., 266, **408**
Ride, S. K., 190, **376**
Ritchie, M. L., 163, **408**
Rivlin, A. M., 354, **408**
Robert, J. M., 291, **380**
Roberts, L., 79, 94, 200, 201, 202, 204, 205, **408**
Roberts, L. A., 219, **399**
Roberts, N., 150, **408**
Roberts, P. O., 11, 168, **408**
Roberts, T. L., 256, **408**
Robertson, R., 132, **409**
Robinson, S., 315, 316, **393**
Rockart, J. F., 17, 32, 313, **398**
Rodhe, H., 48, 90, **408**
Roland, E. F., 249, 278, **412**
Rolf, S. R., 94, **392**
Rollenhagen, C., 252, **416**
Ronovn, A. B., 78, **380**
Rose, A. H., 207, **408**
Rosen, J. J., 197, **388**
Rosen, J. M., 94, **392**
Rosenberg, N., 365, **408**
Rosenberg, N. J., 78, 97, 102, **383**, **408**
Rosenberg, S. A., 200, **408**
Rosenfeld, A. H., 54, 55, 56, 71, **378**, **409**
Roske-Hofstrand, R. J., 254, 255, **405**
Ross, M. H., 54, 57, 58, 208, **409**
Rossi, P. H., 11, **409**
Roszak, T., 147, 285, 307, **409**
Rothenberg, J., 266, **409**
Rotty, R. M., 79, **376**
Rouse, W. B., 182, **409**
Rowell, L. B., 321, **409**
Rowles, G. D., 130, **409**
Rule, J., 303, **376**
Rumberger, R., 307, **397**
Rumelhart, D. E., 284, **409**
Runnion, A., 73, **409**
Rupp, B., 249, **409**

Russell, M., 119, **409**
Ruth, J. C., 249, 278, **412**

S

Salthouse, T. A., 354, **409**
Salvendy, G., 17, 242, 249, **388**, **393**, **411**
Samdahl, D. M., 132, **409**
Sameh, A. H., 233, **396**
Sander, M., 266, **409**
Sanders, A., 266, 319, **393**
Sanderson, P. M., 17, **409**
Sanford, D. W., 245, **409**
Sanghvi, M. J., 51, **388**
Sassin, W., 47, **409**
Sayer, M., 19, **409**
Scace, J. L., 223, **384**
Scardamalia, M., 149, **409**
Schahn, J., 132, **409**
Schein, E. H., 34, **409**
Schlesinger, L. A., 35, **410**
Schlesinger, M. E., 85, **410**, **417**
Schlesinger, W. H., 102, **410**
Schneider, C., 127, **410**
Schneider, S. H., 78, 81, 82, 83, 88, **410**
Schoolland, K., 143, **410**
Schoonard, J. W., 248, **410**
Schultz, T. J., 132, **386**
Schulze, E. D., 100, **410**
Schurick, J. M., 252, **391**
Schurr, S. H., 50, **410**
Schwab, E. C., 249, **410**
Schwartz, A., 350, **410**
Schwartz, J., **410**
Schwartz, J. P., 284, **410**
Schwartz, S. E., 91, **410**
Schweizer, E. K., 23, **385**
Schwing, R. C., 175, **386**
Scrimshaw, N. S., 332, 363, **410**
Segal, H. P., 6, **410**
Segal, J. W., 145, **381**
Seidel, R. J., 249, **394**
Seiler, W., 79, 93, **402**
Seinfeld, J. H., 69, 91, 92, 93, **410**
Seitz, F., 83, **410**
Selfridge, O. G., 202, 234, **410**
Seligman, C., 47, 72, **377**, **410**
Seligman, M. P., 198, **410**
Sellers, P., 99, **411**
Seltzer, S. E., 197, **414**
Seppala, P., 249, **410**
Shaiken, H., 302, **410**

Shalin, V. L., 242, **411**
Shanklin, J. D., 94, **386**
Sharit, J., 17, 249, **411**
Shaw, R. W., 53, **411**
Shea, S., 279, 280, **377**
Sheil, B., 258, **411**
Sheldrick, M. G., 171, **411**
Sheppard, H. L., 301, **411**
Sheridan, T. B., 240, 182, 283, 319, 350, **411**
Shiba, S., 37, 38, **411**
Shinar, D., 174, **411**
Shleifer, A., 35, **411**
Shneidermann, B., 249, 284, **410**, **411**
Short, J., 289, **411**
Shugart, H. H., 102, **385**
Shukla, J., 99, **411**
Shulman, G. L., 253, **391**
Shulman, H. G., 249, **411**
Shulman, S., 113, 114, **411**
Shumate, P. W., Jr., 222, **411**
Sieber, J. E., 330, **411**
Siegenthaler, U., 81, **388**
Silverberg, R. F., 191, **390**
Silvestro, K., 242, **411**
Simmons, H. E., 207, **411**
Simpson, C. A., 249, 278, **412**
Simpson, R. W., 319, **410**
Simutis, Z. M., 155, **412**
Sind, P. M., 209, 210, **412**
Singer, J. E., 71, 113, **377**
Singleton, W. T., 319, **412**
Sirevaag, E. J., 319, **395**
Sleeman, D., 148, **412**
Sleight, A. W., 61, **412**
Slichter, W. P., 19, **412**
Slovic, P., 67, 176, 335, 336, 337, 339, **382**, **387**, **397**, **398**, **412**
Slowiaczek, L. M., 278, **412**
Smith, E. E., 145, **403**
Smith, M. J., 227, 252, **412**
Smith, R. A., 97, 320, **412**
Smith, R. L., 187, **389**
Smith, R. P., **412**
Smith, T. J., 237, **412**
Sobey, A. J., 169, 171, 172, **412**
Soldo, B., 353, **412**
Solomon, N. A., 207, **384**
Solow, R. M., 14, 15, 27, 28, 30, 35, 152, **378**, **384**
Sommer, R., 199, **413**
Soures, J. M., 66, **383**

Spence, D. A., **408**
Spencer, R. W., 85, **413**
Sperling, D., 54, 87, 179, 180, **390**
Spesock, G. J., **413**
Sproull, L., 288, 289, 290, 314, **413**
Sreenivas, K., 19, **409**
Stahl, F. W., 200, **413**
Stammerjohn, L. W., 252, **412**
Stanislaw, H., 173, **413**
Stark, L. W., 252, 253, **400**
Stata, R., 24, 139, 141, **379**
Stauffer, B., 81, **388**
Stefik, M. J., 229, 232, **378**
Steinberg, M. A., 18, **413**
Steinhart, P., 103, **413**
Steinmeyer, D., 54, 57, 58, 208, **409**
Stephens, J. C., 204, **413**
Stephens, S., 326, **413**
Stern, P. C., 129, **413**
Sternberg, R. J., 145, **377**
Sterns, H. L., 357, **413**
Stevenson, M. P., 102, **385**
Stever, H. G., 190, 192, **413**
Stewart, D., 23, 223, **413**
Stoddard, S., 348, 349, **395**
Stokes, A. F., 166, **413**
Stokols, D., 129, 130, **413**
Stolarksi, R. S., 94, **413**
Stoll, C., 226, **413**
Strassman, P. A., 304, **413**
Streeter, L. A., 171, 256, **413**
Streff, F. M., 173, **413**
Stuart, M. A., 237, **412**
Studt, T., 109, **413**
Suess, H., 78, **408**
Suhrbier, J. H., 92, 332, **413**
Sullivan, T. R., 223, **384**
Summers, L. H., 10, 12, 13, 26, 27, **391**
Sun, M., 89, **413**
Sundstrom, E., 40, **414**
Swallow, J., 149, **409**
Swets, J. A., 148, 196, 197, **386**, **389**, **414**
Szekely, J., 14, **414**

T

Takahashi, T., 80, **414**
Tans, P. P., 80, **414**
Taylor, F. W., 39, **414**
Taylor, G. A., 256, **413**
Taylor, G. J., 186, **380**
Taylor, J. C., 300, 302, **397**, **414**

Taylor, J. J., 67, **414**
Taylor, K., 121, 127, 134, **414**
Taylor, L., 332, 363, **410**
Tchobanoglous, G., 105, **414**
Tesler, L. G., 218, 227, **414**
Theisen, G. H., 105, **414**
Theologus, G. C., 321, **414**
Thompson, C. B., 39, **414**
Thompson, R. H., 267, **414**
Thornton, D. C., 319, **416**
Thurow, L. C., 10, 11, 27, 30, 152, 299, **378, 414**
Tichy, W. F., 215, **384**
Tideiksar, R., 358, **414**
Tiegdje, J. M., 208, **414**
Toch, T., 142, **414**
Todreas, N. E., 67, **389**
Toffler, A., 7, 15, **414**
Topalian, S. L., 200, **408**
Toynbee, A. J., 115, **414**
Treshow, M., 330, **383**
Treurniet, W. C., 252, **401**
Triggs, T. T., **407**
Trimmer, W. S. N., 23, **392**
Tuck, A. F., 94, **407**
Tugend, A., 142, **415**
Turkle, S., 285, **415**
Turnage, J. T., 314, **415**
Turner, J. A., 320, **415**
Turner, L., 39, **415**
Turoff, M., 248, **415**
Tushingham, A. M., 87, **406**

U

Udall, J. R., 97, 98, **415**
Uhlig, R. P., 304, **415**
Usher, C. E., 297, **399**
Uttal, Ql, W. R., 238, **415**
Utting, K., 273, **415**

V

Vachon, R. A., 349, **415**
Van Allen, J. A., 190, **416**
Vanderheiden, G. C., 351, **416**
Vaughan, R., 139, 297, 306, **377**
Vaughan, R. J., 138, 152, 154, **416**
Vaydanoff, P., 308, **416**
Vehar, G. A., 201, **396**
Verma, I. M., 201, **416**
Verplanken, B., 336, **416**
Vicente, K. J., 131, 319, **416**

Vidaver, A. K., 336, **400**
Vining, D. R., Jr., 362, **416**
Vining, J., 108, 132, 383, **416**
Virginia, R. A., 102, **410**
Vishny, R. W., 35, **411**
Vitello, D., 171, **413**

W

Waern, Y., 252, **416**
Wagner, A., 214, **394**
Walden, D., 219, **391**
Waldrop, M. M., 277, **416**
Wall, T. T., 17, **416**
Wallace, D. R., 103, **416**
Wallich, P., 40, 41, 80, 133, **382, 416**
Walsh, P. W., 105, **404**
Walton, E., 35, **378**
Waltz, D. L., 277, **416**
Walzer, P., 168, **378**
Wapner, S., 131, **384**
Ward, J. S., 155, **412**
Warren, H. E., 74, **397**
Warriner, G. K., 60, **416**
Watson, J. D., 73, 202, 205, **408, 416**
Watson, R. T., 94, **416**
Weaver, L., 120, **378**
Webb, J. W., 284, **416**
Weinberg, A. M., 6, 52, 54, 67, 117, 339, **416**
Weinberg, C. J., 62, 63, 69, 70, **417**
Weinberg, R. A., 195, 207, **417**
Weiner, J., 81, **417**
Weinstein, C., 250, 277, 280, 281, 292, **398**
Weinstein, N., 133, **382**
Weischedel, R., 279, 280, 282, **417**
Weiser, M., 218, 264, **417**
Weitzer, R., 321, 349, **419**
Welter, T. R., 135, **417**
Weybrew, B. B., 188, **417**
Wheaton, G. R., 321, **414**
Wheelon, A. D., 329, **403**
White, M., 129, 131, **394**
White, R. M., 79, 88, **417**
White, R. W., 130, **417**
Whiteside, G. M., 214, **417**
Whitfield, D., 319, **412**
Whitford, W. G., 102, **410**
Wickens, C., 319, **393**
Wickens, C. D., 166, 183, 320, 354, **379, 413, 417, 418**
Wiener, E. L., 167, 168, 249, **417**

Wiener, J. M., 354, **408**
Wiener, N., 58, **417**
Wierwille, W. W., 249, 319, **388**, **418**
Wigley, T. M. L., 78, 79, 85, **394**, **417**
Wilder, C. S., 347, **417**
Wilensky, R., 279, 280, 282, **417**
Wilfong, G. T., 17, **418**
Will, J., 129, 131, **394**
Williams, D., 244, 252, **408**, **418**
Williams, E., 289, **411**
Williams, R. H., 62, 63, 69, 70, **417**
Williams, S., 62, 74, 127, **418**
Williamson, R. A., 190, **398**
Williges, B. H., 249, 278, **412**
Williges, R. C., 319, 349, 351, **381**, **382**, **390**, **418**
Wilson, E. O., 103, 104, **418**
Wilson, H. J., 65, **417**
Wilson, J. R., 17, **393**
Wilson, J. W., 168, **417**
Wilson, K. S., 140, 236, **418**
Wilson, W. J., **418**
Winett, R. A., 59, **403**
Wise, J. A., 187, **418**
Wisniewski, E. J., 242, **411**
Witten, I. H., 242, **418**
Wivel, N. A., 336, **400**
Wolkomir, R., 170, **418**
Wollard, K., 64, **418**
Wolman, M. G., 97, **412**
Wood, P. H. N., 347, **418**
Wood, R. C., 277, **418**
Woodruff, E., 149, **409**

Woodruff, W., 367, **418**
Woods, D., 330, **418**
Woodwell, G. M., 78, 82, **392**
Woteki, T., 71, **418**
Wright, K., 21, 54, 172, 180, 219, 325, **418**
Wright, P., 79, **394**
Wright, R. F., 91, **398**
Wrighton, M. S., 61, **418**

Y

Yaeger, K. E., 68, **376**
Yam, P., 115, **418**
Yang, J. C., 200, **408**
Yankelovich, D., 300, **418**
Yankelovich, N., 150, 269, 273, 415, **418**
Yanskin, A. L., 78, **380**
Yeh, Y. Y., 320, 417, **418**
Young, J. A., 26, 30, 299, **419**
Yourdon, E., 236, **419**

Z

Zacharias, G., 182, **377**
Zalusky, J. L., 316, **381**
Zammuto, R. F., 308, **381**
Zimring, C., 321, 349, **419**
Zodhiates, P. P., 146, **403**
Zoltan, E., 283, **419**
Zorpette, G., 64, **418**
Zue, V., 250, 277, 280, 281, 292, **398**
Zwally, H. J., 83, **419**

Subject Index

A

Accidents, 355–359, *see also Fatalities; Human error*
  of elderly, 357–358
  highway, 357
  and major causes of death, 356–357
  motorcycle, 358
  prevention of, 357, 358–359
Achievement tests, 145–146
Acid rain, 77, 89, 91–92, 100
Actual risk, 334–335
Adaptation, 130, 372
Adult-education programs, 139
Aerocar, 172
Aerosol propellants, 95
Aerospace operations, 68
Age
  chronological vs. functional, 354
Aging. *See Elderly*
Agriculture, 97–98, 133, *see also Farming*
  biotechnology in, 207–208
  practice of monocropping, 103
  row-crop, 102
  sustainable vs. conventional, 102–103
  and water contamination, 96–97
Air pollution, 92–93
  and forests, 100–101
Air traffic control, 166–167
  upgrading of, 167

Air transportation, 160–163
  and aviation safety, 163–168
  efficiency improvements for, 180
  importance of to future economy, 160–163
Alar, banning of, 334
Alcohol and driving. *See Driving, while drinking*
Alleles, 202
Amazon forests, 99
Antarctica, 94
Appliance, household
  energy efficiency of, 71
Army, 154
ARPANET, 286, 290
Artificial intelligence (AI), 189, 238–239
Artificial realities, 55, 243–245
Atmospheric Radiation and Measurement program, 86
Automation, 139, 301, 301–305, *see also Mechanization*
  effects on white-collar workers, 302–303
  of flight control, 167
  of highways, 172
  and manufacturing, 17–18
Automobiles, 5, 325, *see also Driving; Transportation*
  accidents and fatalities, 172–174
  alternative fuels for, 69–70
  and carpooling to conserve energy, 73

drinking while driving, 173, 174, 175
electric, 180
energy efficiency of, 54, 56–57, 71,
    179–180
and feedback on driving behavior, 71–72,
    73–74
flying, 172
half-width, 179–180
importance of vision for driving, 174–175
and night driving, 174–175
and older drivers, 174, 175
and problem of congestion on highways,
    170
and rear-end collisions, 173
solar-powered, 64–65
as transportation in the U.S., 168–170
Automotive safety, 56–57
Aviation and automation, 167
Aviation safety, 163–168
Aviation Safety Research Act, 164

B

Back-up, 275
Bacteria, 207
Banking industry, 301
Biodegradable materials, 109, 119
Biodiversity, decreasing, 103–105
Biohydrometallurgy, 208
Biomass
    as alternative energy source, 65
Biotechnology, 195–211
    in agriculture and mining, 207–209
    gene therapy and the Human Genome
        Project, 200–206
    human factors issues on, 210–211
    industrial, 206–207
    technology and medicine regarding,
        196–200
Bird populations, 104
Birth control, 364
Birth rates, 361, 363
Books, electronic, 236
Boredom, problem of, 320, 321, 322
Bore-hole tomography, 48
Brake-lights, high-mounted rear, 173
Broad-sense accessibility, 266–267
Buildings
    energy efficiency of, 56, 324
    heating of, 63–64, 70
Bulletin board, electronic, 224, 286–288,
    290

C

Cancer risks, 122, 123, 338–339
Canned foods, 324
Carbon cycle, 79–80
Carbon dioxide, 48, 49, 78, 79–82, 86, 90,
    101, 120, 126
Carbon monoxide, 92, 93
Carpooling, 73
Cathode ray tube (CRT), 252–253
CD ROM, 115, 216, 269
Cellular telephone, 223
Centimorgan, 203–204
Ceramics, 19
Chemical bonding materials
    regarding medical technology, 197
Chemical energy, 46
Chernobyl accident, 52, 325–326, 336
Chlorine, 94
Chlorofluorocarbons (CFCs), 82, 86, 94–95,
    124, 126
Chromosome, 202–205
Chronological age, 354
Circuit design, very large scale integrated
    (VLSI), 273
Circuits, 214
Clean Water Act, 97
Climate change, 90–91
Climate models
    concerning global warming, 83–87
    validity of, 85–86
Clouds
    effects of, 80, 83
    radiation, 86
Coal, 51, 68
Collaboration, 149
Command language, 254
Command-oriented system, 254
Commercial exploitation
    and the United States' economy, 20–22
Communication, 149, see also Information
        services; Information technology
    face-to-face vs. computer-mediated,
        289–290
    history of, 115
    interperson, 285–292
    risk, 341–343
Communication network. See Network
Competence, 130
Competitiveness, 26, 34–35
    and balance of trade, 12
    and the United States' economy, 12–14

Computer-assisted instruction (CAI), 147–148
Computer-based work monitoring, 227
Computer-based imaging techniques and medicine, 196–197
Computer industry, 301
Computerized databases. See Databases
Computer programmer, 307
Computer-synthesized voice, 171
Computer system
  circuits, 214
  command- vs. menu-oriented, 254–255
  crime, 226–227
  human understanding of, 283–285
  memory, 216–217
  security, 225–227
Computer technology, 5, 155, 218 *see also Database; Information technology; Interface design; Network; Person-computer interaction; Programming; Software; Supercomputer; Terminals; Video display terminals*
  and advance of materials science, 19
  and automobiles, 171–172
  concerning materials research and microfabrication, 23
  conflict resolution, use of in 329–330
  and disabled individuals, 350
  and education, 146–151
  and the elderly, 355
  electronic publishing, 110, 112
  and medicine, 196–197, 199–200, 209
  novice vs. experienced users, 261–262
  and predicting environmental change, 133
  regarding speech and natural language, 277–283
  resistance to, 303
  and space exploration, 191, 193
  and telecommuting, 317–318
  in the workplace, 298, 303–304
Computer virus/worm, 225–226
Computing device trends, 213–218
  costs, 217–218
  new architectures, 214–216
  new materials, 214
  personal computers and integrated computing devices, 218
  storage technology, 216–217
Confirmation bias, 328
Conflict resolution, 328–330
  computers as systems for, 329–330
Conservation era, 127

Contamination
  from industrial accidents, 116–117
  water, 95–99
Conventional agriculture, 102–103
Core memory, 216
Corporate dept, 10
Crop agriculture
  biotechnology in, 207–208
Crops, green manure, 102
Crowding, problem of, 188–189
  effects on health, 362–363
Culture
  definition of, 34
  and desire for material possessions, 42–43
  and its effects on productivity, 34–36
Curtiss Autoplane, 172

D

Databases, *see also Information system; Information service*
  environmental, 120–121
  interacting with large information, 265–270
  and merging of human health and geophysical data, 121
Deaths. *See Fatalities*
Decision-making, 314
  and conflict resolution, 328–330
  and dealing with risks, 333–344
  hidden costs and ubiquity of tradeoffs regarding, 323–326
  need for tools in, 326–331
  personal decision aids regarding, 330–331
  policy evaluation and effect prediction regarding, 331–333
  and policy setting, 323–344
Decomposition, 109
Deconglomeration, 35
Deforestation, 99–101, 104
Dematerialization, 115–116
Desertification, 101–103
Desktop publishing, 235
Detergents, biodegradable, 119
Deuterium, 65–66
Development
  and the United States' economy, 20–22
Dialogue, human-computer, 253–255
Diamond, 214
Digital Equipment Corporation, 240
Direct-manipulation interface, 263–264
Disabilities Act, 349

Disabled individuals, 346–351
  computer technology and, 350
  definition of, 347
  distinguished from elderly, 353
  emotionally disturbed, 347–348
  employment for, 349–350
  and independent living, 349
  laws prohibiting discrimination against, 349
  major types of, 348
  opportunities for, 348–349
Disaster, natural, 118
Disease-causing genes, 200–201, 205–206
Displacement problems, 302
DNA technology, 200–205, 207
Domain-specific knowledge, 241
Drinking water, 95–96, see also Water
  chlorine in, 323–324
Drip irrigation, 98
Drivers
  behavior of, 175–176
  elderly, 174, 175, 352
Driver's licenses, 169, 170
Driving, see also Automobiles; Transportation
  and feedback, 73–74
  at night, 174–175
  and vision, 174–175
  while drinking, 173, 174, 175
Dropouts, 142
Drought, 98

E

Earnings, distribution of, 299
Earth
  mean temperature of, 78
Earth Observing System satellite, 121
Earthquake prediction, 118
Earth's radiation budget experiment (ERBE), 83
Ecological economics, 133–135
Ecological effects risks, 122, 123
Economy
  and competitiveness, 12–14
  development vs. commercial exploitation and, 20–22
  effects of population growth on, 363–364
  environmental, 133–135
  importance of air travel to, 160–163
  importance of chance and positive feed-back in, 21
  and manufacturing, 14–18

  and materials science, 18–20
  and microfabrication, 22–24
  and productivity, 25–43, 44
  regarding industry and productivity, 9–45
  three major sectors of, 298
  and services, 24–25
  world, 11, 365–369
  worrisome indicators concerning, 10–12
Ecosystem, 104
Education, 138–159, see also Employee training
  federal funding of, 143–144
  for future jobs, 305–308
  human factors issues concerning, 156–158
  human tutors vs. computers, 147–148
  and illiteracy, 140, 142
  importance of, 138–139
  Japanese system of, 141, 143
  and minorities, 142
  new developments regarding, 144–146
  problem of public, 28
  programs for training teachers, 145
  regarding industrial and military needs, 152–155
  in science and mathematics, 140–142
  signs of trouble in U.S. concerning, 140–143
  simulation and training, 155–156
  teaching methods, 144–146
  and technology, 146–151
  uncertain direction and commitment in, 143–144
  U.S. vs. non-U.S. citizen, 141
Educational software, 147–148, 150–151
Effectiveness
  and productivity, 25–26
Effect prediction
  and policy evaluation, 331–333
Efficiency
  and productivity, 25–26
  of workers, 39–40
Elderly, 351–355
  common accidents suffered by, 357–358
  and computer technology, 355
  distinguished from disabled individuals, 353
  improving transportation for, 175
  individual differences in, 353–354
  limitations and impairments, 353
  marital status and living situations of, 352
  and performance, 320, 353–354
  special requirements for housing of, 352–353

workers, 301, 309
Elderly drivers, 174, 175, 352
Electrical power, 46, 51
  efficiency of use, 55–56
  geothermal, 65
  produced by nuclear power plants, 67
  from the sun, 64
  and use of coal, 68
  windmills as a means of, 62–63
Electric vehicles, 180
Electronic archive, 268–269
Electronic books, 236
Electronic bulletin board, 224, 286–288, 290
Electronic information storage, 109–110, 112
Electronic license plates, 172
Electronic mail, 55, 224, 286, 288, 290, 314
Electronic map, 171
Electronic monitoring systems
  and feedback for energy conservation, 71–72
Electronic publishing, 55, 110
Electronic storage media, 266
Emissions control, 89
Emotionally disturbed individuals. See Disabled individuals
Employee training, 38, 152–155, 300, see also Education; Work
Endonucleases, 204
Energy, 6, 46–75
  alternative automotive fuels, 69–70
  alternative sources of, 52, 61–68
    biomass, 65
    fission, 67–68
    fusion, 65–66
    geothermal, 65
    sun, 63–65
    water, 63
    wind, 62–63
  chemical, 46
  electrical, 46
  individual behavior and use of, 70–74
  nuclear, 6
  and rail transportation, 177
  and superconductivity, 60–61
  traditional sources of, 47–49
  and transportation, 179–181
  U.S. situation concerning, 49–53
  world-wide demand and distribution of, 47

Energy-balance models, 84
Energy conservation, 43, 53–60
  automotive efficiency, 56–57
  and carpooling, 73
  college vs. noncollege populations, 59
  and efficiency of electrical power use, 55–56
  and feedback, 60, 71–72, 73–74
  and monetary incentives, 72
  quality control and recycling as, 57
  reliability of self-reports on, 59–60
Energy-efficient products, 58–59
Engineers, 306
English, computer understanding of, 279
Environment
  definition of, 129
  effects of on health, 130–133
  energy production and protection of, 325
  living and working, 129–130
  military involvement in issues concerning, 127
  and scenic beauty, 132
  that supports learning, 144–146
Environment, hospital, 198–199
Environmental change, 76–137, see also Pollution; Natural disasters
  acid rain, 77, 89, 91–92, 100
  contamination from industrial accidents, 116–117
  decreasing biodiversity, 103–105
  deforestation, 99–101, 104
  desertification and wetland loss, 101–103
  and ecological economics, 133–135
  and environmental psychology, 129–133
  global warming, 77–91, 125
  natural disasters, 118
  and need for better understanding of problem, 118–120
  and perceived environmental risks, 121–123
  political action concerning, 123–128
  and responsible individual behavior, 128–129
  stratospheric ozone thinning, 93–95
  waste, 105–116
  water contamination and depletion, 95–99
Environmental databases
  and research tools, 120–121
Environmental ergonomics, 131–133
Environmentalism, three stages of, 127–128
Environmental laws, 125
Environmental press, 130

Environmental Protection Agency (EPA), 95
  task force, 122–123
Environmental psychology, 129–133
  ecological vs. organismic approaches, 131
  minimalist, instrumental, spiritual views
    of, 130
  origin of, 130
Environmental risk
  perceived, 121–123
Environmental risks
  four types of, 121–123
Ephemeralization, comprehensive, 115,
  116
Ethanol, 69
Etherialization, progressive, 115, 116
Europe
  and highway safety, 171–172
Expert-system, 189, 229, 240–243
  building requirements of, 241
  definition of, 241
  motivation for development of, 241
  psychological factors involved in the devel-
    opment of, 242–243
  rule-based, 242
Extinction, 104

F

Farming, 297–298, *see also Agriculture*
Fast-food industry, 299
Fatalities, 355–359, *see also Accidents; Human
    error*
  automobile accident, 172–174
  aviation, 163
  motorcycle, 173
  pedestrian, 173
  railway, 178–179
Feedback, 134
  on driving behavior, 71–72, 73–74
  and energy conservation, 60, 71–72
  and population growth, 363
  of residential energy use, 72
Fertilizers, 96, 97
Fiber networks, 222–223
First-personness, 263
Fission
  as alternative energy source, 67–68
Food
  canned, 324
  distribution and policy decisions, 332
Forests, destruction of, 99–101, 104
Formaldehyde, 69

Fossil fuels, 6, 47–49, 81, 93, 97
  in the U.S., 50–52
Functional age, 354
Fusion
  as alternative energy source, 65–66

G

Gallium arsenide, 214
Gasoline, 69
Gender, *see also Women in workforce*
  differences in earnings, 299, 310
  discrimination, 309–310
General Motors, 35–36
Gene therapy, 200–206
Genetic linkage map, 203–204
Genetics, 195, 207
Geothermal
  as alternative energy source, 65
Glaciers, 81
  melting of, 83, 87
Global carbon cycle, 79–80
  effect of humans' use of land on, 80
Global-Change Program, 120
Global modeling, 86
Global networks, 221–222
Global warming, 77–91, 125
  carbon dioxide and other greenhouse gas
    accumulation, 81–82
  climate models for, 83–87
  greenhouse effect, 77–80
  and the no-regrets policy, 88–89
  oceans and, 80–81
  reasons to worry about, 87–88
  regarding ice sheets, 83
  and water vapor, 82–83
Global Warming Prevention Act, 126
GNP (gross national product), 10–11, 26,
    42–43, 54, 363
Governmental dept, 10
Graphics, 269
Greenhouse effect, 48, 77–80
  and greenhouse gas accumulation, 81–82,
    90
Green manure crops, 102
Ground transportation, 168–172
  and highway safety, 172–177
Guided-evolution interface design, 257–260

H

Handicapped individuals. *See Disabled individ-
    uals*

Hanford Military Reservation, 113
Health, *see also Medicine*
  effects of environment on, 130–133
  and personal safety, 355–359
  risks, noncancer, 122, 123
Helicopters, 165–166
High-definition television, 20
Highway, *see also Automobiles; Driving*
  accidents on, 357
  automated, 172
  congestion, 170, 171
  risky behavior on, 175
  safety, 171–177
Holistic top-down approach, 332
Homelessness, 11
Hospital environment, *see also Medicine*
  and dehumanization, 198
  design of, 198–199
Housing, elderly, 352–353
Hubble Space Telescope, 191
Human behavior
  as a cause of air pollution, 93
  and climate change, 90–91
  and environment, 135–136
  and extinction, 104
  and upsetting the global ecosystem,
    76–77
  while driving, 175–176
Human-computer dialogues, 253–255
Human error, *see also Accidents; Fatalities*
  in assessing risks, 337
  automobile accidents and fatalities con-
    cerning, 172–174
  aviation safety and, 164
  and industrial accidents, 116–117
  in medicine, 209–210
  in nuclear power plants, 67
  railway fatalities, 178–179
Human factors issues, 372–374
  aviation safety, 164–166, 168
  biotechnology, 210–211
  computer interface design, 250–252
  concerning environment, 131–132
  concerning microfabrication and nanotech-
    nology, 23–24
  disabled individuals, 346
  on education, 156–158
  on energy generation, 52–53
  and equipment safety, 22
  helicopters, 165–166
  information technology, 110–112,
    245–247

  in nuclear power plant operation, 52, 114,
    327–328
  person-computer/machine interaction,
    210, 248–249
  population growth, 364–365
  regarding economics and productivity,
    43–45, 368–369
  and services, 25
  space exploration, 184, 193–194
  speech and language recognition systems,
    281–283
  transportation, 181–183
  work, 318–322
Human Genome Project, 200–206
Human scheduling, 17
Human tutors, 147–148
Hurricanes, 118
Hydrochlorofluorocarbons (HCFCs), 86
Hydrofluorocarbons (HFCs), 86
Hydrogen, 69–70
Hydrogen-powered automobiles, 69–70
Hydrogen-rich fuels, 49
Hydropower, 63
Hypertext, 252, 271, 273, 293

I

Ice, glacial, 81
  melting of, 83
Illiteracy, 140
  unemployment and prison regarding,
    142
Immigrants
  effects of on U.S. population, 361–362
  in the workforce, 309
Incineration, 106
  and medical waste, 112
Incrementalism, 52
Individual behavior
  and energy use, 70–74
  responsible, 128–129
Individual differences
  among the elderly, 353–354
  and genetics, 203
Industrial accidents
  contamination from, 116–117
  and human error, 116–117
Industrial biotechnology, 206–207
Industrial mass-production techniques, 323
Industry, role of in education and training,
    152–154, 155

Information
  access and use, 265–277
    and interacting with large information
      stores, 265–270
    and navigating through information,
      272–274
    on-line help and advice, 276–277
    personal information systems regarding,
      274–276
  filter, 233
  narrow- vs. broad-sense accessibility of,
    266–267
  transmission of, 55, 75, 180
Information presentation techniques,
  270–272
Information refinery, 233
Information services, 151, 235–237
Information storage, 109–112, 115–116
  electronic, 109–110, 112
  reliance on paper for, 111–112
Information systems, medical, 199–200
Information technology, 41, 110–112, 149,
  150, 212–247, *see also Computer technology*
  human factors issues concerning, 245–247
  networks, 218–227
  and productivity, 31–33
  software, 228–234
  some expectations regarding, 234–245
  trends in computing devices, 213–218
Information workers, productivity of, 40–41
Infrared radiation, 77–78
Insurance industry
  and automation, 301–302
Insurance-policy strategy, 88
Integrated computer circuits, 23
Integrated-services digital network (ISDN),
  221–222
Intelligent vehicle/highway systems (IVHS),
  172
Interface, 293
  definition of, 250
  direct-manipulation, 263–264
Interface design, *see also Computer technology*
  evaluation of, 260–261
  evolutionary iterative, 257–260
  intuition in, 255–257
  invisible, 262–264
  and needs of novice and experienced users,
    261–262
International Thermonuclear Experimental
  Reactor

(ITER), 66
Interperson communication, 285–292
Inter-referential I/O, 263
Introns, 202
Invention revolution, 115
Irrigation, 97, 98
Iterative interface design, 257–260
  task analysis and rapid prototyping, 259

J

Japan, 172, 217, 259
  education in, 141, 143
  and total quality management (TQM),
    37–38
Job, training on the, 152–155
Job-related stress, 321, 321–322
Jobs, *see also Employee training; Work*
  quality of, 299–300, 320
  productivity and, 29
  skill requirements of future, 305–308
Job satisfaction, 320–321
Job Skills Education Program, 154

K

Key-word descriptors, 268–269
Knowbot, 270

L

Labor productivity, 26, 27
Lactic acid, 109
Land
  degradation of, 101–103
  semi-arid, 102
Language understanding
  and computers, 279–280
  and semi-automatic translation, 280–281
Lasers, 66
Lead, 92
Learned helplessness, 198
Leisure, problem of, 322
Life space, 130
Light-emitting-diode (LED), 224
Light Helicopter Experiment (LHX),
  165–166
Living space, 188–189
Local network, 221–222
Loneliness
  among elderly, 355

M

Machine memory, 216
Macroergonomics, 310–311
Magellan spacecraft, 191
Magnetic-levitation systems, 177
Magnetohydrodynamic (MHO) topping cycles, 51
Management information system (MIS), 110–111
Manufacturing, 28
  and automation, 17–18
  and changing organizational patterns in U.S. industry, 15–16
  productivity in, 15
  and the United States' economy, 14–18
Mars, 78, 185
Material possessions, desire for, 42–43
Materials
  biodegradable and photodegradable, 109
  chemical composition of, 18
  for computer technology, 214
  controlling microstructure of, 18
  and dematerialization, 115–116
  science and the United States' economy, 18–20
Mathematics, education in, 140–141
Measurement techniques, 7
  for productivity, 39–43, 44
Mechanization, 297, 301, see also Automation
  and productivity, 31
Medical imaging processes, 196–197, 209
Medical information systems, 199–200
Medical waste and incineration, 112
Medicine, see also Health; Hospital environment
  preventive, 358–359
  and technology, 196–200
Meiosis, 203
Memex, 265
Menu-driven systems, 254–255, 261–262
Methane, 48, 82, 90
Methanol, 69, 180, 325
Methanol-powered automobiles, 69
Microergonomics, 311
Microfabrication, 22–24
Microorganisms, 96, 207
  use of in removal of unwanted substances, 208–209
Military
  aviation safety, 164–165
  and involvement in environmental issues, 127

  role of in education and training, 154–155
Miniaturization. See Microfabrication
Mining, biotechnology in, 208–209
Minorities
  and education, 142
  and population growth, 361–362
  in the workforce, 309
Monetary incentives
  for energy conservation, 72
Monitors, 223
Monocropping, 103
Moon, 185, 186
  as spot for astronomical observation, 186
  surface of, 186
Motorcycle fatalities, 173
Multi-factor productivity, 26
Multimedia, 150, 252, 273–274, 293
Multiple choice tests, 145
Multiprocessor system, 215
Municipal Solid Waste Source Reduction and Recycling Act, 126

N

Nanotechnology, 23
Narrow-sense accessibility, 266–267
National Forest Management Act, 104
National networks, 221–222
Natural disasters, 118
Natural gas, 48, 49, 50, 180
Natural language processing (NLP), 282
Network, 218–227, 286, 290, 314, see also Computer technology
  fiber, 222–223
  local, national, and global, 221–222
  packet-switching, 219–220
  promises and problems regarding, 224–227
  satellite, 223
  terminals, 223–224
Newspapers, electronic, 109–110
New United Motor Manufacturing, Inc. (NUMMI), 35–36
NIMBY conflicts, 340
Nintendo as an educational tool, 148
Nitrates, 97
Nitric acid, 91
Nitrogen, 89, 97
Nitrogen dioxide, 92
Nitrous oxide, 90
Noise and rail transportation, 177
Noise pollution, 132–133

No-regrets policy, 88–89
Nuclear fission, 67–68
Nuclear fusion, 65–66
Nuclear power, 6, 48, 325–326
  assessing risks of, 336
  and industrial accidents, 116–117
  and risk assessments, 339
Nuclear power plants
  human error at, 67
  problem of disposing of, 114
  policy changes regarding, 331–332
Nuclear propulsion system, 68
Nuclear war/weapons, threat of, 327, 329
Nuclear waste disposal, 113–115
Nuclear Waste Policy Act, 113
Nursing homes, 354

O

Object-oriented programming, 232
Occupation. *See Employee training; Job: Work*
Occupational hazards, 356
Ocean and global warming, 80–81
Office work and dependency on computers, 298
Oil, 49, 67, 98, 179, *see also Petroleum*
  in the United States, 50
Older individuals. *See Elderly*
On-line help, 276–277
On-line information system, 151
  medical, 200
Optical fiber, 222
Organisms
  decrease in species of, 103–105
  transgenic, 195
Organization, workplace, 313–315
Organizations
  effects of on productivity, 34–36
  and patterns of change in U.S. industry, 15–16
Outreach Program, 186
Overconsumption, 13
Oxygen, 89, 186
Ozone, 90, 92, 93
Ozone hole, 94
Ozone thinning, stratospheric, 93–95

P

Packet-switching network, 219–220
Pandemonium, 234
Paper, *see also Newspapers*
  recycling of, 107
  reliance on for information storage, 111–112
Paper entrepreneurism, 13–14
Paper waste reduction, 109–112
Parallel-architecture system, 215, 232–233, 246
Participation, 37
Particle bed nuclear reactor, 68
Particulates, 92, 96
Pathfinder experiment, 171
Pedestrian fatalities, 173
People-driven vs. driving systems, 300
Perceived risks, 334–335
Performance
  and age-related changes, 320
  of the elderly, 353–354
  productive, 30
Personal computers, 218
Personal decision aids, 330–331
Personal information system, 274–276
Person-Computer/machine interaction, 210, 248–295, *see also Computer technology; Database; Information technology; Programming; Software*
  human factors issues concerning, 248 249, 281–283
  human understanding of, 283–285
  information access and use during, 265–277
  interface design for, 250–264
  and interperson communication, 285–292
  speech and natural language concerning, 277–283
Pesticides, 96, 324
Petroleum, 48, 50, 67, 179, *see also Oil*
Photodegradable materials, 109
Photonic device, 222
Photonic storage media, 266
Photovoltaic power, 64
Physical mapping, 204
Physician's Data Query (PDQ), 200
Piecemeal approach, 332
Pilots, 162–163, *see also Air travel; Aviation*
  general aviation, 163–164
  helicopter, 165–166
  role in flying planes, 167–168
Pioneer-10, 191
Plan, Do, Check, and Act cycle, 38
Planes. *See Air transportation; Aviation; Transportation, air*
Planets, exploration of, 191

Plankton, 80
Plastics, 18–19, 109, 325
  biodegradable, 109
  as waste, 107
Plutonium, 113
Poland, 134
Polar ice caps
  melting of, 87
Policy decisions, 53
Policy evaluation
  and effect prediction, 331–333
Political action
  on environmental change issues, 123–128
Pollution, *see also Environmental change*
  air, 92–93
  noise, 132–133
  point vs. non-point, 97
  and policy decisions, 332
Polymers. *See Plastics*
Polyvinylchloride (PVC), 324–325
Population growth
  and immigration, 361–362
  and increasing urbanization, 359–365
  negative, 363
  world, 360–361, 362, 364–365
Positive feedback. *See Feedback*
Positive reinforcement
  and recycling, 108
Precipitation, acid. *See Acid rain*
Preventive medicine, 358–359
Primary memory, 216
Probabilistic risk assessment, 339
Procognitive systems, 265
Product design
  and waste disposal, 108–109
Productive performance, 30
Productivity, 11, 44, 316
  agriculture and technological innovation
    concerning, 30–31
  definition of, 25, 26
  determinants of, 29–34
  and GNP statistics, 42–43
  and information technology, 31–33
  labor, 26, 27
  measurement of, 39–43, 44
  and mechanization, 31
  multi-factor, 26
  organizational and cultural effects on,
    34–36
  participation, responsibility, and quality
    regarding, 37–39
  of programmers, 231–232

  in relation to effectiveness and efficiency,
    25–26
  six systematic weaknesses concerning U.S.,
    30
  and slowing gains in, 26–29
  and the United States' economy, 25–36
  white-collar, 33–34, 44
Programmers, productivity of, 231–232
Programming, 228–234, *see also Computer
    technology; Software*
  environments and tools, 230–231
  of parallel processors, 232–234
  several styles of, 229
  and software evaluation, 230
PROMETHEUS, 171
Proteins, 195
Psychology, environmental, 129–133

Q

Quality
  control of and recycling as energy conser-
    vation, 57
  and productivity, 37–39
Quality-control circles (QCs), 37
Quality of life, 345–369
  effects of population growth and increasing
    urbanization on, 359–365
  for the elderly, 351–355
  equity and international collaboration re-
    garding, 365–369
  and people with disabilities, 346–351
  personal health and safety regarding,
    355–359
Quantum-tunneling, 214

R

Radiation, 93–94
Radiation-trapping effect, 90
Radiative connective models (RCMs), 84
Radio, use of for broadcasting, 5
Radioactive waste, 113–115
Rail transportation, 177–179
  and fatalities, 178–179
  high-speed magnetic-levitation systems of,
    177
  and safety, 178
Rain, acid. *See Acid rain*
Rain forest, tropical
  burning of, 93
Random-access memory, 216–217

Rapid prototyping, 259
Rational design, 209
Recycling
    demographical differences in programs for,
        108
    paper, 107
    and positive reinforcement, 108
    and quality control as energy conservation,
        57
    of waste, 106, 107–109
Reforestation, 101
Rehabilitation Act, 349
Research
    environmental databases and tools for,
        120–121
Resource Conservation and Recovery Act
        Information
    System (RCRIS), 114
Risk assessment, 334–339
    and communication, 341–344
Risks, 333–344
    actual vs. perceived, 334–335
    reducing and managing, 339–341
Roads, safety on. See Highway safety
Robots, 17, 239, 350
    knowledge, 270

S

Safe Drinking Water Act, 95
Safety, 117
    aviation, 163–168
    health and personal, 355–359
    highway, 172–177
    railway, 178
Salinization, 97, 101–102
Salt, 97
    in soil, 101–102
Sampling techniques, 7
Satellite networks, 223
Satellite office, 315
Savings accounts
    problem of the low rate of, 13
Scenario fulfillment, 328
Scenic beauty, 132
Scheduling, human, 17
Science
    education in, 140–142
Scientists, 306
Seismic monitoring, 118
Seismic profiling, real-time, 48

Self-reports
    on energy conservation, 59–60
Separation technology, 107
Services, 40
    and the United States' economy, 24–25
Signal detection theory and analysis tech-
        niques, 196–197
Silicon, 23, 186, 214
Simulation and training, 155–156
Single-processor machines, 215
Sky writing, scholarly, 288
Small businesses, 311–313
Smog, urban, 92–93
Social traps, 128–129
Software, 228–234, see also Computer technol-
        ogy; Programming
    educational, 147–148, 150–151
    evaluation of, 230
    for general consumer, 235
    and programmer productivity, 231–232
    programming, 228–230
        environments and tools regarding,
            230–231
        of parallel processors, 232–234
Soil, 101–103
Soil erosion, 101
Solar panels, 63–64
Solar power, 63–65, 70
Solidification, 18
Solid waste, conventional, 105–107
Soviet Union
    space program of, 189–190
Space exploration, 68, 184–194
    human factors issues on, 184, 193–194
    human's role in, 192–193
    moon base for, 186
    plans and prospects concerning, 185–186
    space station regarding, 187–189
    status and future of U.S. program for,
        189–192
Space station, 187–189
    characteristics of, 189
    hard constraints concerning, 188
    problem of crowding regarding, 188–189
    safety and habitability of, 187–188
    scientific and commercial functions of,
        187
Species of organisms
    decrease in, 103–105
Speech
    computerized production and recognition
        of, 277–278

Steel, 18
Storage technology, 216–217
Stratosphere, 93–94
Stratospheric ozone thinning, 93–95
Stress, job-related, 321–322
Sulfur dioxide, 89, 92
Sulfuric acid, 91
Sun
  as alternative energy source, 63–65
  and electrical power, 64
Supercomputer, 215–216, *see also Computer technology; Information technology*
Superconductivity, 60–61
Supervisory control situations, 322
Sustainable agriculture, 102–103
Switching speed, 213–214
Switzerland
  use of medical waste incinerators, 112

T

Table of Contents (TOC), 273
Task analysis, 259, 318–319
Teaching methods. See Education
Technological advances, 371
Technology and work, 300–305
Telecommunication system, 183
Telecommuting/telework, 315–318
  reasons for resistance of, 316–317
  telecommunications and computer technology regarding, 317–318
  and white-collar workers, 316, 317
Teleconferencing, 55, 318
Telegraph, 5
Teleoperator, 237–238, 350
Telepresence, 237–238
Television
  as an educational tool, 148
Temperature, mean, 78
Terminals, 223–224, *see also Computer technology; Computer systems; Video display terminals*
Thermodynamics, second law of, 46
Thinking ability, development of, 145–146
Three-dimensional global circulation models (GCMs), 84
Three Mile Island, 52, 113, 117
Top-to-bottom approach, 37
Total quality control (TQC) movement, 37–38
Total quality management (TQM), 37–39, 57

and Japan, 37–38
in U.S. Department of Defense, 38
Tourism, 162
Toxic chemicals, 96
Trade, balance of, 10, 11, 12
Trade, international, 367
Trade deficit, 10, 12
Tradeoffs
  hidden costs and ubiquity of, 323–326
Tragedy of the commons, 128
Training. *See Education*
Trains. *See Railway transportation*
Transgenic organisms, 195
Transistors, 214
Translation, semi-automatic language, 280–281
Transmission Control Protocol/Internet Protocol (TCP/IP), 220
Transportation, 160–183
  air, 160–163, 180
  alternative fuels for, 180
  and aviation safety, 163–168
  and energy, 179–181
  ground, 168–172
  and highway safety, 172–177
  human factors issues concerning, 181–183
  rail, 177–179
Trellis architecture, 233–234
Trends, 2–3
  occupational, 297–299
Tritium, 65–66
Tropical Forestry Action Plans, 100
Tropical rain forest, 104
  burning of, 93
Troposphere, 93, 119
Tutors, human, 147–148

U

Ultraviolet radiation, 93–94
Unemployment rates, 301
Unionism, 300
United States
  air travel and, 160–163
  annual rate of economic growth of, 10–11
  automobile accidents and fatalities, 172–174
  balance of trade in the, 10, 11, 12
  current state of education in, 140–144
  economy of, 9–45

education and economic future of,
    138–139
energy situation of, 49–53
environmental laws of, 125
industrial and military training, 152–155
medical services in, 196
population growth in the, 361–362
rail transportation of, 177–179
solid waste disposal in, 105–107
and space exploration, 185–186, 189–192
three stages of environmentalism in,
    127–128
and unionism, 300
use of automobiles as transportation,
    168–170
use of medical waste incinerators, 112
Uranium, 68, 113
Urbanization
    and population growth, 362–363
U.S. Department of Defense
    and total quality management (TQM), 38
Uselessness, sense of
    among elderly, 355

V

Venus, 78
Very large scale integrated (VLSI) circuit de-
    sign, 273
Video display terminals, 252–253
Vinyl, 324–325
Virtual travel, 236
Virus/worm, computer. *See Computer virus/
    worm*
Vision and driving, 174–175
Visual display technology, 223–224
Visual fatigue, 252
Volcanic eruptions, 79
Voyager spacecraft, 191

W

Waste, 105–116
    and biodegradable and photodegradable
        materials, 109
    conventional solid, 105–107
    incineration, 106
    management, 19
    medical, 112
    radioactive, 113–115
    recycling of, 106, 107–109
    reduction of paper, 109–112

sociobehavioral research on reduction of,
    107–109
Waste disposal, 43
    and product design, 108–109
Waste recovery system
    high- vs. low-technology, 107
Water, *see also Drinking water*
    as alternative energy source, 63
    conservation, 98–99
    contamination, 95–99
    depletion, 95–99
Water vapor
    and global warming, 82–83
Weather models, 84
Welfare effects risks, 122, 123
West Germany
    use of medical waste incinerators, 112
Wetlands, 103
    destruction of, 101–103, 104
Wheel-rail trains, 177
White-collar worker
    effects of automation on, 302–303
    productivity, 33–34
    and telecommuting, 316, 317
Wind
    as alternative energy source, 62–63
Women in the workforce, 308–310, *see also
    Gender*
Wood, 49
Word processors, 291–292
Work, 296–322, see also Employee training;
    Job
    and boredom, 320, 321, 322
    and changing composition of the work-
        force, 308–310
    contexts, 310–313
    and disabled individuals, 349
    and displacement problems, 302
    and effects of aging, 320
    human factors issues concerning, 318–322
    job quality, 299–300
    and job-related stress, 321–322
    and job satisfaction, 320–321
    occupational trends, 297–299
    skill requirements of future jobs, 305–308
    and technology, 300–305
    telecommuting and telework, 315–318
    training at, 152–155
    workplace organization, 313–315
Worker efficiency, 39–40
Working environment, 129–130, 131
Workload, measurement of, 319–320

Work-monitoring, 227
World Commission on Environment and
     Development, 124
World economy, 11, 365–369
  human factors issues on, 368–369
World Environmental Policy Act, 126
Worldnet, 222

World population, 360–361, 362, 364–365
Writers, 291

Y

Yucca Mountain nuclear waste disposal site,
     113